Education Division 2014

Core Programming Area at the 2014 AIChE Annual Meeting

Atlanta, Georgia, USA
16-21 November 2014

ISBN: 978-1-5108-1256-7

Printed from e-media with permission by:

Curran Associates, Inc.
57 Morehouse Lane
Red Hook, NY 12571

Some format issues inherent in the e-media version may also appear in this print version.

Printed by Curran Associates, Inc. (2015)

For permission requests, please contact AIChE
at the address below.

AIChE
120 Wall Street, FL 23
New York, NY 10005-4020

Phone: (800) 242-4363
Fax: (203) 775-5177

www.aiche.org

Additional copies of this publication are available from:

Curran Associates, Inc.
57 Morehouse Lane
Red Hook, NY 12571 USA
Phone: 845-758-0400
Fax: 845-758-2634
Email: curran@proceedings.com
Web: www.proceedings.com

TABLE OF CONTENTS

392003 Control and Automation of Fluid Flow, Mass Transfer and Chemical Reactions in Microscale Segmented Flow

Abstract:

Owing to the constant increase of the worldwide energy demand due to the growth of populations and development of countries, the emissions of greenhouse gases (mostly carbon dioxide (CO_2)) from fossil fuels will constantly increase. Currently, carbon capture and sequestration is one of the most efficient solutions to control and reduce the concentration of the released greenhouse gases in the atmosphere. However, the underground storage of the captured CO_2 is expensive and limited to some specific locations. On the other hand, the current rate of energy consumption along with the inevitable global warming makes energy as one of the crucial worldwide issues of the near future. Therefore, there will be a huge demand for new sets of sustainable energy substitutes. In 2008, it was found that CO_2 in the presence of a frustrated Lewis pair (FLP) and hydrogen can be quickly converted to methanol and water. It has been proposed that CO_2 fuel conversion using the captured CO_2 from power plants (or atmosphere) has the potential to be one of the key solutions to address the energy demand and global warming problems. However, measurement and characterization of the kinetics of CO_2 conversion reactions (i.e., to methanol, ethanol or CO) in the presence of metal-free catalysts (FLPs) are currently challenging or in some cases impossible, owing to the mass transfer and time-resolution limitation of the conventional batch scale characterization tools.

During the past decade, microfluidic (MF)-based technologies have shown promising solutions as a high throughput characterization tool for investigation, screening, and optimization of gas-liquid reactions.

In my PhD, I was working on a multidisciplinary project, collaborating between Mechanical Engineering and Chemistry Departments of University of Toronto. During this time I have made significant contributions in flow chemistry and microfluidic (MF) research fields specifically related to screening gas solubilities in liquids, measuring thermodynamic characteristics of gas-liquid reactions associated with CO_2 sequestration, characterizing CO_2-governed liquid-liquid phase separation process (switchable water) and on-demand preparation of high quality colloidal nanomaterials (e.g. CdSe/CdS).

During this time, I designed and developed the first automated MF platform that allowed the routine screening of solubility data for a wide range of CO_2-physical solvent mixtures. The platform allowed screening of approximately 300 different conditions (temperature and pressure) during one day. In the next step, the developed automated MF strategy was utilized to measure, for the first time, thermodynamic characteristics of CO_2-FLP reaction. I have also designed a two-phase MF platform for high temperature (240-330 C) synthesis of high quality (Fwhm ~30 nm) nanocrystals such as CdSe, CdS and ZnS. The silicon-based microreactor could produce high quality core or core/shell nanocrystals with desired spectral characteristics at a throughput of 0.6 ml/min.

As an independent scientist, I will focus on the development and utilization of MF platforms for the carbon capture in the presence of nanoparticles along with fuel conversion process. I will also study the underlying transport mechanisms in the presence of chemical reactions with experimental and numerical methods.

389363 Mechanistic Correlations Between Material Properties and the Dynamics of Catalytic and Photocatalytic Redox Processes

Abstract:

Chemical reactions catalyzed by materials with electronic band gap (e.g., metal oxides) proceed via redox cycles involving kinetic coupling between elementary steps that reduce and re-oxidize the catalyst materials. Such redox cycles involve transfer of electrons between the catalyst and substrate molecules, with or without photo-excitation of electrons within the material to facilitate the electron transfer, and their dynamics depend on positions of filled and empty electronic bands. Here, I will discuss the use of kinetic, isotopic and spectroscopic methods in combination with quantum mechanical calculations to understand the mechanism of such reactions and to derive rigorous relations between material properties and their catalytic functions.

389825 Understanding of Polymers in Confined Thin Films and Bulk Membranes: Fluorescence Based Approach

Abstract:

My research interest is to develop core understanding of material properties of functional polymers in thin film and bulk membrane format using fluorescence and other complementary surface characterization techniques (spectroscopic ellipsometry, quartz crystal microbalance, neutron reflectometry).

Bulk membranes have potential applications in bioseparation, water purification, fuel cell applications and so on. Virus filtration membranes work based on size exclusion mechanism and are used to remove viral contaminants during the processing and commercial production of therapeutic proteins. The biggest challenge in these processes is lack of sufficient understanding of virus retention mechanism. This knowledge is pivotal to understand the trends in flux and log reduction values (LRVs) of virus filtration membranes with increased volumetric throughput. The application of fluorescence confocal microscope to visualize the virus retention zones along the thickness of the membranes is a very exciting and new concept. This has been taken one step forward by adopting a multiple dye approach to track model bacteriophage retained inside virus filtration membranes (DV20, Vireolsve pro and Viresolve NFP) at different stages of membrane operations with intermittent pressure releases (Post-doc at ChE, Penn State, PI: Prof. Andrew Zydney). Fluorescence based techniques offer unique opportunities to distinguish between virus captured before and after process interruptions. This, in turn, gives interesting insights about the effect of pore structure and morphology on virus retention.

Fluorescence, when used in conjunction with other surface characterization techniques, provides a powerful platform to characterize transport properties and mobility of polymer and water in bulk polymer membranes as well as thin films. Bulk free-standing and confined supported systems behave very differently at hydrated state in terms of material properties. When the film thickness approaches to several tens of nanometers, the optical, mechanical and transport properties are dominantly controlled by the interfaces. The mobility of the polymer chains is constrained due to severe confinement effect. Although a great deal of work has been done so far to explore the material properties of ionomers in bulk membranes, the mobility, hydration and transport properties of functional ionomers in nm-thick films are relatively unexplored. The challenges in characterizing thin films start when the conventional bulk characterization techniques fail to predict the nanoscale changes in materials properties. Fluorescence based techniques can overcome many of these issues. Fluorescent probes sensitive to local viscosity and proton concentration were incorporated into ionomer thin films and membranes and the changes in fluorescence properties of the dyes as a function of film thickness and hydration were studied (Post-doc at MatSE, Penn State, PI: Prof. Michael Hickner). The changes in fluorescence offered valuable insight into stiffness and proton mobility of hydrated bulk and confined systems. The work clearly demonstrated the differences between confined and bulk systems. The study has prime importance in nanoscale understanding and advanced level designing of polymer-catalyst interface in fuel cells.

In addition, fluorescent p-conjugated oligoelectrolytes were synthesized and characterized in solution and solid state utilizing their light-harvesting properties. This was a part of my PhD work. Layer-by-layer self-assembled thin films of fluorescent oligoelectrolytes were investigated to develop fluorescence based bioassay platforms.

The multi-disciplinary work experience enabled me to innovate effective techniques to understand functional materials more deeply for a range of applications. As a future faculty, I would like to perform independent as well as collaborative research focusing on structure-property relationships and material development for (1) energy applications and (2) chemobiosensing. The application of fluorescence based techniques in fuel cell research is quite new, has already started offering interesting insights and has a lot of scope of work. An array of confined and bulk ionomeric systems will be probed using fluorescent functional sensory molecules and water-polymer-substrate interaction will be further explored. Other surface characterization techniques can help to validate the hypothesis from fluorescence based analysis and strengthen the arguments about unknown scenerios in confined systems. This will solve a lot of mysteries hidden inside these geometries. Techniques will be developed to probe zone-specific density, water/proton transport and mobility. This will be of specific interest and the knowledge earned from materials studies will be employed to develop new materials with desired hydration and transport properties. The study will be extended to light harvesting molecules in confined systems with varied geometry and microenvironment. Typical light induced energy and charge transfer phenomena will be tuned for better and cheaper solid state biodetection platforms with improved sensitivity and selectivity.

388999 Protein-Protein Interactions in Disease and Therapy

Abstract:

Protein-protein interactions mediate a wide range of biological processes including receptor-ligand binding, protein signaling cascades, antibody binding, cell adhesion and metabolic regulation. Dysfunction of protein-protein interactions can lead to a variety of diseases including cancers and neurodegenerative diseases. Antibody therapeutics also make use of highly specific protein interactions to provide targeted intervention. The ability to generate, characterize and evolve protein interactions provides a basis to understand disease pathogenesis and to develop potential therapies.

I am interested in how the arrangement of different amino acids leads to highly specific interactions between two proteins and to the overall disease state of an organism. In my work with David Colby at the University of Delaware we have investigated prion pathogenesis and therapy using high throughput technologies such as fluorescence activated cells sorting and deep sequencing and at multiple scales from recombinant DNA techniques to animal models. Prion diseases are a group of rapidly progressing, lethal neurodegenerative diseases that are propagated by the conversion of the native form of the prion protein (PrP^{C}) to an alternatively folded, disease causing form (PrP^{Sc}) catalyzed by the interaction of PrP^{Sc} with PrP^{C}. We used mutagenesis and directed evolution approaches to generate high affinity antibodies to disrupt prion pathogenesis and to characterize prion protein self-interactions. Future applications of the methodologies developed include investigation of allosteric enzyme sites and antibody interactions with non-antigen targets relevant to the biopharmaceutical industry. I am also interested in engineering education and have pursued several opportunities to lead students in classroom and laboratory settings.

Doped Metal Oxides – Increased Functionality for Energy Applications

James Dorman- University of Konstanz, Konstanz, Germany

Metal oxides are commonly used in current technologies due to the numerous properties that can be achieved with slight variations of the metal ion(s). Furthermore, the properties of these compounds can be engineered to increase their applicability through the incorporation of low concentrations of substitute ions into the crystal lattice or the addition of thin shell layers. Such doped metal oxide structures are implemented into state-of-the-art technology, such as catalysts, optical displays, semiconductor devices, etc., due to the following properties:

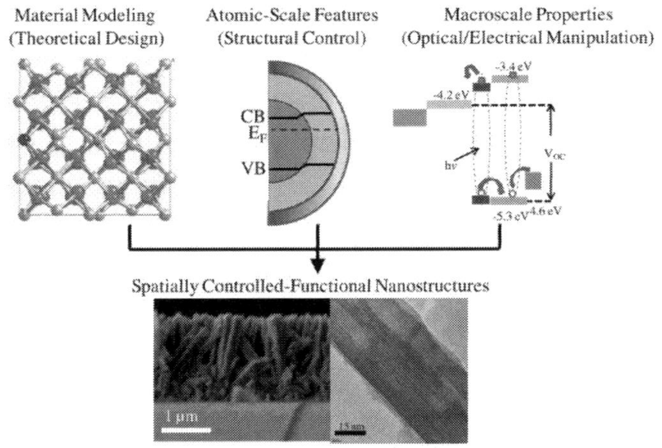

The schematic shows how material modeling, spatially control, and manipulation of the physical properties will be implemented for high functionality nanostructures.

- Predictable luminescence due to allowed transitions after crystal field splitting,
- Charge mobility control proportional to the electron/hole density,
- Electronic state addition around the conduction and valance bands,
- Increased chemical stability in the presence of harsh conditions, and
- Control over stress/strain of the system during chemical reactions.

However, the combination of host and dopant structures and their associated applications are far from optimized or used to their full potential. My research interests focus on the doping of metal oxides to control their physical and electronic properties with a specific focus on energy applications. Specifically, my research focuses on the engineering of spatially controlled doped metal oxide nanostructures for optical/optical (lighting and displays)[1,2], optical/electrical (photovoltaics)[3-5], and electrical/chemical (energy storage) energy conversion. The schematic highlights the implementation of these material modeling, spatial control, and the macroscale properties for the production of functional nanostructures for energy applications. The role of the dopant on the material properties can be confirmed through a combination of theoretical modeling[6] and experimentally[7] as discussed in my pervious publications. Specifically, the position and concentration of the dopant plays a critical role in both lighting and photovoltaic applications.[3,6] Furthermore, the process can be precisely controlled using a combination of wet chemical synthesis methods, i.e. hydrothermal growth and sol-gel chemistry, and low pressure deposition methods, such as atomic layer deposition.

(1) Dorman, J. A.; Choi, J. H.; Kuzmanich, G.; Chang, J. P. *The Journal of Physical Chemistry C* **2012**, *116*, 10333.
(2) Dorman, J. A.; Choi, J. H.; Kuzmanich, G.; Chang, J. P. *The Journal of Physical Chemistry C* **2012**, *116*, 12854.
(3) Dorman, J. A.; Weickert, J.; Reindl, J. B.; Putnik, M.; Wisnet, A.; Noebels, M.; Scheu, C.; Schmidt-Mende, L. *The Journal of Physical Chemistry C* **2014**.
(4) Bretschneider, S. A.; Weickert, J.; Dorman, J. A.; Schmidt-Mende, L. *APL Materials* **2014**, *2*.
(5) Weickert, J.; Zimmermann, E.; Reindl, J. B.; Pfadler, T.; Dorman, J. A.; Petrozza, A.; Schmidt-Mende, L. *APL Materials* **2013**, *1*.
(6) Dorman, J. A.; Choi, J. H.; Kuzmanich, G.; Bargar, J. R.; Chang, J. P. *Journal of Applied Physics* **2012**, *111*, 083529.
(7) Dorman, J. A.; Mao, Y.; Bargar, J. R.; Chang, J. P. *The Journal of Physical Chemistry C* **2010**, *114*, 17422.

379648 Multimodal Nanomedicines for Solid Tumors: Rational Drug Combinations, Targeted Cancer Therapies, Photo-Activated Drugs, and Synthetic Lethal RNA Interference

Abstract:

Nanoscale drug delivery is a power full tool for the treatment of solid tumors. Although a range of approaches have been described over the past several years, the vast majority seek to improve cytotoxic chemotherapies that have been in use for decades. Nanoscale delivery strategies integrating rational drug combinations, synthetic lethal RNA interference, stimuli-responsive drugs, or so-called molecularly targeted cancer therapeutics are and remain underexplored. Here, we present work from our laboratory describing the development of a range of such nanoscale platforms for drug delivery to solid tumors. These include: (i) self-assembled peptide co-polymer nanoparticles for synthetic lethal RNA interference of DNA damage response in ovarian and lung cancers, (ii) multivalent anti-androgen and anti-estrogen gold nanoparticles for the treatment of hormone-refractory breast and prostate cancers, (iii) photo-responsive nanoparticles that deliver drugs or achieve photothermal ablation in response to laser exposure, (iv) nanoscale delivery platforms targeting cancer stem cells, the hypoxic tumor microenvironment, or the tumor stroma, and (iv) Layer-by-Layer (LbL) nanoparticles for horizontal signal blockade in breast carcinoma.

383097 Optical Methods to Probe Biological Questions

Abstract:

The convergence of technical advances in optical imaging, biology (specifically genetic modification), and chemical synthesis provide a wealth of avenues to explore complex biological systems with fluorescent microscopy. My research interests focus on applying novel optical, chemical, and/or biological approaches to better understand medically related processes at a fundamental engineering level in cells and *in vivo*. Using these approaches we can measure biological processes, medical intervention success or failures, and monitor disease state or progress. During my graduate work I developed novel optical nanoparticle sensors to fluorescently measure ions and small molecules. These sensors were used to measure sodium flux during action potentials in cardiomyocytes and interstitial concentrations as a tattoo like minimally invasive clinical monitor. During my postdoc I have focused on combining polarized fluorescence microscopy and fluorescently labeled anticancer drugs to measure target engagement and binding activity in cells and at the subcellular level in *in vivo* tumor models. This approach, discussed in detail in this poster, allows us to better understand small molecule behavior in cells *in vivo* and assess tumor heterogeneity during therapeutic treatment both spatially and temporally. Building on these pursuits, I hope to develop a research program that explores difficult to measure biological questions in complex systems using novel fluorescence imaging approaches.

380477 Discovery of Disease-Specific Antibody Biomarkers and Their Targets

Abstract:

Discovery of biologic molecules specific to a diseased state, or biomarkers, can lead to diagnostic development, therapeutic target identification, and improved understanding of disease pathogenesis. Antibodies remain an attractive class of biomarkers given their amplification by the immune system, stability, and current clinical use. While the antibody repertoire represents a rich source for biomarker discovery, it has been difficult to impartially identify which molecules from this repertoire are associated with disease. This work demonstrates three molecular discovery processes centered on the utility of bacterial-displayed peptide libraries and fluorescence activated cell sorting (FACS) for identifying novel antibody biomarkers. We applied these methods to identify candidate molecular diagnostics for pre-eclampsia (PE), a condition with unknown etiology that affects 5-8% of pregnancies. Previously characterized autoantibodies that bind the angiotensin II type 1 (AT1) receptor in PE patients [1][2] are difficult to detect, vary in prevalence amongst studies, and most importantly, lack specificity. Thus, we sought to identify additional PE-specific antibody biomarkers.

Applying three quantitative screening strategies against a set of PE and healthy outcome pregnancies (HOP) identified unique disease-specific peptides from a 15 amino acid peptide library. With a two-color screening method, we used antibody fractions enriched from plasma to isolate significantly PE cross-reactive and specific peptides distinct from the AT1 receptor epitope. We used a panel of these antibody-detecting peptides to train and validate an Adaptive Boosting classification algorithm that achieved 80% diagnostic accuracy [3]. In a second screen, the library underwent sequential enrichment for PE antibody (immunoglobulin G) binding followed by removal of HOP antibody binding peptides against unprocessed, diluted plasma. This screen more closely replicated the native environment for the antibody binding interaction and yielded a strong consensus motif. Directed evolution expanded this motif for which BLAST searches identified a number of environmental antigens and human proteins that shared similarities.

Importantly, by evaluating antibody binding activity of bacterial-displayed 15-mer fragments from the strongest candidates, we linked this motif to a region of a common viral antigen (VA) and a human G protein-coupled receptor (hGPCRa). Furthermore, we verified that the synthesized VA fragment efficiently blocked antibody binding to the hGPCRa fragment and to HEK293T cells expressing the full length protein. Thus, this method enabled unbiased identification of a disease-specific antibody and characterization of its targets. In a third method, we combined library screening with next-generation sequencing to profile the antibody repertoire of four individual PE patients and four HOP and identify motifs present amongst these PE.

This work demonstrates that we can now probe the antibody repertoire using unbiased screening of bacterial-displayed peptide libraries to identify new markers of disease. These disease-specific antibody-detecting peptide reagents can be used for molecular diagnostics and/or characterizing antibody target(s), potentially elucidating novel therapeutic targets. Building upon this work, we can investigate how the antibody repertoire changes with age or after exposure to a particular virus or environmental factor.

[1] Wallukat, G. *et al.* Patients with Preeclampsia Develop Agonistic Autoantibodies against the Angiotensin AT1 Receptor. *The Journal of Clinical Investigation* **103,**945–952 (1999).

[2] Zhou, C. C. *et al.* Angiotensin Receptor Agonistic Autoantibodies Induce Pre-eclampsia in Pregnant Mice. *Nature Medicine* **14,**855–862 (2008).

[3] **Elliott, S.E.** *et al*. Characterization of Antibody Specificities Associated with Preeclampsia. *Hypertension* **63,**1086-1093 (2014).

389947 Programmable Dynamic Surfaces As Information Carriers

Abstract:

Tuning surface chemistry to achieve selective material properties has been an object of research for decades. Subtle variations in chemical environment at the surface of particles and materials often result in drastic changes in surface reactions, self- assemblies, surface toxicity, and more. Fine-tuning surface chemistry of materials (and/or nano particles' surfaces) in response to environmental signals could be an efficient way to dynamic surfaces capable of adjusting their surface properties and also manipulating material objects resting on such surfaces. Change of the surface properties could be triggered by alteration or gradient in chemical potential, external fields such as electrical or magnetic fields, light, or other environmental stimuli. Potentially applications of such smart surfaces might include: transferring matter (at different scales), capture and release of ions, self-cleaning materials, "smart" delivery carriers, etc. As an example, nanoparticles functionalized by photoactive ligands might undergo conformational changes upon light irradiation ultimately altering surface properties. Careful selection of grafted functional groups at the correct ligand density could yield photoactive surfaces of the particles having dynamic surface charge density or hydrophobicity. Since the relaxation time of light-induced processes is fairly short, such photoactive particles could be applied to trap and release ions or other charged moieties. Yet another type of dynamic surfaces could be fabricated by applying charged ligands on flat substrates of macroscopic extent. Under the influence of externally applied electric fields, surface ligands could actively rearrange in regions of high-low filed. If the fields were then changing in time over the surface, they could create "waves" of surface properties that could alter effective surface properties and ultimately move around objects placed on such surfaces. My research on the design and, later, implementation of such surfaces can be facilitated by multiscale modeling with which one can gain further insights into the structures of the switchable molecules and the overall surface properties that these molecules give rise to. I am also interested in understanding such surface-dynamic phenomena at the level of fundamental thermodynamics of nonequilibrium actuation, characteristic response times, and energetic efficiencies.

379483 Improved Treatment of Fungal Infections through Targeted Drug Delivery: How Can Biomembranes Help?

Abstract:

Fungal infections are a significant cause of morbidity and mortality especially in immune-compromised individuals. Fungal diseases are hard to eradicate as fungal cells develop resistance to most available antifungal drugs and are hard to target inside mammalian cells. Several life threatening fungal infections such as cryptococcosis and histoplasmosis are contracted by inhalation of spores, which then proliferate inside the macrophages of the alveolar region and disseminate throughout the body. The macrophage-pathogen interaction is the first step in these diseases and provides a suitable target for early treatment of the disease. Recent evidence suggests that microdomains composed of saturated phospholipids and sterols (i.e. lipid rafts) play an important role in macrophage-pathogen interaction. Abolishing lipid rafts in fungal cells is shown to drastically reduce their virulence. Also, many pathogens use lipid rafts as the site of entry to the host cell, suggesting that rafts can be used for targeted treatment of fungal diseases. However, understanding the role of lipid rafts in modulating infections and utilizing them for targeted therapy of fungal infections is a complicated research problem requiring knowledge in various fields such as engineering, biophysics and microbiology.

My graduate and post-doctoral experiences in chemical engineering and microbiology have uniquely prepared me to investigate this challenging research problem. In my graduate work, I have studied the interactions between phospholipid microdomains and polymeric nanoparticles. I have examined the effects of particles on microdomain size and surface properties. I have also studied the interactions between aerosolized particles and lung surfactant interfaces to gain knowledge on the potential adverse effects of inhaled particles used for pulmonary delivery. In my post-doctoral research, I have studied the role of plasma membrane biophysics in the virulence properties of *Cryptococcus neoformans* (the causative agent of the disease cryptococcosis). Using a number of fluorescence spectroscopy methods, I have shown that the structure of lipids in the plasma membrane of *C. neoformans* plays an important role in its ability to form a rigid membrane and cause life threatening infections. I have also examined the relationship between lipid rafts in mammalian cells and their infectivity by *C. neoformans*.

My future research interests combine my expertise in characterizing plasma membrane physical properties and the interactions between lipid rafts and nanoparticles. Through modulating the physical properties of plasma membrane, I aim to design strategies to increase the uptake of drugs or drug containing nanoparticles in mammalian cells. Also, I aim to use particle engineering to target fungal cells inside macrophages and treat the disease at the site of infection. I plan to use the knowledge provided by these research directions to develop new and more effective inhalable drugs against fungal infections for pulmonary drug delivery.

365051 New Approaches to Interrogate and Engineer Complex Biocatalytic Networks

Abstract:

I hope to run a multidisciplinary research group that employs novel theoretical and experimental approaches to interrogate and engineer biochemical systems of relevance to human health, energy, and the environment. My research program will have three broad goals: (i) to develop methods for controlling, studying, and altering enzyme activities of metabolic relevance, (ii) to employ those methods to answer fundamental questions of human disease, cellular metabolism, and molecular recognition, and (iii) to apply those methods, and correspondingly evolved theories, to develop novel enzyme inhibitors and to engineer metabolic networks.

My lab will employ concepts and techniques from protein engineering, catalysis, nanotechnology, synthetic biology, and applied mathematics. Our work will range in scope from highly applied (e.g. the development of a new biochemical sensor or an improved metabolic pathway) to highly theoretical (e.g. the use of dynamical systems theory to identify patterns in social networks or ecosystem dynamics that apply to the differentiation of cells or to the behavior of systems of mutually interacting biocatalytic species).

With this presentation, I will describe a few proposed projects consistent with these plans. Examples include (i) optical systems for spatiotemporal control of enzyme activity *in vivo*, (ii) metamaterials-based strategies for evolving enzymes for synthetic metabolic networks, and (iii) enzymatic evolution of small molecules to inhibit pathologically relevant enzymes.

378090 Multiscale Modelling and Process Simulation

Abstract:

Reliable modelling of chemical processes depends on the accuracy of the physical properties used to describe each aspect of the combined system. However, most systems of interest are sufficiently complex that normal operations are data poor. Optimization increases the need for data and compounds the challenge. Using numerical simulation to estimate both the importance of physical properties and the properties themselves provides a way to meet that challenge. Connecting atomic and microscale behavior accessible through electronic structure calculations (commonly with intermediate descriptions using force fields) to continuum, macroscopic behavior requires a familiarity with multiple approaches and computational techniques. One of the most promising avenues of investigation is using molecular simulation techniques, either molecular dynamics or Monte Carlo, to calculate excess energy data in both stable and unstable regions. Correlating these data provides physical property relations analogous to traditional constitutive models that can be implemented at minimal cost to provide atomic scale insight for system level implementation.

Design of heterogeneous catalysts for the conversion of biomass-derived compounds into fuels and chemicals

Jean Marcel R. Gallo, Department of Chemistry at the Federal University of São Carlos, Brazil.
email: jean@ufscar.br

There has been an increasing interest in renewable sources of carbon for the production of transportation fuels and chemicals due to the diminishing petroleum resources. Cellulose and hemicellulose, components of biomass, can be depolymerized into the monossacharides glucose and xylose, which yields a variety of important platform molecules for the production of fuels and chemicals. For the economical viability and sustainability of biomass conversion processes, development of selective and stable solid catalysts is necessary. My future research studies will focus on the context of biomass processing, such as preparing novel heterogeneous catalysts to address current process problems (such as low stability, low yields, and leaching), devising pathways for biomass conversion into chemicals, and studying reaction kinetics and mechanisms. Between these topics I can mention four examples: (1) bioethanol conversion into valuable chemicals; (2) hydrothermally stable catalysts for the conversion of monosaccharides into platform molecules; (3) conversion of furfuryl alcohol into levulinic acid; and (4) upgrading of crude glycerol obtained as biodiesel byproduct. I believe that my research interests lie on strategical areas for developing sustainable and economically viable biomass conversion processes that will be well received by the scientific community and attract fundings.

I hold a Ph.D. in Chemistry (in a joint program between the University of Eastern Piedmont in Italy and the State University of Campinas in Brazil under the guidance of Prof. Leonardo Marchese and Prof. Heloise O. Pastore) followed by three years of postdoctoral research in Prof. James A. Dumesic's group from the Department of Chemical and Biological Engineering at the University of Wisconsin-Madison. My research in Prof. Dumesic's group involved the design and synthesis of heterogeneous catalysts and reaction kinetic studies on the transformation of biomass-derived molecules to platform chemicals. This postdoctoral training has complemented my knowledge in the field of catalysis and reaction kinetics, while I have brought to the group a new field of expertise focused on materials chemistry. In September 2013, I left Prof. Dumesic Group and joined Prof. Jose Maria C. Bueno group at the Federal University of São Carlos, in Brazil. Since then I have been working in the conversion of ethanol into compounds of interest of the chemical industry.

JMR Gallo thanks FAPESP for the fellowship (process number 2013/15861-1).

387445 Climate Change, Air Quality and Public Health: An Uncertainty Analysis

Abstract:

The central goal of this research is to explore the connections between climate change, air quality and public health through numerical modeling, with an emphasis on uncertainty in simulations of human-climate interactions. Climate change is expected to significantly influence meteorology and impact future air quality. In addition, climate change mitigation strategies will also affect emissions and concentrations of air pollutants. Earth system models are vital to advance our understanding of atmospheric processes and are used as key tools to steer environmental policies with enormous economic and social implications. However, there are large uncertainties associated with climate modeling that further propagate into estimates of future air quality, as well as projections of associated health and economic impacts. These uncertainties must be identified and quantified to adequately assess the changes brought about by different climate policies. Furthermore, characterizing uncertainty across the complete climate change problem is essential to target the largest potential reductions to uncertainty in existing modeling systems and allow for improved impact and adaptation analyses.

Towards this end, a modeling framework to simulate air quality and associated health impacts under different climate scenarios has been developed using global earth systems modeling with atmospheric chemistry capabilities. Future pollutant concentrations are simulated with the Community Earth System Model (CESM), while the MIT Integrated Global System Model is used to predict climate change during the 21st century and beyond. The framework can be used to explore the most important sources of uncertainty in climate projections: emissions of climate forcers and conventional air pollutants under changing policy scenarios; model response due to structural uncertainty; and natural climate variability. By using this modeling framework to simulate changes in air quality for ensemble simulations of 21st century climate change, the propagation of climate uncertainties into atmospheric pollutant concentrations predictions, as well as associated health and economic impacts can be evaluated through probabilistic analyses. This information will assist in developing more meaningful climate impacts and risk assessments and better guide decision-making.

Dielectrophoresis (DEP) is an electrokinetic phenomenon used widely in microfluidic devices to manipulate, concentrate and separate particles. A polarizable particle in a nonuniform electric field experiences a physical force called the DEP force. In one format, called 'insulator-based DEP" (iDEP), particles flow through a microchannel which has zones where the cross section changes due to insulating obstacles, channel constrictions or other structures, while a DC electric field is applied across the channel. When an electric field is applied across the microfluidic channel, these structures generate regions of strong electric field gradients near themselves, which causes polarizable particles to experience a DEP force. iDEP has been used for more than a decade, and its advantages include relatively simple fabrication, and virtually eliminated electrode fouling and electrode reaction problems. Manipulation of a particular set of analytes in iDEP is dependent on a combination of the applied electric signal, channel features (design of insulating obstacles etc), suspending medium composition and microfluidic channel material [1]. We have investigated iDEP with <100 Hz AC signals instead of DC signals, and found that a given particle could be trapped under identical conditions in the same microchannel with 100 V_{RMS} weaker electric fields compared to DC signals [2]. Later, we compared several insulating obstacle geometries and found that under otherwise identical conditions, the DC voltage required for trapping the same particle could vary by 300 V between "long" circle shaped and "short" diamond shaped posts. By changing the electric signal or insulating obstacles it is possible to reduce the power requirements of iDEP devices without changing the sample and at no additional device complexity or cost. These studies are expected to pioneer more rationally designed iDEP devices [3].

1. Gencoglu, A., et al., *Particle Manipulation in Insulator Based Dielectrophoretic Devices.* Journal of Nanotechnology in Engineering and Medicine, 2013. **4**(2): p. 021002.1-7.
2. Gencoglu, A., et al., *Dynamic microparticle manipulation with an electroosmotic flow gradient with low frequency alternating current dielectrophoresis.* Electrophoresis, 2014. **35**(2-3): p. 363-373.
3. LaLonde, A., et al., *Effect of insulating posts geometry on particle manipulation in insulator based dielectrophoretic devices.* Journal of Chromatography A, 2014.

365893 Crystal Polymorphism and Texture Control during Solution Growth for Organic Electronics and Pharmaceutical Flow Chemistry

Abstract:

Solution deposition of organic semiconductors (OSC) is a leading contender for producing large-area, inexpensive, and flexible organic electronics. Using a new solution processing technique called solution shearing, we modify the molecular packing of a wide range of OSCs through processing alone. We quantitatively measure the molecular ordering resulting from different deposition conditions and show effective tuning of molecular packing that yields higher charge carrier performance without changing the underlying chemical structure. We investigate how solution flow and evaporation affects the morphological features and their impact on charge carrier transport. We also develop a novel high speed in-situ technique that uses X-ray diffraction to study how the polymorphs grow as a function of solution shearing parameters. This in-situ technique shows that the confined growth of OSC in a thin film promotes formation of a metastable crystal polymorph. We pattern high performance organic electronics through the use of surface selective patterning and nucleation control. The concept of nucleation control and crystal growth is also utilized to perform and extend microfluidic flow chemistry for pharmaceutically relevant reactions. Inorganic and organic solid formation can be processed without clogging the reactor by understanding the characteristics of crystals in flow. My proposed research will expand on the concept of metastable crystallization of OSCs and other organic molecules in order to reliably generate new polymorphs while controlling crystal morphology.

386728 Computational Design of Crystalline Materials Toward Storage and Efficient Use of Energy

Abstract:

One of the greatest challenges that mankind is facing is to continue the development of technologies that improve our quality of life without depleting our energy resources. At the center of addressing this challenge is the need to discover novel materials that allow us to develop processes and systems that make efficient use of energy. The advent of nanotechnology has given us unprecedented access to the atomistic structure of matter and enabled the rational design of materials by precisely controlling their nanostructure. My primary research interest is to use quantum and classical mechanics molecular simulations to (help) design the materials that will play pivotal roles in overcoming the bottlenecks related to energy consumption and/or storage in emerging technologies in areas as diverse as transportation, the chemical industry, and biomedicine. The goal is to capitalize on the link between properties/performance and nanoscale features.

This poster will provide an overview of my computational research work for the design of materials towards energy storage and/or efficient use of energy. During my doctoral work, I sought to design a catalyst/support system that could selectively produce single-walled carbon nanotubes (SWCNTs) of the same chirality. SWCNTs are tubular crystalline materials with outstanding chirality-dependent properties. Therefore, chiral-selective nanotube synthesis is eagerly pursued to avoid energy-intensive separations methods to obtain the desired chirality for applications such as biomedical imaging and photovoltaics. I investigated the nanotube growth mechanism to elucidate and discover ways that chirality could be controlled during synthesis, using reactive molecular dynamics (RMD) and density functional theory (DFT) methods. I investigated the feasibility and relevant conditions under which one could take advantage of the catalyst nanoparticle structure to act as a template to the nascent nanotube, and thus provided useful synthesis guidelines to experts in nanotube synthesis.

During my first postdoctoral project, I sought to design metal-organic frameworks (MOFs) that could be used in vehicular tanks to store fuels such as natural gas and hydrogen in high enough quantities to allow practical driving ranges. These gaseous fuels have specific energy contents (MJ/kg) superior to gasoline, but significantly lower energy density (MJ/L) at standard conditions, and thus their use currently requires energy-intensive compression. However, natural gas and hydrogen represent promising cleaner alternatives to reduce dependence on oil, so their adsorption on porous materials is being actively explored. MOFs are novel, remarkable, crystalline materials whose nanopore structure can be finely tuned by appropriate selection of essentially limitless possible combinations of inorganic "nodes" and organic "linkers." I used a variety of strategies for generating "crystal structures" and Monte Carlo simulations to calculate their adsorption properties. The goal was to establish performance limits (through data mining) for this class of materials and identify the best materials within these boundaries. Through close collaboration with synthetic chemists, some of the best MOFs identified via simulation have been synthesized, and their promising adsorption characteristics have been experimentally confirmed.

My current postdoctoral research focuses on the computational component of a collaborative project with experts in synthetic chemistry and catalysis. We aim to design hybrid MOF/nanoparticle catalysts for regioselective oxidation and/or hydrogenation toward the synthesis of commodity chemicals. Designing highly selective catalysts can result in "greener" reactions by reducing the need for post-reaction separation processes, which typically account for the bulk of the energy cost of operating a chemical plant. Mimicking the working mechanism of enzymes, the nanopores of the MOF component are expected to provide control on how the molecule "attacks" the surface of the nanoparticle component. In this research, I integrate insights obtained from classical molecular dynamics (MD) about the dynamics of the reactants inside MOF nanopores with insights from DFT calculations about the reaction energetics for the transformation of reactants into products on the catalyst surface. These calculations stand out from other DFT calculations in the literature in that the reactants are exposed to steric constraints using a novel faux-pore approach.

Building on my expertise in materials science, catalysis, computational methods, and data mining, I plan to mainly focus my academic career on the computational design and discovery of materials and catalytic systems that will either enable us to reduce energy consumption through highly active and selective catalysts (likely drawing inspiration from nature's enzymes), or directly use the required energy to convert reactants to products from abundant renewable sources such as in metal-oxide photocatalysts.

Doctoral Advisor: Professor Perla B. Balbuena, Texas A&M University, College Station TX

Postdoctoral Advisor: Professor Randall Q. Snurr, Northwestern University, Evanston IL

368013 Optimization of Valuable Intermediates By 11 Alpha-Hydroxylation of Steroid DHEA By Solvent-Enhanced Biocatalyst Beauveria Bassiana

Abstract:

This research evaluates the oxidative capacity of fungus *Beauveria bassiana*; a versatile whole-cell biocatalyst used in the biotransformation of chemicals. The goal of this research is to standardize, control and optimize the biocatalysis capacity of *B. bassiana* in the biotransformation of DHEA. To extend use of *B. bassiana* to commercial applications, the optimization of reaction conditions and accurate prediction of biotransformation products are necessary. This work enhances the selective hydroxylation capacity of Strain ATCC 7159, resulting in a cost effective and eco-friendly process for the synthesis of valuable 11α-hydroxy steroids. Our work establishes the metabolic pathway of DHEA and defines the optimum reactor arrangement, substrate concentration, reaction temperature and pH. NMR and LC-MS analysis were performed to characterize reaction products. Higher substrate conversion, selectivity and yield of desired product was achieved in a "Resting Cells" arrangement. As well, higher volumes of growing medium compared to reaction buffer, increase the activity of the biocatalyst. The apparent rate of reaction fits a Michaelis-Menten kinetics model with a maximum reaction rate of 4.45mM-day. Interestingly, when a diluted amount of substrate is used, a higher yield of 11α-hydroxy steroids is achieved. Also, reactions at 26°C with pH ranges between 6.0 and 7.0 resulted in the highest conversion (70%) and the higher product yield (45.8%). *B. bassiana* has the capacity to metabolize DHEA and similar steroids in different set-ups and has a promising future as biocatalyst to be used in the production of drug metabolites.

Porous Materials for Energy Applications: Shale Gas Recovery, CO_2 Sequestration, and Lithium-Ion Batteries

Gennady Gor
Department of Civil and Environmental Engineering, Princeton University, Princeton, NJ

As a faculty member, I will establish a world leading group specializing in porous materials for energy applications. My unique expertise is a combination of modeling chemical and mechanical processes in porous materials, with my toolkit spanning various computational techniques that cover both micro and macro scales. I am particularly interested in phenomena where fluids affect the mechanical properties and behavior of solid or soft porous materials. My areas of practical interest are shale gas recovery, CO_2 sequestration and lithium-ion batteries.

St. Petersburg State University: I received fundamental education in theoretical physics at St. Petersburg State University, Russia. My research was in the field of first-order phase transitions. Particularly, I studied how a system consisting of liquid droplets and vapor evolves in time. In my Ph.D. thesis, supervised by Prof. A. P. Grinin and Prof. F. M. Kuni, I developed a novel approach to the kinetic theory of non-isothermal nucleation using methods of thermodynamics, mathematical statistics, and transport phenomena.

Rutgers University: During the following two years I worked as a postdoctoral researcher at the Department of Chemical and Biochemical Engineering at Rutgers University (group of Prof. Alexander Neimark). My main focus was on modeling adsorption in nanoporous materials. I achieved the following results: I developed macroscopic and molecular theories of the deformation of nanoporous materials during adsorption. By means of Classical Density Functional Theory and Monte Carlo simulations I modeled nitrogen adsorption in novel types of nanoporous carbons and developed methods for characterization of these materials. These methods have been further successfully implemented in commercial instruments by Quantachrome Inc. I also modeled adsorption and transport of chemical warfare agents in polyelectrolyte membranes.

Princeton University: Currently I work at Princeton University with Prof. Jean Prevost (Department of Civil and Environmental Engineering) and Prof. Craig Arnold (Department of Mechanical and Aerospace Engineering). My first research project at Princeton was on modeling CO_2 leaks from geological storage. It involved the development, implementation, and application of methods of thermodynamics of fluid phases and mechanics for modeling the compositional flow of CO_2-brine mixtures. Then switch my focus to developing a multi-physics model for lithium-ion batteries to study the mechanical reasons of batteries aging and degradation.

Apart from research activities, I have always enjoyed teaching. I have taught a number of courses to a variety of different audiences: from high-school students to Princeton graduate students.

At the poster session, besides giving an overview of my past and current research, I will discuss my plans in research and teaching as a prospective faculty.

Selected publications (from 26 total):

[1] Cao Q., **Gor G. Y.***, Krogh-Jespersen K., Khriachtchev L. Non-Covalent Interactions of Nitrous Oxide with Aromatic Compounds: Spectroscopic and Computational Evidence for the Formation of 1:1 Complexes ***J. Chem. Phys.*** 2014, 140, 144304
[2] **Gor G. Y.***, Stone H. A., Prevost J. H. Fracture Propagation Driven by Fluid Outflow from a Low-permeability Aquifer ***Transport in Porous Media*** 2013, 100 (1), 69-92.
[3] **Gor G. Y.***, Paris O., Prass J., Russo P. A., Ribeiro Carrott M. M. L., Neimark A. V. Adsorption of n-Pentane on Mesoporous Silica and Adsorbent Deformation ***Langmuir*** 2013, 29 (27), p. 8601-8608
[4] Landers J., **Gor G. Y.**, Neimark A. V. Density Functional Theory Methods for Characterization of Porous Materials (Review) ***Colloids and Surfaces A*** 2013, 437, p. 3-32.
[5] **Gor G. Y.**, Elliot T. R., Prevost J. H. Effects of Thermal Stresses on Caprock Integrity During CO_2 Storage ***International Journal of Greenhouse Gas Control*** 2013, 12, p. 300-309
[6] **Gor G. Y.**, Thommes M., Cychosz K., Neimark A. V. Quenched Solid Density Functional Theory Method for Characterization of Mesoporous Carbons by Nitrogen Adsorption ***Carbon*** 2012, Vol. 50 (4), p. 1583-1590
[7] **Gor G. Y.**, Neimark A. V. Adsorption-Induced Deformation of Mesoporous Solids ***Langmuir*** 2010, Vol. 26 (16), p. 13021-13027

362104 Research and STEM Outreach at a Primarily Undergraduate Institution

Abstract:

There is a need for novel multifunctional nanoscale materials in the area of medicine and environment. My goal is to develop a successful research program at an undergraduate institute to focus on these needs by establishing an interdisciplinary well-equipped laboratory that encourages women and diverse groups in science and engineering. My research interests are, but not limited to, i) Pickering emulsions (PE) for biomedical and environmental applications, ii) Synthesis of spider-web peptides for biomimetic materials.

In modern undergraduate STEM education research has become an integral tool that provides training in innovation and collaboration in addition to preparing students for advanced degrees and careers in science and engineering. One of the challenges in performing research at an undergraduate institution involves designing simple projects that can be conducted with limited resources and involves undergraduates at all stages. The above proposed research activity intends to improve and increase research opportunities in Chemical Engineering Department at PUIs. Typically undergraduates are available for short durations - one semester or sometimes during the summer. This will be addressed by designing short, cumulative projects from the proposed work that can be completed quickly and the results generated will contribute towards answering the research questions.

In addition to scientific research, I am also interested in developing K-12 STEM outreach activities that involves active participation of the high school schools students, pre-service and K-12 educators. Some of my recent initiatives include: 1) a Chemical Engineering based workshop for pre-service K-12 teachers, 2) acting as a faculty advisor to local high schools for the Performance Standards Project Science Fair, 3) hosting a STEM day workshop for 7th grade and high school girls

380715 Molecular Engineering of Crystallization in Pursuit of More Potent Functional Materials, Cleaner Sources of Energy and Better Climate

Abstract:

Crystallization is a versatile phenomena in nature. For millennia, it has been widely used for extracting salt and other minerals, and is perhaps the oldest of the traditional chemical engineering separation processes. The importance of crystallization is however not confined to purification only. Crystallization is a vital component of many important natural phenomena, such as precipitation. It can also be employed for making functional materials, as many of these materials have crystalline, quasi-crystalline, or liquid-crystalline long-range order, and their functional properties usually emanate from their underlying long-range order. For instance, optical band gaps can only be realized in certain crystalline and quasi-crystalline arrangements of atoms, molecules or colloidal building blocks. Consequently, studying crystallization processes with the aim of discovering new crystallization routes, or engineering the existing crystallization processes, is of broad scientific and technological interest. Molecular engineering of crystallization processes is however not possible without having a thorough understanding of the kinetics and the mechanisms of the underlying processes. Computer simulations are invaluable tools in this endeavor since the existing experimental techniques lack the necessary spatiotemporal resolution for studying molecular mechanism(s) of crystallization. Typically, crystallization proceeds through a nucleation and growth mechanism, with the nucleation stage being the rate-limiting step. Historically, several advanced simulation techniques have been developed for studying rare event phenomena such as nucleation. However, there are various technical difficulties in applying these generic algorithms to crystallization processes.

One such difficulty arises in studying crystallization in systems in which structural relaxation is slow. In such circumstances, developing more efficient computer software can assist in expanding the range of accessible timescales. However, the existing advanced simulation techniques must also be modified to cope with the issue of correlated configurations/trajectories. A notable example is the supercooled water. Water transforms into hexagonal ice at ambient pressures. However, its crystallization can be suppressed for temperatures as low as 40 K below its melting point. Therefore, the ability to predict the extent and the rate of crystallization is key in understanding the behavior of supercooled aqueous systems. For instance, the fraction of icy droplets in a cloud is a very important parameter to many climatological models, and determines the light-absorbing properties and the precipitation propensity of a cloud, but the extent of freezing in atmospheric droplets and aerosols can not be determined from thermodynamics only. However, computational studies of ice nucleation are very challenging and the problem of calculating the rate of homogeneous nucleation of ice in realistic models of water has been dubbed as one of the most challenging open problems in computational statistical physics, besting the efforts of generations of computational scientists. Another aqueous system that is of utmost interest in the context of clean sources of energy is the water-gas mixtures that can transform into clathrate hydrates. Clathrates are crystalline solids in which water molecules organize into polyhedral cages that encompass small guest molecules such as hydrocarbons, carbon dioxide and nitrogen. Methane hydrates are present in the ocean bed and in the permafrost and it is believed that the methane trapped in the form of hydrates in those environments is orders of magnitude larger in quantity than the methane in known oil and natural gas resources. Methane hydrates can become sources of clean energy in the future. However, the mechanism and the kinetics of hydrate formation are not fully understood, due to the very same reasons that make computational studies of ice so challenging. In my postdoctoral years in Princeton, I developed a coarse-grained variant of the forward-flux sampling algorithm, and successfully used it for calculating nucleation rates in a fully atomistic model of water. As an assistant professor, I intend to expand on my postdoctoral work and study the problem of ice nucleation under atmospherically relevant circumstances, e.g. in droplets and films of supercooled water. There are several interesting questions to be asked in this regard, such as the effect of vapor-liquid interfaces on crystallization, or the role that inorganic impurities play in modifying the rate and the mechanism of crystallization. The findings of this investigation will be pivotal to the area of atmospheric and environmental sciences and can help us gain a better understanding of phenomena such as global warming and climate change. It can also assist us in devising guidelines for engineering our climate, e.g. by engineering the radiation budget and the precipitation patterns across the biosphere via changing the properties of clouds. I am also interested in the problem of hydrate nucleation, mostly at realistic supersaturations that have been mostly overlooked in existing computational studies of hydrates. Understanding the mechanisms of nucleation in those systems can assist us in developing better technologies for energy storage and carbon sequestration. It can also help us prevent the problems that hydrate formation can cause in production and transportation facilities in the petroleum industry.

Another difficulty in studying crystallization is the issue of nonconventional symmetries that can make the characterization of the disordered and the crystalline states a nontrivial task. The most notable examples are the complex structures that assemble from anisotropic nano- and colloidal building blocks. Many of these nonconventional symmetries can result in interesting functional properties, such as photonic band gaps, negative refractive index. During my doctoral years at the University of Michigan, we established the striking capably of anisotropic building blocks to assemble into structures with complex crystalline and quasicrystalline order. For instance, we showed that hard tetrahedra can assemble into a dodecagonal quasicrystal, a discovery that was followed by further discoveries of other intriguing disorder-order transitions in other hard and soft particle systems. Today, it is relatively straightforward to study the thermodynamics and the self-assembly propensities of different building blocks. However, studying kinetics of such transformations are less straightforward. There are two types of difficulties in applying the existing advanced simulation techniques to these crystallization processes. First of all, many of these structures have nonconventional symmetries, and can thus not be properly characterized by the existing order parameters that are widely used for simple atomic and molecular systems. It is necessary to emphasize that the problem of identifying a good oder parameter for these advanced simulation techniques is not just an exercise in geometry, but instead requires a large deal of experience and a lot of effort. The second difficulty is that no efficient algorithms exist for following the dynamics of systems that are comprised of building blocks with singular surfaces e.g. hard polyhedra. Combining my graduate experience in studying such nontrivial symmetries, and my postdoctoral experience in using and modifying advanced sampling techniques, I intend to tackle these two problems to obtain a mechanistic understanding of the processes that give rise to such nontrivial structures.

HIGHLIGHTS [AS OF MAY 2014]:

-Five first-author publications.

-One submitted manuscript.

-Five to be submitted within the next few months.

Multifunctional Polymer Nanocomposites towards Electromagnetic Interference Shielding and Flame Retardation

Qingliang He

Integrated Composites Laboratory (ICL), Dan F. Smith Department of Chemical Engineering, Lamar University, Beaumont, TX 77710 USA

ABSTRACT: Electromagnetic pollution and fire hazard are two major problems that compromise the sustainable development of high technologies in our modern society. Although extensive efforts had been made trying to solve both the two problems separately, the tentative for exploring materials which can solve both the two aforementioned problems simultaneously has rarely been reported. Polymer matrix nanocomposites reinforced with nanofillers possessing unique filler architectures with easily tuned composition are ideal for targeting desirable functionalities. However, the major challenge by using polymer matrix nanocomposites for both the electromagnetic interference (EMI) shielding and flame retardation is the unknown relationship between each other. Hence, the effective improvement for both the two functions by one material is hard to design.

Based on the mechanisms of EMI shielding (magnetic and dielectric materials, which may consist of elements possessing catalytic function for flame retardancy) and flame retardation (carbon and ceramic nanomaterials, which may contribute as dielectric components for EMI shielding), multifunctional polymer matrix nanocomposites with high efficiency in both EMI shielding and flame retardation have been developed. To be specific, polypropylene nanocomposites reinforced with nanofillers consists of magnetic (iron, cobalt and its core shell metal @metal oxide nanoparticles), dielectric materials (carbon nanomaterials including carbon nanotubes and graphene nanosheets, ceramic fillers or coating such as SiO_2) were synthesized using a facile one-pot wet chemistry approach.

Regarding the structure design, hierarchical 3-D filler systems with different dimensionalities (0D, 1D, 2D and 3D) were integrated to achieve unique architecture in polymer matrix; and surfactant with unique functional groups will also be introduced in order to improve the polymer-filler interfacial bonding strength. Meanwhile, how the filler composition, dispersion quality and assembly patterns in polymer matrix will affect the physicochemical properties and the corresponding functionalities for the as-prepared nanocomposites will be studied and the structure-property-performance relationship will be revealed.

SELECTED PEER-REVIEWED PUBLICATIONS (12 out of 40):

1. **Q. He**, T. Yuan, S. Wei, N. Haldolaarachchige, Z. Luo, D. P. Young, A. Khasanov, Z Guo, Morphology and phase controlled iron oxide nanoparticles stabilized with maleic anhydride grafted polypropylene, ***Angewandte Chemie International Edition*** 51(**2012**): 8842 -8845
2. **Q. He**, T. Yuan, J. Zhu, Z. Luo, N. Haldolaarachchige, L. Sun, A. Khasanov, Y. Li, D. P. Young, S. Wei, Z. Guo, Magnetic high density polyethylene nanocomposites reinforced with in-situ synthesized Fe@FeO core-shell nanoparticles, ***Polymer*** 53 (**2012**): 3642-3652
3. **Q. He,** T. Yuan, X. Zhang, Z. Luo, N. Haldolaarachchige, L. Sun, D. P. Young, S. Wei, Z. Guo, Magnetically Soft and Hard Polypropylene/Cobalt Nanocomposites: Role of Maleic Anhydride Grafted Polypropylene, ***Macromolecules*** 46 (**2013**): 2357-2368

4. **Q. He**, T. Yuan, Z. Luo, N. Haldolaarachchige, D. P Young, S. Wei, Z. Guo, Morphology and phase controlled cobalt nanostructures in magnetic polypropylene nanocomposites: the role of alkyl chain-length in maleic anhydride grafted polypropylene, ***Chemical Communications*** 49 (**2013**): 2679-2681
5. **Q. He**, T. Yuan, S. Wei, Z. Guo, Catalytic and Synergistic Effects on Thermal Stability and Combustion Behavior of Polypropylene: Influence of Maleic Anhydride Grafted Polypropylene Stabilized Cobalt Nanoparticles ***Journal of Materials Chemistry A,*** 1 (**2013**): 13064–13075
6. **Q. He**, T. Yuan, X. Yan, Z. Luo, N. Haldolaarachchige, D. P. Young, S. Wei, Z. Guo, One-pot Synthesis of Size- and Morphology- Controlled 1-D Iron Oxide Nanochains with Manipulated Magnetic Properties , ***Chemical Communications,*** 50 (**2014**): 201-203
7. **Q. He**, T. Yuan, X. Yan, D. Ding, Q. Wang, Z. Luo, T. D. Shen, S. Wei, D. Cao, Z. Guo, Flame-Retardant Polypropylene/Multiwall Carbon Nanotube Nanocomposites: Effects of Surface Functionalization and Surfactant Molecular Weight, ***Macromolecular Chemistry and Physics,*** 215 (**2014**): 327–340
8. **Q. He**, T. Yuan, S. Wei, Z. Guo, Unusual Function of Modified Polyolefins for Manipulating Magnetic Nanostructures, ***JOM*** (Invited review contribution), 66 (**2014**): 655-659
9. J. Zhu, **Q. He**, Z. Luo, A. Khasanov, Y. Li, L. Sun, Q. Wang, S. Wei, Z. Guo, Property manipulated polypropylene-iron nanocomposites with maleic anhydride polypropylene, ***Journal of Materials Chemistry*** 22 (**2012**): 15928-15938
10. X. Zhang, **Q. He**, H. Gu, H. S. Wei, Z. Guo, Polyaniline Stabilized Barium Titanate Nanoparticles Reinforced Epoxy Nanocomposites with High Dielectric Permittivity and Reduced Flammability, ***Journal of Materials Chemistry C*** 1 (**2013**): 2886-2899
11. X. Zhang, **Q. He**, H. Gu, H. A. Colorado, S. Wei, Z. Guo, Flame retardant electrical conductive nano-polymers based on bisphenol F epoxy resin reinforced with nano polyanilines, ***ACS Applied Materials & Interfaces*** 5 (**2013**): 898-910
12. H. Gu, J. Guo, **Q. He**, Y. Jiang, Y. Huang, N. Haldolaarachchige, Z. Luo, D. P. Young, S.Wei, Z. Guo, Magnetoresistive Polyaniline/Multi-Walled Carbon Nanotubes Nanocomposites with Negative Permittivity, ***Nanoscale,*** 6 (**2014**): 181-189

AWARDS/HONORS:

1. 1st Place Award, *International Polyolefins Conference 2012*, Houston, TX, USA, 2012
2. Scholarship Award, *Society of Plastics Engineers, Lamar University Chapter*, Beaumont, TX, USA, 2012
3. Scholarship Award, *South Texas Section, Society of Plastics Engineers, College of Engineering, Lamar University*, Beaumont, TX, USA, 2012
4. 2nd Place Research Excellence Award in the *2012 Chinese-American Chemical Society Southwest Chapter (CACS-SW) Annual Poster Competition*, Houston, TX, USA, 2012
5. Scholarship Award, Polymer Modifiers and Additives Division (PMAD) of the *Society of Plastics Engineers*, Houston, TX, USA, 2013
6. 2nd Place Research Excellence Award in the *2013 Chinese-American Chemical Society Southwest Chapter (CACS-SW) Annual Poster Competition*, Houston, TX, USA, 2013

Understanding and Predicting Chemical Reactivity in Supported Metal Catalysis

David Hibbitts

Metal-catalyzed reactions occur on polycrystalline particles, often at high coverages of reaction intermediates and (in some instances) with H_2O present. These complicating factors, however, have only recently been addressed by theoretical treatments of chemical reactions, due to the methodological and computational challenges they present.

Aerobic oxidation of alcohols (including diols) on Pd, Pt, and Au catalysts is active for the formation of carboxylic acids with H_2O and hydroxide present. Density functional theory (DFT) calculations performed with discrete H_2O molecules found that alcohol oxidation is a base- and metal-catalyzed reaction in which hydroxide acts as a Brønsted base in solution and at the H_2O-metal interface to deprotonate alcohols and nucleophilically attack aldehyde intermediates to form carboxylic acids. These conclusions were consistent with kinetic data, which showed that rates are proportional to the pH and pKa of the alcohol and isotopic data, which demonstrated that the additional O-atoms in the product acids originated from the H_2O solvent rather than from the O_2-coreactant. This oxidation process consumes hydroxide and produces electrons, a process balanced by oxygen reduction reactions (ORR) in which O_2 abstracts H-atoms from H_2O (to replenish hydroxide) in subsequent steps to form OOH* and *HOOH* species prior to O-O cleavage. These reactions combine to form a short-circuited electrochemical cell, permitting insights into the two half-reactions to be applied to electrochemical applications.

H_2O also plays a role during gas-phase Fischer-Tropsch synthesis (FTS), which forms large hydrocarbons and H_2O from synthesis gas (CO-H_2) derived from biomass, coal, or natural gas. Kinetic and isotopic data coupled with DFT calculations suggest that CO activates following initial reactions with chemisorbed H-atoms (H*) to form *HCOH* preceding C-O cleavage rather than via direct CO* dissociation, which has been shown to have much larger barriers on both low-index terraces and defect sites of 201- and 586-atom Ru and Co nanoparticles modeled at high CO* coverages. H_2O increases turnover rates on Co and Ru catalysts and DFT calculations on such models show that H_2O acts as a H-shuttle in which H* reacts with nearby H_2O to form a short-lived H_3O^+ intermediate, which then protonates CO* (to form COH*) or HCO* (to form *HCOH*) both of which lead to reduced barriers for *HCOH* formation, increasing the rate of CO activation.

The H*-assisted mechanism of CO activation via *HCOH* intermediates on Co and Ru is reminiscent of O_2 activations via OOH* and *HOOH* species on Pd, Pt, and Au described above and was also observed during NO-H_2 reactions in which NO reacts with H* to form *HNOH* species prior to N-O cleavage on Pt. Kinetic, isotopic, and theoretical studies at high NO* coverage of NO-H_2 reactions on Pt indicated that NO activates through that H*-assisted mechanism to form *HNOH*, which then dissociates into NH* and OH* intermediates that undergo fast reactions to give N_2O, NH_3, N_2, and H_2O products.

This work has indicated the necessity of rigorous kinetic studies and isotopic data to be combined with DFT calculations on complex catalyst models, in which the effects of H_2O, high coverages of co-adsorbates, bimolecular reactions, and polycrystalline surfaces may be required to correctly identify reaction mechanisms. Future plans are to continue such comprehensive studies through research that combines gaseous and aqueous-phase kinetic measurements with isotopic investigations and DFT calculations to understand the factors that affect metal catalyst reactivity and selectivity.

387872 Bridging the Gap Between Academia and Industry: Designing Curricula to Prepare Students for the Profession

Abstract:

One purpose of an undergraduate engineering education is to prepare students to become practicing engineers. However, in recent years, industry has expressed concern that while graduating students are technically competent, they are less prepared for other aspects of engineering jobs. Curricula are well-designed to convey technical content, but focus less on skills such as communication, project management, creativity, and teamwork. Although capstone design courses emphasize the importance of working in teams and communicating progress, students would be better prepared for industry if this emphasis were woven throughout the entire undergraduate engineering curriculum. My research interests lie in developing innovative ways to equip students with these skills early on and throughout their entire undergraduate education, preparing them to attack open-ended problems, develop innovative solutions, communicate ideas, interact with peers, and manage their projects and time.

My doctoral work at Purdue University involved working closely with industry partners to develop an automation and control system for a pharmaceutical manufacturing process that utilized inkjet printing technology. This project not only gave me a solid foundation in chemical engineering fundamentals, but also convinced me of the importance of closely linking curricula with industry principles in order to better prepare students to serve the needs of the profession. During my graduate studies, I sought out many leadership opportunities, including being president of our Graduate Student Organization and heading Future Faculty seminars for interested graduate students. I also devoted a large portion of my time to teaching, receiving two teaching awards for my semesters as a teaching assistant. Through my research, leadership and teaching experiences, I realized that there was an opportunity to improve curricula and pedagogy to better prepare future engineers, leading to my current focus on engineering education research.

My current post-doctoral research at Oregon State University focuses on developing technology to instill conceptual understanding in students through different pedagogical approaches. Through the use of tools such as the AIChE Concept Warehouse and the Industrially-Situated Virtual Laboratory Project, we focus on methods that challenge students to grasp fundamental concepts, as opposed to teaching students to memorize content and algorithmically solve problems. These approaches help to improve knowledge retention, conceptual understanding and transfer and also help to better prepare students to work with the open-ended problems they will encounter as practicing engineers. As a professor, I plan to continue in this area of research by investigating student learning and creating ways to bridge the gap between education and industry. I want to help students develop the skills vital to a career in engineering throughout their entire course of study, better preparing them to work in a challenging professional environment. I look forward to transforming engineering curricula while educating and developing future engineers.

387005 Engineering Biomimetic Cues to Restore Musculoskeletal Tissue Function

Abstract:

Musculoskeletal injuries and diseases, which include meniscal tears and degenerative disc disease, are a significant health concern in the United States. Current treatments, however, typically rely on donor tissues (either allo- or autografts) and suffer from poor availability. Furthermore, donor tissue does not always adequately restore function, integrates poorly with surrounding tissue, and can have a high morbidity. Tissue engineering aimed at replicating and/or restoring the biological and mechanical cues required for tissue function offers an advantage to current treatments and prevents further musculoskeletal degeneration. My research has focused on the intelligent design of hydrogels towards this aim.

During my Ph.D., under the advisement of Anthony Lowman and Giuseppe Palmese, I developed and characterized a fiber-reinforced hydrogel composite for meniscal replacement aimed at replicating the anisotropic mechanical properties of the native meniscus. The hydrogel composite was characterized in compression (0.1 - 0.8 MPa) and tension (0.1 - 250 MPa) for a number of formulations demonstrating fine control over mechanical properties within the range of the human meniscus. Furthermore, the fiber-hydrogel interface was tailored for optimal adhesion using a novel biocompatible grafting technique to form a direct covalent linkage between the fiber and hydrogel, where grafting resulted in a 20-fold improvement to interfacial shear strength. Following implantation into human cadaveric knees, ex vivo gait simulations showed improved contact areas and stresses along the tibial plateau for the hydrogel composites in comparison to a meniscectomy.

Currently, I am a postdoctoral fellow in Jason Burdick's laboratory researching synergistic growth factor delivery for improved osteogenesis using proteolytically degradable hydrogels. A critical-sized calvarial defect was used to determine the effect of combined SDF-1α, a known chemokine, and BMP-2 delivery on bone formation in vivo. The treatment group with combined delivery showed significantly higher bone formation when compared to hydrogels loaded with either growth factor alone or the empty defect group. More importantly, SDF-1α, when delivered in combination with BMP-2, effectively reduced the required BMP-2 dose for similar bone formation by increasing cellular invasion into the defect site. Ongoing work evaluating the activation of other osteogenic signaling pathways is being investigated to further improve BMP-induced osteogenesis.

In the future as a faculty member, I will use my experience with and knowledge of polymer science, biomechanics, and growth factor signaling to develop finely tunable materials that replicate and restore the mechanical, structural, and biological signals needed for function in musculoskeletal tissues.

384727 Developing Novel Biocatalysts Using Synthetic Biology

Abstract:

Synthetic biology is an emerging field to develop new biological systems that perform novel functions for useful purposes. The field has been rapidly growing by designing synthetic genetic circuits, demonstrating biological principles, and developing artificial cells. Advances made have inspired a variety of new applications in industrial and clinical processing to address current needs in sustainable energy and food supply as well as novel therapeutics development. However, it is still early in the development of real-world applications of synthetic biology, and progress is limited by the complexity and the unexpected responses of biological systems. My faculty research program seeks to address this limitation. We will develop biocatalysts for a variety of applications by using interdisciplinary approaches that analyze, understand, and control cellular processes on a molecular and systems level. Specifically, I am interested in three areas: i) engineering biofilms for disease treatment and therapeutics production ii) cell-free synthetic biology for developing artificial metabolic pathways that are impossible, or otherwise difficult, to generate using cell-based systems, and iii) microbial electrosynthesis using engineered biofilms and cell-free systems to convert electricity to biochemical, therapeutics, and fuels. I will apply a unique and diverse set of skills in synthetic biology, protein engineering, genome engineering, and cell-free engineering for the development of novel biocatalysts and biosynthesis pathways for clinical and industrial uses. My work will allow me to build a leading research program for compelling applications in medicine, basic science and quantitative applied biotechnology.

377619 Engineering the Flow Behavior of Colloidal Materials through Surface Modification and Shape Anisotropy

Abstract:

The idea that colloidal materials can be designed with enhanced rheological properties by incorporating shape and surface anisotropy is motivated by our finding that structural rigidity can be used to predict the nonlinear elasticity in gels that have undergone yielding. We demonstrate that the predictive power of microscopic theories can be improved when both the microstructural rigidity and the dynamical heterogeneity induced by yielding are taken into account. Because thermal rupture forces play a critical role in yielding, we develop a model gel system in which rheological measurements can be carried out in conjunction with microscopy experiments and direct force measurements using optical tweezers. The validity of structural rigidity is also tested by synthesizing colloidal spheroids and roughened particles. We find that colloidal oblate spheroids self-assemble into tilted structures in a specific range of volume fractions and attraction strengths. These structures could pave the way for maintaining the elasticity of colloidal gels even at exceedingly low particle loading. Finally, we observe a significant increase in the shear thickening behavior of rough colloids in concentrated suspensions. These results collectively provide experimental support for the applicability of structural rigidity as a guiding principle in engineering the flow properties of a broad range of soft matter.

390384 Microfluidic Biosensors to Monitor Health

Abstract:

Many deaths in the U.S. are attributed to chronic diseases. Diabetes, heart disease, and stroke are all examples of preventable and/or manageable chronic diseases that affect many people under the age of 60. This means that on a yearly basis large amounts of money are spent on health care for treatment. Along with this expense comes compromised quality of life (in some cases disability), and escalating health care costs. Healthcare cost can be alleviated if there is technology available that allows early disease detection, so that diet and exercise can be implemented as a treatment option. Microfluidics and dielectrophoresis (DEP) are technologies applicable to disease detection. Specifically microfluidics manipulates small volumes of fluid (μL) and DEP uses electric fields to detect differences and separate biological cells. Together DEP and microfluidic devices are useful for characterizing biological cell properties and distinguishing between healthy and unhealthy cells. Microfluidic devices are fabricated at a low cost making them suitable in healthcare.

The theme of my research will be improved healthcare using microfluidic biosensors with a focus on (1) health and (2) exercise assessments. I would like to use my knowledge in DEP to achieve these assessments in microdevices. Health care assessments such as blood screens completed in microfluidic biosensors can give indication of disease on a shorter time scale. Wearable microfluidic biosensors that monitor a person's health during exercise are also important in healthcare especially for person's overweight. I have experience using DEP technology to study red blood cells, polystyrene beads, and human mesenchymal stem cells (hMSCs). The focus of my doctoral work was on the quantification of hMSCs dielectric properties (membrane permittivity and capacitance) for the design of a continuous cell sorting microfluidic device. I've also worked on the development of a new DEP experimental technique that rapidly determines the DEP response spectra of biological particles via frequency sweep rates. This work was optimized using polystyrene beads and extended to red blood cells. My plan is to continue research on DEP and biological cells with expansion into the area of exercise science to help improve healthcare.

388149 Hierarchical Hybrid Assembly of Functional Nanomaterials

Abstract:

Functional nanomaterials will play critical role in many future technologies related to energy, environment and health. Next generation hybrid or plug-in-hybrid vehicles would require high-power, high-energy density energy storage devices with lower emission for sustainability. To achieve these goals, it is essential to design, identify and sythesize materials that can accommodate efficient transport of ions and electrons in addition to high surface area. Energy storage devices based on nanoscale materials (nanoparticles, nanotubes or nanofibers of electroactive polymers, metal oxides and carbon nanotubes) recently gained significant interest to afford high-power, high-energy density without compromising cycle life. During my postdoctoral research at MIT, I've developed expertise in the design of novel electrode assembly for energy storage devices. I focused on developing hybrid electrodes based on conjugated polymers (Polyaniline, PANi), metal oxides (TiO_2), graphene oxide (GO) and multi-walled carbon nanotubes (MWNT); these nanostructured hybrid electrodes performed exceptionally well (compared to graphitic electrodes) delivering high-power, high-capacity simultaneously for electrochemical pseudocapacitors that suitable for electric vehicles or plug-in electric vehicles. Along with conventional electrode development, I designed alternative layer-by-layer assembled electrodes based on electrostatic interaction of nanomaterials. The film architecture, assembly technique and post treatment were optimized to create high performance electrodes. These electrodes were engineered to achieve high electronic conductivity and high porosity for improved ion transport of aqueous or organic electrolytes rendering higher capacitance than graphite electrodes with excellent mechanical and electrochemical stability over thousands of cycles. The development of hybrid electrodes of TiO_2, MWNTs and PANi nanofibers with controlled aspect ratio and morphology exhibits attractive potential in terms of improved electronic conductivity and porosity that are competitive with the expensive state-of-the-art RuO_2. However, these multilayer electrodes of nanoscale materials would reach a performance limit unless we optimize the selection of electrolytes which affect ion conduction and thereby overall electrochemical performance. Polymer-based gel electrolyte could be a solution to address this challenge that could be part of the separator membrane in the overall electrochemical cell design. In addition, my Ph.D research experiences on polymeric membrane design and fabrication would allow me to develop such novel separator membranes. In future, as an independent researcher, I am interested in the design and facile synthesis of nanomaterials and their assembly for hierarchical electrodes and electrolytes. Using thses electrodes, my lab will focus on fundamental aspects of charge storages, ion transport which could be ultimately translated to manufacturing. These nanostructured materials along with polymeric/ionic gel electrolyte would have tremendous potential to impact the overall electrode design and performance for improve energy storage devices like metal-air batteries and hybrid supercapacitors.

Searching for epigenetic patterns as biomarkers for cancer diagnosis

Isabel Jimenez-useche[1,2]

[1] School of Chemical Engineering, Purdue University, West Lafayette, IN, USA

[2] School of Engineering Education, Purdue University, West Lafayette, IN, USA

DNA methylation plays an essential role in various biological processes such as stem cell differentiation, imprinting, X-chromosome inactivation, etc. Increased DNA methylation levels have been associated with chromatin compaction leading to gene silencing. For example, abnormal DNA methylation is associated with silencing of tumor suppressor genes and is observed in the onset of tumorigenesis. There is evidence suggesting that not all methylation events are relevant in chromatin compaction and the initiation of cancer. It seems that methylation at certain locations of the DNA might be key to start chromatin compaction and gene silencing, but the location of this methylation sites is still unknown. In order to identify DNA methylation locations that could potentially be involved in chromatin compaction and gene silencing, my research focus has been on studying the effects of different DNA methylation patterns in the modulation of chromatin compaction.

Using multiple molecular biology techniques, I constructed chromatin-like molecules representing three levels of chromatin organization: 1) Naked DNA, 2) Nucleosomes and 3) Nucleosome arrays. These *in vitro* systems allowed the evaluation of the direct effect of DNA methylation on the conformation and dynamics of chromatin without the influence of other proteins found in the cell nucleus. In addition, I have introduced different DNA methylation patterns to my *in vitro* systems to study the effects of the position of the methylation sites in the chromatin structure.

My results showed that compactness of chromatin-like molecules with a stretch of CG dinucleotides in the central dyad of the nucleosomes or at the major groove locations showed a dependence on their methylation status. Specifically, methylation of a stretch of dinucleotides decreased the relative compactness of nucleosomes and increased tetranucleosome compaction. The opposite effect was observed for nucleosomes and tetranucleosomes with CG dinucleotides located at the major grooves. These findings confirm that the presence of methylation in certain locations within chromatin lead to distinctive effects on the compactness of chromatin-like molecules. These results suggest that these two methylation patterns could potentially be biologically relevant for gene expression regulation. Although the biological significance of these methylation locations is still to be determined, the results of this research are instrumental in understanding the mechanism of chromatin compaction by DNA methylation and could be applied in the identification of new and more accurate DNA methylation biomarkers for early detection of cancer.

In addition to my work with epigenetic modifications, I have always been fascinated by the ability of instructors to facilitate the process of learning, in particular in engineering disciplines. This interest lead me to complement my previous research experience with a new research focus in the area of Engineering Education. The goal of my current research is to assess the experience of first-year international undergraduate students in engineering programs in American universities. I have been able to identify four different challenges that international students face when coming to American Universities. These challenges are: 1) cultural differences, 2) communication challenges, 3) personal challenges and 4) academic challenges. Each of these challenges has different implication in the students' academic performance. I intend to assess the impact of these challenges using multiple assessment instruments such as surveys, questionnaires, interviews, focus groups, students' grades and other kind of evaluations. The results of this research will be instrumental in identifying relevant factors that might be addressed to improve the experience of international students in American universities.

393664 Active colloids at the oil-water interface: The biophysics of microbial oil degradation

Abstract:

In natural habitats, microbes regularly encounter a wide variety of interfaces that influence their behavior and activity. Bacteria often accumulate near interfaces and will eventually attach to them. In favorable conditions, bacterial will grow and divide leading to colonization of the interface. Dense colonies formed at the liquid-solid interface are referred to as biofilms while those formed at the liquid-gas interface are referred to as pellicles. Bacterial colonies also form at the liquid-liquid interface, for example at the oil-water interface, although the process is not well understood. Biofilms can be undesirable and destructive causing infection and biofouling, however, they can be beneficial in the production of biofuels and environmental processes such as waste water treatment. Bacterial attachment to an interface is the first step in colonization, therefore understanding the mechanisms of bacterial attachment will provide controlled methods to prevent or enhance biofilm formation.

My research interests are broadly in the field of experimental soft condensed matter physics. My doctoral research (with Richard Lueptow and Julio Ottino, Northwestern University) focused on mixing and segregation dynamics of granular mixtures in rotating tumblers. My postdoctoral research (with Paulo Arratia, UPenn) focused on establishing a link between the molecular microstructure and the macroscopic rheological properties of complex polymeric fluids. My current research (with Roman Stocker, MIT) focuses on understanding the fundamental interactions between microbes and liquid-liquid interfaces, specifically the biodegradation of crude oil by marine bacteria. The degradation of oil in the ocean is a process of immense societal and ecological importance whose efficiency ultimately hinges on fundamental problems in colloidal and interface physics because degradation occurs when individual bacteria attach to individual oil drops.

I will discuss experimental and theoretical results on the effect of oil drop diameter on the attachment of marine bacteria to the oil-water interface. Using phase-contrast and fluorescence microscopy, we directly observed the dynamics of marine bacteria in the presence of oil droplets of varying diameter within microfluidic devices. We find that microbial growth and degradation become strongly limited for drops smaller than a critical diameter. A theoretical model of attachment supports this finding, suggesting that rendering oil droplets too small by the excessive addition of dispersants, as often done in oil spills, might be a counterproductive strategy. This research aims to pursue a deeper understanding of the fundamental oil-microbe interactions and promises to open the door to a broad set of other problems in interfacial physics, hydrocarbon microbiology, and other oil spill management related questions.

381161 Design of Multifunctional Materials for Energy Related Applications

Abstract:

The materials design by first-principles modeling provides an opportunity to study new materials for energy related applications. Today, the importance of this approach rapidly increases in chemical engineering and other disciplines. The atomic level of information, such as the properties of matter, will be obtained by the computational methods like density functional theory (DFT). The specific systems we will study are (a) crystalline membranes for gas purification (b) proton-conducting perovskites for fuel cells (c) Li-ion electrolytes, and (d) ferroelectric and multiferroic materials.

Post-doctoral advisor: Professor Craig J. Fennie, School of Applied & Engineering Physics, Cornell University

Doctoral thesis advisor: Professor David S. Sholl, School of Chemical and Biomolecular Engineering, Georgia Institute of Technology

Design of catalysts for energy conversion and storage

Yijin Kang, Materials Science Division, Argonne National Laboratory

My research interests and achievements focus on fundamental understanding of catalytic processes and development of highly active (or selective) and durable catalysts for the reactions of industrial importance, especially for energy conversion and storage. I establish a research system that connects fundamental investigation on well-defined extended surfaces (e.g. single crystal surfaces), extrapolation onto nanocrystals with highly controlled shape and size, exploration of interfacial interaction using novel nanocrystal superlattices as platform, and finally design of high performance catalysts in which all the possible beneficial properties from complex functional structures are implemented.

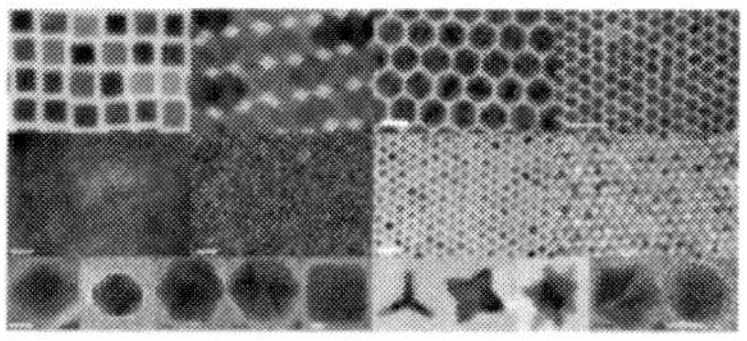

Figure 1. High quality Pt nanocrystals with various shapes. The size is tunable in the range of 2-30nm.[1]

The fundamental understanding of correlationship between materials structure and catalytic properties is the key to design new catalysts with desired properties. The very first step would be to understand the most simplified surface processes using well-defined extended surfaces and computer-powered theoretical simulations. However, the real catalysts used in industrial processes are mainly nanomaterials that take advantage of their high surface-to-volume ratio. It is necessary to confirm that the desirable properties found in simplified cases are actually functioning in the nanomaterials. Therefore, well-defined nanocrystals with exposed surfaces mimicking the simplified surfaces which are used in theoretical studies are highly valuable and appreciated. My work at this level focuses on providing high quality model materials at nanoscale. For instance, in order to study the surface sensitive reactions on Pt surface at nanoscale, I prepared high quality Pt nanocrystals with various controlled shapes[1,2], exposing selective surfaces (e.g. {111} on octahedra and icosahedra, {100} on cubes, and {110} on rhombic dodecahedra). The library of such highly controlled nanocrystals with desired surface is not limited to single element metals[3], but also include alloys[4,5] and oxides[6].

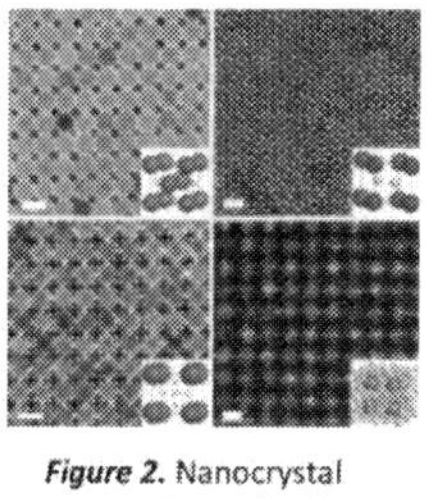

Figure 2. Nanocrystal superlattices, allowing programmable inter-particle contacts.[7]

Besides desired surfaces, the interactions among the components within the catalysts are also important. For example, in the Pt-Ni alloy Ni atoms modify the electronic structure of Pt, weakening the adsorption of OH_{ad} spectator species on Pt surfaces, thus enhancing the oxygen reduction reaction (ORR) activity. Because of certain desired metal-oxide interfacial interactions that promote catalysis or introduce synergistic effect, metal catalysts supported on oxide supports are the other class of important catalysts that exploit beneficial interactions. In additional to well-controlled size and shape, the structures rendering the desired interactions are needed to carry out the investigation of interfacial effects. I established a platform using binary nanocrystal superlattices (BNSLs) to construct tunable interactions[7,8]. For example, by constructing Au-FeO_x BNSLs with various structures (AB, AB_2, AB_{13}), I was able to control the population of Au-FeO_x contacts. In the investigation of CO oxidation, a linear correlation between population of contacts and activity was found. Such correlation provided strong evidence that the interface of Au-FeO_x contacts is responsible for the extraordinary activity toward CO oxidation, rather than small sized Au[7].

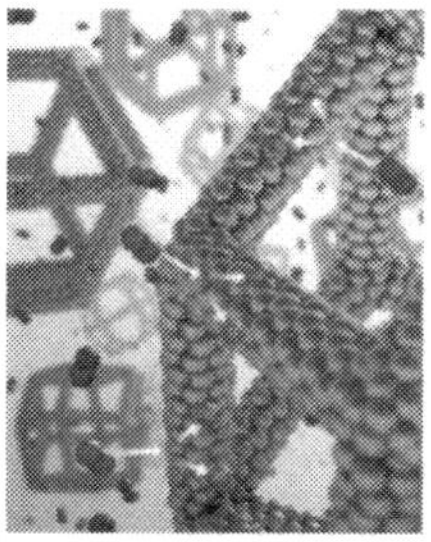

Figure 3. Pt_3Ni catalysts for ORR (oxygen electrode of fuel cells) that incorporated all beneficial properties in one nanostructure.[9]

Finally, with all the knowledge learned from theoretical studies and investigation on model materials, I am able to design catalyst that incorporates the beneficial properties in a single nanostructure. For example, my Berkeley collaborators observed a shape evolution from solid $PtNi_3$ nanocrystal to hollow Pt_3Ni nanoframes[9]. I took advantage of the hollow structure and Pt_3Ni composition of the nanoframes. I treated the surface of nanoframes to promote the surface segregation, obtaining the surface structure mimicking the optimal $Pt_3Ni(111)$-Pt-Skin for ORR. The hollow structure of nanoframes not only substantially increases the active surface area, but also provides a capillary force to hold a specific ionic liquid in which oxygen solubility is much higher than that in regular acid electrolyte. Therefore, combining all these beneficial properties for ORR in surface segregated Pt_3Ni nanoframes, I was able to achieve a 36-fold enhancement in activity in comparison to state-of-the-art Pt/C catalysts[9]. The similar approach is applicable to other catalyst systems and for other important reactions. For instance, I made Pt-Pb core-shell structure with desirable Pt_3Pb intermetallic structure, in order to overcome the poisoning problem thus to enhance the overall activity for formic acid oxidation[10].

In summary, I have established an effective catalyst designing route that extracts knowledge from model study and theoretical computation, fabricates desirable surfaces and constructs beneficial structures, and ultimately directs the production of real catalysts of industrial interests. I am seeking for a tenure-track faculty position at R1 universities at the cycle of 2014-2015. I would like to focus my future research on energy conversion and storage, especially on solar-to-fuel conversion and advanced battery technologies.

Notes and references:
1. Kang *et al.* ***J. Am. Chem. Soc.***, **2013**, 135 (7), 2741
2. Kang *et al.* ***ACS Nano***, **2013**, 7(1), 645
3. Kang *et al.* ***Angew. Chem. Int. Ed.***, **2010**, 49(35), 6156
4. Kang *et al.* ***J. Am. Chem. Soc.***, **2010**, 132(22), 7568
5. Kang *et al.* ***ACS Nano***, **2012**, *6* (6), 5642
6. Kang *et al.* ***Angew. Chem. Int. Ed.***, **2011**, 50(19), 4378
7. Kang *et al.* ***J. Am. Chem. Soc.***, **2013**, 135 (4), 1499
8. Kang *et al.* ***J. Am. Chem. Soc.***, **2013**, 135 (1), 42
9. Kang *et al.* ***Science***, **2014**, *DOI: 10.1126/science.1249061*
10. Kang *et al.* ***ACS Nano***, **2012**, 6 (3), 2818

393556 Investigating the Selective Removal of Anions and Cations in Electrodialysis/ Electrodialysis Reversal (ED/EDR)

Abstract:

Fresh water resources are inherently limited by their quantity, and it has drawn the world's attention. Electrodialysis and electrodialysis reversal (ED/EDR) are two closely-related water purification processes which use an electrical field to propel ions through ion exchange membranes. The most effective use of these technologies has proven to be the production of potable water from brackish water resources. In brackish water resources as well as industrial water streams, water compositions differ, and removal of particular monovalent or divalent ions may be more desired than others. Therefore, removing ions at different rates under more controllable conditions is valuable. Selective removal of ions in an ED / EDR process can decrease the energy due to removing the target ions rather than unwanted ones. This experimental study, which investigates the nature of ion removal and the effects of operating parameters on the process, was conducted to explore the dominant empirical model on the selective removal of monovalent ions and divalent ions in the laboratory-scale ED process.

During my PhD studies, I have conducted the experiments in pilot scale EDR at the Brackish Groundwater National Desalination Research Facility (BGNDRF) in Alamogordo, NM. This set of experiments provided me a general idea regarding the effect of operating parameters on the removal of different ions in the EDR process. Utilizing this data, I designed, and built a complete laboratory scale ED system equipped with a complete data acquisition system. The reason that the data acquisition system was designed and used in this system was to provide continuous and accurate monitoring and recording the operating conditions during the entire experiments. The collected data was used in developing an empirical model to predict the removal of monovalent and divalent ions in the ED/EDR process at various operating conditions. The developed model can be widely used for full-scale electrodialysis operations for which removal of some of the specific ions is desired.

As an independent scientist, I will focus on improving the ED/EDR as well as other membrane-based desalination technologies in order to reduce the energy and cost of the process without impacting performance of these processes. Additionally, I will work on using renewable energies, specifically solar energy, to provide off-grid desalination systems for rural, remote off-grid areas to provide water for people who are suffering from fresh water scarcity.

390958 Combining Computation and Experiment to Uncover Environment Friendly Solutions to Energy Problems

Abstract:

The world's dependence on fossil fuels has lead to the need for alternate sources of energy as supplies dwindle, as well as a growing need to remove harmful compounds from the air. Solar energy and hydrogen energy are promising candidates for supplanting fossil fuels while novel adsorbents like metal-organic frameworks (MOFs) are promising materials for removing harmful gases. To date, my research has been concerned with both aspects of the fossil fuel problem. Specifically, my M.S. and Ph.D studies were concerned with solar energy and hydrogen energy. In order for solar cells to be a feasible alternative to fossil fuels, they need to have a competitive power conversion efficiency. By adding partially sulfonated ionomer to the active layer of the solar cell, my research led to an increase in power conversion efficiency. Regarding hydrogen energy, my research was focused on understanding the thermodynamics of metal hydride reactions for hydrogen storage applications. Specifically, the goal was to screen thermodynamically promising metal hydride reactions from a full database of metal hydride mixtures using first-principles calculations. The large-scale screening ultimately provided a number of promising single-step or multi-step metal hydride reactions. My postdoctoral research at Northwestern University was related to investigating MOFs for the removal and separation of harmful gases. I used computational methods to screen and assess functional groups that could be incorporated into MOF ligands to preferentially adsorb harmful gases under humid conditions. In addition, I was also involved in assessing new methods for characterization of MOFs, as well as assessing MOFs for various gas separations. Currently, I am interested in lithium ion battery issues as a research director and postdoctoral fellow at Georgia Institute of Technology. Inspired by my research background, my future research will be focused on producing energy sources, such as methane or methanol, through carbon dioxide reduction on nanosized clusters or other shapes of metals and metal-based compounds. The nanoscale particles will have different chemical properties from bulk-based surfaces, and thus understanding the effect of the particle size, shape, and type on the carbon dioxide reduction would be an interesting topic for environment-friendly energy applications. Specifically, I have a plan to examine the changes in the mechanisms and activation barriers of the carbon dioxide reduction, increasing the particle size or changing the particle shape and the type of metals.

389054 Design of Biopolymer Building Blocks for Novel Self-Assembled Materials

Abstract:

Bioinspired and biomimetic materials, in particular hydrogels, are currently one of the most interesting areas in academic and industrial research, encompassing both biomedical (drug delivery, tissue engineering) and industrial applications (filters, sensors, and actuators). To meet these emerging demands, studies have focused on tailoring a material's mechanical properties (e.g. mechanical strength, extensibility, toughness, injectability, self-healing, and lifetime) and mimicking the physical/chemical properties of natural materials (e.g. extracellular matrix and muscle). Biopolymers, in particular natural proteins and synthetic polypeptides, have received a great deal of attention as elements in these materials. Amino acid sequences of polypeptides can be precisely controlled and easily modified through recombinant DNA technology while large batches of high purity and monodisperse polypeptides can be cost-effectively biosynthesized. More importantly, different sequences can produce unique folded structures that result in distinct biophysical and biochemical properties at the molecular level including mechanical properties, molecular recognition abilities, and intelligent responses to external stimuli. If these exceptional molecular properties of biopolymers are incorporated into macroscopic materials, multifunctional biopolymer-based materials that exceed those attainable from purely synthetic polymer-based materials can be designed.

Mechanical behaviors of muscle can be potentially mimicked by engineered hyperelastic materials/hydrogels constructed by crosslinking folded protein structures. Using a custom-designed atomic force microscope (AFM) for single-molecule force spectroscopy, structure-mechanical property relationships of α-helical repeat proteins and globular proteins have been characterized, demonstrating that similarly structured proteins can be used as building blocks for hyperelastic materials due to their reproducible large extensibility and energy dissipation, and fast response to external forces. In addition, to maintain proper folding and to preserve the desired mechanical properties in macroscopic materials, end-crosslinking of the proteins was investigated. Simple physical crosslinking was performed by fusing subunits of protein complexes to their termini and it was confirmed to successfully construct the mechanical protein-based supramolecular self-assemblies as designed. A new technique was also developed to directly quantify the interfacial rupture strength of the subunits in their individual assemblies, permitting the investigation of various protein complexes that can be used as molecular "screws" or "bolts" in novel biopolymeric materials.

Protein (nucleoporin)-based biological hydrogels with selective filtering properties have been mimicked by constructing biosynthetic hydrogels built from newly designed nucleoporin-like polypeptide (NLP) building blocks. The biological hydrogels in the nuclear pore have been attractive due to their capability of selectively filtering $< 0.1\%$ of all proteins in a eukaryotic cell while translocating over 1,000 molecules per second. Development of similarly behaving biopolymer gels can potentially impact separation technologies and defense applications but is limited by the low synthesis yield of proteins and long gel processing time. Artificially engineered high-yield NLP building blocks that can mimic the structural properties of nucleoporins have been successfully developed, and rapid gel construction is achieved. Using a combination of selective permeability fluorescence assays, shear rheology, and Raman spectroscopy, the mechanism of artificially engineered NLP hydrogel function is studied.

380028 Development of Microporous Membranes and Thin Films for Energy and Environmental Systems

Abstract:

My research interest is in the areas of inorganic membranes and thin films of nanostructure, particularly the nanoporous zeolites and zeolite membranes. I have focused on developing methodologies for synthesizing materials of desired microstructure and chemical composition and understanding the materials' structure-property relationships. My research has been particularly concentrated on the ceramic membranes for chemical separations and catalytic reactions with direct applications in the emerging clean energy technologies and environmental management systems.

During my PhD program, I have been involved to develop a new method of controlled modification of zeolite pore structure that can make the membrane for simultaneously achieving high selectivity and high permeance in the separation of similarly-sized small molecules such as H_2/CO_2. The modified zeolite membrane was able to realize the long-anticipated zeolite membrane reactor for high-temperature catalytic reactions for the pre-combustion CO_2 capture and H_2 production from fossil fuels and biomass. I have successfully established and validated a one-dimension model to simulate the membrane reaction performance beyond the experimental conditions. In my post-doctoral study, I have continued to work on the development of gas separation membranes and membrane reactor systems for major commercial applications in catalysis, adsorption, and ion exchange. In addition, I have been involved to validate the modeling works by obtaining experimental data via synthesis, characterization and evaluation of nanoporous ceramic membranes.

Based on my knowledge of membranes/catalysts, instrumentation design, and reaction modeling, my future research will focus on the fundamental issues including synthesis of novel materials and membranes and understanding their structure-property-performance relationships. Also, by developing innovative materials/membranes associated with clean and sustainable energy technologies, I am going to establish strong research for studying molecular transport, diffusion, gas-solid reaction mechanisms in the nanostructured materials and thin films.

393444 Solid-State Self-Assembly : Fundamentals and Applications

Abstract:

Nanoscale science and technologies has been developed tremendously during the last two decades, introducing a variety of nanomaterials with unique properties. However, incorporation of the properties into macroscale functional applications has been limited. An essential challenge is the integration of such unique properties into assemblies for micro- and macroscale devices and discovering appropriate conditions under which this can be accomplished. Here I explore the self-assembly of nanoparticles (NPs) in solid-state under external stress for discovering fundamental understandings of mechanisms and dynamics for various engineering applications. 1. Yoonseob Kim et al. ***Nature***, 2013, 500, 59-63 first demonstrated an example of excellent stretchable conductors from self-assembly of spherical NPs in solid-state. Free-standing stretchable conductors were assembled by layer-by-layer assembly. High conductivity and stretchability were observed and the properties originated from dynamic self-organization of NPs under stress. Modified percolation theory to incorporate the self-assembly of NPs gave excellent match with experimental data. 2. The current study demonstrated the chiroptical activities of NPs in solid-state for the applications of biosensing devices and optoelectronics. Unconventional transfer of macroscale stresses shaped nanoscale assemblies with handedness. Chiroptical responses were reversibly tunable by controlled stresses. Computational simulation supported the chiroptical properties originated from self-assembly of NPs in solid-state.

Based on my past experience, I will develop a research plan that includes, but are not limited to, more fundamental studies about 1) initiation force and main driving force of self-assembly, 2) relation between nanoparticles and external stresses, 3) form of probable assemblies not only in chain and cell but for higher dimensionality, and 4) optimization to the commercially available materials. Understanding the fundamentals has great potentials to lead to improved material designs for advanced applications. Specific objectives include the following: 1) stretchable semiconducting nanoparticle composites for reversibly tunable photocurrent with discovery of unique pathways, 2) electro-tunable mechanical properties of angioplasty balloons with enhanced performance including in-vivo experiments, 3) reversibly stretchable Li-ion batteries with high performance during many repeating cycles of huge mechanical stress.

The works were funded, in part, by Rackham Predoctoral Fellowship, University of Michigan, STX foundation, Seoul, Korea, and US Air Force Office of Scientific Research.

Revealing Molecular Mechanisms by Single Molecule Fluorescence Microscopy

Younghoon Kim

[1]*Department of Bioengineering, University of Illinois at Urbana-Champaign, Urbana, IL 61801*

Research Interests:
Single Molecule Fluorescence Microscopy is an emerging multi-disciplinary field that currently finds numerous applications in many diverse industries ranging from Biological Diagnostics to Traditional Reaction Engineering. Single Molecule Fluorescence Resonance Energy Transfer (smFRET) has rapidly become an integral part of nanobiotechnology, and enables researchers to gain insights into systems that otherwise would have been inaccessible. Direct monitoring proteins, nucleic acid, and individual movement of biomolecules provide rapid, deep understanding of molecular mechanisms, that cannot be seen in mass or ensemble state. For example, using smFRET, nucleic acid processing enzyme, helicase motion was monitored to understand the liquid-gas like transition state in sub-cellular level. smFRET allowed to observe milli-second to minute time scale molecular motions. Perhaps the main advantage of this experimental approach is that the unmatched detection level and real-time motion in simple robust manner. Also it gives direct, intuitive results that does not need any over-simplifications or complicated mathematical process and direct monitoring of heterogeneous populations.

Postdoctoral Project: "*DEAD-box helicase in the processing body*" and "*Bright and Photostable Fluorescent Labels for Single Molecule Microscopy*" under supervision of Sua Myong, Bioengineering and Charles M. Schroeder III, Chemical and Biomolecular Engineering, University of Illinois, Urbana-Champaign
PhD Dissertation: "*Cure for Muscular Dystrophy by Self-Assembled Polymer Vesicles and Worm-Micelles*" Under supervision of Dennis E. Discher, Chemical and Biomolecular Engineering, University of Pennsylvania

Research Experience:
My academic career path has been a blend of many fields of biology, biotechnology and chemical engineering. My formal training was in biophysics, which I have learned from many different experimentalist by the projects that I have worked on. I was always close collaborations with other engineers and scientists in the fields of rheology, medicinal chemistry, drug delivery, material science and molecular biologist. As a result of these collaborations I have acquired invaluable and versatile experience in many research areas: biophysics and biochemistry, biomedical imaging, material science, biology, and engineering design. Moreover, as a result of engineering mind-set, I have published two patents covering world and US, which has been actively discussed to commercialize the product.

Industry Experience:
It would be unique experience for a chemical engineer to be a traditional process engineer and later to be a biotech engineer. After bachelor's degree, I have worked at the traditional cement plant as a senior process engineer, and gained invaluable experience. After PhD degree, flourishing environment of biotech company enticed me and realized that how the new technologies can nurture small company to mid-scale or big one. I would like to pass my experience to the next generation chemical engineers how two industries have common points and what are our strengths.

Teaching Experience:

Aside from my research career, I also have extensive teaching experience. I spent two semesters teaching basic math and science to middle school students as a part of outreach program; I TAed and undergraduate lecture and laboratory courses in Chemical Engineering at the University of Pennsylvania; and, lastly, I have mentored high school student, high school teachers, undergrad students and graduate students, resulting a national-wide award and fellowships from the research I had with.

Towards the Next Generation of Chemical Analysis and Semiconductor Devices: Harnessing Light and Spin

Jonathan King
Department of Chemistry, University of California, Berkeley
Materials Sciences Division, Lawrence Berkeley National Laboratory

Nuclear magnetic resonance spectroscopy (NMR) and imaging (MRI) are two of the most powerful analytical techniques available to engineers, scientists, and medical professionals, enabling a wide range of chemical and structural analysis. However, they require costly infrastructure and are plagued by low sensitivity resulting from small nuclear spin magnetization. My laboratory will harness the highly polarized spin states of photons and electrons to control nuclear magnetization, resulting in greatly enhanced signals from magnetic resonance techniques. We will build upon my previous work with nitrogen vacancy (NV) centers in diamond [1] to develop methods of coupling spin-polarized electrons to nuclear spins in target molecules. This "hyperpolarization" gives the potential for more than 10,000-fold enhancement of NMR and MRI signals, greatly lowering sensitivity limits and reducing experimental timescales. Such enhancements will be transformative, allowing the analysis of dilute biomolecules and microscopic samples over rapid timescales, and generating technology for portable, low-cost magnetic resonance instruments.

While traditional NMR spectroscopy yields a great deal of chemical information, it cannot directly probe the chirality of molecules. We will make use of optically detected methods of magnetic resonance, such as atomic magnetometry, to enable zero-field NMR, where the geometry of spin-spin couplings may be fully resolved. These couplings include the traditionally ignored antisymmetric J-coupling, which is linked to the chirality of the analyte molecule. This chiral recognition, in combination with the chemical specificity of NMR, will provide a powerful new technique for the analysis of pharmaceuticals and biomolecules, where chirality is often a dominant factor in the biological function of a molecule.

Semiconductor devices that harness "spin current" can be faster and consume far less energy than traditional devices using charge current. Magnetized nuclei may be used to manipulate spin currents in semiconductors, enabling a host of "spintronic" devices. We will use a combination of optical and magnetic manipulation of nuclear spins in gallium arsenide [2,3,4] to generate re-writable, 3-dimensional patterns of magnetization without the need for microfabrication or lithographic patterning. These structured regions of magnetism will serve as the architecture for spintronic devices integrated into bulk semiconductors.

1. **J. P. King**, P. J. Coles, J. A. Reimer, "Optical polarization of ^{13}C nuclei in diamond through nitrogen vacancy centers." *Physical Review B*, 81, 073201 (2010).

2. **J. P. King,** Y. Li, C. A. Meriles, J. A. Reimer, "Optically Re-Writable Patterns of Nuclear Magnetization in Gallium Arsenide." *Nature Communications*, 3, 918 (2012).

3. Y. Li, **J. P. King**, L. Peng, M.C. Tamargo, J. A. Reimer, C. A. Meriles, "Helicity independent optically-pumped nuclear magnetic resonance in gallium arsenide." *Applied Physics Letters*, 98, 112101 (2011).

4. Y. Li, **J. P. King,** J. A. Reimer, C. A. Meriles, "Near-band-gap photo-induced nuclear spin dynamics in semi-insulating GaAs: Hyperfine- and quadrupolar-driven relaxation." *Physical Review B*, 88, 235211 (2013).

381902 Optimal Control of Neural and Small Length Scale Dynamical Systems

Abstract:

In the last decade, dramatic advancements in experimental and computational techniques have flushed tremendous opportunities in the study of fundamental questions of science and engineering by taking the approach of stochastic modeling and control of dynamical systems. In parallel, emerging applications through these advances have ignited the development of new technologies by integrating advanced control strategies with stochastic dynamical models. One such example is the brain-machine interface. Over the course of my doctoral and postdoctoral studies, my research interests have mainly been focused on developing optimal control strategies for applications in closed-loop neural prostheses, deep brain stimulations, and closed-loop drug delivery systems. During this poster session, I will highlight examples from both my graduate research and postdoctoral research which demonstrate my capability in advancing these subjects by integrating probabilistic tools with optimal control policies. Building on these results, I will describe my future academic plans to commence a distinct, interdisciplinary and independent research program at the interface of engineering and computational neuroscience. I will explain how my interdisciplinary expertise in the theory of stochastic processes, neuroscience, optimization, dynamical systems and control uniquely positions me to capitalize on emerging opportunities at this area.

I am currently working as a postdoctoral research associate in the department of Electrical and Systems Engineering at Washington University in St. Louis. The major theme of my studies is centered on understanding the effect of system dynamics on the emerged system behaviors such as synchronization in complex dynamical systems, in particular pathological synchronization in Parkinson's disease, and developing optimal control strategies to achieve the desired system behavior in such systems. During the course of my doctoral studies at Lehigh University, I developed optimal control-theoretic framework for investigating the design of closed-loop neural prostheses. In parallel, I developed optimal stochastic control strategies for regulating behaviors of particles driven by Brownian motion.

390372 Optimization of Nonsmooth Chemical Process Models

Abstract:

Nonsmoothness can enter chemical process models through sources such as transitions in phase or flow regime, activation of safety mechanisms, and switches between discrete operating modes. Established numerical methods for simulation, sensitivity analysis, and optimization typically require derivative information, and have limited applicability to nonsmooth systems. Specialized numerical methods exist for nonsmooth problems, such as bundle methods for local optimization, and semismooth Newton methods for equation solving. These nonsmooth methods require local sensitivity information in the form of the generalized derivatives developed by Clarke and Nesterov. However, established methods for evaluation of these generalized derivatives are limited either in scope or in the quality of the obtained derivatives. Nonsmooth dynamic systems introduce particular difficulties, since little is known *a priori* about the behavior of state variables as functions of system parameters, and since even theoretical descriptions of generalized derivatives are lacking for dynamic systems.

For my PhD work with Paul Barton in the Department of Chemical Engineering at the Massachusetts Institute of Technology, I have focused on describing and evaluating generalized derivatives for nonsmooth systems. I developed and implemented a variant of automatic differentiation that evaluates generalized derivatives for composite nonsmooth functions. This method is the first computationally tractable numerical method for accurate generalized derivative evaluation, and is fully automatable.

Several new results were also obtained for nonsmooth dynamic systems that are represented as systems of parametric ordinary differential equations (ODEs) with nonsmooth right-hand side functions. Firstly, I obtained sufficient conditions under which the state variables of such a nonsmooth dynamic system are in fact smooth functions of system parameters. These conditions are computationally inexpensive and simple to verify *a posteriori* during a simulation run, and can sometimes be shown to hold *a priori*. Secondly, I provided the first theoretical description of a useful generalized derivative for a general nonsmooth dynamic system. As in classical sensitivity analysis approaches for smooth dynamic systems, this generalized derivative is expressed in terms of an auxiliary dynamic system. These nonsmooth dynamic sensitivity results were generalized to certain hybrid discrete/continuous systems, which can be used to model systems with discontinuous jumps.

363150 Synthetic Approaches to Control Cell Fate and Function

Abstract:

Cells are complex and transformable machines. In addition to their homeostatic functions, they are capable of sensing and processing diverse biochemical and biophysical signals. Furthermore, isogenic cells are capable of transforming into a multitude of states, including diseased ones, each with distinct sets of active regulatory programs. To decode these complex cellular programs, synthetic approaches provide exquisite quantitative control of the spatiotemporal, physical, and chemical properties of both extra- and intra-cellular regulatory components. My research focuses on two complementary experimental approaches: 1) Engineering synthetic cellular microenvironments with defined spatiotemporal, physical, and chemical properties; 2) Engineering synthetic intracellular signal-transducing machineries that have unique input-output functions and can interface with complex, native regulatory systems such as chromatin and the actin cytoskeleton. The main applications of these synthetic approaches are to: 1) Quantitatively interrogate and control the cellular programs and machines that define distinct stem cell and disease states and their developmental or oncogenic transformations; 2) Elucidate molecular mechanisms underlying cell regulation, including chromatin dynamics and cellular mechanotransduction. My research takes advantage of methods from and collaborations with the fields of synthetic biology, systems biology, materials science and engineering, stem cell and cancer biology, and biophysics. Together this work may be applied to diverse areas in regenerative medicine, cancer biology, and eukaryotic cellular biotechnology.

Past and current work in these areas include: 1) Organic-inorganic surface chemistry (Keung et al 2005, 2007); 2) Biophysical control of neural and embryonic stem cell differentiation and proliferation (Keung et al 2009, 2010, 2011, 2012, 2013); 3) Synthetic, site-specific genome-targeting of chromatin factors to engineer cellular logic, spatial transcriptional regulation, and memory (Keung et al 2014).

390749 Towards a Sustainable Energy Future: Modest – Model Aided Optimization and Design of Energy Systems' Toolkit

Abstract:

Just over a decade into the 21st century and new millennium, it is becoming increasingly evident that energy security and climate change are two of the most defining challenges of our time and generations to come. Unfortunately, there is no *silver bullet* and we need to nurture a range of technologies to address these challenges. To accomplish this, motivated by my own diverse research experiences, several ***clean and green technologies*** need to be developed further. These include: .

- *Bioenergy (including advanced biofuels)*
- *Carbon capture, transport and sequestration (technologies and infrastructure)*
- *Hydrogen driven transportation (technologies and infrastructure)*
- *Clean coal technologies (including integrated gasification combined cycle)*
- *Renewable energy sources (including solar and wind)*
- *Energy systems of the future (e.g., urban & rural energy systems, zero-carbon homes)*
- *Water and healthcare (technologies and infrastructure)*

Together, all these technologies have the potential to alleviate both the energy security and climate change concerns. This presentation will discuss effective use of various *Systems Engineering* tools in a comprehensive framework **MODEST (Model aided Optimization, Design of Energy Systems' Toolkit)** to *foster scientific innovation and facilitate the acceleration of commercial deployment of these technologies. Special emphasis will be on large-scale real world energy systems.*

388324 The Physical Cell: Impact of Mechanics and Rheology on Cellular Function

Abstract:

The internal microenvironment of a living cell comprises an intricate choreography of molecules that must be transported from one location to another, elastic forces that must be overcome or harnessed into useful work, and interdependent chemical reactions whose rates must be carefully controlled. My research focuses on examining the basic physical phenomena that underly biological processes at the molecular and cellular level. I develop multiscale models, employing both analytical and computational techniques, that interface directly with experimental data. My work draws on concepts from continuum mechanics, fluid dynamics, and statistical physics to address the interplay of structure, transport, and biological function.

As a doctoral student with Prof. Andrew Spakowitz, I focused on studying how the mechanical properties of DNA impact the packaging and accessibility of the genome. I demonstrated that the elasticity of the DNA molecule can give rise to tension-mediated cooperative binding between DNA-bending proteins, allowing them to interact across a wide range of length scales without direct protein-protein contact. I also investigated the role of this elasticity in the packing of DNA into compact chromatin fibers, and the connection between packaging fluctuations and force-dependent kinetics of enzymes that access the DNA. My work on the statistics of large DNA-protein complexes spurred the development of a generalized approach for coarse-grained modeling of polymer systems by mapping onto effective elastic chains. This methodology enables simulations of polymer dynamics and statistics on much larger length scales than was previously possible.

As a postdoctoral scholar with Prof. Julie Theriot, I investigate cytoplasmic mechanics in motile neutrophils. These cells exhibit rapid motion accompanied by dynamic changes in shape that arise from large-scale fluctuations at the cell boundary, driven by active polymerization processes. Using a combination of analytical theory and simulations, I study the motion of particles near an actively fluctuating membrane. I apply this model directly to cytoplasmic dynamics in motile cells by treating the cell interior as a viscous fluid driven by boundary fluctuations that are tracked using phase microscopy. Fluorescent tracking of intracellular particles indicates that their relative motion is consistent with the predictions of this simple model. In particular, I demonstrate that fluid flows arising from the cell shape dynamics can contribute significantly to mixing and transport of large complexes and organelles within the cell.

My future research goals center on studying the physical processes underlying the propagation of chemical and mechanical signals within a cell. I intend to use my background in multiscale biophysical modeling as a springboard for addressing fundamental problems of how transport phenomena, active mechanical fluctuations, and intracellular spatial organization impact the structure and kinetics of biomolecular networks. Hijacking cellular machinery to engineer novel circuits or to induce therapeutic effects necessitates an ability to harness the kinetic properties of these reaction networks. The role of cytoplasmic rheology and mechanics in modulating biomolecular processes within the cell thus presents a rich set of questions of direct relevance to cellular engineering.

370573 Controlling the Electronic Properties of Surfaces and Interfaces for Energy Applications

Abstract:

The goal of my research is to develop nanostructured materials for energy conversion in which to investigate the fundamental surface and interface interactions. This poster gives an overview of my past and current research efforts on the design and development of nanostructured materials for various applications, including (i) fundamental research on ferroelectric perovskites for nanodevice applications; (ii) fundamental processing-structure-property relationships that govern organic-inorganic nano-interfaces for organic transistors and solar cells (iii) materials synthesis of transition metal oxides for heterogeneous catalytic systems. I will establish an interdisciplinary research program that lies at the interface of engineering, physics, and materials science. My laboratory will study the fundamental properties of nanostructured materials and focus on developing model systems to design efficient and cost-effective materials for energy applications.

370684 Nanotechnology-Based Breakthroughs in Biology and Energy

Abstract:

In my independent faculty career, my research will focus on nanotechnology-based approaches to understanding and controlling fundamental electronic and mechanical properties of biological materials for advancing both biomedical and energy sciences. I am particularly interested in interfacing biology with functional electronics to yield breakthroughs in drug development, to enable new avenues of powering biomedical devices, and to probe biological changes in a truly non-invasive way. For my doctoral research, I developed a novel method for understanding charge transport at nanoscale metal-oxide interfaces with the use of scanning probe techniques and extended that knowledge to understanding and controlling polarization switching in ferroelectric devices. Ferroelectrics are highly efficient mechanical couplers and are ideal candidates for interfacing with biology for a number of applications. As a postdoctoral fellow, I have developed new techniques to measure tight junction adhesion forces at piconewton resolution. With innovative electrode designs, I have developed a novel bioimpedance system to track changes in cellular dynamics – in-real time – while modulating ion transport gateways. With my background and skills, I am uniquely positioned to launch an interdisciplinary, collaborative Nanotechnology in Biology and Energy program that incorporates the expertise of both scientists and engineers.

388936 Polymers, Colloids and Composites: In the Service of Chromatography with New Porous Materials

Abstract:

Polymer Science, Colloid Science and Chromatography can be interfaced during the development of functional materials for particular applications.

During my doctoral studies in ETH Zurich, I developed a new procedure towards macroporous polymeric microparticles. The major novelty of this procedure is that the porous structure is created and controlled without a template, through the controlled aggregation and coalescence of latex nanoparticles.

These macroporous particles were packed into chromatographic columns and functionalized via controlled radical polymerization with polymer chains of variable composition. Thus, different chromatographic modes were addressed, namely ion-exchange, as well as temperature-responsive chromatography, which relies on an unconventional separation mechanism. Several classes of compounds have been successfully separated, such as small pharmaceuticals, peptides, proteins and monoclonal antibodies. This whole procedure, from the preparation of the base macroporous material, to functionalization, to chromatographic application, has been recently secured by two patents.

Diverging from this project, I also became interested in different approaches for controlling the stability latex nanoparticles. To this end, I synthesized a set of block-copolymer surfactants via controlled polymerization and studied first their self-assembly behavior in an aqueous solution. Subsequently, these surfactants were applied in order to control the shear-induced aggregation of commercial latexes, which is an important issue in the recovery of polymers from a dispersed state.

In my postdoctoral research in Lawrence Berkeley National Lab, I developed porous polymeric monoliths in a thin-layer format for the efficient thin layer chromatographic (TLC) separation of proteins, peptides and fatty acids. Their pore surface was modified either with an in-house synthesized polymer or with metal nanoparticles of different noble metals, namely gold and silver, starting from the corresponding colloidal dispersion. All applied modifications provide a dual function: appropriate modulation of chromatographic interactions, coupled with the assistance in the desorption/ionization of the separated compounds by a laser beam. The latter allowed the analytes detection by MALDI Mass Spectroscopy immediately after separation, without the use of an additional "matrix" substance, which needs to be conventionally applied.

Recently I have initiated a project for the preparation of a composite porous monolith in a capillary format, modified with a layer of a metal-organic framework (MOF), aiming to the liquid-phase separation of small molecules. MOF are an advanced class of porous materials, composed of metal centers interconnected by organic linkers. They are currently intensively investigated, owing to their unique combination of high porosity, ultra-high surface area and tailorable structures. Thanks to their versatility, a range of different interactions between MOF and molecules that can enter their porous network can be potentially exploited for multiple applications, including separations. Nevertheless, little attention has been devoted so far to liquid-phase separations.

Apparently, supports with the appropriate porous structure are necessary, where the MOF can reside. Furthermore, procedures for the controlled formation of the MOF in the confined interior of such a porous support need to be developed, ensuring that it is properly anchored and it also exhibits structural integrity. After resolving those issues, analytical-scale separations can be addressed. As soon as the MOF scalability is tackled, preparative-scale applications could be feasible as well. Since this is an emerging field, I anticipate this type of studies to extent into the future and I hope to be able to undertake them as an independent investigator.

I therefore wish to create a research program, encompassing the following overlapping areas: (i) Developing porous polymeric or composite materials with defined pore structure and surface chemistry that can perform efficient chromatographic separations of biologically relevant compounds, (ii) Understanding and controlling the stability of colloidal systems, such as dispersions of polymer and metal nanoparticles, for both fundamental and applied purposes, (iii) Studying the self-association of polymers with tailored composition into structures, which can be utilized for multiple applications, such as structure templating and encapsulation.

Concerning my teaching and mentoring experience and vision, I have been a teaching assistant in undergraduate chemistry laboratory courses for three years. I have also supervised more than 10 undergraduate interns, assisting my research work. I have indeed learned a lot by working with younger colleagues; I enjoy the challenge of instilling motivation and inspiration for science in people with different personalities. As my scientific background is broad, I feel confident with undertaking teaching duties in several areas, within an engineering or a natural sciences curriculum. Such lectures could include Polymer Science & Engineering, Colloid & Interface Science, Analytical Chemistry, Separation Science & Technology and related subjects.

Single-Molecule Visualization of Corona Phase Molecular Recognition

Molecular recognition is central to the design of therapeutics, chemical catalysis and sensor platforms, with the most common mechanisms involving biological structures such as protein antibodies. However, identifying and isolating an antibody for a particular biological molecule is often costly and time-consuming, and their use is limited to physiological conditions in which protein antibodies remain stable. This work reports the development of a synthesis platform for the design of synthetic sensors, based on the principle of carbon nanotube Corona Phase Molecular Recognition (CoPhMoRe). The design of sensors with CoPhMoRe is based on the adsorptive properties of engineered heteropolymers to carbon nanotubes, which create a unique corona with molecular recognition properties for a target analyte. This phenomenon is shown to be generic, with new recognition complexes demonstrated for riboflavin, l-thyroxine, and estradiol, and confirmed with single-molecule imaging of the nanotube and the corona phase. The complexes can be used as new types of sensors based on modulation of carbon nanotube optical fluorescence, as demonstrated using a synthetic antibody for real time spatio-temporal detection of riboflavin in live murine macrophages. The analyte library for these synthetic sensors could be amplified to identify synthetic sensors for a variety of different molecules for which there currently exists no detection method.

Biofilm and swarming genesis: rotary flagellar motors as mechanosensors

Pushkar Lele
Department of Molecular and Cellular Biology, Harvard University
Cambridge, MA 02138

Mechanosensing in bacterial cells leads to the initiation of biofilms and swarming in multiple species. The mechanisms of bacterial sense of touch, however, remain unknown. I am interested in determining the molecular mechanisms by which bacterial rotary motors sense surfaces and trigger biofilm formation and swarming.

The proposed research builds on recent discoveries of stimuli-dependent changes in motor self-assembly [1, 2]. I developed a novel optical-tweezer based assay to mechanically stimulate single motors, which enabled me to identify the motor components that receive mechanical signals and subsequently regulate function. These findings provide a handle on an important developmental problem: how do mechanical forces initiate cell development and biochemical signaling? My research program will focus on the mechanisms of self-assembly in biological macromolecules, signaling networks, mechanotransduction, and physics of motility. The ultimate objectives of this research will be to determine important adaptation mechanisms in micro-organisms, to identify targets for drug discovery and to develop effective strategies to combat the growing menace of antibiotic resistance.

[1] P. P. Lele, Basarab G. Hosu, H. C. Berg, "Dynamics of mechanosensing in the bacterial flagellar motor", under revision, Proc. Natl. Acad. Sci. USA.
[2] P. P. Lele, R. W. Branch, V. Nathan, H. C. Berg, "Mechanism for adaptive remodeling of the bacterial flagellar switch", Proc. Natl. Acad. Sci. USA (2012), 109 (49), 20018-20022.

385536 Engineering Biomimetic Self-Assembled Materials

Abstract:

My research interests involve the design, characterization, and application of novel self-assembling materials. In particular, I am interested in biomimetic systems in which carefully tuned molecular interactions create functional and responsive materials for applications ranging from nanoscale electronics to drug delivery. Electrostatic interactions play a fundamental role in both creating spatially patterned interfaces and also creating separate polymer rich phases (complex coacervation) for both equilibrium and non-equilibrium processes, and yet are still largely misunderstood. Using rational design of peptides and polymers I intend to explore the role that these interactions play in compartmentalized delivery systems and as a dynamic templates to create hierarchical hybrid materials composed of organic and inorganic constructs. My goal is to thoroughly understand the thermodynamic and kinetic processes driving the self-assembly of these materials and take them from the bench to their functional application.

My postdoctoral work with Professor Matthew Tirrell (The Institute for Molecular Engineering, University of Chicago &Argonne National Laboratory) involves the design, characterization and application of polyelectrolyte complex micellar systems. We have designed polypeptide-based micelles that can self-assemble to have either liquid or solid cores depending on their secondary structure, which provides unique self-assembly and degradation properties. After carefully characterizing model systems, we designed polyelectrolyte complex micelles that contain therapeutically relevant charged molecules such as miRNA and peptides, specifically for the treatment of atherosclerotic lesions and cancer. The modular nature of these assemblies enables the addition of a targeting ligand to increase the efficacy of delivering miRNA, creating polyelectrolyte complex micelles with a targeting ligand outside the corona of the micelle. Our rationally designed micelles have successfully delivered miRNA *in vitro* and are currently being tested *in vivo.*

My graduate work with Professor Raymond Tu (Chemical Engineering Department, The City College of New York) involved the interfacial templating of inorganic materials using self-assembled peptides. In nature, biological molecules form interfaces that assemble patterns of chemical functionality with exceptional precision. The role of dynamics during the assembly of biological molecules appears to be important for mineralization processes. Our work applied model sheet-forming peptides at interfaces to explore the dynamics of assembly in order to template mineral growth. Thermodynamic analysis of structure formation with increasing surface pressure allowed us to understand the nature of self-assembly with iterative changes in the peptide sequence. Additionally, we investigated the dynamics of the self-assembled state, where the organic phase switches between short- and long-range order as a function of surface pressure. This model system allowed us to explore the influence of electrostatic interactions on self-assembly and additionally the influence of short- and long-range order on the nucleation and growth of inorganic material. This is in contrast to a system that starts with a well-ordered preformed template that defines the epitaxial growth of the mineral phase. Versions of our model peptides are modified to include histidine in order to nucleate Au nanocrystals in both the short and long range ordered organic matrix showing that the phase behavior of the peptide influences the crystallinity and shape of the resultant nanocrystals.

Mathematical Modeling and Global Optimization for Planning and Scheduling of Petrochemical Processes, and Natural Gas to Liquid Fuel Energy Systems

Jie Li
Department of Chemical and Biological Engineering
Princeton University
Princeton, NJ 08544-5263

In the last twenty years, the petrochemical industry has succeeded by creating markets and supplying them with suitable products used to create goods such as plastics, cosmetics, lubricants, and paints. Petrochemical production begins in a refinery that separates crude oils mainly into lighter components such as naphtha, light naphtha, top oil, and liquid fuels including gasoline, diesel and jet fuel. The lighter components are further processed into various petrochemicals such as ethylene, propylene, butadiene, benzene, toluene, xylol and some other high-valued products via cracking, butadiene extraction, hydrotreating, etherification, and polymerization processes. Nowadays, tighter competition, strict environmental regulations, and lower-margin profits, drive the petrochemical industry to apply new technologies to improve their planning operations.

In this presentation, detailed nonlinear models are first proposed to predict product yields and properties in production units including crude distillation unit, vacuum distillation unit, hydrocracking units, catalytic cracking units, ethylene-cracking units and other processing units present in a petrochemical plant. The yield and property prediction models for crude distillation and vacuum distillation units are developed using swing cut theory based on crude assay data. Empirical non-linear models are developed for other processing units, including bilinear, trilinear, quadratic, polynomial, and exponential terms. Parameter estimation for all of the developed models is performed by globally minimizing the least squares error between the parametric models and real production data. Moreover, property indices in blending units are linearly additive and calculated on weight or volume basis, which introduce bilinear and trilinear terms. We also introduce binary variables to denote different operation modes for several production units.

The entire planning model is a non-convex mixed integer nonlinear optimization model (MINLP). To solve this large-scale nonlinear model, we propose an optimization-based procedure to obtain the tightest lower and upper bounds for variables especially the variables involving nonlinear terms. Then, we incorporate those tightest lower and upper bounds into commercial solver ANTIGONE to obtain ε-global optimality. Finally, a user-friendly platform is developed to allow the user to modify the planning model by updating model parameters when new data is available, product demands and specifications, cost parameters and many more. Several large-scale industrial examples are solved to illustrate the efficiency of our proposed model and global optimization approach.

381981 Sustainable Fuels and Chemicals from Catalytic Conversion of Natural Gas and Biomass

Abstract:

As petroleum reserves decrease, price increases and the impact of rising atmospheric CO_2 is exacerbated, world production of fuels and chemicals must shift to low-carbon and sustainable resources, such as biomass and natural gas. Conversion of these sustainable resources into fuels and valuable chemicals is very attractive, but also very challenging. Innovative technologies for this conversion could help break the technical barriers and lower the costs for commercial applications. I have been working on developing new technologies and new catalysts for biomass and natural gas conversion since my Ph.D. work.

My Ph.D. study was on fast pyrolysis of biomass and catalytic upgrading to liquid fuels and valuable chemicals. One of the most important contributions is the development of a new electrocatalytic method with a ruthenium electrocatalyst for biomass conversion, converting electrical energy into carbon-neutral drop-in replacement fuels and valuable chemicals. This method provides substantial advantages, including no need of hydrogen gas, operation at mild conditions to reduce coke formation and catalyst deactivation, and a great way to store intermittent renewable electricity as chemical energy in liquid fuels. The other significant contribution is the development of a new screw-conveyor pyrolysis reactor with compact design, enabling flexible application in regional biomass processing depots, showing great potential to solve biomass supply issue.

To deepen and expand my knowledge in catalysis and sustainable energy, I continued my postdoc research with Prof. Fabio H. Ribeiro and Prof. W. Nicholas Delgass at Purdue University. One of my projects is to understand the catalytic active sites on precious metal catalysts (Au on metal oxides) for water-gas shift reaction from the fundamental and atomic perspectives by studying kinetics and applying various detailed catalyst characterization methods. Another important aspect of my work involves synthesis and characterization of heterogeneous lewis acid zeolite catalysts, such as Sn-, Ti- and Zr-zeolites. These catalysts are being tested for condensation of glycerol with aldehydes or ketones to produce fuel additives for diesel fuel, and also to probe the active sites and reaction mechanism. As an expansion on sustainable energy study, I have been studying methane activation via oxidative coupling for ethylene formation using $Mn/Na_2WO_4/SiO_2$. The major work is on understanding of the active sites for methane activation by doing kinetic study and catalyst characterization and using the information gained to design better catalysts.

My future research interests are applying the knowledge of catalyst synthesis, active sites characterization and kinetic studies in biomass and natural gas conversion, such as bio-oil upgrading, catalytic pyrolysis, ethanol and glycerol conversion, and syngas to liquid fuels. Another interest lies in electrocatalytic conversion of biomass and natural gas using renewable electricity (such as photovoltaic and wind energy) and simultaneously storing intermittent renewable electricity as chemical energy in the form of liquid fuels and valuable chemicals. I will focus on designing bimetallic and bifunctional electrocatalysts, developing high temperature electrochemical systems and integrating renewable electricity techniques with electrocatalytic method for biomass and natural gas conversion.

388624 Two-Dimensional Molecular Sieves for Selective Transport

Abstract:

Two-dimensional molecular sieves are highly desired building blocks for the fabrication of molecular selective membranes. Miniaturization of the third dimension opens up opportunities for the synthesis of membranes with a thickness of just a few atoms. Due to remarkably short transport path of the molecules, these membranes will have profound impact in the field of separation technology. Apart from membranes, these two-dimensional porous sheets will find applications in catalysis, sensors, low-dielectric constant materials, barriers, etc.

Zeolite membranes are attractive due to their high performance (permeance and separation factor) and robust chemical and thermal stability. Attempts to synthesize segregated, two-dimensional zeolite sheets have often led to an intergrown or a layered zeolite. For the first time, a purified, dispersed suspension of highly crystalline 1.5 unit-cell-thick zeolite nanosheets was prepared by the exfoliation of layered zeolites followed by solution processing. Compared to time-consuming and expensive coating techniques involving modification of supports by masking and/or functionalization, a simple scalable filter coating technique was developed to fabricate a sub-100 nm-thick, oriented and compact coating of nanosheets. Wetting properties of the nanosheets were tuned, allowing it to deposit on the air-water interface. This allowed a monolayer coating of nanosheets on non-porous supports, opening opportunities for applications in sensors, low-dielectric constant materials etc. Thin, zeolite films prepared by hydrothermal treatment of the nanosheet film showed exceptional performance in the separation of xylene isomers, sparking industrial interest in deploying these membranes for the large scale purification of *p*-xylene. A mathematical model based on adsorption and diffusion kinetics was used to gain insights into the parameters affecting membrane performance at industrial process conditions.

In my future research, I aspire to expand the list of two-dimensional materials for molecular separations. Especially, one atom thick graphene sheets, MoS_2 sheets, aluminophosphate sheets, etc. hold exciting potential towards the fabrication of the thinnest membrane. Strategies for the generation of pores with tunable size in the non-porous atomic sheets will be explored. The transport of molecules with varying size and physical properties will be modeled to explore the full potential of two-dimensional membranes.

Design and Development of Advanced Materials for Energy Storage and Conversion

Qi Lu, Chemical Engineering, Columbia University/University of Delaware

It is now a common realization that our society is moving towards renewable and sustainable energy sources. In witness we are observing an increase in renewable energy production from sun and wind. Due to their intermittent availability however, energy storage and energy conversion techniques are starting to play a more and more important part in our lives.

My research focuses on the understandings of fundamental principles that govern the materials performance using well-defined model systems and first-principles-based calculations, based on which the rational design of high-performance materials for energy storage and conversion applications was carried out. For example, my recent work on the development of supercapacitor electrodes demonstrates my research philosophy: three key parameters that control the behavior of supercapacitor electrodes were identified through a series of mechanistic investigations on model systems[1]. Based on the gained fundamental insights, a novel nanotechnology was developed to simultaneously optimize those parameters and to achieve practical supercapacitor electrodes with both high-energy and power densities[2]. In addition, every fabrication procedure employed in this technology was designed to be cost-effective and industrially scalable. As a result, this technique has been adapted for commercialization in a startup company for grid-scale electricity storage.

Figure 1. A supercapacitor electrode with interconnected metal core and active outer layer.

Figure 2. A nanoporous Ag catalyst that converts CO_2 to CO with high-selectivity and efficiency.[3]

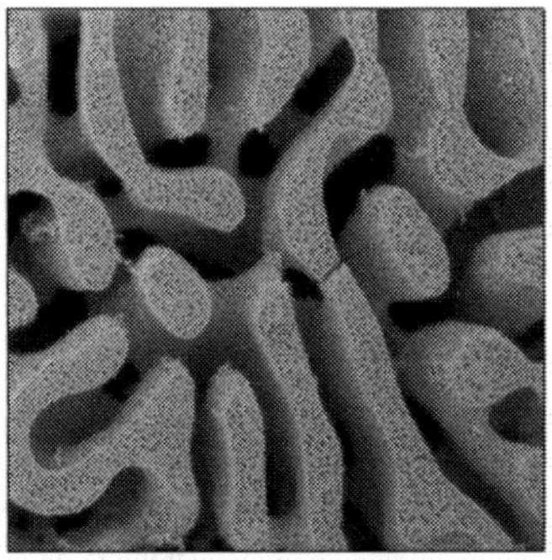
Figure 3. A Bimodal porous CuTi catalyst exhibits a Pt-like surface reaction sites for H_2 evolution reaction.[4]

In two of my most recent works, I developed highly active, non-precious metal-based catalysts for the electrochemical conversion of CO_2 and H_2O to CO and H_2, respectively. The design of this catalyst was investigated using density function theory-based calculations. Dealloy-based techniques were developed afterwards to synthesize practical catalysts with high surface area and density of active sites. The structure of the catalysts was further tailored to maximize the material utilization and electrode conductivity, achieving catalytic materials with optimal over-all performance for CO_2 conversion[3] and H_2 evolution reaction[4].

My future research will focus on the following aspects: 1) battery-capacitor hybrid electrochemical energy storage for solar- and wind- power grid applications; and 2) CO_2 and CH_4 conversion to value-added fuels and chemicals. The overarching goal of my research program is to provide the society advanced materials for future energy usage, which will be pursued via a combination of computational and experimental techniques.

References:

1. Lu Q. *et al.* "Nanostructured Electrodes for High-performance Pseudocapacitors" ***Angew. Chem. Int. Ed.***, 2013, 52 (7): 1882–1889

2. Lu Q. *et al.* "Supercapacitor Electrodes with High-Energy and Power Densities Prepared from Monolithic NiO/Ni Nanocomposites" ***Angew. Chem. Int. Ed.***, 2011, 50 (30): 6847–6850 (Covered by MaterialsViews, ChemistryViews, PhysOrg, *etc.*)

3. Lu Q. et al. "A Selective and Efficient Electrocatalyst for Carbon Dioxide Reduction" ***Nat. Commun.*** 2013, 5:3242 (Covered by Science Daily, R&D, Science World Report, etc.)

4. Lu Q. et al. unpublished results

390292 Next Generation Approaches to Biomolecular Engineering: Synthetic Biology Meets Directed Evolution

Abstract:

Rapidly advancing technologies like multiplex oligonucleotide synthesis and next-generation high-throughput sequencing have opened the door to tremendous progress in synthetic biology. My experience in both protein engineering and strain engineering puts me in an excellent position to forge the next generation of **synthetic biology tools for biomolecular engineering**.

Background

My research career started as an undergraduate at Johns Hopkins University under Prof. Marc Ostermeier, where I helped engineer **allosteric molecular switches via directed evolution**. For my M.S.E. research I created several modifications of chimeras of maltose binding protein and β-lactamase (BLA), the enzymatic activity of which could be allosterically regulated by the addition of maltose (published in *PNAS*). During my PhD with Prof. Matthew DeLisa (Cornell), I developed a **suite of tools for examining protein folding, protein-protein interactions, and post-translational modifications** in *E. coli*. I developed a protein folding reporter for the bacterial periplasm using a BLA-based selection. I showed that we could not only probe intrinsic protein folding, but **factors in the cellular environment** (e.g., chaperones, redox state) that affect protein folding (published in *Protein Science*). Later, I refactored the protein folding reporter to function as a detector for N-linked glycosylation in bacteria as a pipeline for **engineering post-translational modifications** (published in *Biotechnology Journal*).

Current Research

The postdoctoral work I am pursuing with Prof. Ryan Gill at CU-Boulder has provided me a broader view of both **metabolic engineering** and the design of directed evolution experiments. I am currently involved in three projects: (1) Performing selections on **genome-scale libraries** to discover genomic factors underlying **bacterial antibiotic resistance**. (2) Combining multiplex oligonucleotide synthesis and recombineering, I developed a platform for directed evolution of **orthogonal repressors for cellular circuits** on both the DNA and RNA level. (3) **Regulator engineering for ethanol tolerance** in *E. coli*. This work has given me a unique perspective on constructing targeted libraries on the genome scale, generating programmed diversity to complement my expertise in selection design. Specifically, leveraging the fast-developing technologies of multiplex oligo synthesis and next-generation sequencing will be key to my research as a principal investigator.

Future Research Directions

My lab will apply the latest synthetic biology technologies to tackling protein expression in bacteria with applications fpr human health, biofuel production, and synthetic circuit design. **Targeted libraries of expression elements** coupled with selections for protein folding will allow optimization of expression in difficult-to-produce proteins (e.g., membrane, heterologous proteins). **Genome mining** of heterologous proteins leverages the tremendous and ever-increasing amount of metagenomic data to inform and improve selection of enzyme activity. Sub-genome scale libraries of various families of proteins will provide in-depth exploration of cellular factors (and combinations thereof) that contribute to protein production. Finally, expanding into Gram-positive bacteria with **genome-scale libraries** on lactic acid bacteria will provide insight into protein folding and secretion in these clinically-relevant but understudied organisms.

Teaching

Beyond my research accomplishments, I have been fortunate to gain valuable teaching experience. In graduate school I served as a teaching assistant for four different courses, with topics ranging from process control and reaction kinetics (ChemE) to understanding wine and beer (Enology and Viticulture). Throughout my teaching pursuits I strove to implement **active and cooperative learning techniques** to keep students engaged and reinforce concepts learned in lecture. TA evaluations I have received consistently mention my **enthusiasm for the material and creative approaches to teaching** and exam review. I was also employed by Cornell Engineering Learning Initiatives, where I constructed and facilitated Engineering TA Development workshops in Assessment, Academic Integrity, Active and Cooperative Learning, Public Speaking, and Microteaching.

Inspiring students to pursue biomolecular engineering, whether in the classroom or the laboratory, is highly valuable. After all, Niels Bohr once supposedly said: "Biology is too important to be left to the biologists."

379325 Electrochemical Reduction of Carbon Dioxide with Highly Dispersed Metal Nanoparticles

Abstract:

The electrochemical conversion of carbon dioxide into hydrocarbons is an alternative route for synthesizing fuels and feedstocks that are typically derived from oil or natural gas. This route also represents a potential strategy to store electrical energy derived from intermittent sources of clean energy. Although metals in the form of polycrystalline foils are increasingly well-characterized as electrocatalysts for carbon dioxide reduction, their nanoscale counterparts remain poorly understood. Highly dispersed nanoparticle catalysts are ideal for practical electrolyzers due to their high surface area and because they can be integrated into gas diffusion layers in membrane electrode assemblies. The unique ensembles of surface atoms on gold and copper nanoparticle catalysts dramatically alter their activity and stability during carbon dioxide reduction compared to polycrystalline foils.

Our experiments reveal that the primary route by which gold nanoparticles lose surface area during electrochemical carbon dioxide reduction is by forming dendrites on the electrode surface. Random walk simulations coupled with fractal theory establish that the mechanism by which the dendrites form involves diffusion of entire nanoparticles along the support, followed by their collision and fusion to form dendritic structures. Although the mobility of nanoparticles along a support is typically thought to be limited at room temperature, our results demonstrate that nanoparticle catalysts readily move along the support during electrochemical carbon dioxide reduction. We show that an in-depth understanding of the polarization-dependent surface chemistry of gold nanoparticles can be used to limit the dendritic assembly process, thereby increasing surface area for catalysis. We also extend our understanding of surface area loss in gold particles to explain the instability of copper nanoparticles during catalysis.

Both gold and copper nanoparticles exhibit unique activities and selectivities for electrochemical carbon dioxide reduction compared to the polycrystalline foils. We find that there is a continuum in the selectivity for hydrocarbon products between copper nanoparticle and foil electrodes. Detailed electrochemical analysis reveals that nanoscale copper metal particles reduce carbon dioxide to hydrocarbons through a mechanism involving intermediates that are distinct from those observed on polycrystalline foils. Our insights into the structural and chemical factors that influence the activity and stability of copper and gold nanoparticles have enabled us to design advanced catalytic networks for electrochemical carbon dioxide reduction.

379868 Multiscale Modeling to Make Cellulosic Biofuels More Abundant and Affordable

Abstract:

As recognized by the 2013 Nobel Prize in Chemistry, "computer models mirroring real life have become crucial for most advances made in chemistry today."[1]The focus of my research is the application of theoretical and computational methods for uncovering reaction mechanisms that convert non-food biomass into renewable fuels and chemicals. Specifically, my PhD work examines two approaches to cellulose deconstruction: thermochemical and enzymatic processes. In the thermochemical portion of my work, I use quantum mechanics to investigate the elementary steps in cellulose depolymerization under fast pyrolysis conditions. For example, competing mechanisms, specifically homolytic vs. heterolytic, had been debated for many years. Experiments designed to test each hypothesis have been inconclusive. Using computational chemistry, specifically quantum mechanics, I modeled both pathways and discovered a more likely mechanism that better matches for experimental results. My latest work examines the effects of inorganic ions on reaction kinetics, to help understand how these naturally-occurring biomass components change product yield. Additionally, my work has revealed mechanisms that provide targets for catalyst design. The results from this part of my dissertation enable more accurate, mechanistic modeling of biomass pyrolysis that can be used to improve cellolosic biofuel process design and efficiency.

In contrast to biomass fast pyrolysis, which relies on high temperatures to break cellulose's strong bonds, enzymes manipulate cellulose to allow depolymerization at ambient temperatures. Many of the details of how enzymes perform these feats remain unclear. Structural biology has revealed the three-dimensional structures of many enzymes, with and without a substrate present. Molecular simulations allow us to investigate an enzyme's action in the fourth dimension, tracking catalysis over time. These efforts are computationally challenging due both to the system size and the high level of accuracy needed to accurately capture what can be profound stereoelectronic effects resulting from small perturbations in atomic and electronic structure.

Sustainable energy will remain the focus of my future research, and I will continue applying and expanding my expertise in multiscale modeling to make renewable energy more abundant and affordable.

---

[1] "The Nobel Prize in Chemistry 2013 - Press Release". *Nobelprize.org*. Nobel Media AB 2013. Web. 10 May 2014.

385713 Molecular and Mesoscopic Design and Understanding of Energy Materials

Abstract:

The need for novel materials for energy storage and conversion has become increasingly critical as society grapples with limited fossil fuel reserves and the threat of climate change. The challenges and opportunities associated with the discovery and optimization of new energy materials are rooted in a fundamental understanding, and the subsequent control, of how material properties are related over multiple length and time scales. My future research group will seek to design, synthesize, and characterize novel materials for energy applications, with an emphasis on measuring, understanding, and controlling the crucial molecular- and meso-scale phenomena that govern their macroscopic functions. A core aspect of our research strategy will consist of detailed and quantitative characterization of relevant material properties up from the atomic length scale, pushing the limits of current methods when necessary. Among other techniques, novel solid-state nuclear magnetic resonance (NMR) methods will be a cornerstone of such investigations, which enable atom-selective measurements of molecular-level compositions, structures, dynamics, and interactions, as well as transport processes. Multi-scale understanding, combined with a synthesis strategy emphasizing simultaneous control of chemical compositions, material structures, and interfaces, will enable us to more efficiently explore new energy materials composed of low-cost, earth-abundant elements. Initial efforts will focus on materials and devices for electrochemical energy storage, with intentions to also expand to systems for energy conversion. Thematically, we expect to contribute to a broad range of interesting and technologically important problems in chemical engineering, materials science, physical chemistry, and energy.

Within this context, I will highlight my research experiences as a Ph.D. student at the University of California, Santa Barbara, and Marie Curie postdoctoral researcher at the CNRS, France, as well as discuss my future faculty research plans. During my Ph.D., with Profs. Brad Chmelka and Todd Squires, my research broadly focused on elucidating the molecular-level origins of meso- and macro-scale structures and properties in self-assembled and/or hierarchically structured solids, such as mesostructured zeolites, organosilica nanocomposites, and elastomeric organosiloxane foams. In parallel, I modeled electrokinetic flows over surfaces with meso-scale roughness and mass transport to surface-based microfluidic sensors, establishing the dominant physics that influences the transport of ions, analyte, and fluid in distinct asymptotic regimes. At the CNRS, and with the French Network for Electrochemical Energy Storage, my research has focused on measuring disorder and transport properties in novel crystalline electrode and polymer electrolyte materials for lithium-ion batteries, including the development and application of novel solid-state NMR experiments. Collectively, these experiences provide a unique foundation for my proposed faculty research program encompassing materials for energy.

Nanoscale Chirality and Sustainability Laboratory

Kevin M. McPeak

Optical Materials Engineering Laboratory, Department of Mechanical and Process Engineering, ETH Zurich, 8092 Zurich, Switzerland,

The interactions between nanoscale chiral (a shape is said to be chiral if its mirror image is not superimposable on itself) inorganic material, chiral biomolecules and light is leading to a host of new devices with applications in chiral sensing, optical tweezers, optofluidics, nanomechanics, chiral separations and asymmetric catalysis. During my postdoctoral work at ETH Zürich I developed a new method to fabricate chiral nanoparticles and surfaces for visible and ultraviolet plasmonics. The chiral colloids display record molar circular dichroism in the visible and when fabricated from aluminum give a strong chiral optical response in the ultraviolet. This response is resonant with many chiral biomolecules opening up applications for enhanced structural chirality detection in biology, chiral sensing and asymmetric photocatalysis.

The late Nobel Laureate, Richard Smalley, declared "Energy is the single most important challenge facing humanity today." Solar energy is the only clean, renewable energy source available that is capable of supplying the 13 terawatts that are necessary to power the planet. At ETH Zürich I developed a low temperature, all solution-processed periodic nanostructured quantum dot solar cell. Furthermore, during my PhD I studied the chemical bath deposition of nanomaterials for low cost photovoltaic devices. I developed a combinatorial approach for depositing oxide and chalcogenide nanomaterials which enabled rapid and unprecedented understanding of the effect of growth conditions on optical, electronic, and structural properties of the material.

My laboratory will have two primary research areas: 1) fabricating novel inorganic chiral nanomaterials and 2) developing low cost, materials abundant photovoltaic and photocatalytic devices. We will study the interactions between light, inorganic chiral nanomaterials and chiral biomolecules. This fundamental work will allow us to develop new chiral sensors, enantiospecific separation techniques and new routes to asymmetric catalysis which would be beneficial to human health. Lastly, the development of low cost, materials abundant photovoltaic and photocatalytic devices is paramount to energy sustainability and human health in the world.

Selected publications:

Complex Chiral Colloids and Surfaces via High-Index Off-Cut Silicon
McPeak *et al.*, Nano letters ASAP
DOI: 10.1021/nl501032j

Microreactor Chemical Bath Deposition of Laterally Graded $Cd_{1-x}Zn_xS$ Thin Films: A Route to High-Throughput Optimization for Photovoltaic Buffer Layers
McPeak *et al.*, Chemistry of Materials 25 (3), 297-306
DOI: 10.1021/cm3023506

In situ x-ray absorption near-edge structure spectroscopy of ZnO nanowire growth during chemical bath deposition
McPeak *et al.*, Chemistry of Materials 22 (22), 6162-6170
DOI: 10.1021/cm102155m

Research Statement: My long-term vision is to contribute to the development of innovative, impactful biomaterials for fundamental and translational tissue engineering applications and to contribute to the development of the next generation of biomaterial scientists. In particular, as an independent researcher, I will lead a research team in the development and evaluation of cutting-edge biomaterials in the context of novel disease modifying agents as well as for organ replacement using Adult Stem Cells (ASCs). My doctoral and post-doctoral studies have given me strong training in polymeric biomaterial development and characterization, molecular biology techniques, and the culture of primary and stem cells (adult and embryonic). In addition, as a young researcher I understand the relevance of interdisciplinary collaboration, which in my case has resulted in 20 peer-reviewed manuscripts in the area of tissue engineering, biomaterials and regenerative medicine. This expertise will serve as a basis for me to build my own research program. My research will initially be focused on the design of mechanically tunable substrates of varying geometries, chemical and biochemical composition capable of supporting prolonged *in vitro* expansion of ASCs while maintaining their multipotency and therapeutic capabilities. This initial research work will contribute to our understanding of the cellular mechanisms that maintain "stemness" while directly contributing to the improved design of *in vitro* culture environments that enhance the clinical utility of ASCs.

Education:

- PhD, Chemical Engineering; Texas A&M University, College Station TX, May 2011. Advisor: Dr. Mariah Hahn, PhD
- BS, Chemical Engineering; Universidad Industrial de Santander, Bucaramanga Colombia, May 2005.

Research/Work Experience:

- Postdoctoral Research Associate, 6/11-present
  Department of Biomedical Engineering, Rensselaer Polytechnic Institute, Advisor: Dr. Mariah Hahn
- Graduate Research Assistant, 3/06-05/11
  Department of Chemical Engineering, Texas A&M University, College Station, TX
- Lecturer, 10/04-01/06
  Instituto Marco, Bucaramanga, Colombia
- Undergraduate Research Assistant, 01/04-04/05
  Centro de Investigación en Catálisis (CICAT), Bucaramanga, Colombia

Selected Publications (from a total of 20 peer-reviewed publications):

- A.C Jimenez-Vergara, **D.J. Munoz-Pinto**, M.S Hahn. Influence of pressurized cyclic stretch and endothelial cell presence on multipotent stem cell osteogenic commitment. Integr Biol. 5 (7): 983-92, 2013.
- **D. Munoz-Pinto**, X. Qu, L. Bansal, HN. Hayenga, J. Hahn J, MS. Hahn. Relative impact of form-induced stress vs. uniaxial alignment on multipotent stem cell myogenesis. Acta Biomateralia. 8 (11): 3974-81, 2012.

- **D. Munoz-Pinto**, A.C. Jimenez-Vergara, Y. Hou, H.N. Hayenga, A. Rivas, M. Grunlan, M.S. Hahn. Osteogenic potential of poly(ethylene glycol)-poly(dimethylsiloxane) hybrid hydrogels. Tissue Eng A, 18(15-16):1710-9, 2012.
- **D. Munoz-Pinto**, B. Grigoryan, J. Long, M. Grunlan, M.S. Hahn. An Approach for Assessing Hydrogel Hydrophobicity. J Biomed Mater Res A. 100(10):2855-60, 2012.
- **D. Munoz-Pinto**, R. McMahon, M. Kanzelberger, A.C. Jimenez-Vergara, Y. Hou, M. Grunlan, M.S. Hahn. Inorganic-organic hybrid scaffolds for osteochondral regeneration. J of Biomed Mater Res A. 94 (1): 112-21, 2010.

380676 Understanding Diffusion of Small Molecules in Polymer Glasses

Abstract:

The Free Volume Model (FVM) proposed by Cohen and Turnbull over 50 years ago had provided a simple picture that relates the structural properties of amorphous glasses to the diffusing behaviors of small molecules in them. Albeit the semi-phenomenological nature of the model, FVM has met with great success in matching experimental results and providing guidance in designing new materials for membrane separations. However, accumulating experimental data on diffusion of gases through polymer membranes have increasingly suggested inconsistencies of the model. In attempts to correct these deviations within the framework of FVM, effects of activation energy, cavity connectivity and accessible free volume had been introduced. These modifications are largely phenomenological due to the lack of a systematic understanding of the diffusion process in glassy polymers. In this study, using molecular dynamics simulations we aim to provide a coherent picture of small molecules diffusing in polymer glasses. Our results suggest that there are three distinct diffusion regimes depending on the relative size of the diffusant and "cavities". The dominant mechanism that controls diffusion in each regime is identified respectively. Based on the identified mechanisms, deviations from the FVM can now be systematically understood and predicted. Our results provide a basis on which further development of more comprehensive diffusion models will be constructed, a work that is undergoing.

384120 Understanding Artificial Photosynthesis Systems from the Nanoscale to the Device Level

Abstract:

The world's development and the implementation of a sustainable lifestyle depend on significant advances in the alternative energy sector. These advances will require basic scientific developments in energy generation, capture and storage, as well as large scale economic policy drivers to motivate the necessary changes to our current energy infrastructure. I have actively participated in the search for solutions to energy problems with deep fundamental and technological implications in the field of artificial photosynthesis. Artificial photosynthesis devices are complex energy conversion systems that need to absorb sunlight, create and transport charges, and electrochemically generate energy rich molecules such as hydrogen that can be used as a fuel. Given the complexity of the process, integrated systems need to be designed to efficiently carry-out all of these processes in parallel. My research has tackled all of these design aspects ranging from the materials nanostructure level to the overall device design structure, with particular emphasis on the role of hybrid polymeric materials for solar-fuel generation. Here, I will highlight research insights from the basic understanding of functional nanoparticle self-assembly in polymer membranes for solar-energy harvesting, to the structural and transport characteristics of ion-conducting polymers at inorganic interphases for electrochemical water splitting. Lastly, broader aspects on the implementation and device design of cost-effective solar-fuels technologies will be discussed.

Building on the systematic approaches applied on solar-fuel devices, my research program will focus on understanding multifunctional polymer composite systems for energy applications. Specifically, we will first design materials using basic physical models to understand the drivers for performance and potential for a particular application, as well as devise the required morphologies to achieve the desired properties. Then, we will devote the bulk of our research efforts to understanding the self-assembly aspects of model composite systems that lead to the required morphologies by using a wide range of polymer processing techniques and structural characterization tools. Lastly, understanding the effects of composites' nanostructures on their properties (e.g. electrical, electrochemical and physical) will allow us to achieve optimized hybrid materials that can be implemented into devices and will also point out towards directions for improvements in their design. Examples of functional energy systems based on hybrid nanostructured materials include electrochemical catalyst layers, supercapacitor and battery electrodes where the structure and characteristics of the electrochemical interfaces are critical for the device performance.

The Design of Catalytic Materials at the Molecular Level for Sustainable Product Synthesis

Eric G. Moschetta

School of Chemical & Biomolecular Engineering

Georgia Institute of Technology, 311 Ferst Drive NW, Atlanta, GA 30332

My research interests involve the design and implementation of homogeneous and heterogeneous molecular catalysts for production of synthetically useful materials. As a PhD student at The Pennsylvania State University under the guidance of Prof. Robert M. Rioux, I explored the thermodynamics of metal-ligand associations in solution using microcalorimetry for a series of ligands and Pd metal centers. My work showed that ligand binding was enthalpy-driven, that the chelate effect for binding bidentate ligands was enthalpy-driven and not entropy-driven, and that solvent reorganization contributed significantly to the stability of chelated complexes via enthalpy-entropy compensation analysis. I also investigated hydrosulfonation and hydrothiolation reactions using homogeneous and heterogeneous Rh catalysts and examined the structure-function relationships between the active species and their activity and selectivity. This study was the first to use Rh as a hydrosulfonation catalyst and showed that lowering the concentration of the exogenous phosphine ligand that formed the Rh catalyst *in situ* changed the chemo- and regioselectivity in favor of the desired Markovnikov vinyl sulfonate esters. My PhD work allowed me to develop my problem solving skills from the perspective of a chemist as well as the perspective of an engineer.

I am currently a postdoctoral fellow at the Georgia Institute of Technology under the guidance of Prof. Christopher W. Jones where I am expanding my repertoire of catalysis knowledge while delving into new areas such as carbon capture. Whereas my PhD experience imparted a deep knowledge of bonding and catalysis with molecular catalysts, my post-doctoral work emphasizes material design and synthesis. I am probing the impact of Lewis and Bronsted acidic heteroatoms incorporated into amine-functionalized silica materials that act as bifunctional, or cooperative, acid-base catalysts. I am also studying the effects of active sites accessibility and active site dynamics, for example by tuning the alkyl linker length of amine-functionalized silica catalysts. Additionally, I am developing a new procedure for synthesizing amine-functionalized silica adsorbents for applications in catalysis and CO_2 capture. Finally, I am also part of a collaboration through the Center for Selective C-H Functionalization (CCHF, an NSF Center for Chemical Innovation) where I explore enantioselective C-H activation catalysis in flow using a unique hollow fiber reactor that allows for high product yield and enantioselectivity. This collaboration involves organic and organometallic chemists as well as chemical engineers and has been an excellent opportunity for me to expand my areas of research interest.

I plan to apply my synthetic, analytic, and material characterization skills to my future research interests, which include green chemistry, developing sustainable chemical syntheses, and catalyst design. I will emphasize the design of reusable, inexpensive, and environmentally benign catalytic materials for sustainable chemical transformations. My research will focus on the atom efficiency of these chemical transformations and how the structure of each catalyst dictates its activity and selectivity. My goal is to combine the philosophy of a chemist, in terms of materials synthesis and catalyst design, with the philosophy of a chemical engineer, in terms widespread application and sustainability. Ultimately, I will distinguish myself in the catalytic world by considering more complicated organic reactions for heterogeneously-catalyzed systems and applying my current knowledge and skills, such as silica synthesis/functionalization and calorimetry, to other fields such as drug delivery where my background in catalysis will provide a unique perspective in problem solving.

Publications

1. E. G. Moschetta, K. M. Gans, R. M. Rioux, "Elucidating the Roles of Enthalpy, Entropy, and Donor Atom in the Chelate Effect for Binding Different Bidentate Ligands on the Same Metal Center" *J. Catal.* **2014**, *309*, 11-20.

2. E. G. Moschetta, K. M. Gans, R. M. Rioux, "Characterization of Sites of Different Thermodynamic Affinities on the Same Metal Center via Isothermal Titration Calorimetry" *J. Catal.* **2013**, *302*, 1-9.

3. Y. Yang, E. G. Moschetta, R. M. Rioux, "Addition of Sulfonic Acids to Terminal Alkynes Catalyzed by a Rhodium Complex: Ligand Concentration-Controlled Reaction Selectivity" *ChemCatChem* **2013**, *5*, 3005-3013.

4. Y. S. Choi, E. G. Moschetta, J. T. Miller, M. Fasulo, M. J. McMurdo, R. M. Rioux, T. D. Tilley, "Highly Dispersed Pd-SBA15 Materials from Tris(*tert*-butoxy)siloxy Complexes of Pd(II)" *ACS Catal.* **2011**, 1166-1177.

5. E. G. Moschetta, N. A. Brunelli, C. W. Jones, "Reaction-Dependent Heteroatom Modification of Acid-Base Catalytic Cooperativity in Aminosilica Composite Materials" **2014**, *In Press*.

6. E. G. Moschetta, N. A. Brunelli, V. E. Collier, G. I. Lindy, C. W. Jones, "The Effects of Alkyl Linker Length and Pore Size on the Cooperative Catalytic Activity of Amine-Functionalized Silica Materials" **2014**, *In Preparation*.

7. E. G. Moschetta, A. P. Holewinski, C. W. Jones, "Amination of Alkylhalide-Functionalized Silica for Applications in Catalysis, Adsorption and ^{15}N NMR Spectroscopy" **2014**, *In Preparation.*

8. E. G. Moschetta, K. M. Chepiga, N. A. Brunelli, Y. Feng, C. W. Jones, H. M. L. Davies, "Immobilization of Dirhodium Catalysts in Hollow Fibers for Applications of C-H Functionalization in a Flow Reactor" **2014**, *In Preparation.*

387032 In-Situ Polymerization of Functional Materials through CVD Pathway for Energy and Clean Water Resources

Abstract:

Polymers are versatile materials with diverse application in many different disciplines. However the current emerging technologies that rely on nanoscale materials fabrication pose a new challenge in polymer integration into devices. Solar cells, batteries, and capacitors with their mesoscopic dimension are good examples. The solvent intractability makes the integration of polymers into various devices with nanostructured electrodes very cumbersome. Here we rely on Initiated and oxidative chemical vapor deposition (iCVD/oCVD) as unique, solvent free method that enable polymerization and coating in one step to address the current challenges facing device fabrication at nanoscale metric regime.

Here we demonstrate how *in-situ* polymerization using chemical vapor deposition (CVD) can be utilized to integrate polymers within photoanode of a dye senstitzed solar cell. The importance of polymer chemistry investigated by integrating different polymer electrolytes such as poly (2 hydroxyethyl methacrylate) (PHEMA), poly (glycidyl methacrylate)(PGMA), and poly(4-vinylpyridine) (P4VP) into cell architecture. For all the three polymers deposited, we found that fill quality which is defined as the ratio of the volume occupied by polymer over the pore volume realizes an optimum as a function of monomer surface availability. The observed phenomena attributed to the difference in sticking coefficient and surface diffusivity of the monomers. The importance of the interfacial energetic studied using different side chain pendants group and the result compared with the simulation result of our first principal modeling of the cell as a pesudohomogenous medium. We further applied our unique oCVD process to enable step polymerization of conjugated polymer with exceptional and tunable properties. Highly non-soluble unsubstituted polythiophene was deposited with measured conductivity as high as 70 S/cm, which is an indication of low degree of defect. The charge storage capacity of the polymer deposited within nanostructed electrode was evaluated and proved to be enhanced due to the nanostructue effect and nanoconfinement. We used our finding and integrate polythiophene within conventional super-capacitor electrode. A 250% increase in volumetric capacitance was realized when the polymer loading to matrix mass was optimized. We further synthesize highly conductive Poly(3,4-ethylenedioxythiophene) (PEDOT) with near metallic conductivity. The importance of processing parameter on conductivity and the film composition was studied and the reason behind high conductivity of the as deposited film attributed to the overoxidation state of polymer that is stabilized with the oxidant complexes imbedded within the film. The highly conductive film was used as a hole transport media in a all solid state solar cell. A titania mesoporous electrode was deposited on a 50 nm compact layer of titanium dioxide which served as the electron transport layer and thin coating of dedoped oCVD polythiophene was used as absorber layer. The PEDOT with high conductivity was deposited on this assembly and the cell was fabricated using a gold contact on the top of the PEDOT layer. The measured photocurrent efficiency of 0.8% without any optimization was realized. To further benefit from novel iCVD process the potential application of the iCVD in fabricating microporous membrane for membrane distillation (MD) was examined. The exceptional coating ability of iCVD to deposit thin polymer layer on a hydrophilic microporous support and converting the surface to superhydrophic state proved the practicality of this approach for MD membrane fabrication application.

364305 Overcoming Barriers in Structural Biology Though Novel Method Development

Abstract:

The 3-dimensional structures of proteins and other biomolecules are critical to understanding the mechanisms of biological processes. By far the most successful method in structural biology for structure determination is X-ray crystallography, a fact highlighted by the over 88,000 protein and nucleic acid structures deposited in the Protein Data Bank. Despite its remarkable successes, X-ray crystallography suffers from two major bottlenecks: the relatively large quantity of biological material required to screen crystallization conditions, and the need to form large well-ordered crystals. This poster is a presentation of recent research on overcoming these two main roadblocks in order to facilitate the structural study of more difficult biological targets.

The first part of the poster focuses on the strategies used to address the large quantities of protein necessary for structural studies. The project specifically concentrated on improving the overexpression of membrane proteins in *Esherichia coli*. Membrane proteins are extremely important therapeutic targets and hold potential for practical applications in bionanotechnology. However, low yields in normally robust expression hosts such as *E. coli* have hindered progress in our understanding of their structure and function. By engineering the protein folding pathways in *E. coli*, we showed the expression levels of several membrane proteins could be improved 3 to 7 fold. Additionally, expression vectors were improved by directed evolution, resulting in reduced expression-induced toxicity and enhanced functional membrane protein yields in *E. coli*

The second part of the poster details the progress made on overcoming the other major roadblock of protein crystallography: the formation of large well-ordered crystals. For difficult proteins (e.g., membrane proteins and protein complexes), initial crystallization screens will often times only produce small microcrystals which, despite years of attempted optimization, will never grow to suitable sizes required by traditional X-ray crystallography. We demonstrated for the first time that it is possible to use electron crystallography to solve a protein structure from exceedingly small 3-dimensional microcrystals, orders of magnitude smaller than what is required for X-ray crystallography. The development of this new method, which we named 'MicroED', promises to have a large impact on structural biology by allowing structural data to be obtained from protein crystals previously thought to be unusable.

384927 Effect of Physical and Chemical Properties of Nanoparticles on Small Molecule Release from Liposomes

Abstract:

Drug nanocariers such as liposomes are often limited by slow and non-specific release. A novel strategy is to initiate drug release by an external agent such as near infra-red (NIR) pulsed laser light absorbed by encapsulated or tethered nanoparticles (NP). The absorbance of NIR light is desirable because it causes minimal thermal injury to normal tissues and light penetration of several centimeters. We can synthesize Hollow Gold Nanoparticles (HGN) and Sulfide Cupper Nanoparticles (CSN) to absorb NIR light over the range from 700-900 nm. NIR picosecond pulsed laser light absorbed by NP is rapidly converted into thermal energy; the photothermal heating of the NP results in large temperature gradients that leads to the formation of transient vapor nanobubbles in aqueous solution. The collapse of the nanobubbles induces the rupture of the liposome membrane and the subsequent drug release with minimal damage to the contents and/or surroundings. We find HGN show higher specific absorbance of the NIR light than CSN which leads to more efficiently delivery of liposome contents. We have systematically synthesized various sizes of monodsiperse HGN and CSN, and determined the relative efficacy of laser power and duration on the rate of contents release from 200 nm liposomes. A fluorescent dye was encapsulated within a liposome as a model agent to study dye release kinetics triggered by a picosecond pulsed NIR laser irradiation. Subsequently, an anticancer drug was encapsulated within the liposomes to deliver the drug to prostate cancer cell in vitro; near-complete cell killing was observed. The study leads to a better understanding of the capabilities of multi-functional nanomaterials for NIR laser-controlled drug release in order to facilitate the development of optimized more complex therapeutic systems.

390007 Design of Novel Hydrogen and Compressed Natural Gas Fueling Stations

Abstract:

During my PhD at UCLA under Prof. Manousiouthakis, the field of research I have been working on is hydrogen economy, specifically about the distribution of hydrogen fuel to the end user. One of the key challenges for the deployment of hydrogen fuel cell vehicles is the lack of fill-up practices that are time efficient and safe. Moreover, hydrogen fueling station operators do not employ an optimal mass flowrate policy that can satisfy the aforementioned constraints. A hydrogen vehicle user will like that the replenishment of fuel to his/her vehicle take the same time as a gasoline car. On the other hand, the type IV tanks, based on carbon fiber wrapping, have a maximum temperature limitation of 85 degree Celsius, so the fill-up process must ensure that this temperature is not surpassed. The research I have conducted has led to the development novel mathematical models and software that can simulate and optimize the fill-up process of gas fuel such as hydrogen, and even compressed natural gas (CNG). The models are based on a self-consistent thermodynamic model for real gases, transport phenomena principles, and conservations laws, which predict hydrogen's temperature and pressure time evolutions inside the vehicle tank, the station storage tank, the piping connecting the tanks, and the dispenser isenthalpic valve. The model results have proven to match, within 2%, corresponding experimental data from an actual fill-up process. Furthermore, in order to minimize the fill-up time and to determine an optimum mass flowrate policy, I formulated the fill-up process as a minimum time optimal control problem, which through the used of a novel decomposition of the problem into a process simulation problem, that checks for feasibility of the fill-up independent of time, and a simple minimum time optimal control problem, the global minimum fill-up time can be obtained analytically along with an optimal control policy on the mass flowrate. The optimization studies results lead to a fill-up time strategy that can improve current fill-up practices.

In addition to my core research, I have also performed as the safety officer for Prof. Manousiouthakis experimental lab, having to go to intensive safety training and yearly inspections. Additionally, I have participated in the construction of the UCLA hydrogen fuel cell go-kart, and I have served as hydrogen system lead, safety inspector, and project leader. I am a very active member of UCLA Center for Excellence in Engineering and Diversity efforts to motivate students coming from minority populations to attend the UCLA engineering school. I have been participating in the Mathematics Engineering Science Achievement (MESA) program, laboratory tours, panels, talks, tutoring, and volunteering and organizing events since 2006. My research interests are alternative energy sources for vehicles, production of "green" hydrogen, design of hydrogen fueling stations, hydrogen safety, and hydrogen fuel cell vehicles integration.

384783 Understanding the Influence of the Host Environment on Phenotypic Heterogeneity of Bacteria for Development of Effective Therapeutics

Abstract:

Phenotypic heterogeneity in bacterial populations is a common mechanism by which bacteria survive environmental stress. For example, bacterial persisters have the ability to tolerate extraordinary levels of antibiotics due to their phenotypic state. Unlike the resistant bacteria that arise from heritable mutations, persisters are genetically identical to their antibiotic non-tolerant progenies. Persisters are a significant public health problem because they are enriched in biofilms and have been hypothesized to underlie the proclivity of biofilm infections to relapse. Though the medical relevance of persisters is associated with infection sites in the human body, persistence has largely been studied *in vitro* using synthetic media. Pathogens associated with persistent bacterial infections such as tuberculosis, gastrisis, cystic fibrosis and urinary tract infections are developed in host environments including liver, lung and epithelial cells, which are far more complex environments than that attainable *in vitro*. Studies regarding the interaction between persisters and host environment are widely missing, and therefore, it is still unknown how host environments influence bacterial physiology with regard to persistence. Based on the expertise I have gained during my postdoctoral and graduate training, including both computational and experimental experiences for both mammalian and bacterial systems, I am planning to study the effects of host conditions on persister levels and physiology, and develop experimental models including mammalian systems to analyze the effects of host-pathogen interactions on persistence. After establishing experimental models, persister physiology will be analyzed by genetic, biochemical, and flow cytometric assays to elucidate molecular mechanisms responsible for persistent state, which can form the basis of novel therapeutics for the treatment of chronic infections.

384324 Expanding the Chemistry of Life for the Development of Novel Therapeutics and Synthetic Organisms

Abstract:

Synthetic biology has a fundamental scientific focus on developing biological parts to understand and manipulate biological systems. My multidisciplinary research program will focus on utilizing expansions in the chemistry of genetically encoded systems at both the protein and nucleic acids level for the development of engineered biological systems. Aim1: At the protein level, my work will focus on the co-translational incorporation of L-phosphoserine (Sep) into proteins. Our biochemical understanding of protein phosphorylation has been fundamentally limited by our technological inability to manufacture useful quantities of homogenous, site-specific phosphoserylated human proteins. The platform I have developed during my postdoctoral research now allows robust production of active human phosphoproteins. I will use this capacity to address previously intractable questions in cellular signaling associated with epigenetic regulation and disease progression. This work has the potential to add breadth and depth to our understanding as well as provide novel therapeutic windows. Aim 2: While engineered biological components display structural and functional orthogonally to their natural counterparts, they lack chemical orthogonality. To overcome this challenge, I will focus on expanding nucleic acid biochemistry by integrating 2 additional letters to the genetic code. My work will utilize synthetic nucleobases imidazoaminopyrimidone (letter: P) and nitroaminopyridone (letter: Z) in conjunction with G, A, T, and C to develop synthetic DNA and RNA 'parts' for biological engineering applications. P and Z form Watson-Crick pairing like G-C and A-T, do not significantly alter DNA stereochemistry, and are compatible with fundamental biochemical processes such as DNA replication and transcription. The methods I developed during my postdoctoral research allow for assembling P/Z containing synthetic genes. I will use this chemical expansion of the genetic code to develop modular synthetic gene components such as promoters, terminators, ribosome binding sites, riboswitches, as well as chemically-orthogonal translation systems for non-standard amino acid incorporation. In the long run, such biological 'parts' displaying chemical orthogonality provided by the P:Z pair could play an integral role in the bio-containment of synthetic organisms. These efforts will be sustained by my training in RNA-protein biochemistry and augmented by my experience in biological engineering. Furthermore, these aims will provide a framework for training Masters' and PhD level students, and will complement my interest in teaching undergraduate students.

385284 Integrating Computational and Experimental Methods to Discover Disease Causes and Design Protein Therapeutics

Abstract:

With the advent of genetic sequencing, there has been a clear trend in medicine to develop increasingly specialized products. Antibody therapeutics effective for only a specific gene mutation in cancers and individualized genome sequences are just two examples. As more data becomes available, there is a clear need for experimental and computational methods that can identify the most significant facts and utilize them to generate effective results. My vision is to have a combined computational and experimental group that identifies the causes of diseases and develops treatments for them. The graduate and postdoctoral research I have completed has created a strong foundation for achieving this goal.

At Penn State University, my graduate research was on the development of computational tools for protein engineering and *de novo* design, with a specific focus on antibodies. The Optimal Complementarity Determining Regions method (OptCDR) was developed to design the antigen-binding portions of antibodies, known as CDRs. Using Mixed Integer Linear Programming optimization, a rotamer library, and molecular mechanics force fields, libraries of any user-specified size of antibody CDRs against any specified antigen epitope can be generated. Other work focused on creating a database of modular antibody parts that can be accurately assembled to model antibody structures and using these parts in the Optimal Method of Antibody Variable region Engineering (OptMAVEn) to *de novo* design fully human antibodies. As computation speeds increase and force fields are refined, novel methods such as OptCDR and OptMAVEn will be essential tools to design protein therapeutics for specific objectives.

The postdoctoral research I have conducted at the University of California, Santa Barbara is experimental work to find environmental factors associated with diseases. Through a combination of magnetic selection and fluorescence activated cell sorting of a large peptide library and next-generation sequencing, it is possible to identify a large database of peptides bound by a patient's antibodies. In turn, computational analysis of the identified amino acid sequences from multiple disease and control patients can find motifs that are disease specific and cross-reactive. These motifs can be used to hypothesize the environmental factors that are contributing to or causing the disease.

As experimental methods improve, increasingly vast databases will become available. It is imperative that effective computational methods are available to analyze and utilize this data. However, computations alone are insufficient and they must be coupled with experimental techniques to evaluate the generated hypotheses. My future plans are to build upon what I have done and continue to develop computational and experimental methods for identifying causes of diseases and developing treatments for them. The computational and experimental research I have done so far is an excellent foundation for developing integrated methods to utilize the large amounts of data that are becoming available.

Candidate: **Natesh Parashurama MD PhD**

Title: Liver Regenerative Medicine and In Vivo Molecular Imaging for the Study of In Vivo Liver Organogenesis, Liver Disease and Development of New Diagnostics and Therapeutics

The emerging field of liver (hepatic) regenerative medicine encompasses understanding hepatic stem cell biology, organogenesis, tissue regeneration, and tissue homeostasis for the study of liver disease and development of new diagnostic and therapeutic strategies. The liver, the largest internal organ in the body, performs critical metabolic, endocrine, and synthetic physiological functions. As such, it displays elaborate cellular phenotypes and intricate tissue architecture, including hexagonal, lobular structure, complex vascular spaces with specialized endothelial cells, tissue specific fibroblasts and macrophages, complex blood supply (hepatic artery and portal vein) and heterogeneous billiary duct cells which line billiary duct drainage systems. With respect to capacity for organ regeneration, the liver is unmatched. For example, preclinical, rodent studies have indicated acute hepatocellular loss results in compensatory replication of neighboring adult hepatocytes. In contrast, chronic hepatocellular injury, which often leads to end-stage disease, is compensated for by small, liver stem/progenitor cells lying at the junction of intrahepatic bile ducts and the Canals of Hering. Furthermore, circulating and immune cells, like bone marrow mononuclear cells, mesenchymal stem cells, and tissue macrophages, all have major roles in orchestrating repair in liver injury. These repair mechanisms have been translated to established treatments, like human liver transplantation, and promising therapies, such as human hepatocyte, human bone marrow-derived, and human liver stem/progenitor cell transplantation. However, current and new therapies require harvesting of human tissue, underlying mechanisms of tissue homeostasis remain elusive and hard to decipher in vitro, in animals and humans, and new approaches and techniques are needed answer critical scientific questions and take steps towards new diagnostics and therapeutics aimed towards regeneration.

A deeper understanding of human, three-dimensional (3D) liver organogenesis would naturally advance liver regenerative medicine. Unfortunately, crucial steps of liver organogenesis take place in utero, and most research has been focused on making functional hepatocytes rather than re-creating liver organogenesis. Nevertheless, this would lead to more accurate studies of acute and chronic regenerative capacity, metabolic, detoxification, and synthetic functions, and disease modeling. A second point is the need for being able to assess 3D liver organogenesis while it is occurring. Current techniques rely on removing tissue, treating the tissue and analyzing it in a static fashion. However, techniques are also needed to assess dynamic molecular events in real time, in the intact living subject, during organogenesis or regeneration. These needs are addressed by techniques in the field of molecular imaging, which aims to study biological events and pathways in the intact living subjects. Thus the combination of studying 3D liver organogenesis, and applying current techniques and developing new techniques in molecular imaging, will likely further the field of hepatic regenerative medicine.

389491 Characterization of Conducting Polymers for Lithium Battery, Transistor, Thermoelectric Applications

Abstract:

My Ph.D. research at UC Berkeley under the advisement of Prof. Nitash Balsara involved the synthesis and characterization of simultaneous electronic and ionic conducting block copolymers (P3HT-*block*-PEO) for lithium battery electrodes. Block copolymers can self-assemble and form co-continuous nanoscale domains, which are necessary for enabling redox reactions of the active material ($LiFePO_4$). In addition, the block copolymer serves as a binder to hold the redox-active material particles together. This simplifies the electrode design as one material serve both binding and charge transport functions. The application of this material in a lithium battery with $LiFePO_4$ showed specific capacity reaching the theoretical limit and with minimal capacity fade after ten cycles, thus demonstrating feasibility of P3HT-*b*-PEO as a conductive binder material. Furthermore, the ability of the conductive binder to switch between electronically conducting and insulating states in the positive electrode provides an unprecedented route for automatic overdischarge protection within the battery. This is in stark contrast to traditional lithium ion batteries where external electronics is used to provide overdischarge protection.

My current research as postdoctoral researcher at UC Santa Barbara (since May 2013) focuses on structure-property relationships of semiconducting polymers for organic field-effect transistors (OFETs) and organic thermoelectrics. One of the research directions under the advisement of Prof. Ed Kramer has been the structural characterization of aligned semiconducting polymers where the measured mobility has been greater than 20 cm^2/Vs. A combination X-ray scattering and electron microscopy techniques provides evidence for exceptional alignment of the polymer chains. This allows for an efficient intrachain charge transport between source/drain contacts, thus resulting in the measured high mobility. Another research direction under the advisement of Prof. Michael Chabinyc has focused on how various processing conditions effect the thermoelectric properties (Seebeck coefficient, electronic conductivity, and thermal conductivity) of highly doped semiconducting polymers.

My Ph.D. and postdoctoral research provides a unique background in electrochemistry, polymer physics, and solid-state physics which have been applied to lithium batteries, organic transistors, and organic thermoelectrics. My proposed research will build upon my diverse Ph.D. and postdoctoral research background. The foundation of my research group will focus on the versatility of conductive polymers for not only energy storage and conversion applications, but topics branching away from previous research applications.

393540 Syngas Mass Transfer Delivers Fermentation Potential

Abstract:

Bacterial conversion of CO and H_2 from synthesis gas opens a potential supply of energy and chemicals from renewable biomass and waste feedstocks. This potential is being tested in commercial scale plants that use vegetative wastes and industrial waste gases for production of ethanol. Commercial success will hinge on efficiency, selectivity, productivity and stability of the process deployed. The microorganisms that perform the syngas fermentation use the Wood-Ljungdahl biochemical pathway to obtain energy as electrons and protons from the oxidation of CO and H_2 on enzymes inside the cell. The transfer of the substrates CO and H_2 from the gas phase into the cell determines the electrochemical potential inside the cell, and thus the reaction kinetics and the thermodynamic ratio of products. As such, the application of gas/liquid mass transfer is the most important parameter for control of the syngas fermentation process. Analysis of syngas fermentation that employs mass transfer, reaction kinetics, thermodynamics and mass balance derived from the pathway provides a basis for successful control and equipment design in syngas fermentation. Control of mass transfer can be used to effect high energy conservation, high selectivity for desired product and high productivity in stable continuous fermentation.

392845 Establishing Governing Equations for 3D Cell Culture in Perfusion Bioreactors

Abstract:

My research interests are focused on developing new biomedical products such as biomaterials and bioreactor systems that can be used for drug discovery and testing, 3D cell culture and large tissue regeneration. Culturing cells and regenerating tissues *in vitro* on 3D scaffolds involves several challenges, such as efficient nutrient transportation, uniform stress distribution, and the removal of wastes. Perfusion bioreactors not only allow reproducibility but also provide a controlled environment for production of tissues. My goal is to establish fundamental governing equations for the design of tissue engineering bioreactors and scaffolds. My doctoral research involved testing governing equations related to nutrient permeability, mechanical and structural properties of the scaffolds, as well as nutrient consumption kinetics. The strategy was to use a combination of computational modeling and experimental analysis. Large scaffolds with a high aspect ratio were utilized so that the obtained experimental measurements have high signal-to-noise ratio. This allowed the validation of the governing equations used in the computational models with high fidelity. Three different scaffold preparation techniques, freeze drying, salt leaching, and electrospinning were used to fabricate scaffolds with different microarchitecture. Chitosan, gelatin, and polycaprolactone polymers were used to prepare scaffolds. Two types of bioreactor configurations, flow-through and axial-flow, were used in this study. Both were designed to hold same sized scaffolds, but differ in flow configuration, which made them suitable for evaluating and validating the equations. Bioreactors of appropriate flow configuration were constructed in-house for experimental analysis. Computational Fluid Dynamics (CFD) simulations were performed to predict pressure drop, shear stress, deformation, nutrient distribution profile and exit concentration at various operating conditions. These insights help monitor in vitro tissue regeneration, understand the effect of mechanical stimulus on 3D cell culture, and improve quality of the regenerated tissue.

My future research plan will involve developing *a priori* charts for designing systems involving porous media can help simplify the design process for bioreactor systems. The design variables such as flow rate, porosity and permeability can be transformed into dimensionless variables and characteristic curves can be developed to relate these dimensionless variables. This process will significantly reduce the amount of effort required to improve existing bioreactors and develop new ones. Further, I plan to establish other governing equations such as nutrient diffusivity in porous media, kinetic parameter, and non-viscoelastic nature of bioscaffolds to help improve our understanding of the tissue regeneration process and better mimic the native tissue growth phenomenon. Additionally, I would apply my skills and experience from developing the OSU AppCenter at Oklahoma State University in finding ways to use mobile apps to improve productivity of research and development activities. The poster will include examples of my research projects that constitute my research ideas. During the session I will give an overview of the potential and challenges associated with my research plan.

Development of catalysts for energy and environmental applications

Marc D. Porosoff and Jingguang G. Chen

Over the next 30 years, the global energy demand is expected to increase by 60% to maintain pace with current projections of population and economic growth.[1] Fossil fuels will continue to provide a significant fraction of the energy supply; however, cleaner technologies, including solar, wind and nuclear will become critical industries to meet the required demand. As the burning of fossil fuels continues into the foreseeable future, it is necessary to develop more efficient processes to decrease energy usage and the associated CO_2 emissions. With the threat of climate change and ocean acidification becoming more apparent each year,[2] efforts must be put forth to not only develop new forms of clean energy, but also capture and utilize CO_2 as a commodity. These large-scale CO_2 conversion processes require the discovery of effective and low-cost catalysts.

Development of catalysts for CO_2 conversion can be divided into three areas: synthesis, characterization, and discovery, with significant research opportunities available in each area. In catalyst synthesis, the most promising methods include: atomic layer deposition, colloidal synthesis, and incipient wetness impregnation. Although innovative synthesis methods, such as core-shell nanoparticles, can be extremely effective, the complexity of synthesis does not always translate to low-cost catalysts for CO_2 conversion. For example, although relatively simple, uniform bimetallic particles can be synthesized via incipient wetness impregnation. Catalyst characterization requires the utilization of a wide range of techniques. A previous study on CO_2 conversion used extended X-ray absorption fine structure (EXAFS), CO chemisorption, and transmission electron microscopy (TEM) to confirm the formation of well-distributed, bimetallic particles synthesized by incipient wetness impregnation.[3]

Catalyst discovery is the most general area of catalyst development, but perhaps the most important. Previous studies demonstrate that reducible oxide supports, such as CeO_2, exhibit high activity in CO_2 conversion experiments because of their ability to exchange oxygen with CO_2.[3] Molybdenum carbide (Mo_2C), a non-precious metal material, also shows promise as an active and selective catalyst for CO_2 conversion.[4] These materials are strikingly different, but have a fundamental similarity, which is they are each able to uptake and exchange oxygen with CO_2. Because of the importance of oxygen exchange in reactions with CO_2, oxygen binding energy could be a valuable descriptor when searching for CO_2 conversion catalysts. The discovery of such descriptors is an extremely powerful tool that can be applied beyond CO_2 conversion and into other applications of environmental catalysis.

References

[1] U. E. I. Administration, (Ed.: DOE), Washington DC, **2013**.
[2] T. R. Knutson, R. E. Tuleya, *J. Clim.* **2004**, *17*, 3477-3495.
[3] M. D. Porosoff, J. G. Chen, *J. Catal.* **2013**, *301*, 30-37.
[4] M. D. Porosoff, X. Yang, J. A. Boscoboinik, J. G. Chen, *Angew. Chem. Int. Edit.* **2014**, *Accepted*.

367996 Molecular Engineering Strategies for Tunable Assemblies and Hybrid Materials

Abstract:

The intelligent use of processes such as self-assembly, combined with the ability to manipulate the chemical structure of polymers, can lead to a wide array of new materials. Such functional materials can be the solution to many 21st century challenges in fields such as chemistry or bioengineering. Examples of polymer-based materials that combine these two elements will be presented. In the first example, complex coacervation (i.e. a liquid-liquid phase separation phenomenon) is used as a platform for soft material design. To understand the different aspects of complex coacervation (e.g. thermodynamics, affecting parameters, rheological and interfacial properties) polypeptides are used as a model system. More complex molecular design can be utilized wherein polyelectrolyte domains are connected to neutral polymer blocks. Use of such co-polymers can result in the formation of a variety of self-assembly structures with coacervate core domains including nanometer-sized micelles or tunable hydrogels. The second example includes a technique for polymer functionalization of carbon nanotubes (CNTs) that helps circumvent their inherent insolubility, and considerably widens the scope of nanocomposite materials that can be produced. The strategy involves attachment of substituted polymerization initiators onto a CNT surface. With a judicious choice of substitution, initiators of most popular polymerization techniques can be attached. Complete control over grafting percentage of initiator and surface-initiated polymerizations allows synthesis of nanocomposite materials with desired compositions, which is essential for any application. The resulting nanocomposite materials exhibit improved mechanical and thermal properties, compared to pure polymers.

390550 Orthogonal Engineering of Block Copolymers: Tools, Techniques and Applications

Abstract:

Application-directed chemical modification of synthetic and natural polymers can generate bio-functional analogues with diverse topology. Synthetic block copolymers with ordered structural hierarchy are powerful fabrication tools for creating self-assembled systems with applied functionality. Robust and sterically stabilized nanostructures, such as conjugates, micelles and vesicles can be engineered from these synthetic block copolymers to design effective and targeted drug delivery vehicles. However, limited availability of functional monomers, synthetic complexity of polymerization, and lack of orthogonality during post-polymerization processes require efficient and modular synthesis of the polymer backbone. Intelligent application of functional-group specific chemistry, multivalent amplification of linear polymers using dendritic building blocks, and hybridization of natural polymers with synthetic macromolecules can open up a wide range of possibilities for the development of bio-inspired materials with desired form and function.

387753 Catalytic Processes: From Molecules to Complex Reaction Networks

Abstract:

Designing "optimal" processes in terms of yields, selectivity, reaction rates, economics, and environmental impact is an open challenge in the field of heterogeneous catalysis. This can involve multi-dimensional decision-making, including the choice of chemistry, catalyst, and the reactors. My research aims to apply state-of-the-art computational tools and developing new ones for detailed modeling and design of heterogeneous catalytic processes. To this end, I propose to build and expand on my PhD experience at University of Minnesota under the co-advisorship of Profs. Daoutidis and Bhan on developing methods and software for elucidating complex reaction networks and my current research co-advised by Prof. Mavrikakis and Maravelias at University of Wisconsin focusing on using DFT and mathematical optimization for a first-principles-based approach to rigorously model and design catalytic systems.

In this poster, I will present a multi-pronged approach involving computational chemistry, microkinetic modeling, numerical optimization, cheminformatics, graph theory, and network analysis that allows for elucidating molecular events such as surface reactions and relating them to reactor-scale phenomena for systems spanning few tens of species and reactions to several thousands. Specifically, I will discuss: (a) identification of molecular structure-property relationships in competitive adsorption of organonitrogen compounds in gas oil hydrotreating using DFT and vdW-DF methods, (b) leveraging nonlinear programming methods for rational catalyst design, design of experiments, and parameter estimation, and (c) development of a new code, RING, for automated rule-based generation, mechanism analysis, and modeling of complex reaction networks.

Transforming the art of catalyst preparation into a science

Ana C. Alba-Rubio

Department of Chemical and Biological Engineering, University of Wisconsin, Madison, Wisconsin 53706, United States.

Multi-metallic nanoparticles constitute a class of materials offering the opportunity to tune their properties via the composition, atomic ordering and size. In particular, supported bimetallic nanoparticles have generated intense interest in catalysis. However, traditional synthesis methods often lack precise control, yielding a mixture of nanoparticles with varying particle size and composition. This heterogeneity complicates studies to link catalytic properties to structural and/or compositional changes in the catalyst. My experience in both catalyst preparation and characterization puts me in an excellent position to forge the **next generation of tools for design of catalysts**.

My research career started as an undergraduate at University of Málaga (Spain) under the supervision of **Prof. Pedro Maireles Torres** and **Prof. Jose Santamaría González**. I worked on catalysis for the production of biodiesel and I gained preliminary experience on synthesis by traditional methods, characterization and evaluation of the activity of base catalysts.

I thoroughly enjoyed my research experience and I moved to Madrid (Spain) to do my PhD in one of the most prestigious groups in catalysis in Spain, that one lead by **Prof. José Luis García Fierro**, under the direction of **Prof. Manuel López Granados** (Institute for Catalysis and Petrochemistry research, CSIC). My PhD was focused on **base and acid catalysis** for biodiesel production, and the characterization of the catalysts was always a strong point. It helped us to identify the active sites of one of the most active base catalysts for biodiesel production. Characterization also provided us with insight about the causes of deactivation of solid acid catalysts during the reaction. Deactivation was avoided by the development of a novel catalyst that joined the advantages of homogeneous and heterogeneous catalysis. During my PhD, I did two short stays in Brazil and The Netherlands, which helped me to have a more international vision of the science.

My love for the knowledge, the catalysis and the concept of biomass conversion brought me to the **University of Wisconsin-Madison** to conduct **postdoctoral research** under the supervision of **Prof. James Dumesic** and **Prof. Manos Mavrikakis**. Here, I am focused on the **controlled synthesis of bimetallic and bifunctional catalysts** for reactions of interest in biomass conversion and for the direct synthesis of H_2O_2 from H_2 and O_2. I am familiar with many techniques of preparation and characterization of catalysts. I have been trained here for the use of several electronic microscopes, and I really enjoy matching structure with activity. In this respect, I am working to design catalyst architectures for each of my projects at Wisconsin, and I would like to continue this type of work in the future.

My future research directions will combine my experience on characterization and preparation to develop active and selective catalysts for reactions of interest. I love to work on biomass conversion, but since many chemical reactions need catalysts, my future research has diverse applications.

Finally, I have a particular interest in teaching, especially in teaching catalysis and experimental skills. In addition, I would feel confident teaching other subjects to chemical engineering students. When I was an undergraduate student, I taught various subjects to younger students. I developed a teaching technique to keep students engaged and involved in the material. I have also supervised people in the lab and the experience has been very rewarding. Active and cooperative learning will be central to whichever courses I teach, and my students will find from me the encouragement that they need to achieve their goals.

393677 Engineering Approaches to Elucidate Mechanisms of Physiologically Relevant Cellular Behaviors in Cancer

Abstract:

Approximately 90% of cancer related deaths worldwide occur because of metastasis of the primary tumor to other organs. The tumor microenvironment plays a key role in metastatic disease progression; however the associated mechanisms are far from being unraveled. The goal of my research program is develop tools to systematically study these processes and dissect relevant mechanisms *in vitro* and *in vivo*. During my doctoral research at the Ohio State University, I developed brain mimetic biomaterial systems (e.g., hydrogels, electrospun nanofibers) that mimic several features of the *in vivo* tumor microenvironment for investigating migration of patient derived human glioblastoma multiforme (GBM) tumor cells. These three dimensional (3D) cell culture systems recapitulate *in vivo* like morphology, migration, and molecular signatures of GBMs as opposed to traditionally studied 2D cell culture systems. Such biomimetic cell culture models will allow for patient tailored bioassays where patient tumor biopsies can be examined and several drug combinations can tested for personalized medicine, thereby guiding patient care and treatment decisions with the potential to transform current cancer treatment. During my postdoctoral research at Northwestern University, I developed microporous poly(ε-caprolactone) (PCL) biomaterial scaffolds for early detection and treatment of breast cancer metastasis. I also discovered and demonstrated the efficacy a novel therapeutic combination of cyclin dependent kinase-2 (CDK2) inhibitors and Eribulin for the treatment of triple negative breast cancer. To investigate the mechanism of drug action, I utilized a live cellular array that provides high-throughput analysis of dynamic transcription factor activities. This poster will highlight the application of these tools *in vitro* and *in vivo,* which will provide a foundation to elucidate mechanisms of tumor progression and subsequently develop early detection and therapeutic strategies.

383806 Composite Conjugated Polymer/Fullerene Dispersions for Organic Photovoltaic Applications

Abstract:

Controlling the structural morphology of conjugated polymer/fullerene composites is an important aspect in improving the performance of polymer solar cells. The efficiency of exciton dissociation and carrier transport to the working electrodes are both a function of the size and distribution of polymer and fullerene domains within the active layer. In the current solution processing paradigm, these characteristics are intrinsically linked to the detailed process history of the film. This fact compromises the promise of polymer solar cells as an inexpensive renewable energy technology because performance gains made in the laboratory may not translate well to scaled-up manufacturing processes. Aqueous dispersions of conjugated polymer/fullerene nanoparticles have the potential to address this challenge. Similar to their thin-film counterparts, the performance of photovoltaic devices derived from conjugated polymer/fullerene composite nanoparticles (CNPs) is a function of the structural morphology within each nanoparticle. However, whereas the structural morphology of an active layer produced using solution phase deposition is a function of the detailed history of the coating process, the structure of CNPs is fixed during their formulation. Because of this fact, the structure can be identified in the dispersed phase and tied, using single particle characterization, to its photovoltaic performance before it is deposited into a device. Therefore, the optimization of devices derived from CNPs can be broken into two discrete problems: improving the **intrinsic** properties of a nanoparticle as a function of how it is produced and optimizing the **extrinsic** properties of devices through the deposition of optimized CNPs. Our work has addressed both of these challenges. First, using P3HT/PCBM nanoparticles, we show that internal phase segregation is a complex function of the way the particles are produced and their composition. We also show that the resulting structure can be identified and optimized for photovoltaic performance. Second, we outline the principles that affect the deposition of these CNPs onto device relevant electrodes, and provide insight into the production of high-quality active layers from CNP dispersions. The combination of these two separate approaches opens up the potential for novel roll-to-roll processes to produce high performance photovoltaic modules of arbitrary complexity with flexible form factors.

385735 Branched Wormlike Micelles Under Dynamic Flow Conditions Using Spatiotemporally-Resolved Small Angle Neutron Scattering

Abstract:

The nonlinear rheology and segmental alignment of a model series of WLMs is studied to determine the effects of branching on shear flow phenomena. Branching in the mixed cationic/anionic surfactant CTAT/SDBS is controlled via the addition of the hydrotropic salt sodium tosylate. The shear-induced segmental alignment of the micelles is measured for local gap positions and time under large amplitude oscillatory shear (LAOS) in the flow-gradient plane using novel flow-SANS instrumentation [1] as well as temporally in conjunction with rheology in the flow-vorticity plane [2]. Our ability to resolve the microstructure via scattering projections in multiple planes of flow provides new insights into the effects of flow on the WLM microstructure under time-dependent shear flow. Local segmental orientation and alignment in the flow-gradient plane is found to be a complex function of the branching level, radial position, and time during the oscillation. Spatiotemporal SANS demonstrates that LAOS can elicit simultaneous elastic and viscous material responses in localized regions across the gap, thereby corroborating recent theoretical predictions with experiment for the first time [3]. Furthermore, the maximum alignment in the flow-gradient plane under LAOS is found to be significantly higher than that observed in steady shear at the same instantaneous shear rate. This nontrivial hyperalignment under dynamic deformation is interpreted in terms of the dynamics of the microstructural rearrangement during LAOS. This research quantitatively links micellar microstructure and topology to the measured nonlinear shear rheology of WLM solutions, providing a more complete data set to enable interpretation of the rheology and rigorous testing of microstructure-based constitutive equations.

[1] A.K. Gurnon, et al., J. Visual Exp. 84 (2014).
[2] C.R. Lopez-Barron, et al., Phys. Rev. Lett. 108 (2012).
[3] J.J. Stickel et al., Journal of Rheology 57 (2013).

375885 Droplet Microfluidics in Physical and Biological Systems

Abstract:

The exploitation of microdroplets produced within microfluidic environments has recently emerged as a new and exciting technological platform for applications within the chemical and biological sciences. Interest in microfluidic systems has been stimulated by a range of fundamental features that accompany system miniaturization. Such features include the ability to process and handle small volumes of fluid, improved analytical performance and low unit cost. Moreover, microfluidic systems that generate and utilize a stream of sub-microlitre droplets dispersed within an immiscible continuous phase have the added advantage of allowing microscopic investigation of physical properties of complex fluids such as emulsions, suspensions and foams. I will describe here three different studies. All of which use the microdroplet platform to establish fundamental understanding of different physical systems as well as to develop devices for biomedical applications. First I will explore the transition from reversible to chaotic behavior in the oscillatory shear flow of water-in-oil emulsions. The emulsion is being flown through a microchannel and is forced to rearrange due to a central constriction. The surprising results show that while at the Stokes flow limit, the emulsion exhibit behaviors that vary from complete reversible to complete irreversible depending on the volume fraction, velocity and viscosity ratio. I will present here the first direct visualization of the phenomenon. This study is an important step in understanding the microscopic rearrangements of droplets and particles near jamming. The second study deals with the break-up of droplets in a concentrated emulsion during its flow as a 2D monolayer in a microchannel consisting of a narrow constriction. The probability of droplet break-up increases with flow rate and the rate of deformation of the drops. The probabilistic nature of the break-up process arises from the stochastic variations in the packing configuration of the drops as they enter the constriction. The results of this study shed new light on the stability of concentrated emulsions and the resulting applications in droplet microfluidics and two-phase flow in porous media. Finally, I will describe the development of a new method for rapid detection of tuberculosis using droplet-based microfluidics. Ultimately, this tool will be able to detect tubercle bacilli in a sensitive, rapid, specific and quantitative manner at low cost. This tool will be particularly important in resource-limited settings where TB is the most prevalent.

389654 Advanced Materials for Catalysts, Membranes and Sorbents Applications and Renewable Energy Technologies

Abstract:

My research focuses on the interface of chemistry, chemical engineering and materials science. Over the past six years, I have excelled in my research efforts related to the synthesis, characterization and application of organic/inorganic materials for catalysts, membranes and sorbents applications and renewable energy technologies. My PhD research experience at University Putra Malaysia involved development of a cost-effective method for synthesizing novel mixed-metal oxide catalysts being used in light alkane dehydrogenation and oxidation applications, in particular in n-butane to maleic anhydride process, whereby I adapted recent developments in reaction science to advance the development of novel catalysts. My PhD work resulted in 3 patents and 7 papers that have been published in high quality catalysis and chemical engineering journals.

During my previous research at Lulea University of Technology (LTU) in Sweden, I was involved in catalytic production of methanol, dimethyl ether and biofuel from black liquor synthesis gas. The work focused on catalyst synthesis (mainly mixed metal oxides and zeolites), characterization and evaluation into two different catalytic process; (1) conversion of black liquor synthesis gas to methanol (high-pressure hydrogenation reactor), and (2) methanol conversion to dimethyl ether and gasoline-range hydrocarbons (high-temperature dehydration reactor). As part of a multidisciplinary team of researchers and graduate students from LTU university and industry partners, for the first time in the world, we produced bio-methanol from black liquor synthesis gas.

Currently, I am a post-doctoral fellow at Georgia Tech, with Prof. Bill Koros working on membrane gas separations by tailoring membrane characteristics through investigation of new materials and structures to selectively control mass transport at the molecular level. In particular, the current research is focused on producing novel mixed-matrix hollow-fiber membranes and sorbents for post-combustion CO2 capture. This work addresses the economics of CO2 capture with easier scalability.

My research interests in the near future will focus on the fundamentals and applied aspects of catalysis and separation employed in science and engineering for sustainability. The proposed research will seek to design, synthesize and evaluate advanced catalytic materials to allow for revolutionary advances in separation and reaction science to address critical issues in pollution control, waste-to-energy and renewable energy and fuels.

18 Total publications, 13 first author, h-index=9

385802 Evaluating Issues in Environmental Adsorption: Mercury Oxidation and Shale Pore Characteristics

Abstract:

The research presented here will focus on recent work focused around the field of environmental adsorption. The topics are: 1) identification of adsorbed mercury species in applications related to coal-fired flue gases and 2) evaluation of analytic adsorption techniques for the determination of pore characteristics in shale gas. Both of these topics have implications in the field of clean energy: coal-fired power remains the primary source of electric energy in the United States, China and India and the primary anthropogenic source of emitted mercury, while the production of natural gas from non-traditional sources, such has black shale, has outpaced the information available about these sources and the long-term implications of the their development.

1.) Mercury exists in various forms in the flue gases of coal combustion, from its elemental state (Hg^0) at the boiler exit, to various oxidized forms (Hg^{2+}) as it interacts with fly ash particle surfaces as the flue gas cools. The exact species of the oxidized mercury is, at this point, an outstanding question. Thermodynamic calculations suggest that the favored form in coal combustion flue gas is $HgCl_2$, but HgO and $HgBr_2$ could exist under certain conditions, such as the introduction of Br into the combustion process or in cases when coal is burned and contains trace amounts of Br.

An electron ionization quadrupole mass spectrometer (EI-QMS) has been modified for the direct measurement of elemental and oxidized Hg at industrial relevant concentrations ($< 15\ \mu g\ m^{-3}$ or 2 ppbv) from a methane combustion flue gas. Ultimately, the extent of homogeneous and heterogeneous oxidation may be measured, with the Hg species directly identified through mass spectrometry. In addition, the oxidized species created upon exposure to surfaces present in coal combustion, such as fly ash and activated carbon, are identified through the use of packed-bed experiments. A variety of sorbents have been evaluated, with the MS used to track any desorbed species. This information is critical for evaluating the extent of reaction on a sorbent surface and the potential for desorption of various oxidized Hg species. By investigating the desorbed species as a function of temperature provides indication of the relative strength of adsorption of the various Hg species and subsequently their relative stabilities on the various sorbents tested. These types of investigations will also increase the fundamental understanding of the surface reaction of Hg on a variety of sorbents and provide further understanding that can lead to advances in Hg capture.

2.) The rapid ascent of shale sourced gas has created a need for analytical evaluation of shale, in order to determine permeability and gas-in-place. Researchers have focused on the adsorption properties of shale, in order to evaluate potential gas-in-place and understand how gas transports from micropores to macropores to fractures. The literature, to this point, has been unfocused, and pointed towards evaluating as many shale samples from the broadest perspective as possible. However, analytical adsorption approaches appropriate for investigating homogenous samples, such as activated carbon or zeolitic material, has limitations when applied to highly heterogeneous samples. Shale, which has high variability at all scales, from the formation level to the microscopic level, is a particularly difficult challenge.

This work focuses on creating a standard analytical method for the evaluation of gas shale, focused on low-pressure adsorption techniques. The outgas temperature will be evaluated to determine effects on adsorption isotherms, with emphasis on potential changes in pore volume. The samples will be evaluated with 3 probe gases (N_2 at 77 K, Ar at 87 K and CO_2 at 273 K). Isotherms will be evaluated with DFT techniques to determine pore size distribution, while using information from Fourier transform infrared spectroscopy (FTIR), scanning electron microscopy (SEM), elemental analysis and total and inorganic carbon contents to support the gas adsorption observations. The thorough analytical approach presented in this work will present a method that researchers can use to properly evaluate pore characteristics of shale gas, which can be used to support research in flow models and gas-in-place estimates.

389397 Engineered Nanomaterials for Energy Harvesting

Abstract:

Due to a continual rise in demand for energy worldwide, significant attention is being provided to new environmentally clean energy resources such as solar energy on one hand while large efforts are also concentrated on thermoelectric devices which convert thermal energy directly into electrical energy and hold promising contributions in waste heat recovery. For both approaches, strategies are underway to identify materials and device architectures that are inexpensive and scalable for widespread use yet efficient compared to existing technologies. Nanotechnology enables novel approaches to energy conversion that may provide both higher efficiencies and simpler manufacturing methods. I plan to develop a research program which will focus on tuning and improving existing properties and harnessing novel properties through nanostructuring, doping and interface engineering in these materials.

Electronic, optical, mechanical and thermal properties of metals and semiconductors (SCs) in nano-sized systems can be controlled appreciably by varying their size and give rise to potentially new phenomena not observed in their bulk counterparts. Since the properties of the material are directly related to its crystal structure, the ability to control the crystal phase could also lead to new behavior. Nanostructures offer a potential method to obtain additional phases that would not exist in the bulk. Nanostructures can also lead to shifts in phase-transition (PT) temperatures. This effect is not only interesting for fundamental reasons, but the ability to modify the PT temperature allows one to stabilize (or destabilize) desirable (or undesirable) phases, depending on the goals. Because the properties of the nanostructures change through the PT, this can have implications for potential applications.

The introduction of trace intentional impurities (or doping) is central to controlling the behavior of SC materials. Dopants possess the ability to strongly modify the optical, magnetic, and electronic properties of bulk SCs. Modern SC-based technology owes its existence, in large part, to the fact that these materials can be doped. It is the ability to control precisely the number of carriers available in the SC by doping, which has expedited the advance in SC-based electronic and optoelectronic technology. The advantage of doped SCs is that they provide the device engineer with a wide range of mobilities, so that materials are available with properties that meet specific requirements. Hence, it is natural to extend the versatility of nanostructures by adding dopants.

While interfaces are typically viewed as a dividing layer between two neighboring materials, the properties of the interface are in-fact, at times, radically different from the constituent materials. Nanomaterials, in particular, where nearly 20-30% of the total atoms are located at or very close to the surface, provide an ideal platform to study these effects. Rational engineering of the interfaces between two different components in a nano-composite, presents an opportunity for creating materials with novel properties that could not be achieved otherwise. Hybrid nanocrystal/conducting polymer materials have extensively been used in optoelectronic devices such as photovoltaics and light-emitting diodes, and recently for thermoelectric applications as well.

My PhD research focused mostly on zero-dimensional nanostructures (nanocrystals typically less than 10-nm in size). I developed a versatile synthesis that yields high quality silver chalcogenide nanocrystals (NCs) with size-dependent phase behavior, optical and transport properties. However, by changing the surface of the NCs, I was able to modify the phase behavior of these NCs which allowed us to completely tailor both the optical and electronic transport properties of the material. Additionally, I developed a facile approach to lightly dope SC NCs (CdSe and PbSe) with a controllable amount of electronically active impurities. The addition of even a single impurity per NC has a dramatic effect on their optical properties. Furthermore, studies of the electrical transport through the NC films show complex behavior in the Fermi level as a function of dopant concentration. The results demonstrate that dopant behavior in NCs is not as simple as one might expect. Thus, these experiments begin to reveal the properties of a novel class of NC materials in which both interface and impurities play a key role that may be important for future NC devices.

Multicomponent materials allow us to unite the benefits of both components into one hybrid material and design systems that do not necessarily conform to standard mixing theories for carrier transport. My postdoctoral research aims towards designing and characterizing novel solution processable organic/inorganic nanocomposites (inorganic component being one-dimensional materials or nanowires) and gaining a functional understanding and control of carrier transport at these soft/hard interfaces primarily for thermo-electric applications. We observe unusual thermo-electric transport in a high-performance organic/inorganic thermoelectric composite leading to an optimized power factor in the composites that exceed that of the individual components, suggesting non-effective medium behavior. We increase the conductivity in these hybrid systems further by doping the individual components separately. Finally, by changing the nature of the interface, we can completely switch the nature of charge carriers in these material systems which are particularly attractive for building real thermoelectric modules. My results underscore the importance of understanding and controlling interfacial phenomena in hybrid organic/inorganic systems, and provide a general route for enhancing carrier transport in hybrid materials and devices.

Further work will seek to enhance this interfacial phenomenon in layered two-dimensional materials (nano-sheets) where essentially the entire material is at the interface. In the process, we hope to gain a fundamental understanding of these hybrid systems that would provide a general and robust framework to guide materials design and allow us to tailor suitable material systems for energy applications. Future research plans will be discussed in addition to the above topics.

382256 Experimental Fluid Dynamics for Advanced Materials and Biological Systems

Abstract:

Experiments investigating fundamental transport phenomena are the foundation for my past and present research, and they are my primary interest and focus for future research. I have conducted experiments in fluid dynamics in different areas of science and technology including biological systems, advanced materials, and materials processing. During my doctoral work I investigated the properties of lignocellulosic biomass from a rheological perspective. The fluid and mechanical properties of biomass are still of great interest to those processing biomass to biofuels and other bioproducts, and much work remains to link the physiochemical structure of biomass to its rheological properties. My postdoctoral work has focused on transport phenomena in fluid-fluid interfaces like those found in foams, emulsions, and biological membranes. In this area there is enormous opportunity to advance knowledge in synthetic biology by characterizing the fluid and mechanical properties of synthetic cell membranes with microrheological techniques. Most recently I have investigated a novel processing technique for advanced materials called fractal structuring. The technique involves a confined, multi-component flow through a simple but uniquely designed mold of iterative baker's transformations that results in highly structured composites of co-continuous phases with potential applications in photovoltaics, fuel cell membranes, and gas separations. Fluid dynamics simulations and experiments are required here to fully exploit the technique and design the next generation of fractal structuring molds. In this presentation I will highlight important results from my past and present research and indicate where additional investigations of fundamental transport phenomena in each area hold promise for future discoveries.

380066 Engineering Solutions for Systems Biology: Microfluidics and Unbiased Statistical Tools to Uncover Genetic Relationships through Phenotyping

Abstract:

Systems biology focuses on studying living organisms as an ensemble of molecular, dynamic networks that are orchestrated to perform biological functions. The main questions are directed to understanding how all the players interact to accomplish certain goals within a cell or within an organism. This new approach to biology necessitates new experimental paradigms that enable perturbing biological networks and extracting information in a multidimensional scale. Just like systems biology aims at finding how biological ensembles work together (and what happens to them with disease), assays and experimental platforms should focus on extracting ensemble results rather than individual features. Experimental approaches need to be able to look deeper into the information provided by experiments and need to allow better control of the perturbations assayed.

Here I show how we have applied an integrated approach to perform biological studies enabled by microfluidics, automation, mathematical algorithms and machine learning with the nematode *C. elegans* (an excellent model for systems biology). I present how microfluidics can be coupled to unsupervised data extraction tools to obtain deep information and enable high-content studies of complex biological systems. Specifically, I discuss how integrated engineering platforms can be used to elucidate relevant questions in neuroscience, such as synaptic plasticity, learning and the effects of aging on neurological health. We have applied one such system to perform automated genetic screens and to identify differences in synaptic patterns, hidden to the naked eye. These new experimental platforms are promising in our challenging search for a systems-level understanding of biological processes.

389788 Applied Mechanics Studies of Complex Fluids for Pharmacokinetic Application

Abstract:

My research interest lies primarily in non-Newtonian fluid's behavior in response to external field in bulk and near wall. A considerable portion of earlier research work addressed problems related to modeling response of colloidal hybrids, effect of surface charge and contact mechanics, and capillary sintering during drying of colloidal dispersion. My current research involves development of next generation pharmacokinetic models for therapeutic targeted drug delivery.

A class of hybrid complex fluids is made of colloid hard-core with polymer brushes. For battery research they are used for ion transportation. For gas absorption purposes they act as porous support to screen molecules and hold active absorption agents. In drug delivery hybrid complex fluids (which we refer to as nanogels or deformable nanocarriers) are used as vehicles to carry important drugs and target them to specific tissues that are either inflamed or diseased. In this case the core is made of native proteins such as lysozyme and the brush is made out of biocompatible polymers such as dextran or hyaluronic acid. The dimensions of these nanostructures range from 10-100 nm, which puts them below the diffraction limit, making experimental investigations challenging.

Novel Materials for Targeted Drug Delivery: The deformable nanocarrier has a large molecular weight (10^2-10^5kDa) and it is difficult to predict its structural dynamics using molecular dynamics techniques. There is also a lack of mechanistic insight to predict selectivity of binding of the nanogels to inflamed or diseased tissue under physiologically relevant conditions in vasculature. In order to model the deformable gel-based NC, the structure of the deformable gel is coarse-grained based on theoretical models from polymer physics and simulated annealing protocol, which mimics the experimental method of synthesis of this novel material. The Brownian dynamics framework is incorporated to include hydrodynamic interactions, which led to the calculation of equilibrium and steady-shear rheological properties of the nanogel. Brownian dynamics with near-wall interactions along with rheological response in bulk gives important insight that is essential for drug design and predicts the targeting capabilities to specific tissues, which is important for clinical translation.

Novel Materials for Energy Applications: Nano organic hybrid materials are made of silica cores and polyethylene glycol (PEG) brushes. Most of the applications like flow in a semi-confined space is related to complex flows. Primary goal of my research work has been understanding the properties of these materials at Brownian and mesoscale level by developing theoretical model to characterize these complex materials. Based on these models the steady shear rheological properties and transient response[1] have been investigated. A constitutive model for complex flows exhibiting a variety of local linear flows is derived and applied to a complex stochastically varying flow of the fluid through fixed bed of fibers. If the ratio of size of the core to polymer brushes decreases, the structural anisotropy increases and the material shows sign of strain thinning. Such material acts as an additive, which reduces the internal stresses of the fluid in a conduit and the risk of fracturing of underground pipelines (which is a common problem in fracking process) reduces.

Prior research work was primarily on consolidation and cracking in drying colloidal dispersion. The work addressed fundamental understanding related to colloidal films formation broadly desiccation processes of water-soluble colloids. Water-soluble colloids are important ingredient in paint industry (for reduction of volatile organic components) and for continuous manufacturing of drugs in pharmaceutical processing. Drying of water-soluble colloids and its film formation were investigated through experiments and theory. If a dispersion of colloidal particle is allowed to dry, particles packing under a convective field (set by the evaporation) is in compression and the solvent from bulk, pass through porous bed of particles to reach the interface. Charge nature of the particles changes the conventional drying-consolidation profile[3] while the presence of flaws in the packed region is effected due to an interplay between strain energy, stored in the particle network, and interfacial energy.[2] Experiments were performed in rheometer (semi-confined geometry[4,6]) and in microchannels (confined geometry[5]). In semiconfined geometry the pressure required to open a crack from flaw was predicted while in confined geometry the additional effect, due to presence of boundaries and channels, were investigated and hence depicted an overall consolidation behavior.

My research experience is mainly in applied mechanics studies of complex fluids. I would like to further extend the methodologies developed for amphiphilic carriers such as rodlike micelles[7] to include wall-effects and resolve internal and inhomogeneous / anisotropic stresses-strains near wall. In biomedical applications, rodlike micelles can have long circulating times, which are essential in sustained delivery of drugs required for some medical conditions. Rodlike micelles can also be an important self-assembly for drag reduction in oil pipelines and hot water transportation lines. I plan to carry out experiments in 4-roll mill to understand the response of rodlike micelles in a variety of flows and extend the present understanding of soft glassy fluid elements to rodlike micelles, in general FENE dumbbells. I am also interested in packing and sintering of alumina film formation with or without a heat source. Alumina is widely used as refractory coatings, catalyst and more recently transparent optics. I would like to study alumina film formation through theory and experiments to predict packing and its effect on film properties.

I wish to acknowledge funding support for my work through Cornell University-KAUST center and NIH 1R01EB006818-05 & NIH U01-EB016027, made available to me through my postdoctoral supervisors.

References

1. Sarkar, A. and D. L. Koch, "Brownian dynamics simulations of a local-fluid-volume preserving model of solvent-free nanoparticle fluids," The Society of Rheology 84th Annual Meeting (2013).
2. Sarkar, A. and M. S. Tirumkudulu, "Asymptotic analysis of stresses near a crack tip in a two- dimensional colloidal packing saturated with liquid," Phys. Rev. E, 83, 051401 (2011).
3. Sarkar, A. and M. S. Tirumkudulu, "Consolidation of charged colloids during drying," Langmuir, 25(9), 4945–4953 (2009).
4. Sarkar, A. and M. S. Tirumkudulu, "Critical stress required to open a flaw in a two dimensional colloidal packing saturated with solvent," International Symposium on Recent and Emergent Advances in Chemical

Engineering (REACH).
5. Sarkar, A. and M. S. Tirumkudulu, "Delamination of drying nanoparticle suspensions," Soft Matter, 7, 8816–8822 (2011).
6. Sarkar, A. and M. S. Tirumkudulu, "Ultimate strength of a colloidal packing," Soft Matter, 8, 303–306 (2012).
7. Discher Dennis E. and Eisenberg Adi,"Polymer Vesicles", Science 297, 967 (2002).

382321 Bridging the Gap Between Chemical and Biological Catalysis to Produce Biorenewable Chemicals

Abstract:

Biomass is a renewable source of carbon that has the potential to replace fossil-based resources as the feedstock for producing fuels and chemicals. An attractive option for obtaining biorenewable chemicals from biomass leverages the high efficiency of traditional heterogeneous chemical catalysts to upgrade highly functionalized molecules produced via recent advances in metabolic and enzyme engineering. By such coupling of chemical and biological catalysis it is possible to access molecules not easily obtainable by either discipline alone.

My doctoral research has focused on the challenges and opportunities associated with bridging the gap between chemical and biological catalysis. One key aspect of such strategies is to elucidate reaction networks that stem from a particular biologically-derived intermediate. For example, cortalcerone, which can be derived enzymatically from glucose, can be dehydrated over a Brønsted acid catalyst, such as HCl, to yield furylglyoxal hydrate (FH). In turn, FH has appropriate functionality to be converted to furylglycolic acid (FA), a hydroxy-acid amenable for co-polymerization with lactic acid to yield biorenewable polymers reported to have polystyrene-like properties. FA can be produced from FH via an intramolecular Meerwein-Ponndorf-Verley-Oppenauer hydride shift using a Lewis acid catalyst, such as a Sn-containing beta zeolite. Furthermore, FA can be produced in one pot from cortalcerone using a bifunctional Al-containing Sn-beta zeolite possessing both Brønsted and Lewis sties. In another example, triacetic acid lactone (TAL), a highly functionalized molecule obtained via polyketide biosynthesis, can be upgraded to 11 potential end products and intermediates, including bifunctional ketones via decarboxylation of TAL or its hydrogenated analogue, lactones via dehydration of hydrogenated TAL, and unsaturated acids, including the common food preservative sorbic acid, via ring opening of these lactones.

Another key aspect of coupling chemical and biological catalysis is to elucidate the influence of biogenic impurities on heterogeneous catalysts. For example, amino acids can result in deactivation of the metal catalysts used for TAL hydrogenation. This effect can, however, be mitigated by appropriate catalyst design. Overcoating a traditional Pd catalyst with poly(vinyl alcohol) (PVA) results in the formation of microenvironments surrounding the metal nanoparticles that prevent adsorption of such species. For example, the Pd catalyst used in TAL hydrogenation loses 83% of its activity after 14 hours of exposure to methionine (a representative, sulfur-containing amino acid), but a PVA-overcoated catalyst is stable under the same conditions. Indeed, combining the use of a PVA overcoat to prevent adsorption of amino acids with the use of PdAu bimetallic nanoparticles to prevent deposition of carbonaceous residues from TAL decomposition, it is possible to decrease the deactivation rate constant by an order of magnitude when upgrading TAL recovered from spent cell culture medium. My future research will continue to focus on these types of challenges and opportunities that are associated with integrating chemical and biological catalysis, with the ultimate goal of efficiently producing high-value chemicals from biomass.

385612 Computational Design of Advanced Materials to Meet Health, Environmental and Energy Challenges

Abstract:

Society is facing health, environmental and energy challenges, including fast disease diagnosis, affordable supply of clean water, and sustainable energy supply and storage. Rational design of advanced materials is essential to meet these challenges. The fast development of computational technologies and simulation methods enables us to study properties inaccessible to experiments, provide molecular-level principles to guide experiments, and perform large-scale screening of molecular structures for desired properties.

My PhD and postdoctoral research illustrates how computational simulations can be applied in materials design. My PhD work at the University of Washington focused on exploring the differences between zwitterionic and non-ionic materials, and the structure-property relationships of zwitterionic materials. Using quantum mechanical calculations, atomistic simulations and enhanced sampling methods, I have investigated (a) the effects of zwitterionic and non-ionic materials on the structures of chymotrypsin inhibitor 2 and on the association of two hydrophobic sheets, and (b) the hydration, ionic interactions, and self-association of different zwitterionic molecules. My study explained the different effects of zwitterionic and non-ionic materials on bioactivities of proteins, and provided the principles for designing nonspecific protein-resistant zwitterionic materials. A material that I designed computationally was synthesized and tested in my PhD advisor's group.

My postdoctoral research at the North Carolina State University focuses on developing computational approaches to predict the toxicity of different types of nanoparticles. The small sizes of nanoparticles enhance their ability to enter the body, and concerns have arisen over their toxicity. The protein corona on nanoparticles serves as the signature of the nanoparticles in biological systems and determines the interactions between nanoparticles and biological systems. My research aims to develop coarse-grained models that can be used to predict protein-nanoparticle interactions, and to answer fundamental questions on nano-bio interactions.

In my future work, I plan to apply my expertise in computational simulations to develop advanced materials with the aim to address the health, environmental and energy challenges that we face today. I am interested in designing computationally (1) peptide surfaces that resist nonspecific protein adsorption but bind specific proteins for disease diagnosis, (2) zwitterionic porous materials that have high salt-rejection rates and nonfouling properties for desalination, and (3) polymeric electrolytes that have high ionic conductivity for lithium-ion batteries.

Omics-based, Automated Disease Modeling

Jason E. Shoemaker*, Yoshihiro Kawaoka

Department of Microbiology and Immunology
The Institute of Medical Science
University of Tokyo

A systematic understanding of the interactions within biological systems promises to guide the design of next generation disease diagnostics and therapies. In the past decade, systems biologists have successfully applied traditional process systems and controls engineering techniques to analyze the highly complex environments of biological systems; leading to major contributions in the fields of diabetes, drug target prioritization, and metabolite optimization. Yet, many challenges remain for creating medically significant models. The most prominent of challenges is the explosion in data that has occurred since the development of microarray and next generation sequencing techniques. The wealth of data and high degree of biological variance can quickly overwhelm traditional modeling techniques.

My work currently focuses on integrating several computational techniques from the fields of systems engineering, machine learning and bioinformatics for rapidly developing disease models. Specific challenges include:

- Signal deconvolution in *in vivo* whole genome and protein data ('omics' data) to determine the changes local lymphocyte demographics. We have deployed a web tool, CTen, related to this work
- System Inference Microarray Analysis (SIMA) in which we employ nonlinear, biologically inspired models to predict gene expression across experimental conditions
- Applying machine learning to the topological characteristics biological networks to identify safe drugs

Additionally, I am involved in several systems software projects, including Garuda which aims to provide a single platform to unify systems biology tools.

The foundation of my research program will be developing machine learning and artificial intelligence algorithms that can automatically process large scale, dynamic data into effective biological models. In this poster, the successful application of the three tools mentioned above is discussed in the context of influenza virus infection and interferon therapy. Then we review methods and motives to improve these tools and how they can be implemented in long term strategy for developing omics-based, automated disease modeling platforms.

388835 Electrodeposition and Activity of Electrocatalysts in Ionic Liquids

Abstract:

Applications of electrocatalysts can be found in sensors, electro-organic synthesis, waste-water treatment and energy conversion devices. The activity and selectivity of the electrocatalysts can be fine-tuned by applying potential or current. In addition, the electrochemical devices employing the electrocatalysts are not limited by Carnot efficiency. The requirement for triple-phase boundary for electrochemical reactions to occur in these devices necessitate that the electrocatalysts be deposited in a site-specific manner for maximum utilization. Electrodepostion of the catalysts meets this requirement. However, in traditional aqueous system, electrodeposition is limited by the narrow electrochemical window of water (1.2 V). In ionic liquids (ILs), the electrochemical window can be as high as 6 V and the metals having reduction potential negative than that of H_2 evolution could be reduced. Moreover, product selectivity and reaction efficiency can be enhanced in ILs for reactions such as CO_2 reduction. However, studies in these fields are limited. My graduate work in nitrogen-doped ordered mesoporous carbon (NOMC) supported Pt for oxygen reduction reaction (ORR) and my postdoctoral work in electrodeposition of fission platinoids (Pd, Rh, Ru) in ILs puts me in a favorable position to advance these fields.

My graduate work focused on creating strong metal-support interaction (SMSI) between carbon and Pt. Although various types of advanced carbon supports such as carbon black, carbon nanotubes, aerogels, carbon fibers and graphene have been investigated as the support, none have been shown to increase intrinsic ORR activity of Pt. To overcome this, I explored nitrogen-doped carbon supports. Nitrogen has been shown to change electronic and microstructural properties of carbon. However, there had been few works on Pt-supported nitrogen-doped carbon for ORR and most of it was limited to model systems such as highly oriented pyrolytic graphite (HOPG). In my work, NOMC was synthesized by templating on SBA-15 with pyrrole as the precursor. Pt nanoparticles were deposited on the NOMC by electroless reduction in ethylene glycol. I showed how pore-order, graphiticity and nitrogen content plays a role in ORR activity of the Pt. This has provided guidance for designing of high-performing nitrogen-doped carbon supports.

Currently, I am working on recovery of fission platinoids (Pd, Rh, Ru) through electrodeposition in ionic liquids. Current processes for recovery of fission products is based on extraction in highly flammable volatile organic solvents (VOCs) such as kerosene and dodecane. ILs provide safer alternative to the VOCs because of their negligible vapor pressure. ILs also have complexing ability, hence metals could be dissolved in ILs without introducing additional complexing agents. Also the metals solvated could be recovered directly through electrodeposition in ILs. I am studying electrochemical and electroreduction behavior of Pd in many ILs previously not studied. The ILs affect the electroreduction behavior by controlling speciation, electrolyte/electrode interface, and the mass-transfer which in turn affect the thermodynamics, kinetics and morphology of the deposits. This work will expose fundamental underpinnings of electrodeposition of Pd, Rh and Ru in ILs.

In my poster, I will summarize my graduate and postdoctoral work and talk about how I will exploit them to develop the field of electrodeposition of catalysts and their activity in ILs for applications in low temperature fuel cells and CO_2 utilization.

Scalable Nanofabrication of Functional Architectures through Hybrid Lithography

Jonathan P. Singer

Chemical and Environmental Engineering, Yale University, New Haven, CT, 06511

In order to advance patterning capabilities for future applications, innovative lithographic strategies that focus on scalable and cost-effective techniques must be developed. To address this challenge, a growing number of alternative patterning methods to conventional photolithography for the fabrication of complex structures have emerged, all of which possess certain advantages and disadvantages, generally in the form of tradeoffs between speed and control. My research has focused on developing hybrid lithographic nanofabrication techniques to overcome the limitations of individual techniques—specifically, the development of processes to enable the scalable fabrication of previously impossible architectures.

In my doctoral work in the lab of Edwin Thomas at MIT, I designed techniques that combine large area patterning methods (*i.e.* interference lithography and self-assembly) with 3D laser direct write and that can simultaneously exploit emerging developments in enhanced resolution. This research has resulted in hybrid techniques that have enabled (i) the single-shot fabrication of plasmonic resonators, (ii) the local fill-fraction control of complex periodic and quasiperiodic photonic and phononic architectures with deliberate defects, and (iii) the use of thermocapillary forces for the controlled dewetting of polymer thin films and alignment of block copolymer networks for applications in affordable pattern transfer. In my current role as postdoctoral associate in the lab of Chinedum Osuji at Yale, I have continued work on hybrid methods with an increasing focus on functional materials, such as fluorescent, plasmonic, and semiconducting nanoparticles and conjugated polymeric systems, with applications in, for example, organic-inorganic bulk heterojunction photovoltaics. These materials are difficult to deliberately pattern through conventional means, so my work pursues large-scale alternative methods including (i) magnetic field alignment of nanowires, (ii) high-aspect ratio nanoimprint with metallic glass molds, and (iii) durotaxis-directed electrospray of polymers and nanoparticles.

Moving forward, I will continue to couple experiments and simulations in the full cycle of process-to-device development to create new strategies for patterning that incorporate materials with functionalities in their molecular and supermolecular structures into the final complex architectures that are compatible with high-throughput fabrication processes.

Selected Manuscripts

1. Pelligra, C. I.; Singer, J. P.; Huang, S.; Mayo, A.; Mu, R. R.; Osuji, C. O.; " Scalable High-Fidelity Growth of Semiconductor Nanorod Arrays with Controlled Geometry for Photovoltaic Devices using Block Copolymers," *Manuscript in submission.*
2. Hu, H.; Singer, J. P.; Osuji, C. O.; "Morphological Development in Thin Films of a Lamellar PS-b-P4VP Block Copolymer Deposited by Electrospray," *Manuscript in submission.*
3. Shao, Z.; Liu, Z.; Singer, J. P.; Li, H.; Liu, Y.; Osuji, C. O.; Schroers, J.;" Enhanced Non-Polymorphic Crystallization Kinetics of a Bulk Metallic Glass through Shear," *Manuscript in preparation.*
4. Singer, J. P.; Shao, Z.; Gopinadhan, M.; Schroers, J.; Osuji, C. O.; "High-Resolution Nanoimprint of Functional Polymers with Bulk Metallic Glass Molds," *Manuscript in preparation.*
5. Singer, J. P.; Gotrik, K. W.; Lee, J.-H.; Kooi, S. E.; Ross, C. A.; Thomas, E. L.; "Alignment and Reordering of a Block Copolymer by Solvent-Enhanced Thermal Laser Direct Write," *Polymer*, **55**, 7, 1875-1882, 2014.
6. Singer, J. P.; Lin, P.-T.; Kooi, S. E.; Kimerling, L. A.; Jurgen, M.; Thomas, E. L.; "Direct-Write Dewetting of Polymer Thin Films by a Laser-Induced Thermal Gradient," *Advanced Materials*, **25**, 42, 6100-6105, 2013.
7. Lee, J.-H.; Veysset, D.; Singer, J. P.; Retsch, M.; Saini, G.; Pezeril, T.; Nelson, K. A.; Thomas, E. L.; "Dynamical Responses of Layered Nanostructure Composites at High Strain Rate Deformation," *Nature Communications*, **3**, 1164, 2012.
8. Singer, J. P.; Lee, J.-H.; Kooi, S. E.; Thomas, E. L.;" Rapid Fabrication of 3D Terahertz Split Ring Resonator Arrays by Novel Single-Shot Direct Write Focused Proximity Field Nanopatterning," *Optics Express*, **20**, 10, 11097-11108, 2012.
9. Singer, J. P.; Kooi, S. E.; Thomas, E. L.; "Focused Laser Spike (FLaSk) Annealing of Photoactivated Chemically Amplified Resists for Rapid Hierarchical Patterning," *Nanoscale*, **3**, 7, 2730-2738, 2011.

Design, optimization, monitoring and control of continuous pharmaceutical manufacturing plant for QbD and PAT based next generation of efficient manufacturing

Ravendra Singh
NSF Engineering Research Center for Structured Organic Particulate Systems (C-SOPS), Chemical and Biochemical Engineering, Rutgers University, Piscataway, NJ, USA

A novel manufacturing strategy based on continuous processing integrated with inline/online monitoring tools (PAT) coupled with efficient automatic advanced feedback control system is highly desired for efficient QbD based manufacturing of the next generation of pharmaceutical products with optimal consumption of time, space and resources. However, the particulate systems are still poorly understood and there is different level of complexities involved, therefore to achieve this paradigm shift, new techniques and tools need to be developed through extensive fundamental and applied research. This area of research and education are highly encouraged and supported by pharmaceutical companies, academic institutions, regulatory authority (FDA) and national science foundation (NSF). For example, recently national science foundation of USA has established a 100 million dollar research center for structured organic particulate systems (C-SOPS) at Rutgers University New Jersey. Furthermore, more than 40 companies are supporting this center economically and providing scientific inputs so that the research outcomes can be directly applied to the companies.
In this work, a scientific foundation for the optimal design, optimization and control of the continuous process involving structured organic particulate systems has been developed. The mathematical models for continuous manufacturing of active pharmaceutical ingredients and final product tablet, has been developed for virtual experimentation and applied for plant design, optimization and control. Advanced modelling techniques such as PBM, DEM and CFD have been applied for pharmaceutical process modeling. The developed continuous tablet manufacturing plant and process model has been adapted by several pharmaceutical companies. One of the challenges is to monitor and control the process efficiently in real time for QbD based manufacturing. A flexible monitoring and control system incorporating different PAT techniques and tools and control strategies for continuous pharmaceutical manufacturing process has been designed and implemented to the plant situated at Rutgers University. Different control strategies such as PID, PID coupled with dead time compensator, and MPC has been compared and evaluated practically for pharmaceutical process control in terms of set point tracking and disturbances rejection ability, robustness, and stability/instability. A new hybrid MPC-PID control strategy has been developed and implemented into the pharmaceutical tablet manufacturing plant and its performance has been found better than the other control strategies. Principal components analysis (PCA) based additional supervisory control layer has been also integrated with the hybrid feedback control system. For the first time, the closed-loop operation of the tablet manufacturing plant using NIR sensor has been demonstrated, which is a significant advancement in pharmaceutical manufacturing as endorsed by the Food and Drug Administration (FDA) and other regulatory authorities as part of the Quality by Design (QbD) paradigm. A systematic generic framework including the methods and tools through which the cost and profit can be estimated and the Dynamic Real Time Optimization (DRTO) can be integrated with hybrid control system has been also developed. The integrated DRTO provides the optimal operational set points for the plant control system in real time optimizing an objective function. DRTO integrated with hybrid control strategy ensures the maximum possible profit irrespective of the market demand fluctuations.

Prior to this, I worked in a large scale European pharmaceutical project (F^3 Factory project), which involves 25 partners from industry and Academia; where my responsibility was to manage and lead the DTU's parts of the F^3 Factory project and also to coordinate with the other partners associated with it. My doctoral research work, done at Department of Chemical and Biochemical Engineering, Technical University of Denmark, has been awarded for prestigious EFCE Excellence Award 2010, given in

Recognition of an Outstanding PhD Thesis, from European Federation of Chemical Engineering. I have exemplary academic records, previous teaching experiences, and an established trail of valued research accreditations'. I have published several peer reviewed scientific international Journal articles, popular reputed magazine articles, book chapters and presented in more than 47 international conferences.

The objective of this presentation is two-fold. First to give the overview of my research and teaching experience, and second to demonstrate the design, optimization, monitoring and control of continuous tablet manufacturing plant.

Some selected references

1. **Singh, R.,** Sahay, A, Muzzio, F., Ierapetritou, M., Ramachandran, R. (**2014**). Systematic framework for onsite design and implementation of the control system in continuous tablet manufacturing process. Computers & Chemical Engineering, DOI: 10.1016/j.compchemeng.2014.02.029.
2. **Singh, R.,** Barrasso, D., Chaudhury, A., Maitraye Sen, Ierapetritou, M., Ramachandran, R. (**2014**).Closed-Loop Feedback Control of a Continuous Pharmaceutical Tablet Manufacturing Process via Wet Granulation. Journal of Pharmaceutical Innovation, 9, 16-37.
3. **Singh, R.,** Ierapetritou, M., Ramachandran, R. (2013). System-wide hybrid model predictive control of a continuous pharmaceutical tablet manufacturing process via direct compaction. European Journal of Pharmaceutics and Biopharmaceutics, 85(3), Part B, 1164-1182.
4. **Singh, R.,** Ierapetritou, M., Ramachandran, R. (**2012**). An engineering study on the enhanced control and operation of continuous manufacturing of pharmaceutical tablets via roller compaction. International Journal of Pharmaceutics, 438 (1-2), 307-326.
5. **Singh, R.,** Godfrey, A., Gregertsen, B., Muller, F., Gernaey, K. V., Gani, R., Woodley, J. M. (2013). Systematic substrate adoption methodology (SAM) for future flexible, generic pharmaceutical production processes. Computers & Chemical Engineering, 58, 344 - 368.
6. **Singh, R.,** Gernaey, K. V., Gani, R. (2010). ICAS-PAT: A Software for Design, Analysis & Validation of PAT Systems. Computers & Chemical Engineering, Volume 34, Issue 7, 1108-1136.
7. **Singh, R.**, Gernaey, K. V., Gani, R. (2010). An ontological knowledge based system for selection of process monitoring and analysis tools. Computers & Chemical Engineering, Volume 34, Issue 7, 1137-1154.
8. **Singh, R.,** Gernaey, K. V., Gani, R. (2009). Model-based computer-aided framework for design of process monitoring and analysis systems. Computers & Chemical Engineering, Volume 33, Issue 1, 22-42. (Citations>45).
9. Sen, M., Barrasso, D., **Singh, R.**, Ramachandran, R. (**2014**). A Multi-Scale Hybrid CFD-DEM-PBM Description of a Fluid-Bed Granulation Process. Process Journal, 2(1), 89-111.
10. Sen, M., **Singh, R.,** Ramachandran, R. (**2014**). Simulation based design of an efficient control system for the continuous purification and processing of active pharmaceutical ingredients. Journal of Pharmaceutical Innovation, 9, 65-81.
11. Sen, M., **Singh, R.,** Ramachandran, R. (**2014**). A hybrid MPC-PID control system design for the continuous purification and processing of active pharmaceutical ingredients. PROCESS Journal, accepted.
12. Sen, M., Dubey, A., **Singh, R.,** Ramachandran, R. (**2013**). Mathematical Development and Comparison of a Hybrid PBM-DEM description of a Continuous Powder Mixing Process. Journal of Powder Technology, http://dx.doi.org/10.1155/2013/843784.
13. Sen, M., Chaudhury, A., **Singh, R.,** John, J., Ramachandran, R. (**2013**). Multi-scale flowsheet simulation of an integrated continuous purification–downstream pharmaceutical manufacturing process. International Journal of Pharmaceutics, 445 (1-2), 29-38.
14. Sen, M., Rogers, A., **Singh, R.,** Chaudhury, A., John, J., Ierapetritou, M., Ramachandran, R. (**2013**). Flowsheet optimization of an integrated continuous purification-processing pharmaceutical manufacturing operation. Chemical Engineering Science, 102, 56 – 66.
15. Sen, M., **Singh, R.,** Vanarase, A., John, J., Ramachandran, R. (**2012**). Multi-dimensional population balance modeling and experimental validation of continuous powder mixing processes. Chemical Engineering Science, Volume 18, 349-360.
16. Samad, N. A. F. A., **Singh, R.**, Sin, G., Gernaey, K. V., Gani, R. (**2011**). A generic multi-dimensional model-based system for batch cooling crystallization processes. Computers & Chemical Engineering, Volume 35, Issue 5, 828-843.
17. **Singh, R.,** Cameron, I, Gani, R. (2011). Process and Product Modelling: A case study approach. Book chapters 12.2-12.6, Publisher: Elsevier, pp 380-430.
18. **Singh, R.,** Ierapetritou, M., Ramachandran, R (**2013**). Hybrid advanced control of a flexible multipurpose continuous pharmaceutical tablet manufacturing process via direct compaction. Computer Aided Chemical Engineering, Volume 32, 757-762.
19. Muzzio, F., **Singh, R.,** Chaudhury, A., Rogers, A., Ramachandran, R. Ierapetritou, M. (**2013**). Model- predictive design, control and optimization of pharmaceutical process. PharmTech magazine Europe, Volume 37, Number 6, 40-41, 77.
20. **Singh, R.,** Sahay, A., Oka, S., Liu, X., Ramachandran, R., Marianthi Ierapetritou, Fernando Muzzio, F. (**2013**). Online monitoring, advanced control and operation of robust continuous pharmaceutical tablet manufacturing process. BioPharma magazine Asia, Volume 2, Issue 5, October 2013.
21. **Singh, R.**, Raquel Rozada-Sanchez, R., Dean, W., Perkins, J., Muller, F., Godfrey, A., Gernaey, K. V., Gani, R., Woodley, J. M. (**2012**). A generic process template for continuous pharmaceutical production. Computer Aided Chemical Engineering, Volume 31, 715-719.
22. **Singh, R.,** Boukouvala, F., Jayjock, E., Ramachandran, R. Ierapetritou, M., Muzzio, F. (**2012**). Flexible Multipurpose Continuous Processing. PharmPro Magazine, Pharmaceutical Processing, July 2012, Vol. 27, Issue 6, p22-25.
23. **Singh, R.,** Boukouvala, F., Jayjock, E., Ramachandran, R. Ierapetritou, M., Muzzio, F. (**2012**). Flexible Multipurpose Continuous Processing of Pharmaceutical Tablet Manufacturing Process. GMP news, European Compliance Academic (ECE), http://www.gmp-compliance.org/ecanl_503_0_news_3268_7248_n.html.
24. Samad, N. A. F. A., **Singh, R.**, Sin, G., Gernaey, K. V., Gani, R. (**2011**). Systematic Procedure for Generating Operational Policies to Achieve Target Crystal Size Distribution (CSD) in Batch Cooling Crystallization. IEEE, 1-6.
25. Gani, R., **Singh, R.**, PAT Slowly Yielding to Broader QbD (2011). Genetic Engineering & Biotechnology News, March 15, Vol. 31, No. 6, pp 40-41. http://genengnews.com
26. Samad, N. A. F. A., **Singh, R.**, Sin, G., Gernaey, K. V., Gani, R. (**2010**) . Control of process operation and monitoring of product qualities through generic model-based in batch cooling crystallization. Computer Aided Chemical Engineering, Volume 28, 613-618.

27. **Singh, R.**, Rozada-Sanchez, R., Wrate, T., Muller, F., Gernaey, K. V., Gani, R., Woodley, J. M. **(2010)**. A retrofit strategy to achieve "Fast, Flexible, Future (F^3)" pharmaceutical production processes. Computer Aided Chemical Engineering, Volume 29, 291-295.
28. Samad, N. A. F. A., **Singh, R.**, Sin, G., Gernaey, K. V., Gani, R. **(2010)**. Systematic Modeling of Generic Multi-dimensional Model-based System for Batch Cooling Crystallization Operations. Computer Aided Chemical Engineering, Volume 29, 86-90.
29. **Singh, R.,** Gernaey, K. V., Gani, R. **(2009)**. A software tool for design of process monitoring and analysis systems. Computer Aided Chemical Engineering, Volume 26, 321-326.
30. **Singh, R.,** Gernaey, K. V., Gani, R. **(2008)**. Off-line design of PAT systems for on-line applications. Computer Aided Chemical Engineering, Volume 25, 423-428.

386030 Materials and Systems Design for Healthcare and Energy Applications

Abstract:

My research is focused on developing mathematical, computational, and experimental methods for rapid materials screening and systems design for pharmaceutical and solar fuels applications. My research plan in these two different areas is described as follows.

Research Area 1: **Designer Crystals for Healthcare Applications**

Current Collaborators: Prof. Doraiswami Ramkrishna (Purdue U.) and Dr. Hsien-Hsin Tung (AbbVie Inc.)

Description: The crystal structure and morphology are two critical determinants of the physicochemical properties such as solubility, hardness, cleavage, optoelectronic properties, electrical and heat conductivities, and piezoelectric effect of crystalline materials. The polymorphs (crystal structure) are determined in the early stage of crystallization process whereas the morphologies depend on growth, dissolution, and milling conditions. Figure 1 shows a scheme that probes processes at different length scales using a combination of experimental and mathematical tools to control crystal structure, shape, and size by tuning various environmental conditions. The proposed research is fundamental to crystal growth in different areas (*e.g.* the semiconductor industry, catalytic applications, and the pharmaceutical industry).

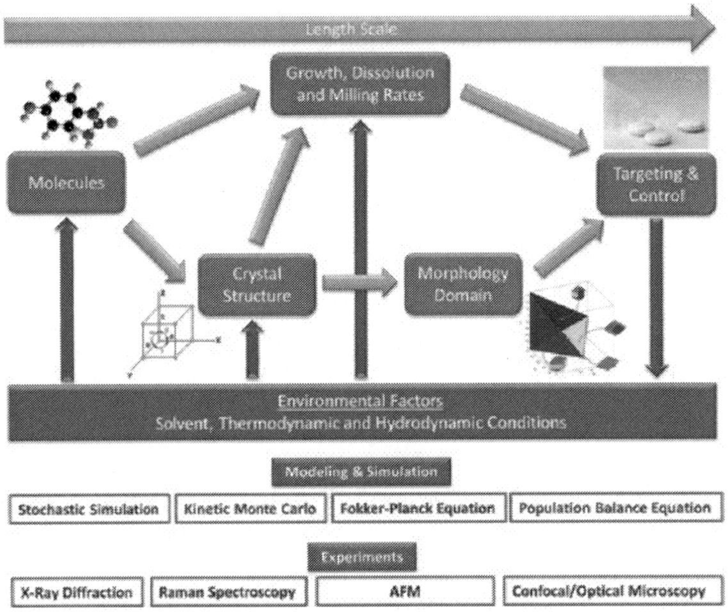

Figure 1: Research Plan for Identification of Structure-Property Relationships and Directed Synthesis of Crystalline Materials

Selected Publications:

Meenesh R. Singh, Jayanta Chakraborty, Nandkishor Nere, Hsien-Hsin Tung, Shailendra Bordawekar and Doraiswami Ramkrishna, "Image-Analysis-Based Method for Measurement of 3D Crystal Morphology and Polymorph Identification using Confocal Microscopy," Crystal Growth & Design, 12 (7), 3735-3748, 2012

Meenesh R. Singh, Parul Verma, Hsien-Hsin Tung, Shailendra Bordawekar and Doraiswami Ramkrishna, "Screening Crystal Morphologies from Crystal Structure," Crystal Growth & Design, 13 (4), 1390-1396, 2013.

Meenesh R. Singh and Doraiswami Ramkrishna, "A Comprehensive Approach to Predicting Crystal Morphology Distributions with Population Balances," Crystal Growth & Design, 13 (4), 1397 – 1411, 2013.

Meenesh R. Singh and Doraiswami Ramkrishna, "Dispersions in Crystal Nucleation and Growth Rates: Implications of Fluctuation in Supersaturation," Chemical Engineering Science, 107 (7), 102-113, 2014.

Doraiswami Ramkrishna and **Meenesh R. Singh**, "Population Balance Modeling. Current Status and Future Prospects," Annual Review of Chemical and Biomolecular Engineering, 5 (1), 2014

Research Area 2: **Artificial Photosynthesis for Solar Fuels**

Current Collaborators: Dr. Nathaniel Lynd (LBNL), Prof. Rachel A. Segalman (UC Berkeley), Prof. Alexis T. Bell (UC Berkeley) and Prof. Nathan S. Lewis (Caltech)

Description: Harnessing solar energy to produce liquid fuels is a promising route to support future energy demands. One possible route is to first produce hydrogen, which can be generated by photoelectrochemically splitting water. The solar fuel generators such as photoelectrochemical cells (PECs) involve a two-step conversion of sunlight to hydrogen using a photovoltaic (PV) cell directly (wireless) or indirectly (wired) connected to an electrochemical cell. My research in this area is focused on identifying materials and methods to develop an efficient, robust, cheap and durable PECs for water-splitting and carbon dioxide reduction. One of the design objectives here is to minimize the potential losses in the device using interface and systems engineering approaches.

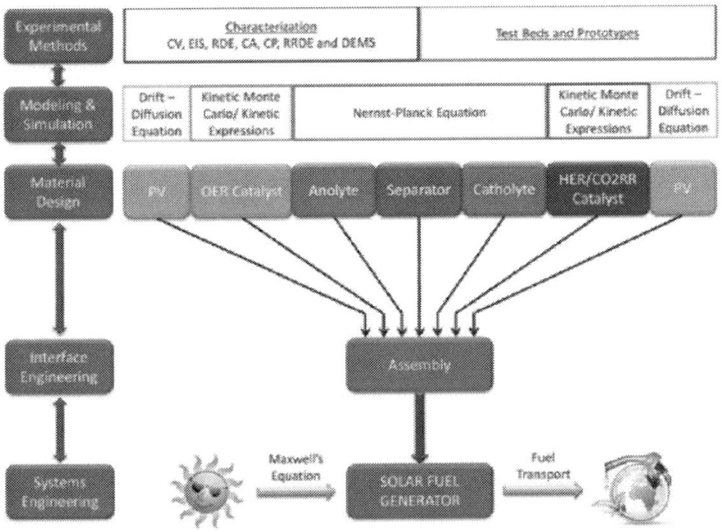

Figure 2: Research Plan for the Design of Cheap, Efficient, and Durable Solar Fuel Generators

Selected Publications:

1. **Meenesh R. Singh**, John C. Stevens, and Adam Z. Weber, "Design of Membrane-Encapsulated Wireless Photoelectrochemical Cells for Hydrogen Production", *Journal of The Electrochemical Society*, 161 (8), 2014

2. Jian Jin, Karl Walczak, **Meenesh R. Singh**, Chris Karp, Nathan S. Lewis, and Chengxiang Xiang, "Experimental and Modeling/Simulation Evaluation of the Efficiency and Operational Performance of an Integrated, Membrane-Free, neutral pH solar-Driven Water-Splitting System," *submitted*

3. Christopher M. Evans, **Meenesh R. Singh**, Nathaniel A. Lynd, and Rachel A. Segalman, "Improving the Gas Barrier Properties of Nafion via Thermal Annealing: Evidence for Dual-Mode Diffusion," *submitted to Macromolecules.*

380723 Engineering Microbial Production Platforms for Efficient Carbon Utilization

Abstract:

Microbial production platforms promise to be a sustainable alternative to today's petrochemical industry. By harnessing the breadth and flexibility of cellular metabolism to process available energy sources, 'living foundries' can be engineered that convert renewable sugar feedstocks directly into a number of enantiopure commodity and high-value specialty chemicals. Despite isolated successes, however, these platforms have been unable to scale to meet the gap created by increasing oil prices due to challenges inherent in driving carbon flux towards the desired product. Drawing on tools from synthetic biology, metabolic engineering, systems biology, and next generation sequencing and synthesis, my research has focused on the development of molecular strategies to increase the viability of these processes, and the elucidation of cellular responses to such intervention.

Here, I present my work towards the engineering of more efficient bacterial and eukaryotic systems. In my doctoral research (with Kristala Prather, Chemical Engineering, MIT), I developed transcriptional and post-transcriptional tools that dynamically divert carbon from central metabolism into a pathway of interest in *E. coli.* These tools pave the way for the development of new product classes within *E. coli* and allow for the optimization of existing pathways by reducing carbon waste. As a postdoc (with Michelle O'Malley, Chemical Engineering, UCSB), I am studying the effects of compartmentalized expression of metabolic pathways on productivity. This approach has tremendous potential to both locally concentrate metabolites and enzymes leading to greater productivity, and limit carbon loss to metabolic side reactions localized elsewhere. While these examples pursue more efficient carbon utilization, there still remains the challenge of sustainable, cost-effective feedstocks for these processes. Towards that end, I am also analyzing next generation sequencing data to identify novel biomass-degrading enzymes from anaerobic gut fungi and elucidate their hydrolytic strategies against diverse lignocellulosic substrates. Such research strives to secure sustainable feedstocks for biotechnology and presents a unique opportunity to examine novel design strategies in a non-model organism.

In the future, my research program will focus on the development of next generation microbial platforms with applications in sustainability, health and energy. My approach is to use the tools of data-driven science to formulate novel hypotheses and elucidate emergent systems-level properties of native systems, and engineer innovative solutions with synthetic biology which exploit robust and versatile features of microorganisms.

388992 Engineering Immunity: Design and Development of Customized Nanomaterials with Controlled Immunostimulatory Effects for Biomedical Applications

Abstract:

Engineered nanobiomaterials (ENMs) are rapidly emerging as a new class of materials that allow tuning of their physicochemical properties for biomedical and pharmaceutical applications, e.g., drug delivery, tissue engineering and vaccines. ENMs could interact with biological entities at the nano-bio interface, which could induce cascades of signaling pathways leading to the activation of immune responses including both innate and adaptive immunity. It has been identified that hydrophobicity, size, and surface charge are the possible parameters affecting the biocompatibility of ENMs. However, the detailed mechanisms by which ENMs could generate immunostimulatory effects at the nano-bio interface are still not fully understood. In light of this, my research will focus on immunoengineering, which will involve developing ENMs with tunable physicochemical properties, investigating the interaction of ENMs with immune systems, and further customizing their applications in vaccine development, tissue engineering, and cancer immunotherapy.

A Spotlight on My Research on Chemical Vapor Deposition

Aravind Suresh
Assistant Professor-in-Residence, Chemical and Biomolecular Engineering Department
University of Connecticut, Storrs, CT 06279
E-mail: aravind@engr.uconn.edu

Background: The research for my doctoral dissertation involved the development of high-temperature oxides with the dual functionality of catalytic activity for hydrogen generation and mixed conductivity for hydrogen purification. Most of my work was done on cobalt-doped barium cerate-zirconates [1, 2]. I also started the investigation on iron-doped barium zirconates, which was then continued by my advisor Prof. Benjamin Wilhite's group. We are now awaiting the publication of the results of the structural and electrochemical characterization of the material.

I have been an Assistant Professor-in-Residence at the University of Connecticut in charge of teaching the junior and senior laboratory courses in the department. During that time, I have also been working on different research projects in collaboration with various faculty members. For this poster, I am focusing the attention on some of my work in the field of Chemical Vapor Deposition (CVD) of polymers.

Spotlight: Initiated CVD (iCVD) and photo-initiated CVD (piCVD) are emerging techniques for creating conformal coatings of polymers on a variety of substrates, providing good control on functionality and minimizing the use of strong solvents [3, 4]. The challenge in iCVD involves decoupling the heated filament and the cold stage, whereas that in piCVD involves the identification of suitable photo-initiators.

During the course of my research on photo-initiated CVD (piCVD), I discovered the property of auto-polymerization in hexyl acrylate vapor when subjected to ultraviolet (UV) radiation. I studied the rate of deposition with respect to process parameters and reported the results through a communication [5]. I am now working on understanding the mechanism of the deposition process, which appears to be unique with respect to the literature on piCVD and also with respect to the polymerization and deposition of the same monomer using iCVD, in that the observed rate of deposition increases with time with an exponential trend and increases with film thickness with a linear trend. The behavior suggests that the deposited poly hexyl acrylate film itself acts as a source of free radicals driving the process. If that is indeed confirmed to be the case, hexyl acrylate vapor or the deposited poly hexyl acrylate film can be used as an initiator in piCVD for depositing polymers whose monomers may not themselves exhibit auto-polymerization under UV radiation.

I am working on the deposition of thermally stable, highly cross-linked polymers using iCVD in collaboration with Prof. Daniel Burkey at the Universtity of Connecticut and Dr. Salmaan Baxamusa at the Lawrence Livermore National Laboratory (LLNL). The focus of the project is to develop CVD polymers that are optically clear and resistant to attack by atmospheric oxygen on aging. CVD techniques that rely on plasmas to initiate the polymerization produced colored films that are susceptible to aging in the ambient atmosphere and iCVD is being explored as an alternative. The preliminary results thus far have been encouraging and testing will continue over the summer of 2014.

I am applying the technique of iCVD to understand the bonding characteristics at the interface between metals and polymers in collaboration with a group at the Institute of Materials Science at the University of Connecticut. They are working more on depositing metals on polymer films; I presented the option of depositing polymers on metal sheets, where the only restriction would be that the monomer of the desired polymer must be vaporizable under the working conditions of the instrument. The project, in its nascent stage, represents situations where CVD can be a valuable tool in a larger research project rather than the focus of the project itself.

Summary: I am looking for tenure-track positions in universities, with a focus on establishing a research direction encompassing two ends of the materials spectrum – thin-film polymer deposition and high-temperature oxides. I would also be interested in a position that combines an emphasis on teaching along with the opportunity of building a research group.

References:

[1] Suresh, Basu, Carter, Sammes, Wilhite, Journal of Materials Science (2010) **45**: 3215-3227

[2] Basu, Suresh, Carter, Wilhite, Journal of the European Ceramic Society (2011) **31**: 1421-1429

[3] Lau, Gleason, Macromolecules (2006) **39**: 3688-3694

[4] Baxamusa, Gleason, Advanced Functional Materials (2009) **19**: 3489-3496

[5] Suresh, Anastasio, Burkey, Chemical Vapor Deposition (2014) **20**: 5-7

390479 Fundamental Studies of Nanoscale Bio-Interactions

Abstract:

Nanoscale bio-recognition is an essential requirement in many biological processes (e.g. enzyme-subtract interaction, immune response, and signaling pathways). Although our knowledge of the molecular and structural basis of these interactions is limited, scientists have found a way to apply these interactions to several applications like affinity chromatography, bio-sensors, and drug delivery. My Ph.D. work focused on affinity membranes for bio-separations. It centered around the development (selection of ligands, modification of membranes) and testing (batch and dynamic adsorption experiments, modeling, estimation of binding capacity, dissociation and kinetic constants, effect of ionic strength and flux) of affinity membranes for the purification of different classes of biomolecules: monoclonal antibodies, lectins and MBP-fusion proteins. More recently, during my post-doctoral studies, I have continued to work on several projects with the common focus of protein-protein interactions, from separation to aggregation. The poster will give an overview of the research conducted in the following topics: (i) specificity of nuclear transport using QCM-D; (ii) insights into the lytic mechanism of antimicrobial peptide using supported lipid bilayer to mimic bacterial membranes; (iii) the use of AFM in force mode to measure the binding forces between different split-protein partners; (iv) biophysical characterization of amyloid aggregates, from dimers to fibrils; (v) charged membranes for the separation of proteins with similar molecular weight and charge. I have broadened my background as a chemical engineer, integrating it with a variety of bio-tools. I plan to continue my career maintaining this interdisciplinary approach in studying protein-protein recognition both as basic science (e.g. better understanding of biological mechanisms) and applied science (e.g. developing bio-sensors).

389776 Reaction Engineering: Bridging Fundamental Engineering Approaches with Biomass Pyrolysis and Transport in Zeolites

Abstract:

Biomass pyrolysis has been widely explored for its potential to generate a sustainable chemical source capable of producing synthetic fuels and chemicals. Lignocellulosic biomass specifically presents great promise with its carbon rich backbone and widespread prevalence. The high temperature thermal conversion of biomass to pyrolysis oil occurs on the order of milliseconds and converts the long chain biopolymers to a carbon-rich liquid crude. The kinetics of biomass pyrolysis is greatly complicated by significant heat and mass transport challenges. The complex fluid dynamics of the reactive liquid intermediate are only beginning to be understood, including the existence of vigorous bubbling and aerosol generation. Unlike traditional crude oil, the condensed bio-oil is highly oxygenated and acidic which often requires significant upgrading prior to end use. The reactive nature of the bio-oil complicates the upgrading process by presenting difficulty in heating and vaporization where the fluid properties can drastically effect the product distributions.

Additionally, the use of zeolites is widely utilized to catalytically upgrade fuels for hydrodeoxegenation, hydrogenation, and cracking. While new materials are being synthesized with transport lengthscales that are increasingly smaller, diffusional transport limitations are presenting in the small particles. The presence of these transport limitations remains a significant technical challenge. However, the use of several experimental and computational techniques is utilized to probe the mechanistic nature of these barriers, and a greater understanding is beginning to be revealed. In this work, original findings are overviewed covering biomass pyrolysis with aerosol generation mechanisms and rate-dominant diffusion limitations in hierarchical zeolites.

Materials for 4D biology: spatial and temporal control of the stem cell niche

Mark W. Tibbitt and Robert Langer, Massachusetts Institute of Technology

Interactions between mammalian cells and the surrounding extracellular matrix (ECM) are critical in the regulation of stem cell differentiation, cell migration, tissue morphogenesis, and disease progression. Cell decisions are often made by integrating these complex interactions, which are dynamic in 4D: three-dimensional space and time. Synthetic polymer-based hydrogels have emerged as an attractive cell culture platform to investigate the influence of these cell-matrix interactions on cell function in a systematic manner. Hydrogels are appealing mimics of the ECM on account of their high water content, tissue-like elasticity, and facile transport of nutrients, waste, and soluble factors. Furthermore, poly(ethylene glycol) (PEG) hydrogels can be formed under mild, cytocompatible conditions and are easily modified to present various biochemical and mechanical signals to cells during culture.

While cell-laden PEG hydrogel constructs have been employed for regenerative medicine and fundamental biological applications, there is still a lack of materials that allow the user to recapitulate the spatially heterogeneous and dynamic environment that cells interact with in vivo. My Ph.D. research has focused on the development, characterization, and application of photoresponsive PEG-based hydrogels that afford the user control of the mechanical and biochemical properties of the scaffold in both 3D space and time. This class of materials facilitates experiments that probe the relationship between spatially heterogeneous and dynamic signals from the ECM and cell function in a well-defined environment to study the fundamentals of the cell-material interactions, stem cell differentiation and plasticity, as well as the cellular response to local changes in soluble factor concentration.

Specifically, we have developed photoresponsive materials for the dynamic modulation of the ECM during 2D and 3D culture. These materials have been used to investigate stem cell response to dynamic changes in elasticity and the plasticity of stem cells during microenvironmental mechanotransduction. Importantly, we uncovered fundamental information about mechanical dosing and memory in mesenchymal stem cells that is regulated by YAP/TAZ signaling. Further, we have utilized photoresponsive hydrogels to develop photodegradable microspheres for the targeted delivery of soluble factors to study tissue morphogenesis and chemotaxis. Thin films of photodegradable hydrogel have been employed for the selective capture and release of cells for diagnostic application.

In my Postdoctoral research, we have extended these paradigms to develop smart and responsive polymeric systems for controlled delivery of therapeutics to intervene during disease progression. Specifically, we are focusing on the use of shear-thinning injectable hydrogels to deliver therapeutics for cardiac valve disease and cancer in a controlled manner.

390144 Utilizing Optimally Designed Peptide-Anchored Liposomes for Specific B Cell Activation and Vaccine Formulations

Abstract:

Lipid-based nanoparticles have been widely characterized for the delivery of small molecules, peptides and proteins. Liposomes containing surface-bound antigens are similarly being developed as vaccines and have been shown to induce strong immune responses when compared to soluble antigens by facilitating cross-presentation and enhancement of humoral responses. Yet so far no studies have focused on how specific liposome design considerations modulate immune responses at the single cell level to enhance CD4+ T cell help and antibody production through interaction and activation of specific B cells. Using an antigen-specific 3-83 Ig transgenic mouse model, we evaluated the role of liposome composition on the direct binding and internalization of liposomes with surface-displayed peptide antigens to antigen-specific B cells. Initially we focused on the role of surface PEGylation and antigen tethering. We hypothesized that while surface PEGylation has been shown to improve liposome stability in serum and enhance lymph node targeting in vivo, it may hinder binding and internalization through specific B cell receptors. Our studies showed that unilamellar PEGylated liposomes containing maleimide-functionalized PEG-lipids conjugated to cysteine-containing antigenic peptides (p31 and p0 peptides with high and no binding affinities, respectively) bound with varying intensities to B cells as a function of PEGylation (10-0% PEG): 5% PEG liposomes bound with the highest intensity to B cells, while 0% PEG liposomes containing peptide antigen anchored directly to the bilayer surface had significantly reduced binding as a result of steric hindrance. These conditions as well as a third condition (5% PEG with surface anchored peptide antigen which resulted in even lesser binding compared to 0% PEG liposomes) were evaluated for their ability to activate antigen-specific B cells in vitro. B cell activation was assessed at the single-cell level using flow cytometry to quantify the up-regulation of early B cell activation markers, such as CD69, CD86, CD40, and MHCII. As predicted, 5% PEG liposomes with tethered p31 resulted in the strongest antigen-specific B cell activation, while the weakest binding 5% PEG liposomes with surface anchored p31 resulted in the lowest levels of activation comparable to unstimulated cells. The incorporation of TLR4 agonist, monophosphoryl lipid A (MPLA), into the liposome bilayer significantly enhanced early activation (day 1) and helped sustain activation and B cell viability at later times (up to 4 days) in an antigen-specific manner. However, MPLA incorporation into liposomes interestingly provided no enhancement on B cell binding despite the presence of TLR4 on the surface of B cells, signifying MPLA-triggered activation is dependent on specific antigen-mediated uptake of liposomes by B cells. Finally, we evaluated the effectiveness of these 3 liposome conditions to induce antigen-specific humoral responses in naïve mice. Antigen-specific antibody titers in mice immunized with antigen-bearing liposomes containing MPLA were found to correlate directly to B cell binding profiles, with 5% PEG liposomes with tethered p31 resulting in titers ~log10-fold higher than 5% PEG liposomes with surface anchored p31. Current studies are focused on understanding the relationship between B cell binding and activation and the ability of these B cells to present helper peptides to CD4+ T cells, as well as how these liposomes can be utilized to probe and expand low-frequency antigen-specific B cells in spleens and lymph nodes post-immunization. We believe the results obtained in these studies will further the development and understanding of effective particle-based peptide vaccines.

New Materials for Chalcogenide Based Solar Cells

Tosun, B. Selin, University of Washington, Chemical Engineering

Global energy consumption is estimated to increase ~50% in the next 30 years (United States Energy Information Administration). Currently, the majority of the world's energy demand is met by burning fossil fuels, generating CO_2, a greenhouse gas that contributes to global warming. Covering only 0.1% of the earth's surface with 10% efficient solar cells can meet the energy demand of the entire world. However, in 2013, solar cells provided only 0.1% of the world's total energy supply. High cost of solar cells still limits their wide-spread use. Currently, the market is dominated by solar cells based on expensive, high purity, single-crystal silicon (c-Si) wafers, which are produced using high temperature processing. Solar cells that are based on thin film semiconductors, such as copper indium gallium diselenide (CIGS), are becoming less expensive than c-Si solar cells because their production consumes less energy and less material. With the excitement generated as a result of their high potential in terawatt-scale energy conversion, thin film solar cells have reached to high overall conversion efficiencies (~20.4%), nearly as high as c-Si solar cells (20-25%). Hence, the thin film solar cell market share is increasing very quickly, from 0.5% to 6% for CIGS during the last 6 years. Innovative products are emerging rapidly, such as Dow Solar's roof shingles, which are expected to increase widespread-use in the coming decade.

During my PhD at University of Minnesota, Chemical Engineering and Materials Science, under the supervision of Prof. Eray S. Aydil, I focused on balancing the industrial needs with long-term solutions, mainly on CIGS solar cells. Thin film CIGS solar cells are built layer-by-layer on substrates (glass or flexible steel foil) through a series of thin-film deposition processes. Typically, substrates are coated with ~1 μm thick molybdenum layer, then by ~2 μm thick light-absorbing CIGS film. Following, a ~50 nm thin cadmium sulfide (CdS) film, also called as the buffer layer, is deposited through chemical bath deposition (CBD) to form the p-n junction. The photogenerated charges are separated at the CIGS-CdS interface to create electric current. While Cd is toxic, the high-efficiency solar cells are obtained only with CBD-CdS. The solar cells are completed by depositing ~100 nm thick zinc oxide (ZnO) window layer followed by a ~200 nm thick conductive transparent tin-doped indium oxide (ITO) film. The CIGS solar cell is very sensitive to moisture; their efficiency degrades with time, particularly when the temperature is high. Sealing with glass increases the weight, the cost, and precludes the possibility of making flexible solar cells. My thesis concerned the layers above the CIGS absorber, particularly developing alternative materials for CdS and ZnO layers that (i) increases their long-term stability, and (ii) minimizes Cd usage. Towards this end, I worked in two parallel directions:

First, I developed an alternative material to ZnO that stops grain boundary diffusion of water to the CdS-CIGS interface: I hypothesized that facile diffusion of water through grain boundaries of polycrystalline ZnO film is the culprit in diminishing the damp-heat stability of solar cells. Accordingly, I worked on using semi-crystalline tin dioxide (SnO_2) films without grain boundaries in lieu of ZnO. The damp-heat stability test emulates the aging of the solar cells under high humidity (85% relative humidity) and high temperature (85 °C) conditions. I used magnetron sputtering to deposit SnO_2 films with various types of microstructure and studied the correlation between the deposition conditions and the film properties. My best results showed over 95% preservation of the peak efficiency even after 240 hours at 85 °C and 85% relative humidity. In comparison, under identical test conditions, the solar cells without the SnO_2 layer lost nearly 80% of their initial efficiency within 24 hours after commencing the test, where 100 hours is standardized as 20-year-lifetime for outdoor efficiency. A patent is filed by the University of Minnesota and Dow Solar Solutions; the results have been published in three journal articles and presented at two international conferences.

Second, I studied CBD of $Zn_{(1-x)}Cd_{(x)}S$ (0≤x≤1) to replace CdS. It is desirable to reduce, or even eliminate, the toxic Cd. CBD of transparent zinc sulfide (ZnS) has been the highest efficiency buffer layer among the alternatives, but is not as efficient as CBD-CdS. A ternary alloy of these two materials, $Zn_{(1-x)}Cd_{(x)}S$, can provide the benefits of both layers to achieve higher efficiencies. I demonstrated a CBD method of $Zn_{(1-x)}Cd_{(x)}S$ where the composition could be changed controllably. I studied the structure and composition variation along the film thickness using a suite of material characterization methods including x-ray diffraction, spectroscopic ellipsometry and depth profiling by Auger electron spectroscopy on these films. An additional distinguishing feature of my work was the addition of ethylenediaminetetraacetic acid (EDTA) in the solution. EDTA chelates metals, slows the deposition and improves the film's crystallinity. Using a combination of Auger depth profiling and spectroscopic ellipsometry, I discovered that a CdS rich layer forms at the film/substrate interface due to the faster reaction of Cd than Zn. In parallel, I also worked on the improvement of the traditional CBD approach by developing a continuous-flow CBD system (CF-CBD) to minimize toxic waste generation. Although the traditional CBD-CdS provides high efficiency solar cells, it is a batch process and wastes materials. Homogenous nucleation and growth of CdS in the liquid solution competes with the deposition reaction. I designed and built a novel CF-CBD reactor that allows the heating of the substrate with minimal heating of the liquid solution. This reduces homogeneous nucleation and growth of CdS particles and thus the precursor waste and the production cost while depositing films uniformly on large substrates, 10 cm in diameter, and also provides excellent control over the compositional depth profiles. These results are presented in three international conferences, and published in three peer review articles.

In the last year of my PhD, I started research on the synthesis of an earth-abundant material that can replace CIGS. While CIGS solar cells have high efficiency, consumption of indium in flat panel display market and the low abundance of this element limits the cost reductions in CIGS solar cells. CZTS is an earth-abundant equivalent to CIGS and has recently shown to yield over 12.6 % efficiency. A thin CZTS film can be deposited using nanoparticles capped with organic ligands dispersed in organic solvents, such as toluene. However, organic ligands are resulting in an undesired C incorporation. I eliminated organic solvents by a ligand exchange procedure and achieved colloidal dispersions of CZTS nanoparticles in water forming environmentally friendly ink. A patent has been filed by the University of Minnesota, published in *Chemical Communications*, and presented in an international conference.

Since my graduation, I have been pursuing my studies in thin films solar cell area as a Post-Doctoral Researcher at University of Washington, Chemical Engineering, under the supervision of Prof. Hugh W. Hillhouse. Here, I focus on 2 projects: (i) improving the optoelectronic properties of solution processed $CZT(S_xSe_{1-x})_4$, and (iii) alternative solution processing methods on Halide Perovskite (HP) based solar cells. Even though, the $CZT(S_xSe_{1-x})_4$ solar cells developed very rapidly, the current device efficiencies still do not come close to single-junction detailed balance (~31%), mostly due the low open-circuit voltages (V_{OC}). I studied improving the optoelectronic properties, and therefore in V_{OC}, at various processing conditions by reducing the potential fluctuations originating from non-stoichiometries in the film. This work is going to be presented in the Materials Engineering and Science Division, and a manuscript is under preparation. Secondly, the recent breakthroughs on solar conversion efficiencies (15%) with HP materials attracted a great attention. Currently, the poor microstructure of HP solar cells is believed to be the limiting factor for higher solar conversion efficiencies. In the light of that, I am focusing on various solution processing methods thought different in-situ and post deposition techniques. In addition, I am working with a graduate student to build and develop a custom-made Multicolor Photoluminescence instrument for identifying the optoelectronic properties of the materials as a function of absorber film depth.

My professional aim is to contribute my research skills toward the improvement of technology, in particular, solar cells as faculty member while training graduate and undergraduate students. During my PhD, I have instructed two core Chemical Engineering classes: Reaction Kinetics and Reactor Engineering, and Mass Transport and Separation Processes; and I have mentored six undergraduate students, three of which joined to PhD programs. As a Post-Doctoral associate, I have been mentoring two undergraduate students, and working with three graduate students on various projects. I believe that my accumulation on topics as Reaction Kinetics, Material Science & Chemistry, Material Characterization Techniques, Semiconductors Fundamentals, Transport Phenomena, Thermodynamics, and experience on industrial collaborations prepare me well for a faculty position in energy related Engineering programs. I consider solar cells as an economically viable energy source for the very near future. As a faculty, I am planning to conduct research on dilute magnetic semiconductors for solar cell applications, especially via solution processing with my background in both PhD and Post-Doctoral studies.

393732 Enabling the Spectroscopic Tools That We Need to Get the Hidden Information We Want

Abstract:

Catalysis lies at the heart of the chemical industry and could be considered as the society's gateway to the sustainable and petroleum-free future for the fuel and chemical production. Since the energy needs of the developed world are currently over-dependent on the utilization of finite mineral resources, enormous research efforts aim to the shift from petroleum supply to natural gas and renewables, such as biomass, for fuel and chemical production in an economic and efficient manner. In that sense, it is common knowledge that catalysts matter and can provide better ways to make new materials from various sources. Spectroscopy is the enabling tool for the knowledge based and hence time-effective catalyst design, which goes beyond the scope of an empirical observation. Monitoring the events taking place in such materials under realistic conditions is crucial for understanding the reaction mechanism of many important chemical processes and would allow the rational design of new or improve the existent catalysts.

Willing to investigate in depth the structural changes of supported metal oxide catalysts, my research as a graduate student relied on the utilization of vibrational spectroscopies under realistic reaction conditions. Operando Raman spectroscopy of MoO_3/TiO_2 catalysts for the Oxidative DeHydrogenation of ethane revealed that the terminal Mo=O functionalities present on the dispersed species could be responsible for the non-selective ethane to CO_x reactions underlining the importance of the anchoring Mo–O–*Support* bonds as the active site. Armed with my already acquired knowledge supported by my inherent eagerness to explore new scientific areas, my postdoctoral research lies on the critical issue of solvent selection in the field of biomass to fuels and chemicals applications. By coupling thorough spectroscopic measurements with ab-initio DFT calculations, we were able to provide an explanation of the enhanced 5-Hydroxylmethylfurfural (HMF) stability if polar aprotic solvents used as a co-solvents.

What I envision in my research plans is the rational utilization of spectroscopic techniques coupled with simultaneous measurements of catalytic reactivity (Operando spectroscopy). In this respect, my ultimate goal is to unravel novel structure-function relationships for heterogeneous and homogeneous catalytic systems. My approach lies first at the design of a multi-spectroscopic reactor that would be able to monitor under realistic reaction conditions the structure of the catalyst by means of Visible Raman, Surface Enhanced Raman and/or FTIR spectroscopies. However, one crucial drawback regarding Operando spectroscopy lies on the fact that cannot be directly applied for the study of homogeneous catalytic systems. For that reason, a new alternative will be proposed namely Operando Solvation that will allow to monitor changes in solvent structure that could be directly correlated with data of the catalytic reactivity for biomass to biofuels and chemical reactions.

Refernces

(1) Tsilomelekis, G.; Boghosian, S. *Phys Chem Chem Phys* **2012**, *14*, 2216.

(2) Tsilomelekis, G.; Boghosian, S. *Catal Sci Technol* **2013**, *3*, 1869.

(3) Tsilomelekis, G.; Josephson, T. R.; Nikolakis, V.; Caratzoulas, S. *Chemsuschem* **2014**, *7*, 117.

385917 Engaging Students in the Capstone Laboratory Using New Processes & Novel Pedagogy

Abstract:

One of the greatest problems facing engineering education in the 21st century is developing ways to engage modern students in the learning process, as these students seem less receptive to the traditional transmission model of lectures and experiments. Further challenges present themselves as engineering disciplines continue to evolve, both in terms of technological advances and scope, necessitating updates to engineering curricula at the university-level. During my PhD research with Prof. Jeffrey R. McCutcheon and Prof. Daniel D. Burkey, I was able to explore possible solutions to these challenges by changing the way the capstone laboratory course was taught at the University of Connecticut, both by building new experiments and by using learner-based pedagogies.

Updating Experimental Curriculum

The capstone laboratory class at the University of Connecticut was a standard unit operations laboratory with fluid mechanics, heat transfer, and pilot-scale separations experiments. In order to better prepare students for the broader applications of modern chemical engineering, several experiments were added that include chemical vapor deposition and bacterial fermentation kinetics in a bioreactor. While I assisted in the development of these experiments, I was principally responsible for the development of several membrane separations experiments, including a reverse osmosis system, a forward osmosis test system, and an osmotic heat engine capable of generating electricity from salinity gradients. All three of these systems serve dual roles as both educational platforms and as research tools. Additive manufacturing elements were also added to the course by the development of a small-scale reactor design experiment, which allow students to test continuous flow reactors they design and print with a 3D printer. For all new experiments, it was important that they all be based on some chemical engineering fundamental, to allow students to see how what they have learned can be used once they leave school.

Improving Engagement with Gamification

In order to improve student attitudes toward the laboratory course, optional game-based elements were introduced. Students were divided into teams that would all complete a common set of experiments. Students were encouraged to complete optional tasks designed to assist with laboratory report preparation. Completion of these tasks gives student teams non-grade points. At the end of the semester, the team that had collected the most points would receive a non-grade based reward. Students have overwhelmingly embraced this system, saying it made the class feel memorable and kept them interested in the course throughout the semester. Other faculty in the University of Connecticut Chemical Engineering Department have adopted some elements of this course into their own.

Teaching & Research Interests

I seek a position that primarily focuses on teaching. I am comfortable teaching a capstone laboratory course, as I have served as a teaching assistant for the course since Fall 2009 and have co-taught the course with the instructor of record since Fall 2011. I also have an interest in teaching courses at other levels, particularly fluid mechanics, unit operations/separations, process design & safety, or first-year engineering courses, as I have recently helped teach a pilot version of a project-based freshman design course. My previous experience also makes me highly qualified to teach specialty courses such as membrane separations, water quality engineering, or engineering ethics. My research interests include applying novel engineering pedagogies such as the ones I have developed as part of my doctoral work. I am very interested in finding collaborators and doing long-term studies with faculty members at other universities. My other research interests are related to membrane separations, with desalination and engineered osmosis process engineering as a major focus.

377937 Nonequilibrium Self-Assembly and Structures

Abstract:

My research focuses on nonequilibrium self-assemblies (NESAs) that require energy supply to maintain their order or, as cells, to "survive". Nonequilibrium assemblies are energy (external field) driven systems that reside in steady states or long-lived metastable structures, some of which can be highly dissipative. DNA transcription machinery, self-propelling micelles, liquid crystals under external fields, cytoskeletal fibers, and molecular motors are all examples of various aspects NESAs. In my research, these complex assemblies are studied using the tools of nonequilibrium thermodynamics. Until now, NESA in artificial systems has been achieved by modifying the interactions between the constituent objects/parts. In the proposed methodology, I will look at the problem from a fundamentally different perspective – namely, I will keep the interactions between the components intact but instead change the agitation mode to achieve different structures. For this purpose, I aimed to control the particle SA in nonequilibrium regime by applying local anisotropic vibrations (for the solvent or particles or both), which effectively translating into anisotropic temperature. According to the equipartition theorem at thermal equilibrium, kinetic energy of the particles is equally divided into all their degrees of freedom translating into isotropic temperature. This theorem does not hold outside of equilibrium and, in principle, one can envision situations in which different degrees of freedom have different "thermal" energies. In my research, I have first studied such hypothetical situations by means of modeling. These experiments validated my hypothesis that the ultimate shape of the assemblies can be controlled by the mode of anisotropic agitation. With this foundation, my current and proposed research will focus both on understanding the fundamental implications of having anisotropic temperature as well as implementing such NESA systems in practice. In terms of theory and modeling, I will implement multiscale simulations that bridge the gap between the molecular (or nano) scale of the non-equilibrium agitating particles moving anisotropically, and the larger particles that are being agitated. During the simulations, I will perform agglomeration tests by measuring the orientational ordering (to probe and quantify the structure of the growing assemblies). I will also calculate energies and entropies at various stages of NESA and compare the results with those of nonequilibrium coarse-grained field theories. Simultaneously, I will implement several system in laboratory practice – these systems will be based on larger assembly components being surrounded by smaller magnetic particles that can be anisotropically "jiggled" by time-varying external fields. These magnetic particles are the non-Brownian particles effectively creating an anisotropic temperature field experienced by the larger particles. The ultimate objective – and, indeed, great hope – of my work, is that I will be able to engineer self-assembling structures of arbitrary shapes and properties by designing appropriate agitation "schedules". Of course, magnetic agitation is but the first step in these directions, and I also envision other classes of systems based on acoustic or pressure waves, and also on rapidly varying external temperature gradients. This work will create a novel experimental test-bed for studying non-equilibrium self-assembly under well-defined conditions. It can also lead to practical development in designing efficient bottom-up micro- and nanomanufacturing schemes.

Molecular dynamics simulation has employed to test the above hypothesis. Several simulations with various settings were performed including change of the long-range interactions between the beads, changing the density of the system, relative diameters. The anisotropy in temperature would act as an external field that can form various patterns of self-assembled (SA) structures including hexagonal and rod shaped considering hydrodynamic and friction effects. It is found that different agitation modes play crucial form in obtaining different SA patterns. These patterns are found in DNA-coated particles, colloids, polymers, and supercooled fluids.

As a faculty candidate, I would apply my beyond equilibrium knowledge to the SA in magnetic fluids in the presence of a magnetic field as an external force and study different effects such as static and alternative field to develop a complet theoretical approach for dynamic self-assembly far from equilibrium. The results of this study would be a basis for more complex NESA systems such as photoswitchable nanoparticles or to control NESA systems that are coupled with chemical reaction networks.

374568 Constructing, Screening, and Evolving Therapeutic Proteins

Abstract:

Protein-based therapeutics are beginning to come of age, but the development of proteins tailored for specific diagnostic and therapeutic applications remains an ongoing challenge. Yeast surface display (YSD) and other display technologies provide powerful approaches to the high-throughput isolation and evaluation of proteins with characteristics suitable for therapeutic settings. However, these techniques require judicious library construction, reformatting of candidate binding proteins prior to most biological characterizations, and are normally limited to screening with the chemical diversity encoded in the 20 canonical amino acids. The work presented here describes the combination of YSD with synthetic antibody libraries and noncanonical amino acids (ncAAs). These efforts have yielded 1) an extremely efficient approach to isolation of binding proteins in conventional YSD format; 2) a switchable display/secretion technology for rapid isolation and immediate production and characterization of candidate binding proteins in soluble form; and 3) a platform for the construction and evaluation of bioconjugates on the surface of yeast.

The discovery of new antibodies remains an important aspect of therapeutic discovery. Yeast surface display (YSD) and other in vitro technologies have emerged as powerful alternatives to conventional immunization and hybridoma technologies, but these in vitro approaches require the construction of antibody libraries with amino acid diversities and loop length variations that enable the efficient discovery of high affinity antibodies. Synthetic antibody libraries described by Sidhu and coworkers have previously been shown to yield high affinity binding proteins when employed in a phage display format. We utilized one of these libraries to construct a billion-member synthetic antibody library in a YSD format in order to take advantage of the quantitative screening capabilities afforded by YSD. Initial screening with model antigens indicated that the Sidhu library yielded full-length binding proteins more efficiently than a YSD nonimmune antibody library previously constructed in the Wittrup laboratory. We have used the synthetic antibody library to isolate a suite of antibodies targeting cancer-associated fibroblasts (CAFs), an understudied component of many solid tumors. We screened the synthetic antibody library for proteins recognizing both the human and murine forms of fibroblast activation protein (FAP), a membrane-bound protease selectively overexpressed on the surface of CAFs. Our screening campaign yielded more than 50 unique antibodies recognizing FAP, with close to 30 exhibiting crossreactivity for both the human and murine form of the protein. Secondary screens for binding enabled efficient identification of candidates recognizing multiple epitopes of FAP, and further development has resulted in a suite of full-length murine antibodies. Pharmacokinetic experiments in mice have confirmed that these antibodies are suitable for in vivo work, and further evaluation of the antibodies as agents for the therapeutic targeting of CAFs is underway. The general antibody isolation and characterization approaches used in this work are accessible to many single-investigator laboratories and should be applicable to the development of antibodies against numerous therapeutic targets.

While the combination of conventional YSD and high-quality synthetic antibody libraries is quite powerful, conventional display technologies are limited by the requirement of converting displayed proteins into soluble form prior to characterization at the end of a screening campaign. Toward this end, we have developed a switchable display/secretion platform that enables the rapid production of soluble reagents immediately following screening efforts without the use of any cloning steps. In our implementation, we use a ncAA-mediated amber suppression event to control the display of an antibody-like scFv-Fc on the yeast surface. Flow cytometry results have confirmed that the presence of the ncAA in the induction media enables display of full-length scFv-Fcs on the yeast surface. This system has the necessary characteristics for screening and soluble reagent production: in model enrichment experiments, a single round of bead-based selections enables approximately 500-fold enrichment of clones binding the target antigen, while yields of secreted scFv-Fcs are consistently upwards of one milligram per liter. Encouraged by these initial results, we have constructed synthetic antibody libraries in the switchable format and screened them against epidermal growth factor receptor (EGFR), a validated cancer antigen. ScFv-Fcs secreted from enriched populations of the library specifically recognize mammalian cells expressing EGFR, confirming both the success of our screening and the recognition of EGFR in its native form. These results suggest that this switchable screening format will enable rapid characterization of additional biological effects of candidate binding proteins (e.g., receptor clustering or downregulation, cellular proliferation, cellular differentiation, or apoptosis) in the future. The implementation of the switchable display/secretion technology enables cloning-free approaches to the generation of specific binding reagents and functional evaluation of these proteins for cancer targeting and other applications.

Although established display technologies have yielded a startling array of engineered proteins with therapeutic properties, the limited chemical repertoire normally found within proteins can impede efforts to tailor proteins for use in specific therapeutic applications. For example, conventional display technologies are rarely suitable for the characterization or engineering of antibody-drug conjugates or other bioconjugates. To overcome this limitation, we have adapted the switchable system described above to enable the incorporation of ncAAs with reactive functional groups into proteins displayed on the yeast surface. We first used the display/secretion system to confirm that several ncAAs, including ncAAs containing ketone or azide groups, can suppress the amber codon in the switchable system in cells expressing an aminoacyl-tRNA synthetase of appropriate specificity; this provides indirect proof of the display of proteins containing reactive side chains on the surface of yeast. Flow cytometry experiments further indicate that cells displaying proteins containing these functional groups can be selectively modified using oxime ligations (ketones), copper-catalyzed azide-alkyne cycloadditions (azides), or strain-promoted cycloadditions (azides) as determined by detection of reaction products on the surface of yeast. Thus, this approach enables both efficient display of proteins containing ncAAs on the surface of yeast and facile, site-specific protein modification, giving this system all of the necessary components for constructing bioconjugates on the surface of yeast. We anticipate that this approach will offer unprecedented throughput for constructing, evaluating, and evolving antibody-drug conjugates and other small molecule-protein conjugates, opening up new possibilities for the development of therapeutics that simultaneously exploit the favorable properties of both proteins and small molecules.

Multiscale Modeling and Simulation of Biological and Polymeric Systems

Balaji Iyer Vaidyanathan Shantha
Swanson School of Engineering, University of Pittsburgh, Pittsburgh, PA-15261

Engineering novel devices with self-regulated intelligent functional response involves seamless integration of materials with multiple components and different properties. Knowledge of structure-function relationships at various length and time scales and harnessing of these relationships to establish specific design criteria is a key requirement to achieve the goal. In view of this requirement, there is an indispensable need for development of sophisticated theoretical and computational models, which could be utilized as valuable tools for designing intelligent multifunctional materials. Recent advances in nanoscale synthesis, functional self-assembly and molecular cell biology has further fueled this demand specifically in macromolecular and biological soft matter systems.

My research interests are in the field of structure-function relationships and multiscale modeling of polymeric and biological systems. This interest stemmed from an intriguing question on dynamics of ring polymers. Notably, the absence of free ends and the complex static structure of ring polymers in presence of topological constraints posed specific challenges different from that of their linear counterparts. I earned my doctorate degree in Chemical Engineering for the study aimed at developing scaling arguments and constitutive relations that connect structure and dynamic response of ring polymeric systems. The experimental verification of some of our theoretical predictions emerged from advances in use of fluorescently tagged DNA strands as model polymers. These advances opened up new exciting avenues for addressing problems in molecular biology from a polymer physics perspective. Specifically, the complex multiscale problem of packaging a meter long DNA within the cell nucleus only a few microns in diameter piqued my interest. I joined Prof. Gaurav Arya's group, Department of NanoEngineering, University of California, San Diego, to devise coarse-grained models for facilitating the study of eukaryotic genome organization within the interphase cell nucleus. Currently, I am seeking ways to combine polymer theory with biologically inspired material design in Prof. Anna Balazs' group, Department of Chemical and Petroleum Engineering, University of Pittsburgh. Computational design of synthetic nanoparticle networks having the desirable features observed in natural dermal armor of animals is the motivation behind this project.

My research goal is to conceive and develop theoretical models and computational tools to establish design criteria for multifunctional and intelligent self-regulated systems at the interface of materials and biology. In this pursuit, I envision the development of multiscale approaches that are inherently multifunctional and can be readily adapted for studying dynamic response. In particular, development of these approaches in synergy with experiments can be valuable to the growth of new technologies that enhance the quality of life. From an academic perspective, the development of multiscale approaches to modeling provides ample opportunities for students to gain a fundamental understanding of interaction between components of a system at different scales, while simultaneously equipping them with the firm knowledge base for developing future technology.

For more information please visit: http://iyerb.weebly.com/

388037 Bioengineering New Solutions for Pediatric Diseases: Platforms to Improve the Treatment of Brain Tumors and Juvenile Diabetes

Abstract:

Pediatric brain tumors and juvenile diabetes (Type I) represent two of the most devastating diseases afflicting children. While significant advances have been made to improve the survival rate of patients afflicted with both diseases these solutions are often at a cost to the overall quality of life for survivors. We foresee tremendous opportunity to utilize the latest advances in Biomedical and Chemical Engineering to develop new solutions that specifically address current challenges faced in the management of pediatric brain tumors and juvenile diabetes.

The first part of my poster will describe the work completed during my Ph.D. at the University of Washington, which was aimed towards the development of multifunctional nanoparticles to aid in the diagnosis and treatment of medulloblastoma (MB) tumors, the most common form of brain tumor afflicting children. Currently, children afflicted with MB require aggressive neurosurgery and the success for tumor resection depends on the accuracy of tumor margin delineation, of which current techniques demonstrate variable success. Biological barriers continue to limit access of therapeutics and contrast agents to tumor cells. To overcome these limitations, we are developing tumor-specific, optical/MR imaging nanoparticles specifically engineered to transverse biological barriers and aid in the diagnosis, staging, resection of brain tumors, and delivery of therapeutics to brain tumors.

The second part of my poster will focus my current postdoctoral research project in Professor Robert Langer's lab at MIT aimed towards the engineering of a bioartifical pancreas. Type 1 diabetes accounts for 10% of all diabetes cases, is typified by a complete deficiency of insulin production, and stems from an autoimmune response in affected individuals that leads to the T-cell mediated destruction of insulin producing β-cells localized in the pancreas. Exogenous cell therapy, in the form of islet transplantation has been intensely investigated as a method to restore glycemic control. However, due to host rejection of transplanted cells and the lack of suitable encapsulation materials or specific immunosuppressive therapies it's widespread clinical application has been limited. Our efforts have been two fold and focused on both understanding the mechanism by which current encapsulation materials become rejected and innovating novel superbiocompatible materials that can mediate foreign body reactions and fibrosis. Significantly, we have developed a prototype device that is able to treat diabetic rodents for over 6 month.

Novel Material Design via Optimized Multiscale/Multiphysical Models

Sesha Hari Vemuri, Myung S. Jhon, Lorenz T. Biegler

Department of Chemical Engineering, Carnegie Mellon University, Pittsburgh, PA 15213

The confluence of engineering and science in the form of nanotechnology has unleashed tremendous opportunities for novel materials and systems bringing about a paradigm shift to the human civilization in the 21st century. New materials and systems have helped us to achieve what was previously though impossible in a broad range of areas including energy production and storage, water purification, information system, and complex fluids systems. Obtaining a thorough understanding of how the atomistic building blocks of the materials influence the overall performance at the systems scale is the holy grail of nanotechnological research. Researchers face a multitude of challenges which include disparate time and length scale interactive phenomena that span across multiple physical realms. Although there have been significant enhancements in experimental and theoretical methodologies over the past few decades to investigate these multiscale and multiphysical processes, it can often be expensive in both time and resources.

In my research, I try to solve the challenge of providing a complete picture of the atomistic to system level integration by developing multiscale/multiphysical modeling framework combined with optimization tools to predict and identify novel materials in a given engineering system via atomistic architecture to provide desired properties [1]. The modeling methodology is provided via case studies in polymer electrolyte fuel cells (PEMFC), micro/nano hydrodyanamic systems, thin films [2], and NEMS. We utilize the mesoscale modeling methodologies originating from the kinetic theory approach called lattice Boltzmann method (LBM) [3] as our centerpiece and integrate the lower level (atomistic/molecular) and higher level (continuum/systems scale) models, thereby obtaining seamless information exchange without losing physical essence of the system. In PEMFC, we were able to identify design criteria such as the atomistic structure of possible membrane materials, minimum catalyst loading & catalyst micro-geometry, and hydrophobicity distribution in gas diffusion layer (GDL). We began by performing computationally intensive atomistic simulation of hydration and proton conduction phenomena in candidate polymer electrolyte membrane materials and combined the information to molecular dynamic simulation via reduced order models (ROM). The complex multiphase interactions between the gaseous diffusing reactants and the liquid water/vapor mixture in complex porous gas diffusion layer are simulated via LBM. Since the kinetic based theory assumes a pseudo-particle picture, these interparticle interactions can be easily realized. We were then able verify previous studies [4] on how liquid water blocks the path for the oxygen to reach the cathode catalyst layer via GDL, thereby introducing further mass transfer limitations to the PEMFC performance. Furthermore, we integrate the atomistic/molecular/LBM model with ROMs and our optimizing solver for providing design criteria of minimum catalyst (Pt) loading while maintaining the same reaction kinetics.

In addition, I will also present our benchmark case study of thin film polymeric systems we also developed hybridized LBM/MD model which can efficiently simulate complex flow in confined geometries via exchanging the velocity and forces at the mesoscale/molecular model interface.

[1] P.S. Chung, R. Smith, S.H. Vemuri, Y.I. Jhon, K. Tak, I. Moon, L.T. Biegler, and M.S. Jhon, *J. Appl. Phys.*, 111, 07B712 (2012)

[2] S.H. Vemuri, P.S. Chung, R. Smith, G.Y. Yeom, Y.I. Jhon, N.E. Lee, L.T. Biegler, and M.S. Jhon, IEEE Trans. Magn., 48, 4277 (2012)

[3] S.H. Vemuri, H.M. Kim, P.S. Chung, and M.S. Jhon, *J. Appl. Phys.*, 115, 17B749 (2014)

[4] L. Hao, P. Cheng, Journal of Power Sources, 195 3870 (2010)

Ab initio discovery of energy conversion pathways

Massachusetts Institute of Technology

Lee-Ping Wang

Ph.D. Advisor: Troy Van Voorhis, MIT. Postdoctoral Advisors: Todd J. Martínez, Vijay S. Pande, Stanford U.

Department of Chemistry, Stanford University. Email: leeping@stanford.edu

STANFORD UNIVERSITY

Discovering new reaction pathways with an *ab initio* nanoreactor

Introduction: Recent methodological and computational advances harken the advent of theory and computation playing a principal role in discovery. I introduced the *ab initio* nanoreactor—a highly accelerated first-principles molecular dynamics simulation of chemical reactions that discovers new molecules and mechanisms without preordained reaction coordinates or elementary steps.

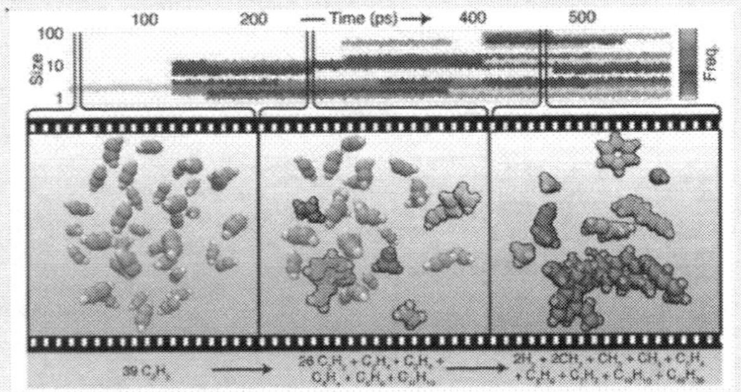

Figure 1: Timeline of a nanoreactor simulation movie. The simulation begins with a collection of acetylene molecules (C, teal; H, white). New molecules are automatically highlighted with molecule-specific colors to indicate the observed reactivity.

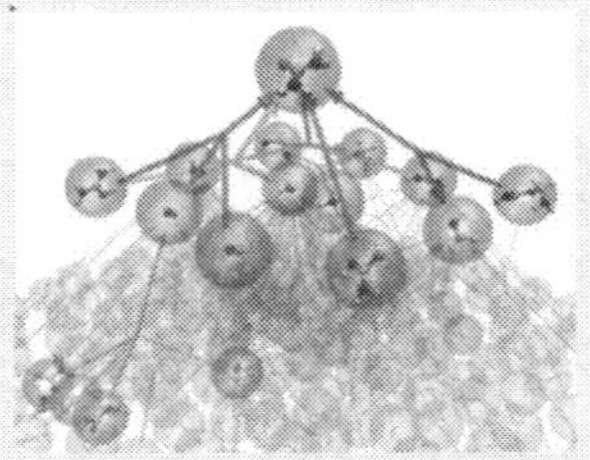

Figure 2: The nanoreactor discovered over 600 compounds, connected by over 1000 pathways. We represent this vast reaction network as a pyramid with a molecule of interest at the top tier (red sphere; $CO(NH_2)_2$). Colored arrows are reaction pathways that link the top molecule with reactants and products on the second tier (blue spheres).

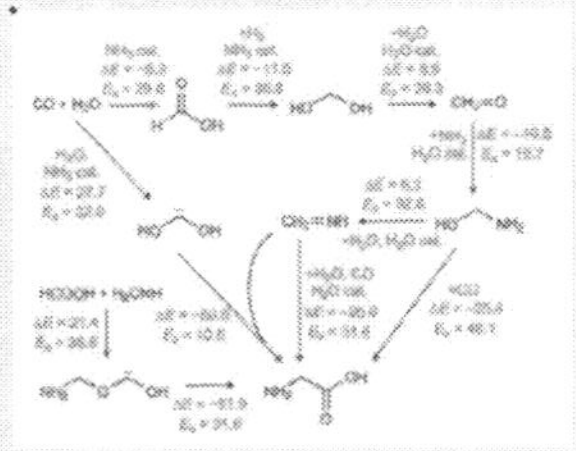

Figure 3: The nanoreactor reveals new pathways for amino acid synthesis from primitive compounds proposed to exist on the early Earth, providing new insight into the classic Urey-Miller experiment. Here the amino acid glycine (bottom) is formed via four different pathways from the starting materials: CO, H_2O, CH_4, NH_3 and H_2.

Outlook: I will use the pathways discovered by the nanoreactor to parameterize kinetic models and formulate design principles for designing improved catalysts for energy conversion. Furthermore, I anticipate that the nanoreactor will contribute to our future understanding of complex reactivity in natural systems by providing novel hypotheses for reaction mechanisms in arenas as diverse as combustion chemistry, atmospheric chemistry and astrochemistry.

Publication: Wang L-P, Titov A, McGibbon R, Liu F, Pande VS and Martínez TJ. "Discovering chemistry with an *ab initio* nanoreactor", *Nature Chemistry* **2014**, doi:10.1038/nchem.2099 (also News of the Week in Nov. 10 issue of *C&E News*.)

Renewable energy conversion mechanisms

Introduction: I have a long-standing interest in renewable energy and the efficient conversion of solar energy into chemical fuels. During my Ph.D. research, I found that ruthenium and cobalt based catalysts for the oxygen evolution reaction (OER) operate via different mechanisms, implying that fundamentally different models are needed to describe the OER kinetics.

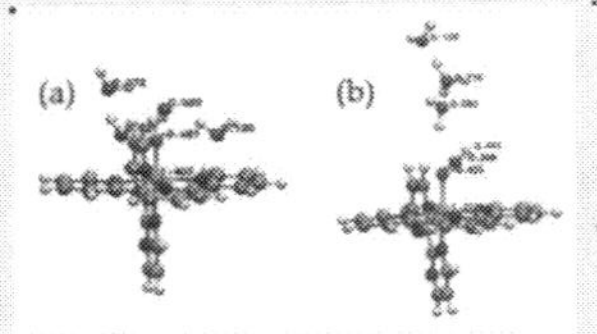

Figure 4: The ruthenium water splitting catalyst operates via an acid-base mechanism where water attacks Ru^{V}.

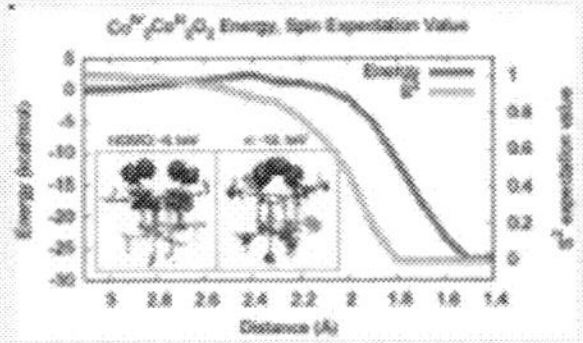

Figure 5: The cobalt oxide water splitting catalyst operates via a direct-coupling mechanism involving two Co^{IV} groups.

Outlook: I am interested in understanding the thermodynamics and kinetics of the most efficient energy conversion catalysts, which are often heterogeneous and nanostructured with many participating pathways. I will approach these challenging problems by applying methods that I developed during my postdoc (left, right, below).

Publications (6 total): Wang L-P, Wu Q and Van Voorhis T. "Acid-base mechanism for ruthenium water oxidation catalysts." *Inorg. Chem.* **2010**, *49*, 4543.

Wang L-P and Van Voorhis T. "Direct-coupling O_2 bond forming a pathway in cobalt oxide water oxidation catalysts." *J. Phys. Chem. Lett.* **2011**, *2*, 2200.

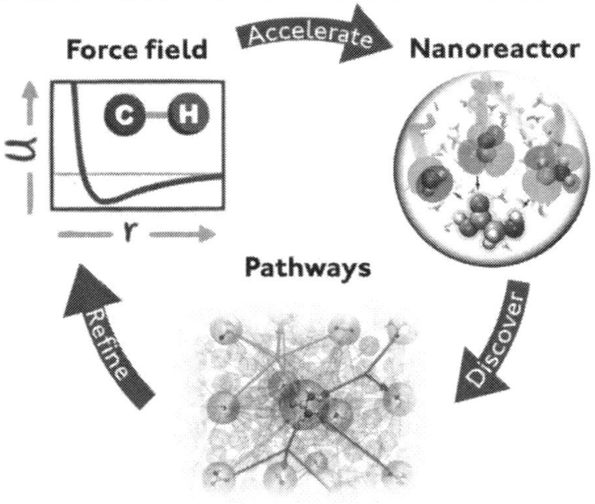

Predict kinetics

Experimental verification

I will take a discovery-based simulation approach toward understanding the complex reaction networks of energy conversion catalysts.

Accelerating simulations via systematic parameterization of accurate force fields

Introduction: *Ab initio* simulations are computationally costly because they treat the electrons explicitly. In protein dynamics simulations and other areas, the electrons are treated implicitly using force fields. During my Ph.D. and postdoctoral research, I introduced the ForceBalance method for systematically deriving force field parameters from combinations of experimental and *ab initio* data. ForceBalance has been applied to improve the accuracy of water, protein, and lipid bilayer models (*in prep.*)

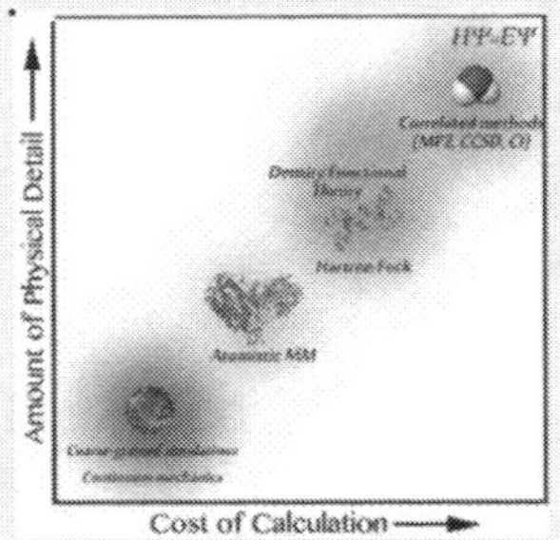

Figure 6: Molecular mechanics force fields can access larger system sizes and time scales than ab initio methods.

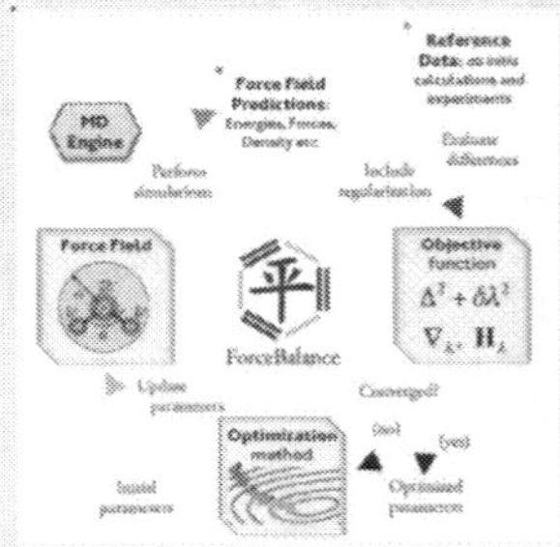

Figure 7: The ForceBalance optimization procedure for systematically building approximate model potentials.

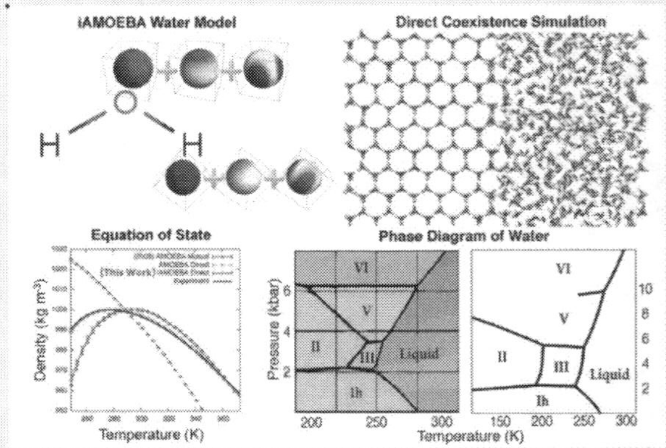

Figure 8: ForceBalance was used to parameterize the iAMOEBA polarizable water model (upper right), which accurately reproduces the equation of state (lower left) and predicts the liquid-ice phase diagram (lower right).

Outlook: I will apply ForceBalance to develop reactive force fields, starting with established models such as ReaxFF and moving towards new functional forms. These force fields will accelerate simulations for discovering new reaction pathways.

Publications (8 total, 3 submitted / in preparation): Wang L-P, Martinez TJ and Pande VS. "Building force fields: An automatic, systematic and reproducible approach." *J. Phys. Chem. Lett.* **2014**, *5*, 1885.

Wang L-P, Head-Gordon T, Ponder J, Ren P, Chodera J, Eastman P, Martinez TJ and Pande VS. "Systematic improvement of a classical molecular model of water." *J. Phys. Chem. B* **2013**, *117*, 9956.

390528 Dynamics of Entangled Rod-Coil Block Copolymers

Abstract:

Polymer science is exploring advanced materials which combine functional domains such as proteins and semiconducting polymers with traditional flexible polymers onto the same molecule. While thermodynamic assemblies of different geometries introduce many interesting new phenomena such as entropic packing and liquid crystalline interactions, dynamic effects are also important to understand for optimal design of material mechanics, processing kinetics, and because of the new physics that directly arises from the motion of multiple domains of dissimilar geometries.

We have recently proposed a theory for the molecular mechanisms of motion in entangled rod-coil block copolymers as a model for this wider class of functional polymeric materials. The large geometrical differences between rigid rods and Gaussian coils cause significant nonlinearities in dynamical behavior as these two motifs are combined on the same molecule. In particular, our theory hypothesizes that the motion of rod-coils is slowed relative to rod and coil homopolymers because of a mismatch between the curvature of the rod and coil entanglement tubes. Dual relaxation mechanisms are predicted in the small and large rod limits, where either the rod or the coil block is expected to determine the motion of the overall molecule.

In this poster we describe the mechanisms of rod-coil block copolymers in detail, while providing significant evidence from both simulation and experiment in support of our theory. We first consider the tracer diffusion of symmetric coil-rod-coil triblocks using molecular dynamics simulation and forced Rayleigh scattering experiments. These results are then generalized to consider rod-coil diblock copolymers, whose asymmetry requires additional analysis. Self-diffusion studies then provide an important step in translating the new physics into practical applications. Finally, additional simulations using coarse-grained models provide a more direct understanding of the underlying mechanisms.

383115 Computer Simulations of Nanoparticle Delivery: From Nanoparticle Design to Cell Membranes

Abstract:

The delivery of therapeutics to disease sites via nanocarriers represents one of the active research areas in the applications of nanotechnology to medicine. It is a rather complicated process involving several different steps ranging from the synthesis of nanocarriers with therapeutics, to the circulation of nanocarries in the body, the interactions of nanocarriers with cell surfaces, the cell membrane penetration and internalization of nanocarriers, and the release of therapeutics from the nanocarriers, etc. Each of these steps, and thus the effectiveness of the overall therapeutics delivery, are affected by different factors, such as the design of nanocarriers, the therapeutics-nanocarriers interactions, the specific composition of the disease cell surfaces and so on. Using computer simulation and modeling, we have investigated the problem from the perspectives of nanoparticles design and cell membranes: 1) the design of the functionalized spherical nanoparticles affects their specific interactions with cell surfaces of interest. Such specific interactions between the functionalized nanoparticles and cell surfaces are achieved using the ligand-receptor interactions. Different nanoparticle design factors, including the characteristics of polymer tethers, nanoparticle size, the ligand density and architectures, the ligand-receptor interaction strength, and so on, are explored in order to generalize their influences on the interaction affinity between nanoparticles and cell surfaces. The 'dilemma' of the interaction affinity and interaction specificity are illustrated and some strategies to overcome the 'dilemma' and achieve both high affinity and specificity of nanoparticles-cell surface interaction are discussed; 2) the architecture of lipid tails influences the formation and the morphologies of water pores inside the lipid membranes. The resultant formation of either hydrophobic or hydrophilic water pores inside the lipid membrane possibly dictates the transport of the materials across of lipid membranes. The knowledge gained from these studies enables the fundamental understanding of the delivery of therapeutics via nanocarries and hopefully serves as guidance for the future experimental design of nanocarriers to obtain optimal therapeutic effects.

387311 Electrochromic Nanocomposites with Endured Energy Storage Properties

Abstract:

Sustainable energies from transient and renewable energy reservoirs such as wind and solar are in urgent demand due to the ever increasing energy crisis arising from the limited reserves of fossil fuels. Given the intermittent nature of these renewable energy resources, reliable energy storage systems are critically needed to store and supply energy in a stable way.[1-5] Meanwhile, electrochromic materials for energy saving have attracted equivalent attention from both the academia and industry, for example, controllable light-reflective or light-transmissive devices for optical information and storage, glare-reduction systems for offices, smart windows for use in cars and buildings.[6-8] In this context, devices with integrated functions of energy storage and electrochromism are highly desirable from the point of view of sustainable development.[9]

Recently, electrically conductive polymers (CPs) have received intensive attention owing to their high pseudocapacitances arising from the rich redox reactions and the corresponding color switching, as well as their facile preparation, environmental stability, high electrical conductivity, and low cost, which make them promising candidates for both energy storage and electrochromic applications. However, structural degradation as a result of the swelling/shrinkage during the intercalation/deintercalation of counterions in the redox reactions remains a challenge for pure CPs thin films.

Targeting to achieve stable conductive polymer nanocomposite thin films with desirable energy storage and electrochromic properties, polyaniline nanocomposite thin films incorporated with different nanofillers, i.e., tungsten oxide and graphite oxides have been prepared by in-situ electropolymerization of the monomers onto the nanofiller coated ITO glass.[10, 11]These nanocomposite thin films have demonstrated much more enhanced stability due to the interactions between the polymer matrix and the nanofillers, which help buffer the stress developed in the polymer matrix. Meanwhile, the nanocomposites exhibit multi-color electrochromism at different potentials, high color contrast, fast switching speed, and high energy efficiency while possessing significant energy storage properties in terms of high energy densities and power densities. The study opens a door for fabricating conductive polymer nanocomposite thin films with integrated energy storage and electrochromism functions and improved stability for future sustainable energy applications via facile electropoloymerization techniques

References

1. J. Zhu, M. Chen, H. Wei, N. Yerra, N. Haldolaarachchige, Z. Luo, D. P. Young, T. C. Ho and Z. Guo, *Nano Energy*, 2014, **6**, 180–192.

2. H. Wei, H. Gu, J. Guo, S. Wei, J. Liu and Z. Guo, *The Journal of Physical Chemistry C*, 2013, **117**, 13000-13010.

3. H. Wei, H. Gu, J. Guo, S. Wei and Z. Guo, *ECS Journal of Solid State Science and Technology*, 2013, **2**, M3008-M3014.

4. H. Wei, H. Gu, J. Guo, S. Wei and Z. Guo, *Journal of The Electrochemical Society*, 2013, **160**, G3038-G3045.

5. J. Zhu, M. Chen, H. Qu, X. Zhang, H. Wei, Z. Luo, H. A. Colorado, S. Wei and Z. Guo, *Polymer*, 2012, **53**, 5953-5964.

6. R. J. Mortimer, *Chem. Soc. Rev.*, 1997, **26**, 147-156.

7. H. Wei, X. Yan, Y. Li, H. Gu, S. Wu, K. Ding, S. Wei and Z. Guo, *The Journal of Physical Chemistry C*, 2012, **116**, 16286-16293.

8. H. Wei, X. Yan, Y. Li, S. Wu, A. Wang, S. Wei and Z. Guo, *The Journal of Physical Chemistry C*, 2012, **116**, 4500-4510.

9. H. Wei, X. Yan, Q. Wang, S. Wu, Y. Mao, Z. Luo, H. Chen, L. Sun, S. Wei and Z. Guo, *Energy and Environment Focus*, 2013, **2**, 112-120.

10. H. Wei, J. Zhu, S. Wu, S. Wei and Z. Guo, *Polymer*, 2013, **54**, 1820-1831.

11. H. Wei, X. Yan, S. Wu, Z. Luo, S. Wei and Z. Guo, *The Journal of Physical Chemistry C*, 2012, **116**, 25052-25064.

SELECTED PEER-REVIEWED PUBLICATIONS (10 out of 33)

1) J. Zhu, M. Chen, **H. Wei**, N. Year, N. Haldolaarachchige, Z. Luo, D. P. Young, T. C. Ho, S. Wei* and Z. Guo*; Magnetocapacitance in Magnetic Microtubular Carbon Nanocomposites under External Magnetic Field, Nano Energy, 6,180–192 (2014).

2) **H. Wei**, D. Ding, X. Yan, J. Guo, L. Shao, H. Chen, L. Sun, H. A. Coloado, S. Wei* and Z. Guo*; Tungsten Trioxide/Zinc Tungstate Bilayers: Electrochromic Behaviors, Energy Storage and Electron Transfer, Electrochimica Acta, 132, 58-66 (2014).

3) **H. Wei**, D. Ding, S. Wei and Z. Guo; Anticorrosive Conductive Polyurethane Multiwalled Carbon Nanotubes NanoComposites; *Journal of Materials Chemistry A*, 1, 10805-10813 (2013).

4) **H. Wei**, H. Gu, J. Guo, S. Wei and Z. Guo; Multiwalled Carbon Nanotubes with Tuned Surface Functionalities for Electrochemical Energy Storage; *ECS Journal of Solid State Science and Technology*, 2(10), M3008-M3014 (2013).

5) **H. Wei**, H. Gu, J. Guo, S. Wei, J. Liu and Z. Guo; Silica Doped Nano-polyaniline with Endured Electrochemical Energy Storage and the Magnetic Field Effects; *Journal of Physical Chemistry C*, 117(25), 13000-13010 (2013).

6) **H. Wei**, H. Gu, J. Guo, S. Wei and Z. Guo; Electropolymerized Polyaniline Nanocomposites from Multi-walled Carbon Nanotubes with Tuned Surface Functionalities for Electrochemical Energy Storage; *Journal of the Electrochemical Society*, 160(7), G3038-G3045 (2013).

7) **H. Wei**, J. Zhu, S. Wu, S. Wei and Z. Guo; Electrochromic Polyaniline/Graphite Oxide Nanocomposites with Endured Electrochemical Energy Storage; *Polymer*, 54(7), 1820-1831 (2013).

8) **H. Wei**, X. Yan, S. Wu, Z. Luo, S. Wei and Z. Guo; Electropolymerized Polyaniline Stabilized Tungsten Oxide Nanocomposite Films: Electrochromic Behavior and Electrochemical Energy Storage; *Journal of Physical Chemistry C*, 116(470), 25052-25064 (2012).

9) **H. Wei**, X. Yan, Y. Li, H. Gu, S. Wu, K. Ding, S. Wei, and Z. Guo; Electrochromic Poly(DNTD)/WO3 Nanocomposite Films via Electropolymerization; *Journal of Physical Chemistry C*, 116(30), 16286-16293 (2012).

10) **H. Wei**, X. Yan, Y. Li, S. Wu, A. Wang, S. Wei and Z. Guo; Hybrid Electrochromic Fluorescent Poly(DNTD)/CdSe@ZnS Composite Films; *Journal of Physical Chemistry C*, 116 (7), 4500-4510 (2012).

AWARDS/HONORS:

3rd Place Award in Chinese-American Chemical Society (CACS) – SW Fall Event,

2012.

APPLICATION AND MECHANISM UNDERSTANDING OF NANO-STRUCTURED CATALYSTS IN BIOFUEL PRODUCTION

Cun Wen, University of South Carolina, Columbia, SC

Department of Energy has set the goal of biofuel production to 36 billion gallons by 2022, while the present production rated at 1.1 billion gallons at 2011. The capacity needs more than 3200% increase in total or 40% increase in annual for the next 10 years. One of the limiting factors in biofuel production is the low throughput of the batch reaction mode. Though it has been long known that heterogeneous processes can accelerate the biofuel production, the actual progress is hindered by poor understanding of reaction mechanisms and requires sophisticated material design and synthesis. The reaction mechanism research is intricate with numerous potential reaction pathways, similar to the possible routes to overcome Himalayas. Our strategy to tackle this issue is to combine the mechanism research with kinetics studies, which serves as loyal contour-mapper and tells us the most feasible way. This strategy has been proven to be effective.[1-12] Herein, I bring in nano-structured heterogeneous catalysts with our strength in advanced nanomaterial synthesis and fundamental reaction-mechanism studies to expedite the biofuel production.

To upgrade biomass to biofuels, the aromatic rings and the C-O bonds need to be broken down and C-C/C-H bonds need to be build up, on which stage heterogeneous catalysts can show off. Firstly, 2-methyfuran acid catalysis reaction is chosen as a model reaction to exemplify our idea of heterogenerizing biofuel production. The 2-methylfuran is a mode molecular representing biomass feed stocks, which usually need acid catalysts for deoxygenating because of oxygen as heteroatom and need chain growth for generating biofuel because of low carbon numbers (<C6). Usually concentrated sulfuric acid is used as acid catalyst to serve the purpose, but it is unfavorable for large scale industrial production. Our results show that the zeolite catalyst can replace the concentrated sulfuric acid with even better production selectivity in terms of cetane number. Furthermore, the cetane number of the product can be increased by decrease the acid strength of the zeolites. To understand the relationship between acid strength and product cetane number, the reaction mechanism on the zeolite catalyst is probed by *in situ* techniques, such as FTIR. The reaction mechanism study reveals that high acid strength increases the adsorption energy, and thus prolongs the resident time of the intermediates on the zeolite. Because longer resident time facilitates the C-C chain growth, the branched products with poor cetane number are favored on zeolites with high acid strength.

Secondly, the products from 2-methfuran acid catalysis reaction need further hydrogenation to upgrade to final biodiesel, and Co is reported to be active for the hydrogenation reaction, whose activity can be further promoted by Cu doping. Thus, bimetallic core-shell nanoparticle Cu@Co with alloy of CoCu on the shell (for high activity) and Cu as the core (for low catalyst cost) would be optimum for this reaction. Traditionally, the core-shell bimetallic nanoparticles are synthesized in two steps. In the first step, the core is formed with one metal, and then in the second step the core is coated with second metal as the shell. Thus, the two-step method not only is poor in economic, but also yields mono-metal instead of alloy on the surface, while alloy is needed for high activity in this hydrogenation reaction. Herein, I demonstrated a one-pot methodology for mono-dispersed

Cu@Co Core-shell nanoparticle synthesis with alloy of CoCu on the shell as catalyst for hydrogenation reaction. Furthermore, I can not only synthesis the nanoparticle in one-step, but also keep a tight control over the size distribution (relative standard deviation 7.9%).

Thirdly, biomass upgrading to biofuel is not limited to the acid and hydrogenation reaction, and other process can also serve the purpose, such as partial oxidation. The partial oxidation has been proved to be an effective way to produce biofuels from biomass aligned with Fischer-Tropsch synthesis. However, the long-standing problem in partial oxidation is the hot spot formation in the catalyst bed, which can lead to the catalyst deactivation or even explosion. It has been known that the hot spot is formed by the side reaction that is the total oxidation taking place beside the partial oxidation. The key factors guiding the reaction path through total or partial oxidation is unknown, which hinders the catalyst development. To address this issue, I looked into the mechanism of partial oxidation production of syngas ($CO+H_2$). Our research has revealed that the lattice oxygen reducibility of the catalyst is the key factors governing the reaction mechanism. With high lattice oxygen reducibility, total oxidation rather than the partial oxidation is favorite. By tuning the lattice oxygen reducibility of Er_2O_3 with Mn doping, I successfully shift the reaction mechanism from total oxidation to partial oxidation, and eliminated the hot-spot formation.

Read Me More:

1. Wen C, Dunbar D, Zhang X, Lauterbach J, Hattrick-Simpers J. Self-healing catalysts: Co3O4 nanorods for Fischer-Tropsch synthesis. *Chem Commun.* 2014;50:4575-4578.
2. Wen C, Barrow E, Hattrick-Simpers J, Lauterbach J. One-step production of long-chain hydrocarbons from waste-biomass-derived chemicals using bi-functional heterogeneous catalysts. *Phys Chem Chem Phys.* 2014;16(7):3047 - 3054.
3. Wen C, Liu Y, Guo Y, Wang YQ, Lu GZ. Strategy to eliminate catalyst hot-spots in the partial oxidation of methane: enhancing its activity for direct hydrogen production by reducing the reactivity of lattice oxygen. *Chem Commun.* 2010;46(6):880-882.
4. Wen C, Liu Y, Guo Y, Wang Y, Lu G. Synthesis of the rare earth compound nanosheets induced by lamellar liquid crystal. *Solid State Sciences.* 2009;11(11):1985-1991.
5. Wen C, Liu Y, Tao F. Integration of surface science, nanoscience, and catalysis. *Pure Appl Chem.* Jan 2011;83(1):243-252.
6. Liu Y, Wen C, Guo Y, Lu G, Wang Y. Modulated CO Oxidation Activity of M-Doped Ceria (M = Cu, Ti, Zr, and Tb): Role of the Pauling Electronegativity of M. *J Phys Chem C.* 2010;114(21):9889-9897.
7. Liu Y, Wen C, Guo Y, Lu G, Wang Y. Effects of surface area and oxygen vacancies on ceria in CO oxidation: Differences and relationships. *J Mol Catal a-Chem.* Feb 2010;316(1-2):59-64.
8. Liu Y, Wen C, Guo Y, et al. Mechanism of CO Disproportionation on Reduced Ceria. *ChemCatChem.* 2010;2(3):336-341.
9. **Meadia Highlights:**Jacoby M. Custom Catalyst Resists Oxidation. *C&EN News.* 2014;92(13):25.
10. **Meadia Highlights:**Pichon A. Catalysis: Lower reactivity means more hydrogen. 2010; http://www.nature.com/nchina/2010/100120/full/nchina.2010.7.html.
11. **Meadia Highlights:**Saxton C. Enhancing catalytic activity. 2010; http://www.rsc.org/Publishing/ChemTech/Volume/2010/02/enhancing_catalytic.asp.
12. **Meadia Highlights:**Chinese New Year. *Chem Commun.* 2010.

391971 Surface and Interfacial Properties in Ceramic and Inorganic Materials

Abstract:

Nano-scale materials have recently attracted great interest due to their ability to advance currently available materials to engineer next-generation technologies and devices. This can lead to lower cost of production, reduced carbon foot print, higher energy efficiency, and improved performance in a variety of applications. Since nanoparticles and nanocomposites have higher surface areas than their traditional micro-scale counterparts, it is becoming increasingly important to understand how surfaces and interfaces interact with their surrounding environments to contribute to the advanced properties and performances. Using a variety of characterization tools, my research focuses on identifying the molecular-scale surface and interfacial structures to explain the mechanisms that control the observed macroscopic properties.

My current and past research includes investigation of interfaces that exist in polymer-derived, nanocomposite ceramics and of surfaces of *di*calcium silicate nanoparticles. I completed my doctoral degree at the University of California, Davis under the advisement of Alexandra Navrotsky and Sabyasachi Sen, where I developed structural models for silicon-based polymer-derived ceramics that were synthesized from pyrolysis of a variety of preceramic polymers. The nano-scale and micro-scale structures are highly tunable and have been proposed for a wide variety of high temperature applications. It was found that the interfaces between the nano-scale domains in these ceramics contribute to high chemical and thermal stability that make these materials a desirable candidate for next generation high temperature ceramics. Furthermore, it was determined that the morphology contains percolation networks that may also contribute to interesting transport and mechanical properties. I am currently a University of California President's Postdoctoral Fellow at the University of California, Santa Barbara, working under the advisement of Bradley Chmelka. My postdoctoral research focuses on understanding the hydration mechanisms in *di*calcium silicates, a minor component in anhydrous cement powders. Sol-gel derived β-*di*calcium silicate nanoparticles were recently synthesized to increase the hydration rate, with respect to conventionally prepared, micron sized β-*di*calcium silicates. Increasing the overall hydration rate of *di*calcium silicate is considered to be an attractive option for replacing a portion of *tri*calcium silicate (the major component in anhydrous cement powders) to achieve similar mechanical strength properties with lower total calcium contents. Correspondingly, this could reduce the carbon foot print of the cement manufacturing process due to the release of carbon dioxide during the refinement of calcium. The molecular-scale surface structures of anhydrous and hydrated β-*di*calcium silicates synthesized by the sol-gel and conventional methods are currently under investigation to identify the surfaces chemistry and interactions that cause the increased hydration rate.

My future research interests include molecular-scale investigations on nanostructured ceramic and inorganic materials with applications in fuel cell electrodes and membranes, battery electrodes, supercapacitors, and high temperature ceramics. My experimental approach to investigate these systems primarily involves the use of high-resolution, solid state nuclear magnetic resonance spectroscopy, which enables identification of molecular-scale structures in inorganic and ceramic materials. Additionally, intermolecular interactions that occur at surfaces and interfaces can be elucidated to provide knowledge of the fundamental chemistry that is involved in the enhanced properties of next generation engineering materials.

393338 Multi-Level Engineering Approaches for Manipulating Plant Metabolism in Culture

Abstract:

Plants are a source of a vast diversity of natural products that have commercial applications as flavors, fragrances, pesticides and pharmaceuticals. These products are often the result of sophisticated secondary metabolic pathways that are specialized to plant cells. Due to the complex structures of these compounds, chemical synthesis is often infeasible, necessitating the use of biological production platforms. In this project, methods were developed and employed to characterize and engineer plant cell systems at both the environmental and molecular levels to decipher mechanisms for promoting plant secondary metabolism. Studies utilized suspension culture of the non-model plant system, *Taxus*. These cultures are utilized for commercial production of the anti-cancer agent paclitaxel (Taxol™, Bristol-Myers Squibb), which is used in the treatment of breast, lung and ovarian cancers, as well as AIDS-related Kaposi's sarcoma.

Environmental manipulation: When plant cells divide, the cell walls do not fully separate, leading to the formation of aggregates, which can range in size from two to thousands of cells. These aggregates result in the formation of subpopulations within a culture. This heterogeneity is believed to be a major source of the inherent variability associated with plant production systems. To aid in the understanding of culture aggregation, a method was developed to reduce aggregation through the application of mechanical shear. A population balance model was established to determine the amount of shear required to reach a target aggregate size distribution. By applying shear over multiple generations of cell growth, a more heterogeneous culture with a constant mean aggregate size was developed with no affect on cell health.

Molecular-level manipulation: Although plant primary metabolism is well characterized and highly conserved across plant species, plant secondary metabolism is often species specific and, as a result, poorly defined. Additionally, the interface between primary and secondary metabolism is poorly understood, limiting metabolic engineering efforts. High-throughput methods were developed to evaluate the flux distribution through specific pathways involved in *Taxus* metabolism. It was found that cultures that produce paclitaxel also divert carbon flux towards the production of phenolics and flavonoids. A long-term study revealed high levels of variability in culture aggregation and secondary metabolite accumulation. These studies are amongst the first to study global secondary metabolism in a non-model plant species and provide valuable insight into the design of effective metabolic engineering strategies to promote production of a particular class of secondary products.

Due to the high level of interaction between cells in these complex cellular systems, successful engineering efforts must look beyond the level of the metabolic pathway. This project is the first of its kind to characterize and manipulate *Taxus* cultures on a multi-scale level to allow for the effective engineering of cultures for increased paclitaxel production. Moving forward, manipulation of culture aggregation through mechanical shearing and cellular metabolism through metabolic engineering could lead to more homogeneous cultures with reduced variability and increased paclitaxel accumulation.

Polyhedral Oligomeric Silsesquioxane (POSS)-Based Hybrid Nanostructured Thermoplastic Polyurethanes (TPUs)
---- Synthesis, Processing and Biomedical Applications

Polyhedral Oligomeric Silsesquioxane (**POSS**) is a class of well-defined, highly symmetric molecules that contain an inner inorganic framework made up of silicone and oxygen $(SiO_{1.5})_x$ and externally covered by eight organic substitutes. The hybrid organic-inorganic framework renders POSS thermally and chemically robust nature, and the eight vertex R-groups feature the versatility in chemistry. In a subtle way, nano-scale POSS moieties were introduced into conventional thermoplastic polyurethanes(TPUs) to form a new generation of hybrid TPUs nano-composites. By tailoring the biocompatible soft segments (e.g. poly(ethylene glycol)(PEG), poly(caprolactone)(PCL), poly(d,l-lactide)[P(d,l)-LA], etc), we have developed a series of novel hybrid POSS-TPUs with a wide variety of biomedical applications, including mechanical robust hydrogels, nano-structural antimicrobial hydrogel webs with long-term duration, enzymatic biostable polyester-based hybrid TPUs for long-term biomedical implantation, hybrid TPUs with high-drug (i.e. Paclitaxel) loading and tailored drug release for biodegradable stent coating, among others.

The cornea is the transparent outermost layer of the eye. Trauma, bacterial and viral infections, and heritable conditions lead to loss of corneal function and visual impairment in over ten million individuals world-wide. Recapitulating the microstructure of native human corneal stromal tissue is believed to be a key feature in successfully engineering corneal tissue. Employing a strategy of surface contact guidance and growth factor supplementation, we demonstrated that on aligned fibrous substrates made from biodegradable poly(ester urethane) urea (PEUU), human corneal stromal stem cells (hCSSCs) could be induced to secrete and organize a type-I collagen-based extracellular matrix (ECM) abundant in characteristic human corneal stromal ECM components. Spatial self-organization of the collagen-based ECM by hCSSCs featured stratified multilayered collagen-fibril lamellae with orthogonal orientation, and uniform fibril size and inter-fibril spacing in a pattern mimicking human corneal stromal tissue. The approach of combining substrate cues with growth factor augmentation offers a new means to engineer well-organized, collagen-based constructs with appropriate nanoscale structure for corneal repair and regeneration.

Heterogeneous Catalysis for Renewable Energy and Green Chemistry

Xiaofang Yang, Chemistry Department, Brookhaven National Laboratory

Exhaustion of fossil-based fuels and the concern of climate change due to carbon dioxide emission have generated growing interests in searching for renewable energy sources and clean energy technologies. The future success of these technologies largely relies on advancing catalyst development for many catalytic processes such as biomass conversion, dry reforming, oxygen reduction and alcohol oxidation in fuel cells, hydrogen production from biomass or water splitting reaction, etc. To meet the challenges in catalyst development, my research is primarily focused on the study of surface science, electrochemistry and heterogeneous catalysis with the aim to discover high performance catalysts in energy-related catalysis and green chemistry.

My doctoral research extensively explored Pt monolayer (ML) and novel non-Pt model electrocatalysts for proton exchange membrane (PEM) fuel cells and electrolyzers. This work sought insight and discovery of new materials by establishing clear structure-activity relationships in electrocatalysis. My study combined the synthesis and characterization of single crystal surfaces in ultrahigh vacuum (UHV) and evaluating their catalytic performance in electrochemical conditions. Single crystal surfaces are characterized by well-ordered surface structures, which serve as ideal models to study fundamental behavior or properties for the related functional materials. Supplemented with theoretical simulations, this approach created an excellent platform to study the relationship between catalytic activity and surface geometric structure and electronic properties of electrocatalysts.

My postdoctoral research focuses on developing high performance catalytic systems for CO_2 conversion and alkene epoxidation and addresses the key scientific issues through theoretical and experimental studies on both model surfaces and powder catalysts. Successful understanding of the principles to control the selectivity and activity allows me to design better catalysts for propylene epoxidation and converting CO_2/CH_4 to syngas. Epoxidation of propylene is an important industrial reaction. However, a lack of selectivity to propylene oxide (PO) is the major challenge in the direct epoxidation of propylene by O_2. My studies on mixed oxides demonstrate the feasibility of designing novel materials as active and selective catalysts which could significantly impact the production of PO in the future. Recently, rising CO_2 concentration in atmosphere has been associated with adverse global climate changes. An alternative way to alleviate this problem is utilizing CO_2 as inexpensive and readily available feedstock to produce other chemicals or fuels. Thus, my research efforts are to identify active catalysts for converting CO_2/CH_4 to syngas. My preliminary findings of the highly active carbide surfaces could potentially lead to the design of high performance catalysts for CO_2 conversion.

My future research will continue in the area of catalysis for novel fuels, green chemistry and fuel cell catalysis, in the context of research and development of alternative energy technologies. Specifically, by engineering model catalysts of alloys, metal oxides and nanostructures, my research will address the key factors such as geometric structures and electronic properties that impact catalytic properties of these materials. The goals of my studies are to provide an improved in-depth understanding of the origin of the enhanced catalytic activity and to tailor the catalyst performance for specific applications.

Ph.D. Advisor: Prof. Bruce E. Koel

Postdoc Advisors: Prof. Jingguang G. Chen and Dr. Jose Rodriguez

PUBLICATIONS

20. "Molybdenum carbide as alternatives to precious metal catalysts for selective conversion of CO_2 by hydrogen, M.P. Porosoff , **X. Yang** and J.G Chen, *Angew. Chem. Internat. Ed.*, **2014**, in press.
19. "Theoretical and experimental studies of ethanol decomposition and electrooxidation over Pt-modified tungsten carbide" T. G. Kelly, A. L, Stottlemyer, **X. Yang**, J. G. Chen, *J. Electrochem. Soc.* **2014**, *161, E3165-E3170*
18. "Nanostructures of Au sub-monolayer: synthesis and its application for reduction of molecular oxygen", **X. Yang** and B.E. Koel, *Nat. Mater.*, **2014**, to be submitted.
17. "Stabilization of catalytically active Cu^+ surface sites on titanium-copper mixed oxide films", A. Baber, **X. Yang**, H. Kim, K. Mudiyanselag, A. Al-Mahboob, J. Sadowski, J. Rodriguez, P. Liu, J. Chen, and D. Stacchiola, *Angew. Chem. Internat. Ed.*, **2014**, in press.
16. "Geometric requirements for catalytic sites on Pt-Sn alloy surfaces", J Gao, H.Zhao, **X. Yang**, B.E. Koel and S.G. Podkolzin, *Angew. Chem. Internat. Ed.*, **2014**, in press.
15. "Opportunities and challenges in utilizing metal carbides in renewable energy applications", **X. Yang** and J.G Chen, *ACS Catal.* **2013**, to be submitted.
14. "Structural origin of anisotropic transport in electrically conducting dichloroacetic acid-treated polymers", J. M. Mativetsky, J. Tarver, **X. Yang**, B. E. Koel, Y. Loo, *Organic Electronics* , **2014**, 15, 631–638.
13. "A New Class of Electrocatalysts for Oxidation of Ethylene Glycol by Supporting Monolayer Pt on Engel-Brewer Alloy Substrates", **X. Yang**, M. Li, B. Koel and J.G. Chen, *Angew. Chem. Internat. Ed.*, **2014**, Submitted
12. "Controlling acetylene adsorption and reaction on Pt-Sn catalytic surfaces", J. Gao, **X. Yang**, S.G. Podkolzin and B.E. Koel, *ACS Catal.*, **2013**, 3, 1149-1153.
11. "WO_3-α-Fe_2O_3 composite photoelectrodes with low onset potential for solar water oxidation" , P, Zhao, C. Kronawitter, **X. Yang**, J. Fu and B.E Koel, *Phys. Chem. Chem. Phys*, **2014**, 16, 1327-32.
10. "Highly stable Pt–Au@Ru/C catalyst nanoparticles for methanol electro-oxidation", Q. He, B. Shyam, M.Nishijima, **X. Yang**, B. Koel, F. Ernst, D. Ramaker, and S. Mukerjee, , *J. Phys. Chem.* **2013**, 117 (3), 1457-1467.
9. " Electrochemical and spectroscopic study of novel Cu and Fe-based catalysts for oxygen reduction in alkaline media", Q.He, **X. Yang**, X. Ren, B. Koel, and S. Mukerjee, *J. Power Sources,* **2012**, 213, 169-179.
8. "Activation of tungsten carbide catalysts by use of an oxygen plasma pretreatment", **X. Yang**, Y.C. Kimmel, J. Fu, B.E. Koel, and J.G. Chen, *ACS Catalysis* **2012**, *2*, 765–769.
7. "Nano-faceted C/Re(11-21): fabrication, structure and template for synthesizing nanostructured model Pt electrocatalyst for hydrogen evolution reaction", **X. Yang**, H. Wang, W. Chen, B.E. Koel and R.A. Bartynski, *ACS Nano.*, **2012,** *6*, 1404-1409.
6. "A novel CuFe based cathodic catalyst for oxygen reduction reaction in high pH environment", Q.He, **X. Yang**, X. Ren, B. Koel, S. Mukerjee, and R. Kostecki, *J. Power Sources,* **2011**, 196, 7404.
5. "Role of surface Fe in enhanced activity for the oxygen reduction reaction on a $Pd_3Fe(111)$ single crystal alloy", **X. Yang**, J. Hu, J. Fu, R. Wu, and B. E. Koel, *Angew. Chem. Internat. Ed.*, **2011**, 50, 10182-10185.
4. "Formation of Pd monomers and dimers on single crystal $Pd_3Fe(111)$ surface", **X.Yang**, J. Hu, R. Wu, B. Koel, *J. Phys. Chem. Lett.*, **2010**, 1(16), 2493–2497.
3. "Influence of phosphate anion adsorption on the kinetics of oxygen electroreduction on low index Pt(hkl) single crystals", Q. He, **X. Yang**, W. Chen, S. Mukerjee, B. Koel, and S. *Chen, Phys. Chem. Chem. Phys.*, **2010**, 12(39),12544-55.
2. "Surface structure of $Pd_3Fe(111)$ and effects of oxygen adsorption", **X. Yang**, L. A. Welch, J. Fu and B. E. Koel, in *Catalytic Materials for Energy, Green Processes and Nanotechnology*, edited by C-Y. Mou, J. Liu, H.H. Kung, S. Dai (*Mater. Res. Soc. Symp. Proc.* Vol. 1217, Warrendale, PA, 2010), 1217-Y08-43.
1. "Improving electrocatalysts for O_2 reduction by fine-tuning the Pt–support interaction: Pt monolayer on the surfaces of a $Pd_3Fe(111)$ single-crystal alloy," W.P. Zhou, **X. Yang**, M.B. Vukmirovic, B.E. Koel, J. Jiao, G. Peng, M. Mavrikakis, and R.R. Adzic, *J. Am. Chem. Soc.*, **2009**, 131, 12755–12762.

389126 Engineering Advanced Membrane Materials for Gas Separation and Biofuels Production

Abstract:

My research work focuses on design and development of novel high performance hollow fiber membranes and sorbents with complex morphologies out of advanced polymers and nanostructured materials, and their applications in membrane-based gas separation, and biofuels purification. During my PhD at University of Chinese Academy of Sciences with Prof. Yinhua Wan, I introduced a new process for improving the affinity between nano-sized particles and polymer matrix through chemical bonds and hence suppressed the formation of microvoids at polymer–inorganic interface, which was used to develop novel organic permselective pervaporation membranes for biofuels purification from biomass fermentation broth. This work has been successfully resulted in a new research program around $2.0M at Chinese Academy of Sciences funded by Chinese Government as well as incubated several collaborations with companies to carry out the pilot scale application for bio-ethanol and bio-butanol production via fermentation/pervaporation coupled process based on cellulose and Cassava renewable feed stock, respectively.

Currently, I am a research engineer and lab manager at Georgia Tech, where I am expanding my expertise in membrane-based gas separation by engineering the economical membrane materials. I also provide mentorship and instruction to the undergraduates, PhD students and post-doctoral fellows in a group of over 20 researchers. This position has allowed me to expend my expertise from flat-sheet membranes to high performance hollow fiber membranes, and also from membrane-based separations to combined hollow fiber sorbents adsorption and membrane-based coupled separations. By using advanced cross-linkable polymers, functional polymers, and several commercially available polymers, I have created high performance hollow fiber membranes for aggressive sour gas separations, CO_2 removal from flue gas, and biofuels purification.

My future research will address both fundamental issues and applied research related to membrane-based gas separations, organic permselective pervaporation membrane technology for biofuels purification, and recovery and/or removal of volatile organic compounds (VOCs) from industrial wastewaters. Using the knowledge of the industrially available material properties developed in the fundamental studies, I will strongly focus on conducting cutting-edge research in engineering the advanced polymer structures and economical processes to produce high performance membranes with the emphasis on global CO_2 reduction, energy conversion, and clean technology which will be met industrial applications.

385622 Production of Furan-Based Products from Lignocellulosic Biomass

Abstract:

Lignocellulosic biomass has great potential for producing alternative fuels and chemicals. In particular, furan-based products such as HMF and furfural are attractive intermediates of biofuels and chemicals. Various conversion technologies for furan-based products have been investigated; however, there are still some challenges including low productivity, formation of by-products, complexity of reaction, high cost of catalysts, and difficulty of product separation. In order to overcome the aforementioned barriers, two novel processes for furan-based products were developed.

The first approach is a biphasic process for production of 5-bromomethyl furfural (BMF). BMF is a furan derivative with high reactivity and easy to be converted to other chemicals, and it is also an intermediate of furan-based biofuels. The reported in this study was a single step biphasic system including an aqueous phase of molten salt hydrate with mineral acid as catalyst and an organic phase of extracting solvent. The BMF was produced in the aqueous phase and immediately extracted into the organic phase to prevent the further decomposition and condensation. The production of BMF from different substrates including monosaccharides, polysaccharides, and real biomass was investigated. Under the optimum conditions, BMF yield from cellulose reached 90.3 %. The effects of reaction factors including acids loading, concentration of halide salt, and volume ratio of organic to aqueous phase were investigated. Different molten salt hydrates solutions (LiBr, LiCl and $CaBr_2$) were also compared to convert cellulose into halidemethyl furfurals (BMF/CMF). The application of biphasic system improved the products selectivity and prevented the formation of byproducts (humins).

The second approach for production of furan-based products from biomass is a one-pot process (HDA) integrating hydrolysis of polysaccharides to monosaccharides, dehydration of the monosaccharides to furans, and aldol condensation of the furans with acetone to produce furan acetone adducts as hydrocarbon precursors. The conversion was completed in a LiBr/acetone system without pretreatment or preprocessing of the biomass. Acetone was not only used as a solvent but also a reactant involved in aldol condensation reactions. Halide salt (LiBr) and hydrochloric acid (HCl) were used as catalysts in the process. According to the experimental results, halide salts play crucial roles in all three reactions above. A series of experiments were performed to optimize reaction factors for efficient conversion of corn stover. A product yield of 78% was obtained under the optimum conditions. In the meantime, lignin was significantly depolymerized and separated from other components during the process as a potential valuable co-product (HDA lignin). Most of LiBr and HCl (99%) were recyclable.

385611 Predictive Theoretical Modeling of Complex Fluids: From Advanced Materials to Engineering Nanomedicine

Abstract:

Theoretical modeling provides insights into fundamental understanding of many complex systems ranging from organic-inorganic hybrid materials to biological fluids. A suitably chosen coarse-grained theory that captures the essential physics of the system can be predictive and guide the materials design for a variety of applications before conducting exhaustive experimental efforts or computational tests of the full complex problem. Here I demonstrate the application of classical theoretical modeling in two nanoparticle systems followed by my proposed research plans.

Nanoparticle-organic hybrid materials (NOHMs) are a new type of complex fluid consisting of 10 nm diameter spherical inorganic core particles surface-functionalized with oligomeric organic molecules with no other solvent. The absence of a solvent, the small size of the nanocores and oligomers, and the incompressibility of the tethered oligomeric fluid make the oligomer-mediated interactions non-pairwise-additive. In my PhD research, I developed a classical density-functional approach for model hard spheres with tethered bead-spring oligomers that allows a direct description of the system free energy as a functional of the probability densities of cores and oligomers without assuming pairwise additivity. Based on the coarse-grained model and my theories, I predicted a striking result of solvent-free NOHMs that the structure factor must go to zero at zero wave number, which can be used to characterize the polydispersity in experimental systems. I also discovered that the carbon capture capacity, disorder-order phase transition, and transport properties of these materials are governed by the oligomer-configurational entropy as the tethered hairs have to uniformly fill the interstitial space. These predictions have been verified by relevant molecular dynamics simulations and experiments.

The use of ligand-functionalized nanocarriers that bind to specific receptor proteins is gaining translational prominence in targeted vascular drug delivery. Owing to the wide spectrum of time and length scales involved in the hemodynamics and nanocarrier-endothelium binding landscape, a unified theoretical framework bridging the microscale many-body hydrodynamic interactions in the vessel tube and the nanoscale nanocarrier adhesion on the vessel wall would provide useful insight into the tailored design of functionalized nanocarriers for specific targeted biophysical environments. As a postdoctoral researcher, I developed a theoretical formalism for studying nanocarrier-cell adhesive dynamics in the presence of red blood cells (RBCs) using a combined framework of generalized Langevin equations and dynamical density-functional theory. Analyzing the RBC-driven margination and the position/velocity autocorrelation functions of the nanocarriers for different vessel diameters, blood flow rates, and binding free energies would allow the quantification of nanocarrier binding affinity and the prediction of biodistribution. The predicted correlation functions for nanocarrier motion have been validated by finite-element direct numerical simulations of Navier-Stokes fluid around a moving nanoparticle at different separations from the vessel wall.

My proposed research plans are aimed at addressing some of the key challenges in engineering nanomedicine and advanced materials design using nanoparticle-based systems. By combining classical theories and computational techniques, my research group will explore how the active mass transport of the magnetic nanoparticle is affected by physiologically relevant interactions in blood flows, how the drug-loaded nanoparticle may overcome biophysical barriers in the extracellular matrix to reach the target site, and how the functionalization of the particle can be optimally designed to achieve self-assembling bioactive nanostructures. Concurrently, I will develop and apply new theoretical and simulation approaches to advance fundamental studies of systematic materials processing, complex networks characterization, and ordered directed self-assemblies for energy applications.

385492 Microfluidic Approaches Towards Metabolic and Genetic Engineering

Abstract:

Microbes provide a promising platform for broad and sustainable chemical production because of their potential to produce a wide range of molecules with high specificity from renewable feedstocks. Although metabolic engineering has existed for several decades, leaps in progress have been driven in part by tangential technological advancements in DNA synthesis, sequencing and automation. These advancements have streamlined editing genomes towards desirable phenotypes and have spurred the development of seminal technologies like trackable multiplexed recombineering (TRMR), multiplexed automated genome engineering (MAGE) and phage-assisted continuous evolution (PACE). Even entire genomes are now possible to synthesize *de novo*. The ever-improving simplicity of manipulating organisms and the demonstrable effectiveness of automation technologies towards this end, suggest that that the future of metabolic engineering and synthetic biology will soon require a drastically different focus, shifting from the basics manipulating and editing genomes to how to effectively engineer and optimize organisms from interdisciplinary approaches. My research effort will utilize an interdisciplinary approach from a fundamental chemical engineering perspective, combining microfluidic and automation technology development to address the needs of metabolic engineering and synthetic biology. This approach will enable more efficient and effective methods to optimize organisms thereby drastically improving the scope, throughput and basic understanding of genetic and metabolic engineering. My planned research efforts are as follows:

Proposed Research Projects

Parallelization and miniaturization of chemical processes as a broad screening approach: Correlating the phenotype of an engineered cell to its genotype by screening limits throughput in the design-build-test phase of genetic engineering. Identifying extracellular products produced, like biofuels, can involve chemical processes in series (production, separation, detection) which reduces the throughput of genotypes that can be tested and thereby limits the effectiveness of engineering organisms. The research proposed here is to develop a broad approach toward parallelizing the production, separation and detection of extracellular products using multiple emulsions in a microfluidic system. This idea can be used to screen for a variety of chemicals and capable of throughputs up to 10^3-fold greater than conventional approaches. By coupling separations and colorimetric reactions in a multiple emulsion format, a method could be developed to perform processes in parallel enabling high-throughput analysis single-genotype analysis. With high-throughput screening, genome-wide search technologies and methodologies will be applied to understand the metabolic engineering of strains whose investigation is currently limited by screening technologies.

Miniaturized, long-write DNA synthesis methodology and development: Many genetic engineering tools and technologies use massive quantities of diverse synthetic oligonucleotides and as such have relied on advancements in DNA synthesis. But, the high and somewhat stable costs of gene synthesis, caused by harsh chemistry and inefficient synthesis, have limited the further applicability of DNA synthesis. The research proposed here is developing a concurrent approach of enzyme and chemistry development with microfluidic system engineering to develop a novel and superior approach to DNA synthesis with respect to the current industry standard. Specifically an enzyme-based DNA synthesis chemistry will be developed in an electrophoresis-driven microfluidic platform allowing for millisecond transport and reaction times, precise temperature control and low reagent use. By coupling molecular biology and protein engineering along microfluidic system development, I envision a bench top system could be developed to improve de novo oligonucleotide and gene synthesis.

Automation of functional pathway discovery: Integrating heterologous genes in non-native hosts can allow for the production of a wide-range of chemicals by either adding a new metabolic pathway or bypassing the host cell's native regulation scheme. However, this is largely a trial and error process and heterologous gene expression typically can result in incompatible enzyme products. To overcome this, I propose an automated system of chemostats, coupled with novel microfluidic transformation technologies in series to continuously create combinations of different pathways of heterologous genes in *E.coli*. This system will be use to rapidly construct entire pathways which can be screened for chemical production. By integrating an automation approach to recursive postivite/negative selection I will develop an open-sourced continuous flow directed evolution system capable of constructing pathways.

Previous Research

I studied under Professor Mark Burns in the Ph.D. program at the University of Michigan to develop technologies enabling active control of flow, particles and chemicals in microfluidic systems. There, I integrated chemical engineering concepts governing microfluidic design (transport and fluid dynamics), system engineering and microfabrication. My Ph.D. resulted in three first author research publications, three contributing author publications and a book chapter contribution. I was awarded the competitive Microfluidics Training Program Fellowship, an NSF graduate honorable mention, and doctoral qualifying exam distinction having the top research and grades for my Ph.D. qualifying exam. After completing my Ph.D. I worked at Becton Dickinson on their BD Max platform - an open-source automated molecular diagnostic robotic system which utilized microfluidic technology developed from the Burns lab.

Currently, I am a research associate in Professor Ryan Gill's research group at the University of Colorado studying methods to relieve throughput bottlenecks in the design-build-test cycle which is core to synthetic biology and metabolic engineering. My most notable work has been developing a method to track populations of combinatorial mutants by utilizing core chemical engineering concepts along with next-generation sequencing technologies. This work has improved the throughput of analyzing combinatorial engineered populations by over 10^4-fold. Another of my publications uncovered implications of a fluid phenomena in microchannels - an effect that arose while developing microfluidic systems for genetic engineering. I have also been involved in authoring several successful grants including the DARPA living foundries grant and DoE biosystems design grant.

Teaching and Professional Activities

Teaching is an important focus of mine and as a professor I will aim to impart students with the knowledge, skills and attitudes to ensure successful transition from education to application. Aside from positively contributing to the undergraduate and graduate educational structure through the ABET requirements, I would also like to apply concepts and ideas published in educational literature like, *Chemical Engineering Education.* I have experience teaching undergraduate discussion classes in Material and Energy Balances and Transport Phenomena. Through these experiences, and other instructional experiences outside of the classroom setting, I've worked closely with a variety of learning styles giving me a good perspective on how best to instruct and interact with students. In addition, I was able to work with many B.S. and M.S. engineers at Becton Dickinson, and was able to observe the skills that I saw promoted success from an engineering perspective. I would like to use my opportunity as a professor to support rich undergraduate extracurricular work. I have had extremely positive experiences in the past with undergraduates and REU students and hope to continue giving opportunities for independent and directed hands-on learning for students. I would also like to support undergraduate students by making them aware of the opportunities available to them for industry, graduate school and various internships (SURF, REU, etc). Finally, I was trained on the Microfluidics in Biomedical Sciences Training Program at Michigan and saw it as a fantastic experience. I would like to support programs like this and initiate the development of collaborative training programs.

386935 Unique Physicochemical Phenomena in Polymer Nanocomposites

Abstract:

Unique physicochemical phenomena in polymer nanocomposites (PNCs), including giant magnetoresistance (GMR), negative permittivity, interfacial polarization and reduced viscosity attracted more attention owing to their significant effects on both the scientific community and the practical applications. However, how to achieve the properties by tailoring the structure of the PNCs is still unknown. Understanding the structure-property relationship is the key solution to the challenge, which strongly relies on mechanism study of these phenomena. Semi-conductive nanocomposites with the aforementioned unique properties have been developed and are applied to explore the working mechanism. In composition optimization and structure design, materials with different magnetic property, electrical conductivity and morphologies will be integrated to achieve the targeting performances and reveal the structure-property-performance relationship.

Targeting the GMR and negative permittivity, a surface initiated polymerization method has been developed to fabricate the polyaniline (PANI) nanocomposites with ferrite and perovskites nanoparticles, the polymerization process takes place on the surface of the nanoparticles and the microstructure study reveals a distinctive structure with the insulated nanoparticles coated by the conductive PANI. Depending on the localization or delocalization function of the nanoparticle plays on PANI, PNCs show different GMR value. PNCs with ferrite nanoparticles show negative GMR and positive real permittivity, however, the PNCs with perovslites nanoparticles have positive GMR and negative real permittivity. The positive and negative GMR were analyzed theoretically from the wave functional shrinkage model and forward interference model, respectively. And the negative permittivity was attributed to the delocalization of charges and the metallic state in PANI.

Targeting the reduced viscosity and interfacial polarization, PNCs with different kinds of conductive nanofillers including polyaniline nanofibers and nanospheres, polypyrrole nanofibers and nanospheres, carbon coated iron (Fe@C) nanospheres, carbon nanotubes, graphene nanosheets and graphene nanosheets decorated with core-shell nanoparticles have been fabricated. The embedded nanoparticles in polymer matrix were found to favor the disentanglement of the polymer chain. And the reduced viscosity of PNCs was attributed to the attraction between the nanofiller and polymer matrix caused by physical or chemical mechanism. The interfacial polarization were found to depending on the specific surface area of the nanofiller, an enlarged interfacial area between the nanofiller and polymer matrix allows more charge carriers blocked at the interfaces and leads to enhanced interfacial polarization.

SELECTED PEER-REVIEWED PUBLICATIONS (7 out of 28):

1. **X. Zhang**, V. Bitaraf, H. A. Colorado, S. Wei and Z. Guo, Vinyl Ester Resin: Rheological Behaviors, Curing Kinetics, Thermo-mechanical and Tensile Properties, ***AIChE Journal***, 60(1), 266-274 (2014).
2. **X. Zhang**, O. Alloul, Q. He, J. Zhu, M. J. Verde,Y. Li, S. Wei and Z. Guo, Strengthened Magnetic Epoxy Nanocomposites with Protruding Nanoparticles on the Graphene NanoSheets, ***Polymer***, 54(14), 3594–3604 (2013).
3. **X. Zhang**, O. Alloul, J. Zhu, Q. He, Z. Luo, H. A. Colorado, N. Haldolaarachchige, D. P. Young, T. D. Shen, S. Wei and Z. Guo, Iron Core Carbon Shell Nanoparticles Reinforced Electrically Conductive Magnetic Epoxy Resin Nanocomposites with Reduced Flammability, ***RSC Advacnes***, 3, 9453-9464 (2013).
4. **X. Zhang**, Q. He, H. Gu, S. Wei and Z. Guo, Polyaniline Stabilized Barium Titanate Nanoparticles Reinforced Epoxy Nanocomposites with High Dielectric Permittivity and Reduced Flammability, ***Journal of Materials Chemistry C***, 1, 2886-2899 (2013).
5. **X. Zhang**, Q. He, H. Gu, H. A. Colorado, S. Wei and Z. Guo, Flame Retardant Electrical Conductive Nano-polymers based on Bisphenol F Epoxy Resin Reinforced with Nano Polyanilines, ***ACS Applied Materials & Interfaces***, 5(3) 898-910 (2013).
6. **X. Zhang**, S. Wei, N. Haldolaarachchige, H. A. Colorado, Z. Luo, D. P. Young and Z. Guo, Magnetoresistive Conductive Polyaniline - Barium Titanate Nanocomposites with Negative Permittivity, ***Journal of Physical Chemistry C***, 116(29), 15731–15740 (2012).
7. **X. Zhang**, J. Zhu, N. Haldolaarachchige, J. Ryu, D. P. Young, S. Wei, and Z. Guo, Synthetic Process Engineered Polyaniline Nanostructures with Tunable Morphology and Physical Properties, ***Polymer***, 53(10), 2109–2120 (2012).

AWARDS/HONORS:

1. 3rd Place Research Excellence Award in the *2013 Chinese-American Chemical Society Southwest Chapter (CACS-SW) Annual Poster Competition,* Houston, TX, USA, October 2013.
2. Honorable Mention Award in student poster competition at International Polyolefins Conference 2013, Houston, TX, USA, February 2013.
3. Scholarship Award, SPE (Society of Plastics Engineers) Thermoplastic Materials & Foams Division, at International Polyolefins Conference 2013, Houston, TX, USA, February 2013.
4. 2nd Place Research Excellence Award in the *2011 Chinese-American Chemical Society Southwest Chapter (CACS-SW) Annual Poster Competition,* Houston, TX, USA, October 2011.

383850 Rational Engineering of Microbes for Metabolite Overproduction

Abstract:

It has been a century since industrial scale microbial fermentation processes were established, which produced simple natural metabolites, such as ethanol, acetone, glycerol and butanol. During the past two decades, advances in DNA technology have enabled transfer of heterologous pathways to microbial host cells and manipulation of a specific step of an endogenous pathway, leading to great improvements in diversity and titer of the metabolites that microbes can make. For example, gram-per-liter titer of plant therapeutic molecules (e.g. paclitaxel and artemisinin precursors) has been achieved in *Escherichia coli* or *Saccharomyces cerevisiae*. However, these amazing successes still heavily relied on trial and error experiments, which are very time-consuming; this is because the interactions between heterologous pathways and cellular metabolism and those among parts of a foreign pathway were not well understood. The faculty candidate is interested in understanding details of these interactions for developing fast and rational approaches to improve the metabolite overproducing strains, and inventing novel practical means of preventing negative interactions among pathway segments - such as spatially separating incompatible pathway modules into different cells or compartments of a cell. During the poster session, the candidate will present our projects that embody these research ideas, and discuss his research plan with search committees.

Computational Systems Biology of Metabolism in Single- and Multi-Species Microbial Systems

Ali R. Zomorrodi

Graduate Program in Bioinformatics and Department of Biomedical Engineering, Boston University, Boston, MA

Metabolism is defined as the set of all biochemical reactions in a living organism required to maintain life. The efficient engineering of metabolism in natural and synthetic microbial systems for biotechnological applications requires a systems-level understanding of the metabolic interactions. This calls for the development of systems-level analysis tools and modeling techniques capable of providing a quantitative understanding of the metabolism as well as quantifying the system-wide impact of engineering manipulations.

Here, I review my recent work on the development of efficient computational systems biology tools to shed light onto the less-understood aspects of metabolism in single- and multi-species microbial systems, and to identify promising network rewiring strategies for biotechnological applications. I first introduce a workflow that combines the ensemble modeling of metabolic networks with customized optimization-based algorithms for creating large-scale kinetic models of metabolism. This is exemplified by constructing a large-scale kinetic model of *E. coli* core metabolism satisfying multiple sets of mutant flux data. I next introduce k-OptForce, a computational strain design protocol integrating the available kinetic information with the genome-scale stoichiometric-based models of metabolism to identify system-wide metabolic interventions leading to the targeted overproduction of a biochemical of interest. Results and lessons learned by the application of k-OptForce to overproduce a number of biochemicals and biofuels (e.g., fatty acids, triacetic acid lactone, L-serine etc.) in *E. coli* and *S. cerevisiae* will be shared.

The focus is then shifted from single-species systems to microbial ecosystems, where multiple microorganisms are involved and communicate through the exchange of biochemical cues. Here, I introduce a multi-level and multi-objective optimization framework, called OptCom, for the metabolic modeling and analysis of microbial communities using genome-scale metabolic models. Next, I introduce an extension of this procedure, called d-OptCom, which enables capturing the impact of dynamic changes in the environment on the function and structure of microbial communities. This is followed by a discussion of ongoing efforts on the game-theoretic analysis of microbial communities using genome-scale metabolic models in dynamic spatially structured environments. Overall, the wide array of computational and systems modeling tools introduced in this poster highlight their utility in model-driven biological discovery and in complementing metabolic engineering efforts for biotechnological applications.

393772 Plasma Catalysis for Materials and Energy Challenges

Abstract:

Low-pressure plasmas can be used for gas phase activation via electron-molecular reactions producing ions, radicals, metastables and photons that can be beneficial for catalysis of several technologically important reactions. First of all, when used along with heterogeneous catalysts, the plasma activation could lower the activation energy further. Such synergistic effect is termed here as plasma catalysis Plasma catalysis is emerging as one of the most promising alternatives to address energy and environmental challenges.

However, it is not clear whether one needs the same type of catalyst materials even for plasma environments or not. For example, the low-melting metals in the presence of plasmas can catalyze the growth of one-dimensional materials. In the absence of plasma activation, the molten metals do not exhibit any catalytic activity towards selective dissolution even with application of high temperatures. The reasons for such synergic effect with plasma activation and metal clusters suggest that such metals could be interesting for plasma catalysis research to address environmental/energy challenges in addition to materials growth. In this research program, using results from our prior research, new catalysts will be designed for studying carbon dioxide reduction using plasma catalysis approach. CO_2 is a cheap, non-toxic and non-flammable, renewable feedstock compared to oil or coal for producing valuable chemicals and transportation fuels. However, the existing methods for CO_2 reduction are not economically viable. The production of chemicals from CO_2 could have a significant positive impact on the global carbon balance.

Hydrogen electrocatalysis: from mechanistic study to advanced catalyst development

Wenchao Sheng

ws2440@columbia.edu
Department of Chemical Engineering, Columbia University, New York, NY 10027

My research interests are centered on addressing challenges in renewable energy research, and mainly focused on electrochemical energy storage and conversion applications such as electrolyzers, fuel cells, flow batteries, and electrochemical synthesis of high-value chemicals. I aim to bridge the fundamental understanding in electro-catalysis and surface science constructed on model electrode systems to the design and development of high-performance nano-scale materials and structures for clean energy applications.

One of the intriguing topics in renewable energy research is the utilization and generation of hydrogen energy. Hydrogen oxidation and reduction reactions (HOR/HER) are two of the key components in low-temperature fuel cells and electrolyzers. The pH values of the electrolytes have a great impact on the reaction kinetics such that the HOR/HER on the state-of-the-art Pt catalyst in alkaline media decreases by two orders of magnitude than that in acids. The slower HOR kinetics on Pt is projected to impose an additional overpotential (~130 mV at 1.5 A/cm^2_{MEA} with a Pt loading of 0.05 mg_{Pt}/cm^2_{MEA}) in alkaline or alkaline membrane fuel cells (AFCs/AMFCs), thus the use of Pt catalyst at the anode in AFCs/AMFCs requires a higher catalyst loading.[1, 2] With the development of many non-precious metal catalysts for the oxygen reduction reaction in both acid and alkaline electrolytes, showing comparable activity to Pt, the demand for highly active and cost-effective HOR catalyst in alkaline medium is arising and becomes one of the crucial challenges to make AFCs/AMFCs more practical.

Rational design of highly efficient HOR/HER catalysts requires identification of the reaction descriptor for the HOR/HER in alkaline electrolyte. It is shown that the HOR/HER exchange current density in alkaline media on a wide range of monometallic surfaces follows a volcano type of relationship with the calculated hydrogen binding energy (HBE) of the metals with Pt having the optimum HBE value.[3] This discovery indicates that HBE values can serve as the designing principle for developing efficient electrocatalysts for the HOR/HER in base. Motivated by this finding and the fact that Ni exhibits certain HOR/HER activity in base, we calculated the HBE of NiMo bi-metallic and CoNiMo multi-metallic surfaces and found that they have similar HBE values as Pt. We subsequently synthesized NiMo and CoNiMo alloys by electrodeposition and performed HOR/HER tests in 0.1 M KOH at different temperatures. The results show that the HOR/HER on CoNiMo outperforms electroplated Ni by 20 times.[4] At sufficient catalyst loading, this promising material will alleviate the catalyst cost issue raised by higher Pt loading in alkaline electrolytes, and will facilitate the development of the AFCs/AMFCs technology.

References:

1. K. C. Neyerlin, W. B. Gu, J. Jorne and H. A. Gasteiger, *J Electrochem Soc*, 2007, 154, B631-B635.
2. W. C. Sheng, H. A. Gasteiger and Y. Shao-Horn, *J Electrochem Soc*, 2010, 157, B1529-B1536.
3. W. C. Sheng, M. Myint, J. G. G. Chen and Y. S. Yan, *Energ Environ Sci*, 2013, 6, 1509-1512.
4. W. C. Sheng, A. P. Bivens, M. Myint, Z. B. Zhuang, R. V. Forest, Q. R. Fang, J. G. Chen and Y. S. Yan, *Energ Environ Sci*, 2014, 7, 1719-1724.

393802 Polymer Supported Catalysts for Heterogeneous Catalysis

Abstract:

Research Interests:

Selective and efficient conversion of oil or gas feedstock into highly valued bulk and fine chemicals is one of the major focal points in heterogeneous catalysis. Traditional inorganic catalyst supports suffer from mass transfer limitations, coking and low interaction with organic molecules. In particular, catalytic reactions in fine chemical synthesis require a suitable environment of ligands and solvent, which provide the required geometry and stabilize the transition state of the reaction. Polymers create a 'solvent'-like organic reaction environment, onto which ligands and metal complexes can be grafted, or metal nanoparticles can be loaded to provide the catalytic functionality. Polymers also swell when placed in a solvent enabling new mass transfer pathways. However, most of polymers do not offer high thermal stability and resilience to mechanical stress, which consequently leads to a limited use in industrial reactions. My core competence is to design and synthesize thermally and mechanically stable polymer supports, to introduce functional groups into these polymer supports, to test them in model reactions and to apply them to practically relevant industrial processes.

PI Experience:

In November 2012, I joined Prof. Ferdi Schüth's group as a postdoc. I discussed the idea of ultrasonic spray pyrolysis as a method to synthesize catalysts with him and he agreed to give me the financial support but I had to find a lab facility to conduct the research. With his help, I was offered laboratory space from Prof. Christof Schulz and Dr. Hartmut Wiggers at the University of Duisburg-Essen. I applied and obtained an Alexander von Humboldt fellowship to add to the financial support given by Ferdi Schüth. After proposing the idea, three PhD students were very eager to join the lab and work with me on this interesting work. Currently, mesoporous sulfonated carbon spheres have been synthesized via one step ultrasonic spray pyrolysis using cheap precursors and will be tested in dehydration reactions.

Research Experience:

My PhD thesis focused on controlled synthesis of colloidal nanoparticles, heterostructures and their applications in catalysis. For instance, plasmonic-catalytic nanostructures have been synthesized to utilize solar energy in heterogeneous catalysis. Entering my postdoc, I broadened my field into the catalyst-support interaction. I have developed a polyphenylene support as an excellent platform for heterogeneously catalyzed reactions, which are normally carried out under homongeneous conditions. High thermal stability of the polyphenylene support further enables gas phase continuous partial oxidation of benzyl alcohol to benzaldehyde. I have also built an *in situ* electron paramagnetic resonance setup to study the redox properties of CuO-CeO_2. Using this set-up combined with online mass spectroscopy, the three step redox mechanism could be directly observed for the first time.

Teaching Experience:

I have two years' experience as a teaching assistant in Chemical Principles and three years' experience as a teaching assistant in Astronomy. I taught the exercise courses and tutorials, off-class activities, and also main lectures. I have also tutored 4 PhD students. Currently, I supervise an exchange student, Jacob Oskar Abildström, from The Technical University of Denmark in our group.

Future Research:

My core competence as a faculty member in chemical engineering department is to design polymer supports for heterogeneous catalysis. In the short term, my research comprises of polymer synthesis, studing their chemical and thermal stability, mass transfer, functionalization, and activity in model reactions. Cross coupling polycondensation, cyclotrimerization, radical or acid/base catalyzed polymerizations are the major routes to obtain conjugated and non-conjugated polymers. Metal nanoparticles are impregnated via swelling-impregnation method developed by myself. In addition, the spray drying technique offers a continuous way to synthesize polymer catalysts. Chemical stability of polymers can be studied by treatment with redox gas or liquids at reaction conditions. Their relations with polymer structure will be investigated by varying the ratio between aromatic and aliphatic units. Selective hydrogenation of alkynes with different molecular sizes will be used to investigate the mass transfer limitation. The polymer supported catalysts will be tested in several model reactions both with loaded metals, ligands or a combination of both. The focus for the application of the reactive polymers will be cross-coupling reaction, selective oxidation and hydrogenation. If selected to obtain tenure, I will aim for more challenging tasks with polymers, such as inorganic-organic hybrid catalysts, gas phase heck reactions, and benzene to phenol conversions.

Selected Publications:

(14 total, 7 first author, 427 total citations, 1 patent – ResearcherID F-6893-2014)

1. **Feng Wang**, Jerrik Mielby, Felix Herrmann Richter, Guanghui Wang, Gonzalo Prieto, Takeshi Kasama, Claudia Weidenthaler, Hans-Josef Bongard, Søren Kegnæs, Alois Fürstner, Ferdi Schüth "A Polyphenylene Support for Pd Catalysts with Exceptional Catalytic Activity" *Angew. Chem. Int. Ed.* **2014**, DOI: 10.1002/anie.201404912.
2. **Feng Wang**, Si Cheng, Zhihong Bao, Jianfang Wang "Anisotropic overgrowth of metal heterostructures induced by site-selective silica coating" *Angew. Chem. Int. Ed.* **2013**, *52*, 10344-10348.
3. **Feng Wang**, Chuanhao Li, Huanjun Chen, Ruibin Jiang, Ling-Dong Sun, Quan Li, Jianfang Wang, Jimmy C. Yu, Chun-Hua Yan "Plasmonic harvesting of light energy for Suzuki coupling reactions" *J. Am. Chem. Soc.* **2013**, *135*, 5588-5601.
4. **Feng Wang**, Ling-Dong Sun, Jun Gu, Wei Feng, Yi Yang, Jianfang Wang, Chun-Hua Yan "Selective Heteroepitaxial Nanocrystal Growth of Rare Earth Fluorides on Sodium Chloride: Synthesis and Density Functional Calculations" *Angew. Chem. Int. Ed.* **2012**, *51*, 8796-8799.
5. **Feng Wang**, Chuanhao Li, Ling-Dong Sun, Chun-Hu Xu, Jianfang Wang, Jimmy C. Yu, Chun-Hua Yan "Porous Single-Crystalline Palladium Nanoparticles with High Catalytic Activities" *Angew. Chem. Int. Ed.* **2012**, *51*, 4872-4876.

6. **Feng Wang**, Chuanhao Li, Ling-Dong Sun, Haoshuai Wu, Tian Ming, Jianfang Wang, Jimmy C. Yu, Chun-Hua Yan "Heteroepitaxial Growth of High-Index-Faceted Palladium Nanoshells and Their Catalytic Performance" *J. Am. Chem. Soc.* **2011**, *133*, 1106-1111.
7. **Feng Wang**, Ling-Dong Sun, Wei Feng, Huanjun Chen, Man Hau Yeung, Jianfang Wang, Chun-Hua Yan "Heteroepitaxial Growth of Core–Shell and Core–Multishell Nanocrystals Composed of Palladium and Gold" *Small* **2010**, *6*, 2566-2575.

393894 Electrokinetic Instability in Microchannel-Nanochannel Devices: A Comparison of Slip-Velocity and Full-Formulation

Abstract:

formulation

In systems containing ion-selective membranes or nanochannels, concentration polarization (CP) under DC voltage beyond the classical Levich limit leads to the breakdown of local electroneutrality over micron or larger scales at the salt-depleted interface. This manifests itself in the growth of an extended space charge (ESC) region, which becomes unstable above a critical voltage drop. The instability results in the formation of a fast-flowing vortex system with complex, often chaotic, dynamics. In unconfined systems, i.e. large electrolytic cells, this contributes strongly to the overlimiting conductance (OLC) of the system. However, both the role of the instability in OLC as well as its origin and onset become more complicated in highly confined systems such as microchannel devices. The problem of instability under geometric confinement has been studied both analytically and numerically using two different approaches. The first approach considers the ESC to be negligibly thin. Thus it is replaced by a slip boundary condition for the hydrodynamics and boundary values for the concentration and potential obtained from asymptotic analysis. This yields directly a marginal stability curve and critical parameters for the onset of instability. The second approach employs a direct numerical calculation of the stability problem resulting in a growth-rate surface from which marginal stability and critical parameters may be extracted. While both show suppression of the instability with increasing confinement, there are some interesting discrepancies between the features which merits further consideration.

392943 DNA-Controlled Partition of Carbon Nanotubes in Polymer Aqueous Two-Phase Systems

Abstract:

Sorting single-wall carbon nanotubes (SWCNTs) of different chiralities is both scientifically interesting and technologically important. Specific short single-stranded DNA sequences have been identified to show recognition abilities for specific chiralities of SWCNTs from their synthetic mixture. In addition, polymer aqueous two-phase (ATP) extraction has been demonstrated recently as a very effective way to achieve nanotube sorting. In this work, we report a systematic study of sequence-dependent partition of DNA-wrapped SWCNTs in several polymer ATP systems. This unique approach of combining DNA sequence and polymer ATP system enabled the purification of many single-chirality SWCNT species at high purity. In addition, the DNA-controlled partition of SWCNTs in ATP system suggests the sensitive dependence of hydration energy on the exact spatial distribution of hydrophilic groups which are determined by the DNA sequence and the SWCNT chirality (n,m).

393908 Design of Biomimetic Functional Nanomaterials to Study Integral Membrane Proteins

Abstract:

Integral membrane proteins (IMPs) reside within the cellular membrane and play a key role in modulating a wide range of physiological processes, from neuronal signaling to immune and hormonal response. Their malfunctioning has been implicated in various diseases, including cancer and cardiovascular diseases, which makes them an important pharmaceutical target. The native lipidic environment is imperative for the stability and activity of IMPs, which makes *in vitro* studies of these proteins very challenging. The overarching goal of my research has been to develop novel biomimetic interfaces for reconstituting IMPs to facilitate their structural and functional studies. For example, during my postdoctoral research at the NIST Center for Neutron Research, I fabricated a functionalized nanotubular membrane for reconstituting IMPs, and used this platform in combination with neutron scattering techniques to determine the IMP's structure. Additionally, during my graduate research, I developed a microarray-based platform to selectively immobilize IMPs for high-throughput target screening. Currently, my research at the University of Delaware is focused on studying the role of surfactant micelles in the stabilization and crystallization of IMPs. Surfactant micelles are widely used as a bilayer mimic for reconstituting IMPs in protein-detergent complexes (PDCs). We are employing nuclear magnetic resonance and small-angle neutron and x-ray scattering (SANS and SAXS) techniques to study the stability of PDCs over molecular and microscopic length scales, respectively, which we expect will enable the rational design of PDCs and IMP crystallization trials. Using my interdisciplinary research background in biomaterials, nanotechnology, neutron scattering techniques, and IMPs, I intend to develop an independent research program on *biomimetic functional nanomaterials* for: biophysical characterization of IMPs using neutron scattering; biosensors to monitor IMPs; and high-throughput drug screening platforms.

394054 A Systems Level Approach Towards Rational Strain Development for Microbial Bioprocesses

Abstract:

Microbial based biochemical and/or biofuel production presents an attractive and sustainable potential source of industrial chemicals and transportation fuels have been the research focus in the recent past. The attractiveness of this approach is greatly enhanced by the use of renewable feedstock such as lignocellulosic materials, waste product and byproduct streams such as crude glycerol derived from biodiesel production and gaseous effluents (CO_2 and CO) from industrial plants. Generation of value added products from waste streams not only offers an efficient and effective means to recycle, reuse and/or dispose the waste stream but also improves the overall economics of a sustainable biorefinery. However, there are challenges associated with the efficient utilization of the materials such as lignocellulosic compounds and other waste product, thereby lowering the efficiency of the bioconversion process. Moreover, the production of biofuels and biochemicals from microbial fermentation has been severely limited due to the accumulation of products, such as butanol, which are toxic to the cells producing them, resulting in a stress response associated with growth and survival inhibition, and other cellular physiological events such as sporulation observed in the genus of *Clostridia*, thereby limiting the productivity of the desired end product.

The metabolic potential and regulation of the microbial metabolism can be comprehensively interpreted using the tools of systems biology and the information obtained from such study can be used for rational design of the metabolic pathways and regulation. By linking the changes in genome (structural genome) for strains exhibiting desired characteristics from directed evolution and/or chemical mutagenesis to the (functional genome) transcriptome, proteome and metabolome of the cells thereby, enabling targeted genome engineering towards the development of highly productive and/or tolerant strain.

Meet the Faculty Candidate Poster Session

AIChE 2014 Atlanta, GA

Blake N. Johnson, Department of Mechanical and Aerospace Engineering, Princeton Institute for the Science and Technology of Materials, Princeton University, Princeton, NJ, USA

The development of a biofabrication method which offers the ability to precisely assemble biologically functional multi-component devices and materials could allow for both revolutionary fundamental biological discoveries using advanced *in vitro* models for tissue regeneration and could ultimately lead to new tissue engineering-based applications for treatment of disease and complex injury.

Along these lines, my postdoctoral research has been dedicated the development of novel three-dimensional (3D) bioprinting approaches for creating anatomically conformal nerve guidance channels capable of regenerating complex peripheral nerve injury. Additionally, the approach was adapted to develop complementary nervous system-on-chip models which remain highly faithful to the composition and structure of peripheral nerve tissue. A light scanning-based model acquisition technique known as 3D structured-light scanning was used to generate high resolution computer models of the sciatic nerve tibial branch which served as the template for the final 3D printed nerve guide. The bioprinting approach also facilitated incorporation of supporting physical and biochemical cues which included longitudinal guidance cues and gradients of neurotrophic factors. The selective regeneration of the sensory and motor nerve pathways bifurcating from mixed nerve was also examined by incorporating internal hydrogel matrix which contained guiding biochemical cues in the different pathways. In order to elucidate the critical factors affecting regeneration of the bifurcating mixed nerve injury, we have developed complementary nervous system-on-chip platforms for the studying the influence of surface topology and biochemical cue spatial distribution on various important factors critical to successful peripheral nerve regeneration, including the support and guidance of axonal growth and the migration of Schwann cells.

My PhD research was focused on the developing ultra-sensitive biosensor technology for real-time monitoring of protein-protein, DNA-DNA, and DNA-RNA interactions in complex backgrounds, such as circulating fluids and environmental settings. Specifically, the research involved the modeling of sensor physics, as well as the fabrication and characterization of piezoelectric-based hybrid mechanical and electrochemical label-free biosensors. Significant biosensing achievements included the label- and preparation-free detection of a class of important circulating biomarkers, known as microRNA, directly in human serum and the detection of toxin-producing cyanobacteria, also known as blue-green algae, in environmental source waters. In addition to biosensing applications, my research also developed means of using sensor-based approaches for modulating the surface binding energy of immobilized molecules and cells using electromechanical and electrochemical energy.

As a new faculty member, I would continue to develop novel lab-on-chip devices for neuroregenerative therapy in the peripheral nervous system (PNS) as well as attempt to expand towards solving problems targeting fundamental biological questions related to central nervous system regeneration and degradation. Using the nervous system-on-chip platforms, I would also plan to examine new approaches for sensor-based stimulation of neuroregenerative response, as such devices provide a controlled setting for examining cellular responses to mechanical, electrical, and drug stimuli.

394061 Packings and Assemblies for Continuous Families of Polyhedra

Abstract:

Nanoparticles and colloids of various polyhedral shapes are synthesized and used as building blocks for self-assembly. It has been shown in simulations that a plethora of complex crystals (including quasicrystals) can form entropically solely due to the anisotropic shape of the particles. At the limit of maximum density, the densest packings, as have been studied in math for centuries, becomes the relevant quantity. However, a general understanding of how changes in shape affect packings and assemblies remains largely unexplored. Here, we study continuous families of polyhedra to demonstrate richness and complexity in behavior as a function of shape. We investigate connections between assemblies and densest packings, discuss the possibility of predicting one from the other and outline general guidelines for experiments.

394727 Co-culture based modular engineering for aromatic and aromatic-derived compound production in Escherichia coli

Abstract:

Traditional metabolic engineering studies focus on engineering target metabolic pathways in one single cell. Despite having gained many successes, this approach suffers from several disadvantages which limit its wide application in the future metabolic engineering research. Here, we developed a novel bacterium *E. coli-E. coli* co-culture approach to completely modularize and optimize the biosynthetic pathway in two separate cells. In particular, a membrane-bound transporter was utilized to facilitate the intermediate mass transfer between the two modules/cells so that the two cells worked as an integrated whole. The co-culture system was shown to produce value-added aromatic and aromatic-derived compounds with high titer and yield. Moreover, the co-culture system can be engineered to efficiently utilize sugar mixtures, which is challenging for the single cell strategy. The established method in this study offers a novel perspective for conducting metabolic engineering studies in the future.

396659 Power and Remediation with Photovoltaics

Abstract:

My research interests are focused on the fundamental understanding and practical application of photovoltaics. The prospective work will focus on the synthesis of new materials and better understanding of the underlying properties that determine the effectiveness of different materials for photovoltaic and photoelectrochemical processes.. I believe that the projects described below are feasible and will provide new insights in their respective fields.

Remediation of carbon dioxide is a problem without a clear practical solution. Sequestration is a temporary answer and may not be providing even the minimal results originally predicted. Electrochemical reduction of carbon dioxide is possible but requires such a significant energy input that it is not practical at this point. Using "free" energy from sunlight to reduce carbon dioxide with a photoelectrochemical system is an appealing solution. P-type semiconductors are able to directly reduce molecules in solution. Accordingly, although the direct reduction of CO_2 requires approximately 1.9V of input, this could be reduced with co-catalysts and the use of strategic electrolytes (acetonitrile, ionic liquids).

Oxynitride photocatalysts are known in the literature. While their synthesis is highly energy and time intensive, they show promise in both photovoltaic and water splitting applications. Synthesizing oxynitride photocatalysts at lower temperatures and shorter times would enhance the ability to seek out new oxynitride photocatalysts. The use of a local exothermic reaction will meet these goals. The use of a localized exothermic reaction will generate the heat needed to synthesize the oxynitride materials. This procedure may also allow for high-throughput synthesis techniques that will speed the search for new and better materials even more.

Hybrid organic-inorganic perovskite materials have recently garnered a great deal of interest due to their low cost and high efficiency in photovoltaic applications. All of the focus, however, has been on Pb-halide based materials. Although cheap, using lead in the material could be problematic. This work would focus on finding other perovskite-type photovoltaic materials by substituting the metal and halide. Experiments would include full physical and spectroscopic (time-resolved ultrafast) characterization of the material as well as application testing for photovoltaic, water splitting, and remediation purposes.

These projects will allow for collaboration in a number of fields, including materials synthesis techniques, energy storage methods (batteries, flow cells, etc.), and reaction kinetic measurements (steady-state and transient).

396658 First Principles-based Multiparadigm, Multiscale Strategies for Simulating Complex Materials Processes

Abstract:

My research focuses on, (1) development of advanced computational algorithms for atomistic modeling, and (2) their applications to current problems in chemistry and chemical engineering. The research lies at the interface of physics, chemistry, chemical engineering, and numerical and algorithms analysis. The aim of my research is to study processes that are of current interest from both small- and large-scale perspectives. The research is motivated by diverse applications including energy technologies, nanotechnology, biomaterials, catalysts, and organic materials where experimental approaches are inaccessible and/or insufficient. In principle, any material is a collection of particles including nuclei, electrons, and phonons. To engineer alternative materials for improved or new functionalities, it is crucial to understand chemistry and energetics of the materials, which govern their mechanical, electronic and thermal properties. I use *ab-initio* quantum mechanics (QM) and density functional theories (DFT) to achieve the goals of my research. Such methods fail, however, when more realistic systems must be studied. For instance, micro-second studies of systems involving large number of particles and high temperature are far beyond the scope of QM and DFT methods. Much of my research is focused on bridging first-principle simulations, normally feasible for small systems, and large-scale simulations necessary for studying systems of practical interest. I have extensive experience in development of reactive force fields, such as ReaxFF as well as invoking reactive force fields to study real-life problems.

I will present an example of such problems by discussing a process-based molecular model of nano-porous silicon carbide (SiC) membrane. A broad class of important materials, such as carbon molecular sieves, SiC, and silicon nitride, are fabricated by temperature-controlled pyrolysis of preceramic polymers. In particular, the fabrication of SiC membranes by pyrolysis of a polymer precursor that contains Si is quite attractive for separation of hydrogen from other gases. It has been quite difficult to extract atomistic-scale information about such SiC membranes since they are amorphous. I will discuss development of a reactive force filed and its application for the thermal decomposition of hydridopolycarbosilane (HPCS) to form SiC nanoporous membranes. I will also highlight my future research plans which I believe are of substantial scientific merit and national interest which fit well with a range of research programs launched by different funding agencies.

Directed Self-Assembly of Sub-10 nm Particles: Role of Driving Forces and Template Geometry in Packing and Ordering

Shafigh Mehraeen[1,2], Mohamed Asbahi[3], Wang Fuke[3], Joel K. W. Yang[2,3], Jianshu Cao[1], Mei Chee Tan[2]

[1]Department of Chemistry, Massachusetts Institute of Technology, Cambridge, Massachusetts 02139, United States
[2]Pillar of Engineering Product Development, Singapore University of Technology and Design, Singapore 138682
[3]Institute of Materials Research and Engineering, Agency for Science, Technology and Research (A*STAR), Singapore 117602

Abstract

By comparing the magnitude of forces, a directed self-assembly mechanism has been suggested[1] previously in which immersion capillary is the only driving force responsible for packing and ordering of nanoparticles, which occur only after the meniscus recedes. However, this mechanism is insufficient to explain vacancies formed by directed self-assembly at low particle concentrations. Utilizing Monte Carlo and Brownian dynamics simulations, we developed a theoretical model, and proposed another mechanism. In our proposed mechanism, the competing driving forces controlling the packing and ordering of sub-10 nm particles are (1) the repulsive component of the pair potential, and (2) the attractive capillary forces, both of which apply at the contact line. The repulsive force arises from the high particle concentration, and the attractive force is caused by the surface tension at the contact line. Our theoretical model also indicates that the major part of packing and ordering of nanoparticles occurs before the meniscus recedes. Furthermore, utilizing our model, we are able to predict the various self-assembly configurations of particles as their size increases. These results lay out the interplay between driving forces during directed self-assembly, offering a better template design now that we know the importance and the dominating driving forces in each regime of particle size.

References

1. Liddle, J. A.; Cui, Y.; Alivisatos, P. *J Vac Sci Technol B* **2004,** 22, (6), 3409-3414.

396344 Microfluidics for Polymer Self-Assembly in Confined Quasi-2D Geometries and Enzyme Design towards Plastic Recycling and Biofuel Production

Abstract:

My postdoctoral work with Prof. David Weitz (Harvard University, 2010-present) involves high throughput microfluidics for performing chemical reactions and biochemical assays in picoliter drops. Taking advantage of material behaviors at liquid-liquid interfaces, enables me to engineer complex materials of hierarchical orders with novel functionalities. For example, I created triggered-release systems using stimuli responsive polymers and enzyme-based functional materials in collaboration with Advanced Energy Consortium and BASF companies. Additionally, in collaboration with New England BioLabs, I genetically engineered proteins and also performed single molecule assays using DNA templates and an *in vitro* transcription-translation system. Moreover, part of my duties as a postdoc included mentoring and supervising the research of undergraduates, graduates and postdocs on a large number of projects.

Thus, I believe that my training and expertise position me advantageously for becoming a faculty member.

As a new faculty member, my first thrust is to take advantage of confinement to engineer novel structures that otherwise are unattainable in bulk. It has been demonstrated that confinement is a powerful tool to break symmetry on the scale of single molecules, allowing polymers to form new phases that are not possible using bulk self-assembly. However, the phase behavior of polymer blends under confinement has not been characterized experimentally in detail due to the lack of available techniques. We recently developed a novel microfluidic method for fabricating monodisperse double emulsions with ultra-thin polymer shells, less than 100 nm in thickness. Due to the quasi-two dimensional physical confinement this method offers exquisite control over the phase behavior of the polymer blend enabling non-equilibrium assembly to create membranes with controlled porosity and permeability. In addition, I will synthesize new polymers and investigate how changing their molecular structure affects self-assembly to obtain quantitative structure-function correlations rules. Applications range from encapsulation and controlled release to membranes for separation and purification.

My second thrust addresses the huge environmental challenges arising from plastic waste and energy production. The current approach for dealing with plastic waste is incineration, which releases large amount of CO_2. In contrast, enzymatic degradation is a possible solution for efficient transformation into high-value fine chemicals without significant greenhouse gas emissions. However, naturally occurring enzymes are inefficient at processing plastic substrates. To increase their efficiency for plastic degradation controlled mutations can be utilized; this approach necessitates generating an enormous library of enzyme mutants. However, developing reliable high-throughput screening platforms to screen for functional variants in libraries with large size is very challenging.

I will use drop-based microfluidics to develop a high-throughput screening platform to overcome these difficulties by using new assays based on *in vitro* transcription and translation (IVTT) system. I will adapt this approach to engineer enzymes and enzyme-polymer conjugates through mutagenesis and study their interaction with natural and non-natural substrates. The results of this study will be applied to the design of enzymes for new applications: such as the biochemical recycling of plastics and production of renewable energy through degradation of cellulose.

Bio-inspired Engineering: Formulation Principles and Design Strategies for First-in-class Therapeutics

Rahul K. Keswani, PhD
Research Fellow, University of Michigan, Ann Arbor

Recent advances in pharmaceutical engineering have entailed the development of multifunctional materials that produce tailored pharmacological responses. Concurrently, scientists have taken advantage of naturally occurring materials to design components for incorporation with synthetic agents to produce super-efficient, targeted and intelligent therapeutics that provide sustained therapy. In this poster, I will provide proof-of-concept and design strategies for *two novel, efficient and biologically relevant therapeutics*.

1. Typically, drugs have unique sites of action and mechanisms that are dependent on formulation and administration routes. The associated chemical interaction of drugs with the human body leads to either pharmacologically relevant metabolites or alternative biochemical forms of the compound. One such drug is the riminophenazine dye, clofazimine that is known to have biocrystallized forms. These have been clinically reported but the mechanism of formation has never been investigated. I will describe the *first-in-class biocrystal-based theranostic agents* that mimic the stabilized chemical drug formulation of clofazimine and that pharmacologically, is both a potent anti-inflammatory medication and a fluorescence and photo-acoustic signaling-based diagnostic imaging agent. Most importantly, the design principles are easily translatable for other pharmaceuticals with varied chemical properties such as basicity, lipophilicity and solubility.
2. Most gene therapeutics rely on using either non-viral vectors for packaging genetic cargo or employ modified viruses for their natural ability to insert genetic cargo into cells. I will provide proof-of-concept and design principles of combining enhanced functional capabilities of naturally sourced biomaterials with the remarkable efficiency of viruses into *first-in-class superlative hybrid gene therapeutics*.

Selected References:

1. **RK Keswani**, K Su & DW Pack, 2014, "Efficient In Vitro Gene Delivery by Hybrid Biopolymer/Virus Nanobiovectors", Journal of Controlled Release, 192, 40-46.
2. **RK Keswani**, IM Pozdol & DW Pack, 2013, "Design of Hybrid Lipid/Retroviral-Like Particle Gene Delivery Vectors", Molecular Pharmaceutics, 10(5), 1725-1735.
3. **RK Keswani**, J Baik, C Hitzman, L Yeomans, N Rodriguez-Hornedo, KA Stringer, GR Rosania, 2014, "A concentrative chloride transport pathway parallels massive accumulation of a weakly basic drug in mouse macrophages", Journal of the American Chemical Society, submitted.

396061 Engineering synthetic microbial cell factories for fuels and green chemicals

Abstract:

Global energy demand, rising petroleum prices and environmental concerns have stimulated increasing efforts to develop more sustainable and cost-effective approaches to produce carbon-neutral fuels and green chemicals. Exploitation of the diverse microbial pathways based on metabolic engineering and synthetic biology frameworks provides a promising solution to synthesis of fuel and pharmaceutical molecules that are previously derived from the petroleum-based resources. As an emerging discipline, synthetic biology is becoming increasingly important to design, construct and optimize metabolic pathways leading to desired overproduction phenotype. We will highlight some of our most recent efforts in applying synthetic biology tools such as combinatorial pathway engineering strategies to fine tune the expression level and balance various sections of a metabolic pathway. Some of the synthetic biology techniques that will be highlighted include the construction of the *ePathBrick* vectors that support the modular assembly of multi-gene metabolic pathways and development of synthetic malonyl-CoA switches that allow dynamic tuning of metabolic flux in *E. coli*. Efficient genome engineering tools such as CRISPR-cas9 directed multiplex combinatorial pathway engineering will also be covered. Applications of such strategies have resulted in the robust production of important compounds that are of particular interest to the pharmaceutical and biofuel industry.

Swelling in polymeric membranes for natural gas applications

Leslie Schulte, Chemical Engineering, Kansas State University
leslieschul@gmail.com

Natural gas production in the US has increased dramatically with the expansion of fracking. These operations require separations to remove CO_2, H_2S and higher hydrocarbons from CH_4 before natural gas can be placed in the pipeline and delivered to the consumer. Polymeric membranes are an attractive technology for this separation because they are modular in nature, have a small footprint and do not require a phase change for separation. Unfortunately, the different components in the raw natural gas mixture swell polymers. This swelling, or plasticization, limits the separating ability of the membrane. Many researchers have worked to develop an understanding of why swelling occurs and there are several solutions including thermal and chemical treatments. These treatments, while suppressing swelling, also decrease the productivity of the membrane. *My research goals are to understand how plasticizing molecules affect membrane materials. I'll use this understanding to develop novel techniques that prevent the decline in selectivity when membranes are exposed to plasticizing environments.*

My PhD project focuses on fabricating and characterizing flat sheet membranes. The primary goal of this research was to design a more chemically and thermally stable polymeric membrane with gas separation properties. This was accomplished by blending commercially available polymers. My work is part of an NSF-funded IGERT program. This program brings together an interdisciplinary group of engineers, agronomists, and sociologists working on the conversion of biomass to fuel.

Thermochemical Conversion of Biomass to Renewable Fuels

Paige A. Case

University of Maine Department of Chemical and Biological Engineering

Doctoral Advisors: William J. DeSisto and M. Clayton Wheeler

I am passionate about research and enjoy working in groups with a diverse set of people. My current research focuses on development of simple, robust processes to deoxygenate and stabilize bio-oils to produce drop-in transportation fuels. I plan to continue research in the field of sustainable fuels and chemicals.

As part of a multidisciplinary, multi-institute research group working on thermochemical conversion strategies for production of fuels and chemicals from woody biomass, I developed methods and techniques that have been adopted by other researchers. Over the course of my research career, I have published four peer-reviewed journal articles and have given several papers and posters at conferences. Each of my poster presentations has been awarded special recognition.

I have extensive experience in non-traditional education and outreach. I assisted with grant writing and was responsible for coordinating UMaine's curriculum for the Renewable Energy minor. I have enjoyed mentoring several undergraduate students, and my work with an REU student last year resulted in a peer-reviewed publication about value-added products from tannins. I have been active in K-12 education outreach as well, and have developed activities for several organizations and represented our department at recruiting events. Perhaps most notably, I have spent three weeks each summer developing design projects and leading teams of high school juniors in laboratory experiments for UMaine's Consider Engineering program.

Publications

"Formate Assisted Pyrolysis of Pine Sawdust for in-situ Oxygen Removal and Stabilization of Bio-oil"
Case, Paige A., Wheeler, M. Clayton, DeSisto, William J. Bioresource Technology. In Press.

"Effect of Residence Time and Hot Gas Filtration on the Physical and Chemical Properties of Pyrolysis Oil"
Case, Paige A., Wheeler, M. Clayton, DeSisto, William J. Energy & Fuels. 28 (6): 3964–3969 (2014)

"Pyrolysis of Pre-treated Tannins Obtained from Radiata Pine Bark"
Case, Paige A., Bizama, Constanza, Segura, Christina, Wheeler, M. Clayton, Berg, Alex, DeSisto, William J. Journal of Analytical and Applied Pyrolysis. 107: 250-255 (2014)

"Liquid hydrocarbon fuels from cellulosic feedstocks via thermal deoxygenation of levulinic acid and formic acid salt mixtures"
Case, Paige A., van Heiningen, Adriaan R.P., Wheeler, M. Clayton. Green Chemistry. 14: 85-89 (2012)

Awards
Third place, TCS2014 Frontier Young Scientist Poster Award
"Proposed mechanisms for pyrolysis of calcium pretreated pine"
Poster presentation at TCS2014, Denver, CO (September 2014)

Second place, TCBiomass2013 Student Poster Challenge
"Formate-Assisted Pyrolysis of Biomass"
Poster presentation at TCBiomass2013, Chicago, IL (September 2013)

395739 Functional Nano-Surfaces for Transformative Materials in Biological & Energy Applications

Abstract:

My research interests are in surface science, thin films, and nanostructures. Surface science has a very broad impact on industrial, environmental, electronic, biotechnological, biomedical, and military applications. The focus is to modify and design synthetic functional nano-surfaces in two different areas: (1) biological, and (2) energy-related applications. The first area of research includes designing biologically active surfaces for tissue engineering, regenerative medicine, drug/nanoparticle delivery, and antimicrobial surfaces. Our aim is to design biocompatible surfaces in which nanoparticles and drugs are embedded in various biodegradable polymers for combinational therapy. Our main goal is to create controlled staggered-release delivery vehicles to embed and deliver various biologically active constituents. The second area of research is to modify surfaces to create smart materials for energy applications. We will study the structures in complex fluids in terms of aggregation, precipitation, and adsorption to modify surfaces against corrosion and fouling in complex fluid transportation. We will employ microfluidics to study these complex systems.

First-Principals Modeling of Methanol Fuel Cells: Kinetics and Catalyst Design

Glen R. Jenness

The decreasing supplies of petroleum resources coupled with an increase in energy demand has created a need for an alternative power supply. Proton-exchange membrane fuel cells (PEMFCs) are an attractive prospect, due to the high energy density and low emissions. Previous PEMFC technology relied on the oxidation of H_2, but due to the difficulties of storing H_2, this route is not economically favorable. By replacing H_2 with an alternative renewable proton source we can alleviate the cost of storing and transporting H_2, as well as making the PEMFC technology more environmentally friendly. Methanol is an attractive proton source due to low storage costs, good thermal stability, high energy density, and is readily produced from biomass resources. While methanol fuel cells offer a high energy density, the efficiency remains low, limiting the use of methanol fuel cells. As a consequence, there has been an increased effort in recent years to find an efficient catalyst for the oxidation of methanol. Previous catalysts employed a platinum nanoparticle with an oxide support, such as γ-Al_2O_3. However, recent experiments indicate that a platinum nanoparticle supported on a carbon-based support, such as carbon nanotubes (CNT), have led to an increase in the efficiency of methanol fuel cells[1].

Several challenges still remain to be resolved in order to increase the efficiency of methanol fuel cells. Firstly, the CNT itself can carry an intrinsic reactivity, with the reactivity being determined by the metallic or semi-conducting nature of the CNT, arising from the chirality. However, the effect of the chirality on the reactivity of the CNT support is still debated[2]. I have shown that the choice of nanoparticle support can result in a charge state being induced on the nanoparticle, which in turn affects the bonding/anti-bonding interactions between the reactants/products and the catalyst, with the resulting charge state being related to the band gap of the support material[3, 4]. As the band gap of the CNT is driven by the chirality, the Pt–CNT interaction will also be driven by the chirality. Additionally, the efficiency of CNT supported catalysts can be increased via the doping of the CNT with nitrogen[5], which in turn changes the electronic properties of the CNT (i.e. band gap), as well as introduces a Lewis acid center into the support. As the mechanism for the oxidation of methanol on the Pt/CNT catalyst is still in question, the effect of introducing a Lewis acid center is currently unknown. I have also shown that the strength of the Lewis acid center can be related to the atomistically resolved conduction band properties of the Lewis acid center, with the reactivity being related to the strength of the Lewis acidity[6]. Finally, it should be mentioned that one factor preventing the wide spread use of methanol fuel cells is the cost of the Pt metal, and it would desirable to use a base metal such as nickel or iron, without sacrificing efficiency.

Through the use of computational tools (i.e. state-of-the-art *ab-initio* methods such as density functional theory), I seek to resolve these issues. In particular, a) how the metallic vs. semi-conducting nature of the carbon nanotube affects the methanol oxidation process; b) how the effect of doping the CNT affects the electronic properties of the catalytic system; c) how the introduction of a Lewis acid center into the support affects the mechanism and kinetics of the methanol oxidation process. The investigation of these effects from an atomistic level will allow us to understand the origin of these interactions, which enable a rationale and systematic improvement of the efficiency of methanol fuel cells.

References

[1] Sharma, S.; Pollet, B. G. Support Materials for PEMFC and DMFC Electrocatalysts—A Review. *J. Power Sources* **2012**, *208*, 96–119.

[2] Denis, P. a.; Iribarne, F. On the Applicability of Cluster Models to Study the Chemical Reactivity of Carbon Nanotubes. *J. Comput. Chem.* **2011**, *32*, 2397–403.

[3] Jenness, G. R.; Schmidt, J. R. Unraveling the Role of Metal–Support Interactions in Heterogeneous Catalysis: Oxygenate Selectivity in Fischer–Tropsch Synthesis. *ACS Catal.* **2013**, 2881–2890.

[4] Hermes, E. D.; Jenness, G. R.; Schmidt, J. Decoupling the Electronic, Geometric and Interfacial Contributions to Support Effects in Heterogeneous Catalysis. *Mol. Simul.* **2014**, 1–11.

[5] Maiyalagan, T.; Viswanathan, B.; Varadaraju, U. Nitrogen Containing Carbon Nanotubes as Supports for Pt—Alternate Anodes for Fuel Cell Applications. *Electrochem. commun.* **2005**, *7*, 905–912.

[6] Jenness, G. R.; Christiansen, M. A.; Caratzoulas, S.; Vlachos, D. G.; Gorte, R. J. Site-Dependent Lewis Acidity of γ-Al_2O_3 and Its Impact on Ethanol Dehydration and Etherification. *J. Phys. Chem. C* **2014**, *118*, 12899–12907.

395320 Mechanism of non-disruptive membrane fusion by amphiphilic, monolayer-protected gold nanoparticles

Abstract:

The cellular membrane is the biological interface responsible for controlling the transport of material into the cell. Thus, it is a critical barrier that must be overcome to successfully deliver desired molecules to the cell interior. Recently, gold nanoparticles (NPs) protected by a binary mixture of purely hydrophobic and anionic end-functionalized alkanethiol ligands were observed to spontaneously penetrate cell membranes via a non-endocytic, non-disruptive mechanism. This direct penetration process is highly promising for applications in drug delivery, bioimaging, and biosensing as NPs can reach the cytosol without being trapped in endosomal compartments or inducing cytosolic leakage. However, no mechanism has yet emerged to explain the experimental observations, limiting the ability to rationally engineer NPs for optimal internalization.

Inspired by these findings, I present here research conducted at MIT with Prof. Alfredo Alexander-Katz in close collaboration with Prof. Darrell Irvine at MIT and Prof. Francesco Stellacci at EPFL. I used a combination of coarse-grained and atomistic simulation methodologies to show that these amphiphilic NPs can spontaneously insert into and fuse with lipid bilayers to obtain configurations resembling transmembrane proteins. The full insertion pathway is remarkably similar to the mechanism of vesicle-vesicle fusion due to the physicochemical similarity between these synthetic and biological systems, allowing us to take cues from biology to understand NP behavior. Predictions from my simulations were confirmed experimentally leading to the development of design guidelines to optimize fusion efficacy. Furthermore, experimental results demonstrated that the same NPs predicted to fuse with bilayers are internalized by the non-endocytic process described above, establishing a link between favorable bilayer interactions and intracellular localization. These findings yield significant physical insight into behavior at this important nano-bio interface and illustrate the power of such simulations in guiding NP design for biomedical applications.

In my future research, I will utilize detailed simulations across multiple time and length scales to generalize this strategy to the study of various membrane-active biological and synthetic systems in close collaboration with experimentalists. In particular, I am interested in understanding mechanisms of post-translational membrane protein insertion, capturing the role of lipid interfaces in membrane binding, and using novel optimization algorithms to tune NP-membrane interactions. My goal is thus to both study fundamental questions in membrane biophysics and develop novel engineered nanomaterials for biomedical applications.

395023 Characterizing the Structure and Function of a Protein Fusion Complex of Photosystem I and Hydrogenase Enzyme

Abstract:

Proteins are large, complex molecules, and their interactions within even the simplest of microorganisms are a complex and intriguing process, requiring a mechanistic understanding at the single molecule level that often defies the capabilities of current science and technology. Gaining insight into proteins and their interactions is crucial to our understanding of the biological world, and could provide technological advancement in alternative energy, drug discovery, and a wealth of other areas. Membrane proteins in particular comprise ~30% of open reading frames, yet represent only a small fraction of known protein structures due to a lack of understanding of the effects of protein-environment interactions on their structure and stability. Here, we are interested in the study of structural effects on the stability and activity of redox proteins. In particular, we are focused on generating protein fusions between the integral membrane protein photosystem I (PSI) of *Synechocystis* sp. PCC 6803 and the membrane-bound hydrogenase (MBH) of *Ralstonia eutropha* H16, a unique *in vitro* complex capable of harnessing solar energy for the production of molecular hydrogen (H_2). We present our study of the detergent-solubilized solution structure of PSI using small-angle neutron scattering (SANS) experiments and all-atom molecular dynamics (AAMD) simulations, and discuss our results in applying a similar approach to studying PSI-MBH protein fusions. By utilizing a combination of computational modeling and experimental characterization of these fusion complexes, we hope to provide a novel method for probing the structure-function relationship that governs redox proteins.

394917 Self-assembly, self-organization, and the design of new materials

Abstract:

My research focuses on understanding and controlling the self-assembly and self-organization of nanoparticle-based materials, which find applications ranging from photonics and plasmonics to industrial formulations such as coatings. A central aim of my work is to *develop quantitative design strategies for controlling dynamic pathways to self-assembly*. At the intersection of physics and engineering, this research utilizes synthetic chemistry, advanced imaging techniques, micromanipulation, and statistical physics. I will discuss the development of a suite of non-equilibrium pathways to self-assembly that are implemented using DNA reaction networks, including self-limiting growth of finite-sized structures, assembly of novel periodic lattices due to specific next-nearest-neighbor interactions, and design of smart materials that can reconfigure themselves in response to an external cue. These pathways can be understood quantitatively using core chemical-engineering concepts, including chemical kinetics and reaction engineering, transport, and thermodynamics.

Engineering human pluripotent stem cell-derived neurovascular systems to model neurological disease and explore therapeutic strategies

Ethan Lippmann, Ph.D.

My research focuses on engineering human pluripotent stem cell (hPSC) fates for applications in developmental and disease biology, drug screening, and regenerative medicine. During my graduate studies, I developed a process for differentiating human pluripotent stem cells (hPSCs) to endothelial cells possessing a blood-brain barrier (BBB) phenotype. In conjunction with renewable sources of pericytes, astrocytes, and neurons, we created a fully-human, scalable, BBB co-culture model that possesses physiological barrier properties and is being evaluated for its ability to predict brain uptake of small molecule drugs. For my postdoctoral work, I have been developing methods to improve the diversity and scope of neural differentiation from hPSCs. We devised a novel platform for generating pure neural stem cells (NSCs) from hPSCs in a fully defined and scalable manner and then used this platform to uncover biological cues that deterministically regulate posterior neural patterning, thereby generating uniform populations of NSCs possessing fixed positional identities corresponding to specific hindbrain and spinal cord domains. We are currently incorporating these defined NSCs onto custom surfaces that facilitate spatiotemporal control over cell behavior in an effort to improve the derivation of neuronal subtypes such as motor neurons. As an independent investigator, I will integrate the technology developed during my graduate and postdoctoral studies to engineer hPSC-derived, disease-specific *in vitro* neurovascular models that will serve as tools for dynamically assessing drug efficacy, identifying novel druggable targets, and studying progressive neurodegeneration in a dish.

Education

Ph.D., Chemical Engineering, 2012, University of Wisconsin, Madison, WI
B.S., Chemical Engineering, 2006, University of Illinois, Urbana, IL

Selected publications

1. **Lippmann ES**, Al-Ahmad A, Azarin SM, Palecek SP, and Shusta EV. A retinoic acid-enhanced, multicellular human blood-brain barrier model derived from stem cell sources. *Scientific Reports* 4:4160 (2014).
2. **Lippmann ES**, Estevez-Silva MC, and Ashton RS. Defined human pluripotent stem cell culture enables highly efficient neuroepithelium derivation independent of small molecule inhibitors. *Stem Cells* 32:1032-1042 (2014).
3. Sha J, **Lippmann ES**, McNulty J, Ma Y, and Ashton RS. Sequential nucleophilic substitutions permit orthogonal click functionalization of multicomponent PEG brushes. *Biomacromolecules* 14:3294-3303 (2013).
4. **Lippmann ES***, Azarin SM*, Kay JE, Nessler RA, Wilson HK, Al-Ahmad A, Palecek SP, and Shusta EV. Derivation of blood-brain barrier endothelial cells from human pluripotent stem cells. *Nature Biotechnology* 30:783-791 (2012). (*Equal contributions)
5. **Lippmann ES**, Weidenfeller C, Svendsen CN, and Shusta EV. Blood-brain barrier modeling with co-cultured neural progenitor cell-derived astrocytes and neurons. *Journal of Neurochemistry* 119:507-520 (2011).

394808 Rational design of multi-input controlled protein switches towards the development of target specific therapeutic enzymes and biosensors

Abstract:

Protein switches are engineered proteins that are composed of an input domain that recognizes and responds to an input signal and an output domain whose function is regulated by the state of the input domain. Engineered protein switches have a number of exemplary properties for sensing and therapeutic applications including a large dynamic range, high specificity for the activating ligand, and a modular architecture that will facilitate fine-tuning of the desired properties. A modular design of protein switches allows the possibility of developing versatile protein switches, in which different input domains can be coupled to the output domain. Thus, protein switches that can be regulated through exogenous or endogenous inputs have a broad range of biotechnological and biomedical applications such as the next generation therapeutic enzymes and biosensor systems. For enhanced target specific controls and improved specificity of the engineered protein switch, it is desirable to establish a means to regulate protein activity with multiple input controls; yet, engineering of such a complex regulatory system is challenging. Here we describe the design of switchable enzymes that requires multi-input signals such as an effector molecule, temperature, pH, redox agent, or electrochemical signal for activation. First, we inserted an enzyme domain into an effector-binding domain such that both domains remained functionally intact. Second, we induced the fusion to behave as a switch through the introduction of a linker designed to respond to different environmental conditions including temperature, pH, or redox potential. We confirmed the multi-input switching behavior *in vitro* and *in vivo*. We also demonstrated that some of these engineered switches could function as protein logic gates. Structural and thermodynamic studies supported the hypothesis that switching resulted from manipulation of allosteric control mechanism via change in conformational equilibria or dynamics of the fusion domains. These results embody a general strategy for the rational design of complex protein switches with multi-level controls.

Engineering polymer materials for the energy-water nexus: new chemistries, in-situ diagnostics, and directed self-assembly

Christopher G. Arges, Ph.D

Institute for Molecular Engineering
University of Chicago
carges@uchicago.edu; http://ime.uchicago.edu/nealey_lab/people/christopher_arges/

Polymer electrolyte membranes (PEMs) are found at the heart of membrane electrode assemblies (MEAs) and they are central to the performance of electrochemical energy conversion and storage devices in polymer electrolyte fuel cells, water electrolyzers, flow batteries, supercapacitors, and reverse electrodialysis. They are also essential to electrodialysis processes used for water desalination and purifying industrial effluent streams. Two of the most stringent requirements for PEMs are exceptional ionic conductivity, which is paramount to device efficiency, and longevity. This poster will cover research strategies for addressing PEM chemical stability and ionic conductivity through i.) the synthesis of new PEM chemistries, ii.) elucidating material performance and degradation in electrochemical systems using *in-situ* diagnostics, and iii.) the adoption of block copolymer directed self-assembly (DSA) for creating ordered, periodic architectures with high degree of functionality. These strategies will be illustrated through three different case studies. The first two parts of the poster stem from my Ph.D. work with Professor Vijay Ramani at the Illinois Institute of Technology, while the latter part highlights my current postdoc work with Professor Nealey at the University of Chicago.

Part 1 of the poster describes the use of a micro, *in-situ* bifurcated optical fluorescence probe for monitoring the reactive oxygen species (ROS) generation in an operating PEM fuel cell. Chemical degradation of the proton exchange membrane observed by fluorescence response correlated with the macroscopic fluoride ion flux rate collected from the water emanating from the fuel cell. The latter technique substantiated the accuracy of the *in-situ* probe for real-time monitoring of PEM degradation. This part of the poster will also describe mitigation of reactive oxygen species degradation through the incorporation of mixed-metal oxide nanoparticles that scavenge the free radicals in the PEM.

In part 2, the chemical and mechanical stability of anion exchange polymer electrolyte membranes (AEMs) for alkaline fuel cells and water electrolyzers and redox flow batteries will be highlighted. Understanding AEM alkaline stability will be demonstrated through a best practices approach that leverages multidimensional NMR spectroscopy, membrane mechanical analysis, examination of AEM electrochemical cell stability using a variety of electroanalytical techniques, and post-mortem NMR analysis of MEAs. Additionally, chemical structure-property relationships for mitigating chemical degradation of AEMs in alkaline electrochemical systems will be given.

The final part of the poster will showcase research for improving PEM conductivity by orienting periodic, ion-conducting domains in diblock copolymer electrolytes perpendicular to electrode surfaces using DSA. Other extensions of this powerful technique, such as creating periodic patterned arrays of electrocatalyst nanoparticles and fabrication of miniaturized electrochemical devices (e.g., micro-supercapacitors) will be disseminated.

My future academic career seeks to compliment my background in electrochemical engineering and polymer chemistry from my Ph.D training with polymer physics and lithography/nanolithography learned in my postdoc work. From my point of view, amalgamating these different skill sets will lead to new classes of functional and resilient materials for the energy-water nexus and efficient micro-electrochemical systems for portable consumer products. The independent, academic program I am aim to build will cut across a multitude of different disciplines spanning the fields of chemical engineering, materials science, physics, chemistry, and mechanical engineering. My vision is to construct a team with students and research professionals with diverse backgrounds in addition to fostering collaborative alliances with researchers worldwide.

394799 Modeling and Simulation of Corona Phase Molecular Recognition Sensors

Abstract:

Corona Phase Molecular Recognition (CoPhMoRe) has been recently introduced as a means of generating synthetic molecular recognition sites onto nanoparticle surfaces. A synthetic heteropolymer is adsorbed and confined to the surface of a nanoparticle, forming a corona phase capable of highly selective molecular recognition due to the conformational imposition of the particle surface on the polymer. In this work, we develop a computationally predictive model for analytes adsorbing onto a polymer corona phase around a single walled carbon nanotube surface using a 2D equation of state that takes into consideration the analye-polymer, analyte-nanoparticle, and polymer-nanoparticle interactions using parameters determined independently from molecular simulation. The curvature dependence of the interaction energies is determined to be a weak contribution to the overall interaction energy. Overall, the resulting model is able to correctly predict 83% of a 374 analyte-polymer library generating experimental fluorescence responses within 20% error of the experimental values. This modeling framework is an important step to the design of such polymers to target a specific analyte.

Universality of Block Copolymer Melts from Metadynamics and Multi-GPU Simulations

I will discuss how we used massively parallel molecular dynamics simulations on multiple GPUs and the well-tempered metadynamics method to locate the ODT of diblock copolymers. Simulation results for five different coarse-grained models of symmetric diblock copolymers demonstrate both a universal equation of state and parameter $(\chi N)_{\mathrm{ODT}}$ at the order-disorder transition, for invariant degrees of polymerization as low as $\overline{N} \sim 200$. In this regime, Fredrickson-Helfand theory must break down. I will also present HOOMD-blue v1.0, a high performance particle simulation framework, which runs in parallel on over 3,000 GPUs.

Nanotherapeutics for Neurological Disorders

Elizabeth Nance, PhD
Anesthesiology and Critical Care Medicine, Center for Nanomedicine
Johns Hopkins University, Baltimore MD

Neurological disorders account for 13% of the global burden of disease. Due to the complex nature of delivering drugs to the central nervous system (CNS), therapeutics for neurological disorders take 35% longer to develop than drugs for other diseases. The design and manipulation of nanoparticle platforms for drug delivery is an increasingly prominent area of interest for medical applications; however, while many drug delivery platforms have shown efficacious pre-clinical results, this has not translated to a significant number of FDA approved, clinically applied nanoparticle systems. This is especially true for neurodevelopmental disorders, such as autism and cerebral palsy, which are life-long chronic disabilities that begin in early childhood. They have no effective cure, and are often underserved by novel drug delivery technologies, which primarily focus on adults. Therefore, there is great opportunity to bring nanotherapeutic approaches to neurodevelopmental disorders, with results that can be translated to adult brain disorders. *My research goals focus on understanding nanoparticle platform interaction, both as biophysical probes and imaging biomarkers, within disease physiology and pathology, with a focus on neurodevelopmental diseases. This knowledge will then be used to better design and implement therapeutic nanoparticle platforms in clinically relevant models of both pediatric and adult CNS diseases.* My laboratory will provide training for engineering students, applied neuroscience students, and clinical fellows. My long-term goal is to transition my laboratory into a Translational Nanomedicine Center, focused on understanding neurological diseases as they relate to therapeutic applications, and engineering nanomedicine platforms that lead to symptom free survival in patients diagnosed with neurological disease. This multidisciplinary work will involve a team of experts in engineering, nanomedicine, chemistry, materials science, pediatrics, neurology, and neuroscience.

I am uniquely suited to establish a laboratory that integrates training in engineering nanoparticle platforms with an understanding of nanoparticle formulation, biophysics, and physiology and progression of neurological disorders. During my doctoral work with Dr. Justin Hanes, I developed nanoparticles for use as biophysical probes in the CNS, a now patented technology, to better understand drug delivery limitations in both normal and diseased brain. Throughout my graduate studies, I led collaborations with faculty and students in the departments of neurosurgery, oncology, pathology, and radiology. My postdoctoral training with Dr. Sujatha Kannan and Dr. Michael Johnston has built my foundation in neuroscience and pediatric neurology, leading to experience in understanding neuroanatomy, neurophysiology and neural development, as well as development of clinically relevant animal models of pediatric CNS diseases. Based on my combined doctoral and postdoctoral work, I was recently awarded the Burroughs Wellcome Fund Career Award at the Scientific Interfaces, an award that provides $500k over 5 years for transitioning from postdoctoral level training to a tenure track faculty position.

Education

Ph.D., 2012, Johns Hopkins University, Chemical and Biomolecular Engineering
B.S., 2006, North Carolina State University, Chemical and Biomolecular Engineering

Selected Publications

1. **Nance E**, Zhang C, Shih T-Y, Xu Q, Schuster B, Hanes J. Brain penetrating nanoparticles improve paclitaxel efficacy in malignant gliomas following local administration. *ACS Nano*, Under review.
2. **Nance E***, Timbie K*, Miller WG, Song J, Louttit C, Klibanov AL, Shih T-Y, Swaminathan G, Tamargo RJ, Woodworth G, Hanes J, Price RJ. Noninvasive delivery of stealth, brain-penetrating nanoparticles across the blood-brain barrier using MRI-guided focused ultrasound. *J Control Rel.*, 189C:123-132 (2014).
3. Kannan R, **Nance E**, Kannan S, Tomalia D. Emerging concepts in dendrimer nanomedicine: translational aspects. J Internal Medicine, doi: 10.1111/joim.12280. (2014).
4. Balakrishnan B, **Nance E**, Johnston M, Kannan S. Nanotechnology in cerebral palsy: current developments. *Int J Nanomedicine* 8:4183-4195 (2013).
5. **Nance E***, Woodworth G*, Sailor K, Shih T-Y, Swaminathan G, Xiang D, Eberhart C, Hanes J. A dense poly(ethylene glycol) coating improves penetration of large polymeric nanoparticles within brain tissue. *Sci. Transl. Med.* 4, 149ra119 (2012). **Selected Cover Image

394783 Micro- and nano-engineered advanced materials for renewable energy, thermal, and electronic applications

Abstract:

Research Interest

My multidisciplinary research interests are in thin films and nanoscale processing for advanced materials – from materials to devices to circuits, blurring the boundaries between chemistry, electronics, physics, materials science and mechanics. With the state-of-the-art micro/nano fabrication techniques and surface engineering, I will explore both the fundamental and theoretical nature of new materials and devices. As an engineer and scientist, I view the area of micro- and nano-engineered advanced materials and next generation electronics as a diverse discipline full of formidable challenges that will engage the research community for decades to come.

Previous and Current Research

Synthesis of graphene films and organic compounds for developing electronic devices: Taking on the challenge to develop inexpensive materials with enhanced functionalities for electronics and in general energy applications, I have focused on low-cost direct synthesis of wafer-scale graphene on a variety of substrates for touch screens, displays, solar cells, and next generation transistors. Graphene, a one-atom-thick sheet of sp^2-bonded carbon atoms in a honeycomb structure, has attracted significant interest due to its outstanding mechanical, thermal, and electrical properties. I have synthesized graphene films on a variety of substrates including silicon, quartz, and sapphire, through a single-step thermal annealing process of a Cu/Ni bilayer pre-deposited on the substrate in a low pressure environment through chemical vapor deposition (CVD) technique. My method eliminated the need for solution process transfer of graphene to the target substrate, which usually induces defects in its structure. In addition, doping of graphene and organic thin films is important, as it gives rise to p-type and n-type materials, and it adjusts the work function of the film that is necessary to control charge injection and collection in devices such as solar cells, displays, and printed and flexible electronic devices. I developed a low temperature method to fabricate p- and n-type graphene and graphene p-n junctions by modifying the interface between graphene and its dielectric substrate with covalently bonded self-assembled monolayers (SAMs). In a multidisciplinary collaboration work, I showed that surface modifiers based on polymers containing simple aliphatic amine groups and redox-active solution processable metal-organics substantially adjust the work function of conductors, including metals, transparent conductive metal oxides, conducting polymers, and graphene. These methods can pave the way to simplified manufacturing of low-cost and large-area organic electronic technologies.

CVD polymers-surface modification: Conventional polymer deposition (for e.g. ink-jet printing, spin-coating or dip coating) uses solvents that may have adverse effects during the deposition process—such as degradation of the underlying layers or alteration of mechanical/electrical properties—and that cannot be applied for coating of surfaces with high aspect ratio patterns. Chemical vapor deposition (CVD) as a polymer thin film deposition technique offers a versatile platform for fabrication of a wide range of polymer thin films preserving all the functionalities.[5] The solvent-free, conformal vapor phase deposition technique enables the integration of polymer thin films or nanostructures into micro- and nanodevices for improved performance. In my work at MIT, I synthesized thin films of polymers via a substrate-independent and all-dry-initiated chemical vapor deposition (iCVD) technique. Surfaces coated with fluoropolymers exhibited high advancing water contact angle (WCA) and very low WCA hysteresis. These coatings reduce the strength of ice and clathrate hydrate adhesion to the substrates up to five times compared to bare substrates and are very desirable for flow assurance strategies aimed at reducing the occurrence of blockages in oil and gas pipelines. In addition, anti-reflective coatings are significantly enhancing the contrast of today's high resolution displays, but their usage in touch applications suffers from the issue of fingerprint visibility on the surface of such low-reflective substrates. In general, the appearance of fingerprints and oil residues are of major concern for numerous industries involved with maintaining display clarity for applications where handling occurs. To this end, I am working on developing anti-fingerprint and anti-reflective coating using bio-degradable CVD polymers that exhibit high water/oil contact angle. In order to use such coatings in a touch application it must withstand hundreds of thousands of finger touches and rubs. Additional advantages of the CVD technique for polymer deposition is that the mechanical properties of the synthesized polymers against wear and erosion can be improved via in-situ grafting, multilayer composite polymers, and post-synthesis treatment of polymers. Films with mechanical properties comparable to glass were obtained using this technique.

Proposed Research

Development of functional polymers:

Low-cost and scalable CVD polymers for high efficiency heat recovery: Thermoelectric (TE) materials have recently attracted significant interest for converting thermal energy to electric energy in the automobile industry, for recapturing energy currently lost from the hot effluent of power plant smokestacks, and for harvesting the heat currently generated but not used by photovoltaic cells. By understanding how heat and energy flow through materials, energy conversion mechanisms and processes can be integrated into functional devices. The performance of the thermoelectric material is evaluated by the dimensionless figure of merit (ZT) that is a function of electrical conductivity, thermal conductivity, absolute temperature, and Seebeck coefficient—the ratio of the induced voltage over the temperature gradient across the thermoelectric device. A general principle for engineering good TE materials is to separate electrical from thermal conductivity—in particular, to make them dependent on structures at different length scales. Today's thermoelectric materials of choice for low temperature energy conversion are bismuth antimony telluride alloys. The constituting atomic elements have a low natural abundance and are therefore expensive. Additionally, low natural abundance goes hand in hand with some level of toxicity for the environment. Therefore, in today's scenario it is unthinkable to enable a widespread use of thermoelectric installations for waste heat recovery. In that context, widely available, although less efficient, organic thermoelectric materials have become the subjects of interest to the scientific community. Polymers as TE materials have been developed via electro spinning, inkjet printing, and electrochemical deposition. However, these methods are not conformal, use solvents, and require additional treatments to dope the polymer before integrating it into thermoelectric generators. I plan to develop polymer TE materials through oxidative CVD where the conformal, all-dry, substrate-independent, and low-cost characteristics of conducting polymers can be leveraged to make scalable thermoelectric generators. The extent of doping and electrical conductivity of such polymers can be tuned throughout the synthesis process by changing growth conditions. In addition, a state-of-the-art mechanism for grafting such polymers to surfaces provides the robust adhesion required for industrial application. Understanding thermal and electrical properties of such polymers is inherently interdisciplinary, involving backgrounds in synthetic chemistry, polymer physics, condensed matter physics, and materials characterization. Through better understanding of transport in amorphous and disordered materials, it is possible to tailor the thermal and electrical conductivity in synthesized polymers and integrate them into low-cost, scalable, and environmentally friendly organic thermoelectric generators.

Mechanical properties of CVD polymers: CVD polymers fully retain the rich chemistry of organic monomers enabling technology for surface modification of membranes, microfluidic structures, and biomedical devices. These CVD polymeric layers possess mechanical flexibility and conformal coverage attributes and can be integrated into device fabrication schemes for optoelectronics and sensors. While promises have been demonstrated in terms of photovoltaics, organic displays, and sensors, the reliability of these devices ultimately depends on their ability to retain their functionality during operation. In terms of organic electronics, this will present a challenge when considering their need for enhanced mechanical properties as well as the potential for fracture and loss of adhesion at critical interfaces. In this work, I will continue to build on my expertise in studying mechanical properties of ultrathin polymers, synthesized using CVD technique. I will investigate the use of functionalization layers containing hydroxyl, vinyl, and fluorine groups to graft these polymers to the desired substrate to enhance their adhesion. I will also seek to improve the mechanical properties, for example elastic modulus, hardness, and toughness through synthesis of stacked ultrathin multilayer structures, cross-linking of various monomers, and post-synthesis treatment of polymers. The onset cracking strain will be related to film thickness, film composition as determined by XPS/FTIR, mechanical properties will be examined through nanoindentation/nanoscratching, and film defects determined from SEM and optical microscopy techniques. XPS and spectroscopic FTIR will be critical in understanding the early stages of growth of these films and linking this to their formation and mechanical properties. I will also make multilayer stretchable and conductive CVD polymer films and integrate them in flexible and organic electronic/photonic devices and examine their reliability under enhanced flexing and stretching conditions.

394345 Experimental Evolution of Heterologous Pathways in Microbes

Abstract:

Our ability to engineer microbial metabolic pathways is improving rapidly, enabling the efficient and sustainable biosyntheses of both bulk and fine chemicals. Tools from synthetic biology have supported these advances and allowed the use of diverse natural and engineered enzymes. However, novel enzymes frequently require significant optimization to function efficiently in a new production host. More precisely, optimization is a complex co-adaptation process, as the host must accommodate its new pathway and the pathway adjust to its new host. Many metabolic engineers seek to use rational and semi-rational strategies to predictably design highly-productive pathways in heterologous hosts. Unfortunately, these approaches are limited by our inability to account for the complexities involved in genome-scale engineering of microbial metabolism. When knowledge is limiting, evolutionary approaches can be powerful tools, both for forward optimization and as learning opportunities to investigate potential design strategies.

In my graduate research, I used RNA switches to perform directed enzyme evolution in live cells and systems biology to optimize an engineered microbe. My postdoctoral research has applied strategies from metabolic engineering to study natural microbial evolution, looking at how microbes evolve after horizontal gene transfer.

My independent research laboratory will investigate strategies for optimizing the interactions between a host and a novel enzyme or pathway using three related approaches. In my first project, I will extend my postdoctoral research to construct and optimize synthetic mobile genetic elements for bioremediation, effectively performing metabolic engineering in indigenous microbes. Next, I will study adaptation after horizontal gene transfer in a pathogen, looking at how previously-benign bacteria can evolve to become virulent. My final proposal describes a novel selection strategy using synthetic competitive ecosystems in microfluidic droplets. These ecosystems will be used to optimize complex pathways, such as those required to produce antibiotics, while studying the evolution of cooperation and competition in microbial communities. In combination, these projects will offer a novel perspective on important questions in health and ecology while also producing forward design strategies for synthetic biology and metabolic engineering.

388584 Rationally Designed and Deterministically Engineered Electrodes for High-Performance Energy Storage Applications

Abstract:

Over the past two decades, the widespread miniaturization of portable electronic devices has led to the need for similarly scaled energy storage systems. However, the energy storage technology has not improved at the same rapid pace as the portable electronics technology, resulting in limited operation duration of the device on a single charge. The vast majority of the energy storage systems utilized in the portable electronic devices consist of batteries. Considering the currently available battery chemistries, a drastic improvement in the battery performance in terms of capacity is not expected in the near future. Unless a completely new battery chemistry with an extremely high energy density is developed, the only way to achieve longer operation times is through increasing the mass of the active material present within the battery, which would be incompatible with the current shrinking trend of the portable electronics. However, improvements can be achieved in the rate of the energy transfer, i.e. power density, which these batteries can deliver. A battery that can be charged very rapidly would still be quite desirable even if it possesses a relatively low energy density.

The batteries with rapid charge and discharge capabilities can only be realized via simultaneous minimization of certain inherent resistances encountered during the transport of the ionic and electronic species. There are both intrinsic and extrinsic factors affecting the transportation process of ions and electrons. Intrinsic factors, such as diffusivity and conductivity of the battery components, depend on the choice of materials to be utilized within the system. Extrinsic factors, including the surface area, diffusion and conduction path lengths, on the other hand, primarily rely on the size, dimension, and geometry of the system components.

The focus of our research has been on the realization of rationally designed and deterministically engineered three-dimensional electrodes with precisely controlled dimensions that address the aforementioned critical extrinsic factors determining the ultimate performance of the battery. MEMS technologies and electrochemical techniques have been utilized for the fabrication as well as characterization of these scalable, well-ordered, and high-surface-area electrodes with rapid charge and discharge capabilities. Utilization of micro- and nanofabrication techniques also enables the realization of electrodes for high-performance power sources for a wide variety of applications, ranging from autonomous microsystems to macroscale portable electronics. The resultant electrodes demonstrate an improved performance in not only batteries, but also other energy storage systems, such as capacitors and supercapacitors.

394233 Design of Genetically Encodable Protein Switches and Potent Metallobiocatalysts

Abstract:

Precise molecular recognition between proteins and small molecules is critical to cellular function. A rigorous test of our understanding of the physiochemical principles that govern molecular interaction phenomena is the creation of de novo protein-ligand interfaces with programmable properties. I recently developed a generalizable computational/experimental hybrid method for designing proteins that bind small molecules and used it to create selective binders for the steroid digoxigenin. This work and other exciting advances provide a platform for creating novel ligand binding proteins, but fundamental gaps in our understanding of structure-function relationships in proteins need to be addressed in order to robustly and routinely design molecular biodevices that sense the presence of the bound small molecule and respond by performing a specified function. The goal of my interdisciplinary research group will be to uncover the design principles necessary to develop next-generation systems-level protein technologies in which molecular recognition events are coupled with downstream biological or chemical functionalities. Design advancements will be motivated by and provide insight into the biochemical/biophysical principles that promote allosteric changes in proteins and/or enable the coordinated integration of multiple complex chemical events. In particular, we will focus on two major avenues: designing novel molecular switches for regulating metabolic pathways, and developing new metallobiocatalysts for challenging small molecule activation reactions related to green chemistry and energy. Through our design efforts, we hope not only to create novel functional biomolecules but also to uncover design principles that shed light on the underlying structure-function relationships and regulatory mechanisms in natural systems.

394100 Understanding Molecular Mechanisms of Rare Events: Applications in Nanofabrication in Semiconductor Industry, Novel Drug Delivery Systems and Fuel Cells

Abstract:

At molecular level, rare events are events which occur at timescales of a few nanoseconds to several days. These events underlie many observed physical phenomena, such as phase transitions, conformational changes in proteins and polymers, chemical and biological reactions and self-assembly. Rare events often have a rate limiting step, which lies at the maxima of the free energy along the reaction coordinate pathways. Since traditional molecular simulation methods are ineffective in sampling such events, special techniques have been developed to sample them, such as forward flux sampling, variants of string method and transition path sampling. Deciphering molecular mechanisms behind rare events is fundamental to our understanding of physical systems of interest and useful for many technological innovations. In the last four decades, semiconductor industry has seen a relentless scaling down of transistor dimensions in accordance with the Moore's law. The state of the art semiconductor technology is operating at critical dimensions of tens of nanometers. The traditional scaling of transistor dimensions, often called Dennard scaling, is no longer applicable at present length scales, and hence focus has moved to innovations in novel materials and new fabrication methodologies. One approach is to use a high K dielectric material in thin film transistors. Hafnium oxide based inorganic dielectric materials have many problems, such as poor interface properties with silicon causing development of interface traps, which affect reliability of devices. Molecular modeling methods can be utilized to design optimal polymeric substances which are capable of forming a good interface layer with silicon. High-K polymeric materials, such as poly(vinylidenefluride-co-trifluoroethylene) and poly vinyl phenol with varying molecular weight, grafting density and co-polymeric mixtures would be materials of interest for this study. Long relaxation times of these polymers as well as their conformational changes under electric field can be studied as rare events in molecular simulations. Another innovation which has improved performance of semiconductor devices is the introduction of strain in silicon to increase mobility of charge carriers. One way to induce strain is by first depositing thin films and then either curing them or by providing a free surface to allow them to expand or contract. Molecular modeling of structural modifications in thin films via rare event sampling techniques can provide useful insights on the role of chemistry of the film and its interaction with the underlying silicon crystal, which will help in better design of this process. Hydrogels are cross-linked network of hydrophilic polymers. Hydrogels can undergo gel-sol transition upon changes in environmental conditions, such as pH, temperature or light. Using environmental sensitive hydrogels as site specific drug delivery systems is a promising idea. The important parameters of investigation would be the sensitivity of sol-gel transition to stimuli, and the kinetics and thermodynamics of the transition. These investigations can most effectively be undertaken by simulating the sol-gel transition under varying environmental conditions using molecular simulations. Transport properties of membranes are functions of the structure of their polymer network. For fuel cell applications, it is desirable to have high proton conductivity with reduced water cross-over. Molecular modeling of proton transport as a function of the structure of the membranes would help develop more selective membranes with high permeability for protons for fuel cell applications. In my post-doctoral work at Princeton University, I applied forward flux sampling in molecular dynamics simulation of an all-atom water model to study kinetics of evaporation and condensation of water in hydrophobic confinement at ambient conditions, thereby simulating time-scales in order of few hours. Using transition path sampling, I determined transition state ensemble associated with liquid to vapor transition of water in hydrophobic confinement. In my doctoral work at Columbia University, I studied conformational changes in proteins upon adsorption to hydrophobic surfaces. In order to eliminate kinetic traps while sampling different conformational states of proteins, I did Monte Carlo simulation at different temperatures and surface hydrophobicity to study different regions of the phase space. The entire phase space is generated by using methods such as simulated annealing, weighted histograms and parallel tempering. In order to study kinetics of a system which can transition between two states, the phase space can be first sampled by using equilibrium sampling techniques such as parallel tempering, simulated annealing and weighted histograms. Once different minima in the phase space have been identified, rare event sampling can be used to study the kinetics of transition from one minimum to the other.

397143 The Flame Synthesis of Carbonaceous and Heterogeneous Nanomaterials, and Using Soot for the Fischer- Tropsch Synthesis High Value Chemicals

Abstract:

We have a limited understanding of highly desired industrial chemical processes that naturally occur in flames – one reason that soot production remains a global issue. Still, empirical knowledge has driven the development of flame synthesis techniques for widely used products like carbon black, TiO_2, and even specialized carbon nanomaterials. Although the flame synthesis of metal oxide nanoparticles has seen success because this process follows a decent on a Gibbs free energy landscape, carbon particle formation is dominated by chemical kinetics and remains largely unknown to this day. Thus, further understanding it will allow for greater control over carbon nanoparticle formation under flame conditions, which can scale better than other synthesis techniques like the sol-gel method. One of my research plans is the theory-aided design of flame synthesis environments to harvest self-assembled nanoparticles and quantum dots. Within flames, the self-assembly process of carbon particles is a direct result of fuel-rich combustion, the characteristics of which can be tuned by optimizing reagent mixtures and reactor conditions for a desired property. These characteristics include, but are not limited to: size, band gap, surface functional groups, and heterogeneity. Furthermore, reagent fuels may be doped with varying concentrations of metals and/or transition metals to create a new generation of catalysts, sensors, and components of targeted therapeutics, smart materials, and even quantum computers.

Another project of interest relies on the AFM, TEM and FTIR evidence that hydrocarbon chains polymerize on nascent soot surfaces. My earlier computational work strongly suggests that persistent free radicals and combustion byproducts play a role in this process. I plan to exploit this process to produce hydrocarbon compounds that may be used as chemical feedstock or even liquid fuels. A longer term goal is to realize a solar-thermal based energy conversion technology that operates much like reactors employing the Fischer-Tropsch process.

B.A. Chemistry, Virginia Tech

B.S. Engineering Science and Mechanics, Virginia Tech

Ph.D. Mechanical Engineering, University of Souther California (Hai Wang)

Princeton Roving Postdoctoral Fellowship:

High Temperature Gasdynamics Lab, Stanford (Ronald Hanson)

Chemical Engineering Department, MIT (William Green)

397162 Metabolic reprogramming of hypoxic cancer cells

Abstract:

Cancer cells survive under nutrient-limited conditions to surpass normal cell growth. For example, oxygen-limited conditions significantly influence cancer cell metabolism; hypoxia aggravates the Warburg effect by elevating the conversion from glucose to lactate significantly in cancer cells because one oncogene, hypoxia-inducible factor (Hif) increases the expression of metabolic genes in glycolysis. Consequently, it inhibits pyruvate dehydrogenase via activating pyruvate dehydrogenase kinase to decrease oxidative phosphorylation in the TCA cycle. These are key events in tumorigenesis and proliferating cells. Thus, the study of cancer metabolism under hypoxia provides valuable insights in understanding cancer progression and potentially discovering novel therapeutic targets. However, the simple measurement of extracellular metabolites or gene expression levels can provide only limited information on cancer metabolism because most *in vivo* metabolic fluxes consist of reversible reactions and metabolic pathways are also connected to each other. Here, we quantified the metabolic fluxes covering glycolysis, the pentose phosphate pathway and the TCA cycle using ^{13}C-metabolic flux analysis with human cancer cell lines. Furthermore, we validated the observed metabolic behaviors using shRNA knock-down and over-expression cell lines for specific genes. In this course of study, we identified the significant role of hypoxia in regulating cancer cell growth through molecular regulation and metabolic reprogramming by the Hif oncogene.

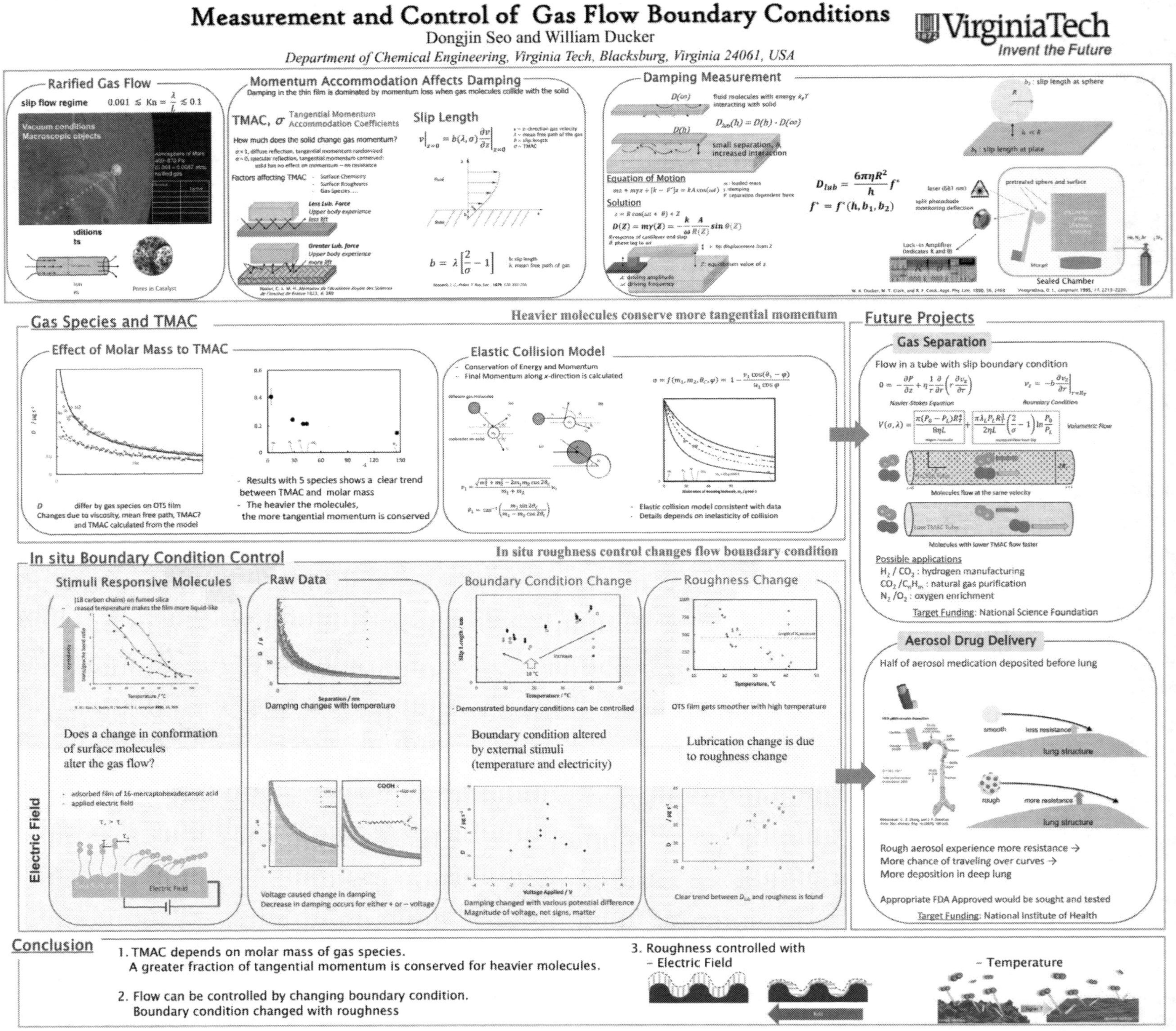
Measurement and Control of Gas Flow Boundary Conditions
Dongjin Seo and William Ducker
Department of Chemical Engineering, Virginia Tech, Blacksburg, Virginia 24061, USA
VirginiaTech
Invent the Future
Rarified Gas Flow
slip flow regime 0.001 ≲ Kn = λ/L ≲ 0.1
Vacuum conditions
Macroscopic objects
Pores in Catalyst
Momentum Accommodation Affects Damping
Damping in the thin film is dominated by momentum loss when gas molecules collide with the solid
TMAC, σ Tangential Momentum Accommodation Coefficients
How much does the solid change gas momentum?
Factors affecting TMAC
Surface Chemistry
Surface Roughness
Gas Species ....
Less Lub. Force
Upper body experience less lift
Greater Lub. force
Upper body experience more lift
Slip Length
Damping Measurement
small separation, h, increased interaction
Equation of Motion
Solution
Sealed Chamber
Gas Species and TMAC
Heavier molecules conserve more tangential momentum
Effect of Molar Mass to TMAC
differ by gas species on OTS film
Changes due to viscosity, mean free path, TMAC?
and TMAC calculated from the model
- Results with 5 species shows a clear trend between TMAC and molar mass
- The heavier the molecules, the more tangential momentum is conserved
Elastic Collision Model
- Conservation of Energy and Momentum
- Final Momentum along x-direction is calculated
- Elastic collision model consistent with data
- Details depends on inelasticity of collision
Future Projects
Gas Separation
Flow in a tube with slip boundary condition
Navier-Stokes Equation
Boundary Condition
Volumetric Flow
Molecules flow at the same velocity
Molecules with lower TMAC flow faster
Possible applications
H2 / CO2 : hydrogen manufacturing
CO2 / CnHm : natural gas purification
N2 / O2 : oxygen enrichment
Target Funding: National Science Foundation
In situ Boundary Condition Control
In situ roughness control changes flow boundary condition
Stimuli Responsive Molecules
Does a change in conformation of surface molecules alter the gas flow?
- adsorbed film of 16-mercaptohexadecanoic acid
- applied electric field
Electric Field
Raw Data
Damping changes with temperature
Voltage caused change in damping
Decrease in damping occurs for either + or – voltage
Boundary Condition Change
- Demonstrated boundary conditions can be controlled
Boundary condition altered by external stimuli (temperature and electricity)
Damping changed with various potential difference
Magnitude of voltage, not signs, matter
Roughness Change
OTS film gets smoother with high temperature
Lubrication change is due to roughness change
Clear trend between D and roughness is found
Aerosol Drug Delivery
Half of aerosol medication deposited before lung
smooth
less resistance
lung structure
rough
more resistance
lung structure
Rough aerosol experience more resistance →
More chance of traveling over curves →
More deposition in deep lung
Appropriate FDA Approved would be sought and tested
Target Funding: National Institute of Health
Conclusion
1. TMAC depends on molar mass of gas species.
A greater fraction of tangential momentum is conserved for heavier molecules.
2. Flow can be controlled by changing boundary condition.
Boundary condition changed with roughness
3. Roughness controlled with
- Electric Field
- Temperature

397401 Advanced catalyst design strategies: Integrating machine learning and probability theory with quantum mechanics and descriptor-based analysis

Abstract:

Chemical processes driven by heterogeneous catalysts are central to a broad range of existing and emerging renewable energy technologies. The ability to engineer novel catalysts which are more active, selective, or inexpensive for a given chemical transformation has the potential to enable or radically improve processes for sustainable energy conversion. A particularly exciting development in the search for novel catalysts is the ability to design catalysts *in silico* using computational methods. A number of promising strategies for computer-aided catalyst design have been developed, including descriptor-based optimizations, genetic algorithms, and high-throughput computational screening. The previous success of *in silico* catalyst design demonstrates the potential power of catalyst prediction through fundamental understanding.

The problem of catalyst design can be considered in terms of two essential challenges: establishing a connection between atomic-scale models and observed reaction rates, and systematically predicting novel catalysts and processes based on these models. *My work will utilize machine learning and probability theory to more efficiently utilize existing theoretical and experimental results to provide fundamental understanding and robust predictions.*

Understanding the connection between atomic-scale models and observed rates is particularly difficult due to the ubiquitous nature of uncertainty in heterogeneous catalysis. The potential importance of transient species, rare events, and sparse active sites makes experimental investigations difficult; moreover, the complex electronic structure of surfaces and vast possibilities of active site geometries presents a challenge for even the most advanced computational methods. Using probability theory as an extended logic, *my research will seek to account for uncertainty in order to develop quantitative connections between atomic-scale models and catalytic activity.*

The development of predictive methods for catalyst discovery relies on access to large amounts of data and innovative approaches to data analysis. The current state of computational catalysis research places the field on the brink of a data-rich era, where research is limited by data analysis rather than data generation. The impact of available information can be maximized by the use of improved methods for data analysis and catalyst optimization. Descriptor-based analyses provide one route towards insight-based catalyst discovery. *My research will build on descriptor methods by using more advanced statistical methods and generalizing the design space beyond surface properties.*

397838 Catalytic Conversion of Transportation Fuel and Cellulosic Biomass Derivatives over Zeolite Catalysts

Abstract:

The current energy policies around the world have encouraged the development of fuel conversion processes and renewable energy sources. With this point of view, my Ph.D. research area can be classified into three regions; transportation fuel conversion, cellulosic biomass conversion into transportation fuel and value-added chemicals, and study of zeolite catalyst materials. For the first subject, various fuels, such as gasoline, diesel, and jet fuel including logistic fuel were investigated for production of hydrogen and light hydrocarbons in attempt to improve fuel efficiency and flexibility in remote area. For example, catalytic cracking of JP-8, logistic fuel in NATO (North Atlantic Treaty Organization) and the U.S. military, was conducted for the purpose of overcoming logical and functional limitation of existing devices such as diesel generator or unmanned aerial vehicle (UAV) in battle field by using liquefied petroleum gas (LPG) as power source of solid oxide fuel cell from JP-8 itself.

As aforementioned, the conversion of sustainable biomass for the production of energy and high value chemicals has been proposed as an enabling technology and received a great deal of interest for decades. Particularly, cellulosic biomass has been recently drawn attention as attractive feedstock due to abundance, low cost, and non-competing with food chain. Hence utilization of cellulosic biomass feedstock was allocated as another major research project. A variety of bio-feedstock, such as furfural, levulinic acid, gamma valerolactone, and pentanediol, was investigated over various zeolite catalysts, and finally, we were able to demonstrate a novel method for the production of LPG-like fuel, benzene, toluene, and xylenes from cellulosic biomass feedstock.

As catalytic fuel process and biomass conversion studies have been carried out using zeolite materials, synthesis of zeolite and zeolite based catalysts are essential part of my doctorial work. Several attempts, for instance varying synthesis temperature, time, organic template, etc., to modify crystal size, pore structure, and acid property of zeolite materials have been conducted. Furthermore, microwave-assisted heating method was compared with conventional hydrothermal heating method for the endeavor to development an efficient method for zeolite material synthesis.

A Route Towards Sustainability Through Multifunctional Hybrid Materials and Interfaces

B. Reeja Jayan, Karen K. Gleason[1], and Arumugam Manthiram[2]
(1) Chemical Engineering, Massachusetts Institute of Technology, Cambridge, MA
(2) Materials Science and Engineering, The University of Texas at Austin, Austin, TX

Areas of Expertise:

Solution and vapor phase synthesis of thin film materials (ceramic, polymer)
Materials discovery and advanced manufacturing
Hybrid (hard-soft matter) interfaces
Polymer photovoltaics
Lithium ion batteries

Abstract for Future Research:

The hallmarks of sustainable development include easy access to clean energy, clean air, potable water, adequate nutrition, and quality medical care. In order to solve the broad challenges we face in achieving a sustainable future, new **"hybrid" material systems** must be envisioned. These materials consist **of soft-hard matter interfaces** between organic and inorganic constituents at varying length scales (molecular to macro scale). Combining these materials incorporates the desirable properties of their organic (e.g. polymers, biomolecules) and inorganic (e.g. metals, ceramics) constituents; further, successfully modifying their compositions and interfaces can give rise to completely new functionalities that did not exist in the parent materials. Examples include organosilicon networks that can selectively conduct ions and block electrons in batteries,[1] and metal-organic frameworks that provide high surface area both to enhance light absorption in photovoltaic devices and to increase the attached fraction of binding biomolecules (proteins) in biosensors.

To enable applications in sustainability, I propose a research program with the following three goals: (1) **design** new multifunctional hybrid materials incorporating organic, inorganic and biological constituents at the molecular or nanometer scale; (2) **process** these materials into nanoscale thin films that form precisely engineered, covalently grafted interfaces with diverse materials like inorganic surfaces, polymers, and biomolecules (proteins); and (3) **investigate** how the morphology and chemistry of these materials and their interfaces influence mass and charge transport. The resulting knowledge will significantly advance technology development in energy generation, energy storage, and biosensing.

Postdoctoral Projects:

1. *Chemical Vapor Deposition of Conformal Nanoscale Polymeric Electrolytes for Three Dimensional Battery Architectures*

2. *Conducting Passivation of Silicon Solar Cells by initiated Chemical Vapor Deposition (iCVD) Polymerization*

Under supervision of **Professor Karen Gleason**, MIT

PhD Dissertation: *Development of Efficient, Stable Organic-inorganic Hybrid Solar Cells*

Under supervision of **Professor Arumugam Manthiram**, (UT-Austin)

Summary of Research Experience:

My **doctoral research** involved chemical and morphological engineering of the interfaces between thin films of organic (polymers) and inorganic (ceramic) materials. I probed the fundamentals of charge transfer phenomena at these interfaces using a combination of microscopic, spectroscopic, and device characterization techniques. My research led to improved hybrid polymer solar cells and lithium ion batteries. I also developed diverse thin film synthesis techniques for inorganic materials. In particular, my solution based process to crystallize ceramic thin films at low temperature using microwave radiation is extremely attractive for several functional material systems with applications in electronics, energy, and medicine.

My **postdoctoral research** addresses the key technological limitation to the development of three-dimensional (3D) battery architectures; the availability of a conformal thin film electrolyte. 3D batteries represent a novel approach to miniaturizing on-board electrochemical energy storage for many applications including sensing, actuation, communications, and medical implants. Using Initiated Chemical Vapor Deposition (iCVD) polymerization, I synthesized conformal nanoscale cyclic siloxane polymer films, which can be made to conduct ions at levels comparable to that of the most explored polymeric electrolyte, poly(ethylene oxide). This demonstration of a low-cost, thin film approach for creating highly stable siloxane-based compositions introduces a new direction to the field of polymer electrolytes, which has been quiescent for several years.

My research work has resulted in 15 publications in leading journals and 4 related patent applications have been filed.

Teaching, Leadership, and Outreach:

Over the years, I have demonstrated strong commitment to teaching, mentoring, and service to the scientific community. At UT-Austin and MIT, I assisted in teaching courses on Materials Engineering, Thermodynamics, Solid State Electronics, and Solar Conversion Devices. Additionally, I serve as reviewer for several scientific journals and as grant proposal review panelist for the *National Science Foundation (NSF).* I am also an Associate Technical Editor of *MRS Bulletin*, a leading publication in materials research. My leadership abilities are demonstrated by my roles as a founding member of the *MIT Think Tank* and as advisory committee member for the *MIT Innovation Initiative*.

Awards and Honors:

- Materials Science and Engineering Nominee for the Michael Granoff Outstanding Graduate Student Award, UT-Austin (2012)
- Graduate Poster Award, Nano Night, UT-Austin (2012)
- Cockrell School of Engineering Student Leadership Award, UT-Austin (2011)
- Travel Grant Award, Graduate Nano Portfolio Program, UT-Austin (2011)
- Doctoral Fellowship, American Association of University Women (AAUW) (2010)
- Mike Hogg Endowed Doctoral Fellowship, UT-Austin (2010)

- Professional Development Award, UT-Austin (2008, 2011)
- H.H. The Maharaja of Cochin Endowment Prize, University of Kerala, India (2002)

Selected Publications: ** denotes shared first-authorship*

N. Chen*, **B. Reeja-Jayan***, J. Lau, P. Moni, A. Liu, B. Dunn, and K. K. Gleason, "Conformal Cyclic Siloxane Electrolyte for Three-Dimensional Battery Architectures", *Nature Materials (under review).*

B. Reeja-Jayan, T. Adachi, R. J. Ono, D. A. Vanden Bout, C. W. Bielawski, and A. Manthiram, "Effect of Interfacial Dipoles on Charge Traps in Organic-Inorganic Hybrid Solar Cells", (2013), *Journal of Materials Chemistry A,* **1**, 3258-3262.

B. Reeja-Jayan*, K. L. Harrison*, K. Yang, Chih-Liang Wang, A. Yilmaz, and A. Manthiram, "Microwave-assisted Low Temperature Thin Film Growth in Solution", *(2012), Scientific Reports,* **2,** 1003.

B. Reeja-Jayan and A. Manthiram, "Understanding the Improved Stability of Hybrid Polymer Solar Cells Fabricated with Copper Electrodes", (2011), *ACS Applied Materials & Interfaces,* **3**, 1492-1501.

Jun Liu, **B. Reeja-Jayan**, and A. Manthiram, "Conductive Surface Modification with Aluminum of High Capacity Layered Li[Li0.2Mn0.54Ni0.13Co0.13]O2 Cathodes", (2010), *Journal of Physical Chemistry C*, **114**, 9528-9533.

Reviews:

B. Reeja-Jayan, P. Kovacik, R. Yang, H. Sojoudi , A. Ugur, D. H. Kim, C. D. Petruczok, X. Wang, A. Liu, and K. K. Gleason, (2014), "A Route towards Sustainability through Engineered Polymeric Interfaces", *Advanced Materials Interfaces*, **1**, 1400117.

398428 Instructing cell bioactivity using programmable DNA-peptide hybrid materials

Abstract:

DNA is one of the most promising molecules for the construction of nanomaterials, with programmable control over both nanostructure and dynamics. Despite this potential, DNA has not been used to a great extent in supramolecular biomaterials for regenerative medicine. My research uses novel DNA-peptide hybrids to create scaffolds for instructing cell behavior in ways not possible with current materials. I will present work on two main projects. The first uses a DNA nanotube as a scaffold for neural stem cell (NSC) attachment and directed differentiation. Modifying one of the strands that comprises the nanotube with a cell adhesion peptide results in a multivalent display of the biological signal on the nanotube surface. These nanotubes are potent scaffolds for NSC attachment compared with unmodified nanotubes, and selectively promote their differentiation to neurons. DNA allows for rational control over the structure obtained, and if nanotube formation is prevented, the NSCs no longer differentiation into neurons. My second project uses DNA as a handle to endow a biologically inert polymeric surface with bioactivity. DNA handles on an alginate coating allow for immobilization of peptides that promote fibroblast adhesion and spreading. By dynamically removing the peptide signal (using several DNA-specific triggers), it is possible to programmably reverse this bioactivity. DNA can also serve as a scaffold to present multiple signals with nanoscale control, promoting a distance-dependent synergy between two bioactive peptides or a peptide and a growth factor. These two projects highlight the power of DNA in both controlling nanostructure and dynamic properties of a cell scaffold, and are the foundation for a new area of DNA-biomolecule scaffolds for regenerative medicine.

399195 Micro and nanostructured interfaces for characterization of disease states in biological systems

Abstract:

The adhesion and aggregation of cells at solid-liquid interfaces plays a key role in biological function in many systems and is often mediated by both surface and environmental cues. For example, the formation of platelet aggregates onto the vascular sub-endothelium during *in vivo* clot formation is mediated by both the surface area available for adhesion and by shear stress, while the development of bacterial communities into biofilms is influenced by physicochemical surface properties and by interspecies interactions. Using innovative approaches in micro/nanofabrication and surface chemistry, we seek to engineer synthetic interfaces that recreate relevant *in vivo* situations in order to better understand how cell-surface and cell-cell interactions lead to disease states. These interfaces can then be used in clinical applications aimed at improving the diagnosis, monitoring, and treatment of disease. Specific applications of our engineered surfaces will include (1) measurement of platelet function for dosing anti-platelet agents, (2) measurement of bacterial interactions that occur during biofilm formation in the cystic fibrosis airway, and (3) high-avidity isolation of bacterial pathogens for rapid detection of bloodstream infections.

399283 Investigating the Interphase Regions Between Inorganic Nanomaterials and Thermosetting Polymer Backbones

Abstract:

The interfaces between a polymer and a metallic or ceramic nanoparticle in a polymer nanocomposite are generally not well understood. Polymer nanocomposites are typically approached on a case-by-case basis using chemistry based functionalization techniques on the particle surface to promote coupling to a tailored bulk polymer matrix phase. A great deal of difficulty lies in practically characterizing the resulting interphase region between the nanomaterial and bulk matrix, which has been demonstrated to play a significant role in the ultimate nanocomposite transient and/or dynamic mechanical response. During nanocomposite synthesis/fabrication, the interactions between the metallic/ceramic and polymeric phases are governed by a number of factors including mass transport from localized concentration gradients and/or differences in surface energy, and the kinetics of the reaction in a thermosetting type system. Practically, issues such as particulate flocculation and agglomeration, along with nanomaterial distribution within the composite, substantially influences bulk structure property relationships. The goal of this research is multifold and includes 1) development of novel characterization approaches to successfully probe the morphological evolution and final state of interphase regions in polymer nanocomposites and 2) fundamentally understanding the correlation between transport, kinetics, and thermodynamics that impacts the formation of the interphase region between polymer and nanoparticle. Ultimately, the goal is to develop predictive control of nanocomposite response as a function of nanomaterial integration approach for broad ranges of materials, particularly with respect to bulk composite dynamic or recoverable transient mechanical response.

399285 Interdisciplinary Engineering Fundamentals: Undergraduate Teaching and Learning

Abstract:

Interfacing nanotechnology and biology, my graduate research aims at understanding the physical and chemical mechanisms associated with the blood-borne spread of cancer which results in over 90% of cancer-related fatalities. Specifically, these studies further elucidate the roles of receptor-ligand interactions during breast cancer dissemination and investigate the efficacy of alternative engineered in-vitro platforms for studying the adhesion, migration and invasion of human cancers. The inherent breadth of this research in addition to my College Teaching and Learning Certificate Program at the University of Kentucky have inspired me to pursue a tenure-track faculty position in engineering education subsequent to my PhD defense in January 2015. Principally, I am passionate about undergraduate teaching and learning. My professional goals are to foster the personal and professional growth of future engineers by providing a solid foundation of introductory chemical engineering and mathematics as well as enhance communication and critical thinking skills. My specific teaching interests lie in early engineering-based coursework on process principles and balances, computational tools and fluid dynamics. In addition to these core courses, I am also interested in teaching elective courses on cell biology for engineers, cross-disciplinary journal clubs and micro/nanofabrication techniques. The biochemically-focused courses and undergraduate research I completed while earning a B.S. in Chemical Engineering at Rose-Hulman Institute of Technology as well as my participation on both the NSF Integrative Graduate Educational Research Traineeship (IGERT) and the NIH Cancer Nanotechnology Training Center (CNTC) grants at UK have shaped my firm views on the need for continued collaborations amongst STEM field majors and clinicians to ensure translational success from bench top experimentation to bedside patient care. In my experiences mentoring undergraduate researchers, I have seen that these opportunities and training can provide hands-on applications to supplement and enrich course materials as well as probe interest in designing innovative solutions to real-world clinically-relevant issues. Accordingly, it is also my desire to lead and participate in outreach and service efforts geared towards inspiring students, especially young women, to obtain degrees in STEM fields with a special emphasis on the importance of interdisciplinary collaborations.

399286 Multiscale Systems Analyses Advancing Drug Discovery Efforts and Therapeutic Development

Abstract:

Systems biology and biomedical control research has guided the rigorous assessment of regulation at the molecular, cellular, and tissue levels. Systems biology approaches — with its combination of computational, experimental and observational inquiry — has extended from model systems to human physiology and now to areas of pharmacology and human health. The Young Lab will combine expertise in systems biology, protein and cellular engineering, membrane biology, and synthetic biology to (a) identify and differentiate cell stress response pathways following local perturbations and (b) advance a mechanistic-understanding of spatiotemporal signaling that influences cell/tissue function. Our research pursuits are motivated by current challenges in drug discovery, therapeutic development, biopharmaceutical production, and limitations of human disease models.

The Young Lab will focus initially on three research objectives:

(1) Characterize diverse, global consequences of system disturbances leading to cell stress in a model eukaryote,

(2) Exploit these host-response pathways to facilitate biopharmaceutical production in yeast, *S. cerevisiae*, and

(3) Investigate the regulation of system perturbations resulting in the altered behavior of multicellular human tissue as a consequence of inflammation, including therapeutic intervention to facilitate drug discovery efforts in disease progression.

Due to the intrinsic complexity of biological systems, integration of experimental and computational approaches is crucial to transform the way we study cell physiology and behavior. To achieve a global understanding of how cellular processes and molecular interactions perform under external and internal perturbations, molecular measurements from many hierarchical cellular levels require integration within appropriate computational frameworks as recently described [1,2]. In the Young Lab, we will foster a research environment that is committed to excellence, encourages innovation and diversity, and advances scientific pursuits in drug discovery and development through multidisciplinary efforts.

[1] C. L. Young and A. S. Robinson, *Protein Folding and Secretion: Mechanistic Insights Advancing Recombinant Protein Production in S. cerevisiae*, **Current Opinion in Biotechnology**, 2014 Jul 14;30C:168-177.

[2] M. R. Ebrahimkhani, C. L. Young, D. A. Lauffenburger, L. G. Griffith, J. T. Borenstein *Approaches to In Vitro Tissue Regeneration with Application for Human Disease Modeling and Drug Development*. **Drug Discovery Today**, 2014 Jun;19(6):754-762.

Infections and diseases of the respiratory system contribute to morbidity and mortality in both the developing and the developed world. The therapeutic advantages of inhalable therapeutic treatment of respiratory diseases include the application of the therapeutic to the site of infection which enables a high localized concentration of therapeutic with low systemic exposure. However, the successful development of inhalable therapeutics requires the ability to successfully formulate the therapeutic into an inhalable form, the successful deposition of the material into specific respiratory regions, and controlled interactions of the material with extracellular and cellular environments. The solution to these challenges requires the expertise and interaction of multiple disciplines including: material science, engineering, molecular biology, immunology, pharmaceutics, and physiology.

My graduate and post-doctoral experiences in chemical engineering and pharmaceutics have uniquely prepared me to tackle these challenges. For drug delivery vehicle design, I have synthesized biodegradable polymers and used these to encapsulate antibiotics and proteins into particles on both the micron and nano scales. I have also used oligosaccharides from the surface of non-typeable *Haemophilus influenzae* and polyethylene glycol to modify the surfaces of polystyrene nanospheres to provide cell targeting and mucosal avoidance, respectively. The biodistribution, extracellular and cellular interactions of particles were evaluated utilizing *in vivo* murine models and *in vitro* models of the upper respiratory tract, mucosal secretions, and alveolar macrophages.

To supplement the *in vitro* experience, translational studies into particle aerosolization and aggregation of dry powder formulations have provided an understanding of deposition and distribution of inhalable therapeutic formulations. Combining this knowledge with my experience with human respiratory physiological measurements provides me with the tools required to address the challenge of designing inhalable therapeutic delivery systems for treatments of respiratory diseases.

374087 Dynamic Modeling of Colloidal Assembly

Abstract:

Understanding concentrated colloidal dynamics in the presence of different pairwise interactions and external fields provides a basis to predict the temporal evolution of colloidal microstructures in diverse phenomena including suspension rheology and colloidal assembly. However, a microscopic theory of concentrated colloidal dynamics does not yet exist that rigorously includes both statistical mechanical (configuration dependent free energy changes) and fluid mechanical (configuration dependent multi-body hydrodynamic interactions) contributions. This work shows the implementation of a novel approach to model colloidal assembly by means of analyzing the temporal evolution of microstructures, that can be observed in experiments and simulation, with a small number of order parameters that describe the state of the system. This methodology is used to analyze the structure evolution of isotropic and anisotropic colloidal building blocks. Knowledge of these dynamic models enable the fundamental understanding of assembly systems, as well as the optimization, design and control of assembly systems to produce low-defect colloidal crystals.

387166 Next Step in Computational Materials Design: Surface Properties of Metal Oxides

Abstract:

In the last decade, metal-oxide (MOs) based materials have found many important applications in batteries, as water-splitting catalysts or as cathodes of solid oxide fuel cells and other uses. Computationally, the bulk properties (i.e. thermodynamic stability, optical response) of these materials have already been qualitatively well captured. However, in real world situations, the processes of interest are often localized to a single site on a surface, reduced to nanostructures, or the bulk is self is altered under the contact with environment (i.e. Pourbaix Diagrams of metal-oxides). While for simple metals some of these challenges have been already addressed, the understanding of metal oxides in reduced dimensions is still in its infancy and offers great promise for future research. Especially important topics are a) types of possible surface terminations, b) relative energetic stability as function of environment, c) modification of properties due to localization and d) role of different active sites such as terraces vs. edge-sites for the process of interest.

My current post-doctoral research with Prof. Jens Nørskov (Stanford U.) and Prof. Alex Bell (UC Berkeley) already resulted in some progress in above topics. Very recently, in a series of papers, I was able to establish importance of types of surface structural motifs for water splitting activity of earth-abundant transition metal-oxides (TMOs: Fe, Co, Ni) [1,2]. As a results, I have identified highly active Fe sites in very promising mixed (Ni,Fe)OOH catalysts [2]. Currently, I continue to work on more general aspects of surface structure of TMOs. In my PhD work, I have also significantly contributed to progress of highly accurate quantum Monte Carlo methods [3,4], which serve as larger framework for widely used density functional theory and quantum chemistry.

My future research interests focus on significantly enhancing the understanding of metal-oxides in reduced dimensions. My knowledge and experience of three classes of state-of-the-art methods: periodical DFT methods, cluster-based quantum chemistry approaches and quantum Monte Carlo methods enables me to obtain high accuracy and cross-validated results, which are necessary to tackle the challenges in materials discovery. My objective is to establish 1) multi-purpose computational electronic structure laboratory including development of more accurate but scalable methods 2) classify and characterize the types of important structural motifs occurring in reduced dimensions such as in surfaces and 3) use these findings to discover new materials with enhanced surface related properties.

Selected Publications (22 total, 10 first author, h-index=10):

[1] Bajdich, M.; García-Mota, M.; Vojvodic, A.; Nørskov, J. K.; Bell, A. T., J. Am. Chem. Soc. 2013, 135, 13521–13530.

[2] Friebel, D.; Louie, M.; Bajdich, M; Sanwald, K.E.; Cheng, Mu-Jeng; Cai, Y.; Sokaras,D; Alonso-Mori, R; Weng, Tsu-Chien; Davis, R; Wise, A. M.; Bargar, J.; Bell, A. T.; Lercher, J. A.; Nørskov, J. K.; Nilsson, A.: Identification of highly active Fe sites in (Ni,Fe)OOH for electrocatalytic water splitting (submitted to Nat. Chem.)

[3] Bajdich, M.; Tiago, M. L.; Hood, R. Q.; Kent, P. R. C.; Reboredo, F. A., *Phys. Rev. Lett.* 2010, *104*, 193001.

[4] Bajdich, M.; Mitas, L. *Acta Phys. Slovaca* 2009, *59*, 81–168.

371890 Controlling Interactions at Nanoscale: A Versatile Tool for Assembling Multifunctional Future Materials

Abstract:

The interactions between nanoscale objects play an essential role in various industrial and biomedical processes. These interactions determine the material's physical properties such as tensile strength, flexibility, electrical and thermal conductivity of the resulting macroscopic material. My research focuses on the fundamental understanding of the interactions at the nanoscale in order to develop new multifunctional hybrid materials. During my doctoral studies at Technische Universität, Berlin in the group of Prof. Gerhard Findenegg, I investigated the effects of modulating the interaction between soft and hard nanomaterials. The influence of surface energy and curvature on the self-assembly of nonionic surfactant at silica nanoparticles[i] and in the nanopores of SBA-15 silica material[ii],[iii] were studied. In addition we also investigated the interaction of biologically relevant entities such as globular proteins with silica nanoparticles[iv]. We revealed how the adsorbed protein induces a pH-dependent completely reversible bridging aggregation of silica particles[v]. Earlier in 2014, I visited the group of Prof. Katsumi Kaneko at Shinshu University, Japan as a JSPS postdoctoral fellow for short period of time. My research in Japan was focused on the development of single walled carbon nanotube (SWCNT) and silica hybrid materials which can further be used as precursors for energy efficient electrodes. In this work we used the approach of dispersing the tubes in liquid and its subsequent drying to form highly porous hybrid aerogel[vi].

Since 2012, I am working in the group of Prof. Orlin Velev at North Carolina State University and Triangle Materials Research and Engineering Center (ΔMRSEC) deals with the programmed assembly of isotropic and anisotropic particles into chains as actuators for soft micro-robotic applications. We demonstrated that heteroaggregating particle pairs can be assembled into rigid permanent chains using dielectrophoresis (DEP) as a structure directing tool[vii]. We proposed a combinatorial based approach for predicting the chain length distribution and established the rules for permanent chain assembly[viii]. Recently, we developed a novel method for assembling superparamagnetic nanoparticles covered by lipid shell into magnetically responsive ultraflexible chains[ix]. Initial application of external magnetic field aligns the particles into linear chains, after which the particles are bound by the soft attractive potential induced by the surface condensed lipid. We demonstrated that the surface wetting lipid governs the formation of nanocapillary bridges and hence enables the ultraflexible chain assembly. We also established that the chains can be assembled in the form of self-repairing networks and a number of other hierarchal structures. In the future, this research field will be extended into the design of new functional materials with unusual optical, electrical and heat transport properties.

In future I plan to continue my research in the field of nanoscience with the aim to fabricate new materials with unusual physical and chemical properties. Based on my previous research experience on self- and directed assembly, I plan to establish my independent research team which would focus on both fundamental and applied aspects nanoscience and nanotechnology. In particular I am very interested in extending my research for assembling carbon based energy harvesting materials. In addition, I would also like to continue my research efforts towards developing a better understanding between nonentities and biological matter.

---

[i] B. Bharti; J. Meissner; U. Gasser; G. H. Findenegg *Soft Matter*, **2012**, *8*, 6573

[ii] B. Bharti; M. Xue; J. Meissner; V. Cristiglio; G. H. Findenegg *J. Am. Chem. Soc.*, **2012**, *134*, 14756

[iii] B. Bharti, T. Shin; B. Deme; G. H. Findenegg *in-preparation*

[iv] B. Bharti; J. Meissner; S. H. L. Klapp; G. H. Findenegg *Soft Matter*, **2014**, *10*, 718

[v] B. Bharti; J. Meissner; G. H. Findenegg *Langmuir*, **2011**, *27*, 9823

[vi] B. Bharti; R. Kukobat; D. Minami; K. Kaneko *in-preparation*

[vii] B. Bharti; G. H. Findenegg; O. D. Velev *Sci. Rep.* (*Nature*), **2012**, *2*,1004

[viii] B. Bharti; G. H. Findenegg; O. D. Velev *accepted,* **2014**

[ix] B. Bharti; A.-L. Fameau; M. Rubinstein; O. D. Velev *submitted*

378500 Design of Injectable Hydrogels for Regenerative Medicine

Abstract:

Cell transplantation via direct cell injection at the target site is a minimally invasive strategy for treating various injuries and degenerative diseases. However, cell viability is typically only 5% in these procedures, and the therapeutic success critically hinges on the survival and subsequent maintenance of the transplanted cells. To protect these cells, we designed an injectable hydrogel that undergoes two different physical crosslinking mechanisms. The first crosslinking step occurs *ex vivo* through peptide-based molecular recognition to encapsulate cells within a physical hydrogel, while the second crosslinking step occurs *in situ* through a thermal phase transition to form a reinforcing network. At room temperature, the physical hydrogel is shear-thinning and self-healing to facilitate gentle cell encapsulation and transplantation by syringe injection. At body temperature, the hydrogel forms a dual-network resulting in a 10-fold increase in shear modulus and significantly reduced erosion rates and prolonged retention time. Human adipose-derived stromal cells (hASCs) injected through a 28-gauge syringe needle were significantly protected from disruptive mechanical forces when encapsulated within the hydrogel compared to medium alone. *In vivo* subcutaneous injection of hASCs in a murine model demonstrated significantly improved cell retention when delivered in a double-network hydrogel compared to a single-network gel or medium alone. These cell-delivery materials are being further investigated for the ability to support functional tissue regeneration at sites of ischemia. In one strategy, the hydrogel mechanical properties are being tuned to enhance the secretion of pro-angiogenic growth factors by hASCs. In a second strategy, the hASCs are being co-cultured with induced pluripotent stem cell-derived endothelial cells (hiPSC-ECs) to promote tissue regeneration in a murine hindlimb ischemia model of peripheral arterial disease.

390757 Multi-Objective Modeling, Simulation, and Optimization for Economically and Environmentally Conscious Decision Makings

Abstract:

Chemical facilities, where large amounts of chemicals and fuels are processed, manufactured, and housed, are at high risk to originate air emission events, including intensive flaring and toxic gas release caused by various uncertainties such as equipment failure, false operation, natural disaster, or terrorist attack. Chemical plant concentrated regions may suffer localized and transient air pollution events that violate national ambient air quality standards (NAAQS). Flaring emissions, especially intensive start-up flaring emissions from chemical plants, have potentially significant impacts on local air quality.Furthermore, geographic allocation of chemical plants significantly affects industrial business sustainability as well as regional environmental sustainability. According to site selection rules, the air quality impact to surrounding communities of a newly constructed chemical plant must be taken into account. Therefore, it is very important to apply modeling, simulation, and optimization to provide economically and environmentally conscious decision makings and valuable quantitative engineering supports for multiple stake holders, including government environmental agency, regional chemical plants, and local communities.

In this paper, the general introduction will be given for our previously developed multi-systematic mathematical simulation and optimization models based on the methodology of multi-objective optimization. The first model will conduct the multi-plant start-up emission evaluation and help to provide control strategy. For any air quality violation is predicted to an AQCR, a multi-objective scheduling problem will be generated and solved to optimize the start-up sequence and start-up beginning time for all chemical plants. The scheduling model minimizes the overall air quality impacts to all of the AQCRs as well as minimize the total start-up time mismatch of all plants, subject to the principles of atmospheric pollutant dispersion. The second model aims to detect the possible emission sources (chemical plants) and identify the abnormal emission profile (emission source location, starting time, time duration, dynamic emission rate and total emission amount) from an accidental emission source responsible for an observed emission event based on an available air-quality-monitoring network so as to support diagnostic and prognostic decisions in a timely and effective manner.It provides valuable information for investigations of accidents and root-cause analysis for emission events; meanwhile, it helps evaluate the regional air-quality impact caused by such emission events as well. The third model evaluates the potential air-quality impacts from candidate sites of a new chemical plant in order to optimize the final site determination by minimizing its air-quality impacts. This study can not only determine the potential air-quality impact for allocation of multiple chemical plants with respect to regional statistical meteorological conditions, but also identify an optimal site for each new chemical plant with the minimal environment impact to surrounding communities. Case studies are employed to demonstrate the efficacy of the developed methodologies.

Engineered Natural Biomaterials for Understanding the Interplay Between Cells and Their Environment

Steven R. Caliari[1]
[1]Department of Bioengineering, University of Pennsylvania

Dynamic new biomaterial tools are necessary to address modern medicine challenges ranging from the repair and replacement of tissues and organs to the treatment of diseases such as fibrosis and cancer. In particular, the ability to engineer biomaterial milieus that re-create the heterogeneous microenvironments of physiological and pathological conditions is critically important. My PhD work in Prof. Brendan Harley's lab at the University of Illinois focused on the design of spatially-patterned collagen biomaterials to engineer orthopedic interfaces and direct cell behavior using approaches with broad applicability to a variety of regenerative medicine challenges. During my postdoctoral work in Prof. Jason Burdick's lab at the University of Pennsylvania I have developed a suite of mechanically dynamic hyaluronic acid hydrogel substrates useful for a range of fundamental mechanobiology studies, focusing in particular on understanding hepatic stellate cell behavior during liver fibrosis. As a principal investigator I will leverage my experience in chemical engineering, biomaterial fabrication and characterization, and mechanobiology to design regenerative templates for tissues including skin and skeletal muscle. On a more fundamental level, I will use these materials to investigate how cells interrogate their environment in disease states such as liver fibrosis and cancer.

Selected Publications (out of 15 first author published/submitted/in preparation):

7) **S.R. Caliari**, M. Perepelyuk, R.G. Wells, J.A. Burdick, 'Mechanically dynamic hydrogels for probing hepatic stellate cell behavior in fibrosis,' In preparation, 2014.

6) **S.R. Caliari**, B.A.C. Harley, 'Structural and biochemical modification of a collagen scaffold to selectively enhance MSC tenogenic, chondrogenic, and osteogenic differentiation,' *Adv Healthc Mater*, In press, 2014.

5) **S.R. Caliari**, L.C. Mozdzen, O. Armitage, M.L. Oyen, B.A.C. Harley, 'Periodically-perforated core-shell collagen biomaterials balance cell infiltration, bioactivity, and mechanical properties,' *J Biomed Mater Res A*, 102(4):917-27, 2014.

4) **S.R. Caliari**, B.A.C. Harley, 'Composite growth factor supplementation strategies to enhance tenocyte bioactivity in aligned collagen-GAG scaffolds,' *Tissue Eng Part A*, 19(9-10):1100-12, 2013.

3) **S.R. Caliari**, M.A. Ramirez, B.A.C. Harley, 'The development of collagen-GAG scaffold-membrane composites for tendon tissue engineering,' *Biomaterials*, 32(34):8990-8, 2011.

2) **S.R. Caliari**, B.A.C. Harley, 'The effect of anisotropic collagen-GAG scaffolds and growth factor supplementation on tendon cell recruitment, alignment, and metabolic activity,' *Biomaterials*, 32(23):5530-40, 2011.

1) T.A. Martin*, **S.R. Caliari***, P.D. Williford, B.A. Harley, R.C. Bailey, 'The generation of biomolecular patterns in highly porous collagen-GAG scaffolds using direct photolithography,' *Biomaterials*, 32(16):3949-57, 2011. *co-first authors

367093 Molecular Simulation As a Tool for Materials Synthesis, Catalysis, and Biocatalysis

Abstract:

The poster displays an overview of my PhD and postdoctoral computational research. Unifying themes include computational chemistry, molecular simulation, rare events, transition states, and structure-function predictions. Currently, at the Pacific Northwest National Laboratory my postdoctoral research is focused on materials and catalysis. In one project the mechanism of metal organic framework (MOF) synthesis is being investigated with computational chemistry focused on establishing detailed reaction schemes from metal ions and organic linkers to MOF building blocks. In a second project the kinetics and dynamics of water on metal oxide surfaces is studied. Using density functional theory (DFT) and *ab initio* molecular dynamics, energy barriers for individual adsorption, reaction, diffusion, and desorption events were obtained, and used in kinetic models, explaining experimental observables. My dissertation work at Iowa State University, as part of the Center for Biorenewable Chemicals, focused on enzymes. A bioinformatics approach was taken to find target enzymes with desired activities among the many ones existing in nature. To that end, a database that classified enzymes by structure was constructed. With experimental collaboration, a statistical approach identified enzymes with novel activity. Molecular dynamics and mixed quantum mechanics/molecular mechanics (QM/MM) simulations were done to understand in detail the mechanism of enzymatic catalysis; revealing transition state conformation, the roles of active site residues, and energy landscape.

380570 Multiscale Simulations of Polymer Dynamics, Polymerization Kinetics, and Polymer Blend Morphology

Abstract:

The availability of fast and large mulitnode computer clusters have made computer simulations of polymers routine. Development of parallel algorithms and availability of accelerators such as graphical processing units (GPUs) further enhanced our capability in performing these simulations. Despite these advances, simulating polymeric systems still remains a challenge because of the multiscale nature of polymers that spans multiple length and time scales. To illustrate I am presenting a vignette of my latest works where I used different coarse-graining techniques and atomistic simulations to simulate: (1) the dynamics of unentangled polymer melts confined in a highly adsorbing cylindrical walls, (2) the polymerization kinetics and chain characterization of bottle-brush macromolecule synthesized through the macromonomer approach, and (3) the morphology and characterization of organic photovoltaic blends with emphasis on the structure of the donor/acceptor interface. All simulation results were qualitatively, and in some cases quantitatively, in agreement with neutron scattering experiments.

For example in the simulations of polymer melts confined in pores, we used a polymer bead-spring model to simulate anodize aluminum oxide (AAO) infiltrated with polydimethyl-siloxane (PDMS) and found excellent agreement in the dynamic and static properties of the chains between neutron spin echo (NSE) and small angle neutron scattering (SANS) experiments, respectively. In the second example, we used a bead-spring model of a macromonomer and simulated ring opening metathesis polymerization (ROMP) by sequentially adding macromoners to form bottle-brush macromolecules. The structure of the bottle-brush was compared to SANS experiments. In the third example, a hindered-rotation model of a polymer was used to model poly(3-hexylthiophene) (P3HT). The simulation results of the P3HT/PCBM blend, in the presence of a substrate, shows qualitative agreement with neutron reflectivity (NR) experiments where a vertical composition profile indicating PCBM rich regions in the vicinity of the substrate and air interfaces. Furthermore, the orientation of the thiophene rings in the donor/acceptor interface was also investigated using atomistic molecular dynamics simulations.

My objective is to perform simulations that would tackle difficult areas in polymer physics such as charged polymers, liquid crystalline polymers, polymers at interfaces and polymer crystallization. In these future works, I will actively search for areas of collaboration with the goal providing insights that will provide a better understanding of these polymeric systems. The above examples illustrate my strategy to meet the challenge.

Graphene Electrode-Based ZnO Nanowire Hybrid Solar Cells

Sehoon Chang, Silvija Gradečak

Department of Materials Science and Engineering, Massachusetts Institute of Technology, Cambridge, Massachusetts, 02139;

Abstract

Growth of semiconducting nanostructures on graphene would open up opportunities for the development of flexible optoelectronic devices, but challenges remain in preserving the structural and electrical properties of graphene during this process. We demonstrate growth of highly uniform and well-aligned ZnO nanowire arrays on graphene by modifying the graphene surface with conductive polymer interlayers. Based on this structure, we demonstrate graphene cathode-based hybrid solar cells using two different photoactive materials – the colloidal quantum dots (CQDs), lead sulfide (PbS) QDs and the conjugated polymer, P3HT – with AM 1.5G power conversion efficiencies of 4.2% and 0.5%, respectively, approaching the performance of ITO-based devices with similar architectures. Our method preserves beneficial properties of graphene and demonstrates that it can serve as a viable replacement for rigid ITO in various flexible solar cell device configurations.

The continuous growth of global energy consumption has led to increasing demand in the development of next generation devices for energy storage and conversion. Proton exchange membrane fuel cells have attracted attention due to their potential as clean, eco-friendly power sources. The ability of a fuel cell to deliver power depends crucially on the rate of proton transport through a hydrated polymer electrolyte membrane (PEM) that is placed between the electrodes. Many people have studied proton transport properties in PEMs, but few have successfully linked properties with the structure and morphology of PEMs, especially in the hydrated states. This is largely due to the technical challenges associated with probing morphology of PEMs in the hydrated states.

My current research as a postdoctoral fellow at Lawrence Berkeley National Laboratory is focused on the fundamental investigation of morphology-conductivity relationships in model PEMs based on ion-containing block copolymers, comprising ionic and non-ionic blocks. Specifically, I study the water uptake and proton conductivity of a model block copolymer, polystyrenesulfonate-*block*-polyethylene-*block*-polystyrenesulfonate (S-SES), as a function of water activity in both humid air and liquid water. I observed a discontinuous increase in both water uptake and proton conductivity of S-SES membranes when nearly saturated humid air is replaced with liquid water. Using humidity-controlled synchrotron small angle X-ray scattering and cryogenic scanning transmission electron microscopy, I discovered the morphological underpinnings of these observations. The discontinuous increase in conductivity coincides with the emergence of heterogeneity in the hydrated channels: a water-rich layer is sandwiched between two polymer-rich brushes. While the possibility of obtaining heterogeneous hydrated channels in polymer electrolyte membranes has been discussed extensively, this is, to our knowledge, the first time that direct evidence for the formation of water-rich subdomains is reported.

This poster depicts the crucial findings of my postdoctoral work in establishing structure-property relationships in PEMs, along with an outline of water management in PEMs by introducing nano-scale pores into the proton conducting phase.

389629 Design of Thermodynamically Consistent Coarse-Grained Models in Soft Matter

Abstract:

The world's population growth is enhancing the need for efficient materials, however, the design, synthesis and processing of novel materials with optimal properties for applications is an important challenge, due to the manipulation of matter at ever decreasing length scales. Computer simulation methods are essential for improving the design of novel materials, because they can provide quick and cost-effective promising directions for materials synthesis, by indicating improvements in process design or identifying reasons for shortcomings. To date, my research has been in the broad area of soft matter and with particular emphasis on methods to control or enhance self-organization on those systems. Through my work I had tackled a series of challenging problems including, but not limited to developing models to successfully reproduce morphologies with long-range order of thin films of diblock copolymers under shear [1-4] (which show promising use in synthesizing nanowires, surface modification, and pattern transfer) and provide a fundamental understanding of the structural as well as the dynamical properties of solvent-free grafted nanoparticles [5-7] (which show promising use in carbon dioxide (CO_2) capture, energy capture and storage, as well as water purification). Most of these problems are multi-scaled ranging from the atomic scale to the macroscale. My current work, focus on the thermodynamics of the vapor-liquid equilibria of pure and mixed solvents of multi-functional amines, which is of great relevance for CO_2 capture processes. By coupling together a predictive equation of state (EoS) for the thermodynamics and phase behavior together with robust coarse-grained models for use in explicit molecular simulation greatly enhance predictive solvent design for CO_2 capture processes. The advantage of such approach is that a molecular-based equation of state can be used to obtain effective coarse-grained models that reproduce the macroscopic experimental thermophysical properties over a wide range of conditions, while conventional molecular simulation with these coarse-grained models can be performed to obtain properties (such as structure or dynamics) that are not directly accessible with the EoS. The realization of these exciting models and methodologies require a new level of understanding of the structure and properties of complex systems, as well as the capability to engineer and optimize their performance. More importantly they are not restricted to one field of research thus making them very versible. For specific molecular structures (e.g., block copolymers) there are well established methods and understanding of their behavior. We still lack a fundamental understanding of self-assembly as the building blocks deviate from the ideal molecular structures that theories are built on. Nevertheless, a more complex molecular structure would offer new possible venues for creating building blocks with selective functionality. One such example is hollow particles, e.g., nano-scale vesicles, protein cages. Such systems can be excellent candidates as nano-capsules for enzyme encapsulation and controlled release of drugs, which would have immediate relevance to many applications in physical and biological sciences. My interdisciplinary research program will therefore be dedicated to the science and engineering soft matter systems with optimal properties.

Selected Publications

[1] A. Chremos, K. Margaritis, A. Z. Panagiotopoulos, Soft Matter 6, 3588-3595 (2010).
[2] A. Chremos, P. M. Chaikin, R. A. Register, and A. Z. Panagiotopoulos, Macromolecules 45, 4406-4415 (2012).
[3] A. Chremos, P. M. Chaikin, R. A. Register, and A. Z. Panagiotopoulos, Soft Matter 8, 3803-3811, (2012).
[4] A. Chremos, A. Nikoubashman, and A. Z. Panagiotopoulos, J. Chem. Phys., 140, 054909, (2014).
[5] A. Chremos and A. Z. Panagiotopoulos, Phys. Rev. Lett. 107, 105503 (2011).
[6] A. Chremos, H. Y. Yu, D. L. Koch, and A. Z. Panagiotopoulos, J. Chem. Phys. 135, 114901 (2011).
[7] A. Chremos, D. L. Koch, and A. Z. Panagiotopoulos, J. Chem. Phys. 136, 044902 (2012).

Charge Carrier Dynamics in Thin Film Solid-State Solar Cells

Tailoring Solution-Processed Semiconductors for Energy Applications

Jeffrey A. Christians

Department of Chemical and Biomolecular Engineering and Radiation Laboratory, **University of Notre Dame**, Notre Dame, IN

The Clean Energy Demand

Global energy demand (currently ~13TW) is expected to more than double by 2050. To prevent dangerous anthropogenic climate interference, renewable and carbon neutral energy will become increasingly important in fulfilling this energy demand.

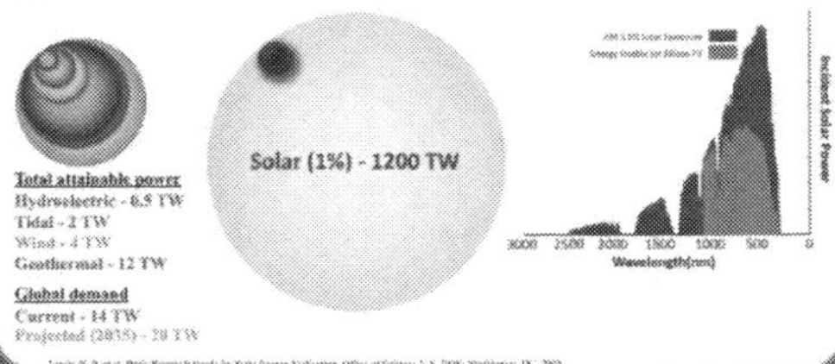

Research Overview

My doctoral research in Prashant V. Kamat's lab at Notre Dame has involved understanding the properties of semiconductor thin films for photovoltaics

(1) Constructing high efficiency, solution-processed photovoltaic devices

(2) Investigating charge transfer across semiconductor heterojunctions

(3) Elucidating semiconductor material properties

(4) Developing new methods for semiconductor thin film fabrication

Next Generation Photovoltaics

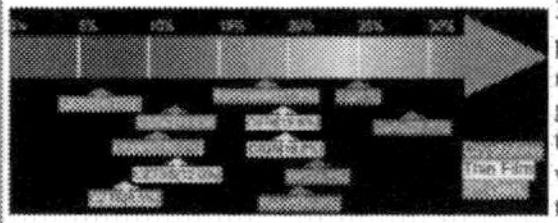

3rd gen. devices can be solution-processed, making them potentially much less expensive than 1st & 2nd gen. PV. Optimizing the charge transfer process in these solar cells will allow further efficiency gains for these materials.

Charge Transfer Processes in QDSCs

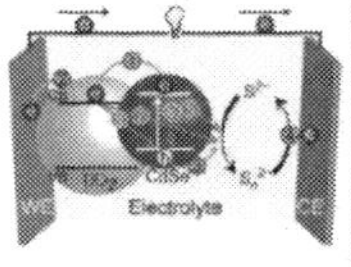

1. Electron transfer from quantum dot to metal oxide
2. Diffusion of electrons through metal oxide film
3. Regeneration of QD by electrolyte (hole transfer)
4. Electron-hole recombination at counter electrode

5-8. Negative electron-hole recombination processes

Extremely Thin Absorber Photovoltaics

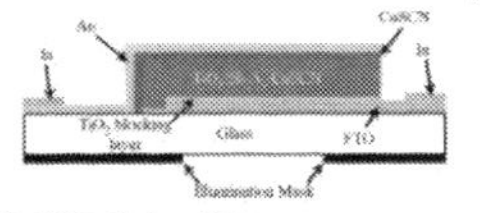

Sb_2S_3 ETA Solar Cells

The near-optimal band gap (1.7 eV), and low-cost processing make Sb_2S_3 a promising material for photovoltaics (below).

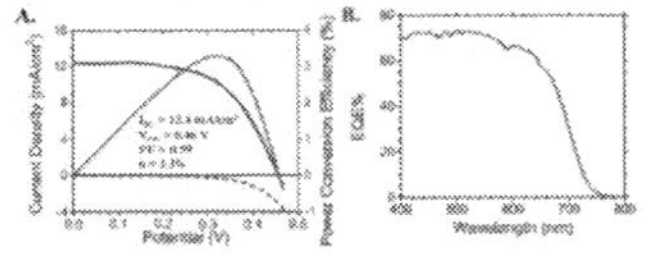

Transient absorption spectroscopy (TAS) offers the ability to study charge carrier dynamics on the femtosecond timescale (10^{-15} s). This can provide information of the rate of charge transfer and recombination reactions, providing valuable information for the further optimization of Sb_2S_3 solar cells.

Hole Transfer from Sb_2S_3 to CuSCN

Hole transfer rates were determined by tracking the sulfide radical induced absorption signal at 550 nm in the transient absorption spectra of Sb_2S_3 with and without the hole acceptor, CuSCN (below, A&B).

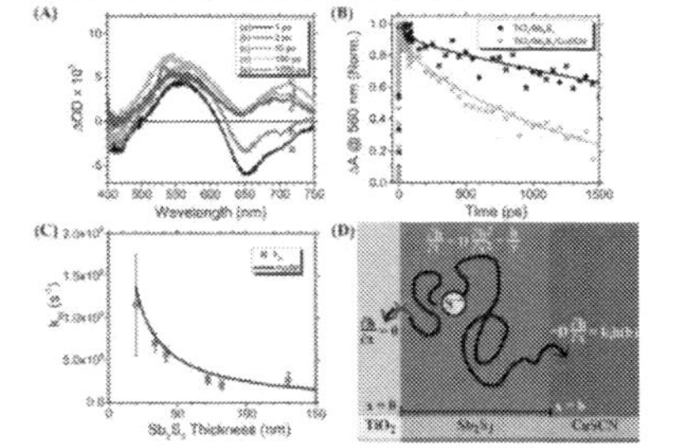

Modeling Interfacial Hole Transfer

By developing a model based on Fick's Laws of diffusion (above, C&D), we resolved the contributions of hole diffusion in Sb_2S_3 and hole transfer across the Sb_2S_3/CuSCN interface. This provides key information about the system and offers a guide for future studies to further improve PV performance.

Organometal Halide Perovskite Photovoltaics

Perovskite Solar Cells

Methylammonium lead iodide perovskites have swiftly advance beyond all other third generation materials, approaching CdTe and CIGS in terms of power conversion efficiency (right), while utilizing earth abundant materials and solution-processing fabrication methods.

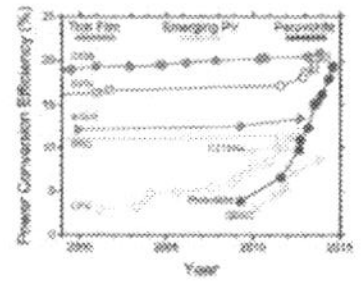

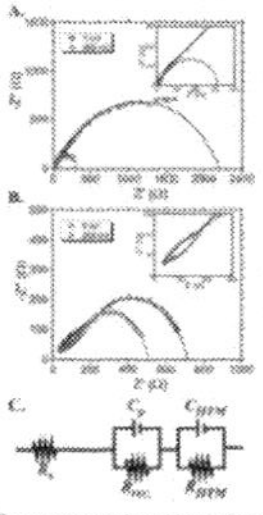

Inorganic Hole Conductors in Perovskite Solar Cells

CuI functions as an inexpensive hole conductor for organo-lead iodide perovskite solar cells, a promising alternative to organic hole conductors such as spiro-OMeTAD. It offers improved conductivity, as demonstrated by impedance spectroscopy (left), and efficiency of 6% compared to 7.9% for spiro-OMeTAD control devices (below).

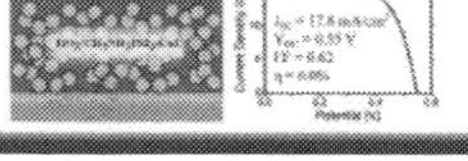

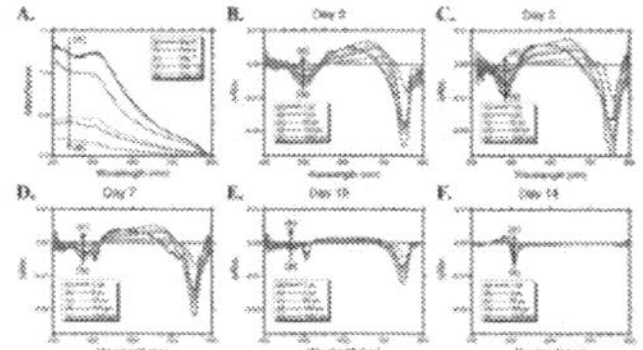

Humidity-Induced Excited State Transformation

TAS of $CH_3NH_3PbI_3$ exposed to humidity (above) elucidates how humidity exposure degrades the performance of perovskite solar cells (right). This work characterizes the nature of this degradation and reveals that changes in the optical and excited state properties occur much more slowly than the loss of PV performance.

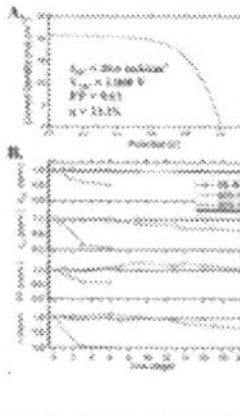

Proposed Future Projects

Solution-processed semiconductor devices have the potential to dramatically simplify the manufacturing of a wide variety of optoelectronic devices. Specifically, I am interested in developing solution-based techniques for the construction of photovoltaics and LEDs.

Solution-Processed Metal Chalcogenides

Recent advances have demonstrated how V_2VI_3 semiconductors can be deposited from solution.[1] This approach offers a facile method for creating $Sb_2Se_xS_{3-x}$ ETA solar cells with the ability to modulate the band gap between 1.7 eV and 1.3 eV.

Lead-Free Hybrid Perovskite PV

Current Sn-based hybrid perovskite show promise for photovoltaics but lack stability. Employing a layered hybrid perovskite structure (right) allows for increased stability and tunable material properties.[2]

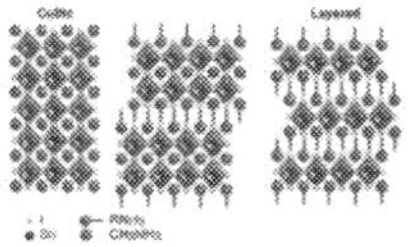

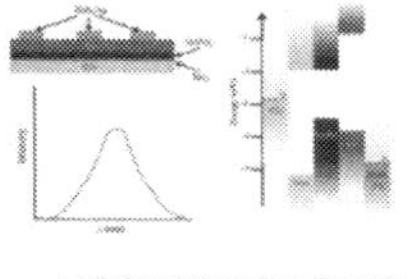

Hybrid Perovskite LEDs

The low contribution from trap-state and Auger recombination makes hybrid perovskites a promising material for LEDs.[3] Employing layered perovskites with high exciton binding energy should increase radiative recombination, leading to even better LED performance.

1. [illegible]
2. [illegible]
3. [illegible]

Acknowledgements

This work would not have been possible without Dr. Prashant V. Kamat, my doctoral advisor, the other members of the Kamat lab and my collaborators, and the members and staff of the Radiation Laboratory. This work was supported by the Office of Basic Energy Sciences of the U.S. Department of Energy, the Jana & Patrick Eilers Energy Research Fellowship, and the Bayer Predoctoral Research Fellowship.

Publications

1. Manser, J. S.; Christians, J. A.; Kamat, P. V. Current State of the Excited State Dynamics of Methylammonium Lead Iodide Perovskite. **2014** In Preparation (Review Article)
2. Christians, J. A.; Manser, J. S.; Kamat, P. V. Best Practices in Perovskite Solar Cell Efficiency Measurements. Avoiding the Error of Making Bad Cells to Look Good. **2014** In Preparation – To be Submitted 12/2014 (View Point)
3. Christians, J. A.; Miranda Herrera, P. A.; Kamat, P. V. Transformation of the Excited State and Photovoltaic Efficiency of $CH_3NH_3PbI_3$ Perovskite upon Controlled Exposure to Humidified Air. *J. Am. Chem. Soc.* **2014** Submitted.
4. Kamat, P. V.; Christians, J. A.; Radich, J. G. Quantum Dot and Nanowire Solar Cells. Hole Transfer as a Limiting Factor. *Langmuir* **2014**, 30 (20), 5729-5738 (Invited Feature Article)
5. Kim, J.-P.; Christians, J. A.; Choi, H.; Krishnamurthy, S.; Kamat, P. V. CdInS Nanowires. Compositionally Controlled Band Gap and Charge Dynamics. *J. Phys. Chem. Lett.* 2014, 5, 1103-1109
6. Christians, J. A.; Leighton Jr., D. T.; Kamat, P. V. Rate Limiting Interfacial Hole Transfer in Solid-State Solar Cells. *Energy Environ. Sci.* **2014**, 7, 1148-1158.
7. Christians, J. A.; Fung, R. C. M.; Kamat, P. V. An Inorganic Hole Conductor for Organo-Lead Halide Perovskite Solar Cells. Improved Hole Conductivity with Copper Iodide. *J. Am. Chem. Soc.* **2014**, 136, 758-764.
8. Christians, J. A.; Kamat, P. V. Trap and Transfer. Two-Step Hole Injection Across the Sb_2S_3/CuSCN Interface in Solid State Solar Cells. *ACS Nano* **2013**, 7, 7967-7974.
9. Opperwall, S. R.; Divakaran, A.; Porter, E. G.; Christians, J. A.; Denhartigh, A. J.; Benson, D. E. Wide Dynamic Range Sensing with Single Quantum Dot Biosensors. *ACS Nano* **2012**, 6, 8078-86.

Molecular Engineering for Regenerative Medicine

Eun Ji Chung

Institute for Molecular Engineering, University of Chicago, Chicago, IL 60637, USA

When designing bioactive materials to direct desired outcomes for regenerative medicine, the appropriate structural and functional properties at all relevant hierarchical levels—including tissue, cellular, and molecular scales—must be incorporated. Furthermore, just as important is the intentional design of the biomaterial to be practical for the clinical setting, complementing available surgical techniques and emerging healthcare technology, while remaining safe and cost-effective.

To understand the biological and engineering criteria for multi-dimensional biomaterials design, my research-to-date has utilized biodegradable polymers and self-assembling assemblies for tissue engineering and theranostic applications. My graduate research at Northwestern University under the direction of Prof. Guillermo Ameer in the Dept. of Biomedical Engineering encompassed developing citric acid-based nanocomposites for orthopaedic tissues. Specifically, I investigated materials parameters that are optimal for stem cell differentiation and *in vivo* bone and ligament regeneration, and developed a novel method termed "low pressure foaming" to fabricate porous scaffolds. Upon receiving the IBNAM-Baxter Early Career Award, I continued my research at Northwestern as a postdoctoral fellow in Prof. Ramille Shah's lab in the Dept. of Materials Science and Engineering, where I fabricated *in situ*-forming, self-assembling structures derived from natural extracellular matrix proteins. Specifically, I focused on novel, collagen-hyaluronic acid membranes to act as bioactive coatings for drug delivery applications that can also complement minimally-invasive procedures. Moreover, I elucidated that the mechanism of self-assembly involved electrostatic interactions and osmotic pressure imbalances. To further diversity my experience in self-assembly and to integrate the rational design of nanoparticles for theranostic applications, I joined Dr. Matthew Tirrell's group at the Institute for Molecular Engineering at the University of Chicago as a recipient of the American Heart Association postdoctoral fellowship. My current research aims to develop peptide amphiphile micelles to target, diagnose, and treat a variety of diseases including atherosclerosis and cancer.

Building on my previous work and harnessing my biomolecular engineering background, I plan to establish an independent research program that aims to apply novel biomaterials strategies that addresses the limitations of medical solutions and laboratory models to enhance tissue regeneration and generate optimal clinical outcomes. Four areas of interest that I plan to pursue include: 1) Molecular assemblies for diagnostic and therapeutic applications, 2) Multifunctional, biomimetic polymers for biomedical applications, 3) Novel sources of stem cells for regenerative medicine, and 4) Design of *in vitro* models of disease.

Selected Publications (14 total):

1. **Chung, E.***, Cheng, Y.*, Morshed, R., Nord, K., Han, Y., Wegscheid, M., Wainwright, D.A., Lesniak, M.S., Tirrell, M.V. Peptide amphiphile micelles for targeting glioblastomas. Biomaterials 35(4): 1249-56 (2014).
2. **Chung, E.**, Jakus, A.E., and Shah, R.N. In situ forming collagen-hyaluronic acid membrane structures: Mechanism of self-assembly and applications in regenerative medicine. Acta Biomaterialia 9(2): 5153-5161 (2013).
3. **Chung, E.**, Sugimoto, M., Koh, J., Ameer, G.A. Low pressure foaming: A novel method for the fabrication of porous scaffolds for tissue engineering. Cover of Tissue Engineering Part C 18(2): 113-121 (2012).
4. Serrano, M.C.*, **Chung, E.***, Ameer, G.A. Advances and applications of biodegradable elastomers in regenerative medicine. Advanced Functional Materials 20(2): 192-208 (2010).
5. Wang, J., Singh, C., Liu, L., Irwin, R., Chen, S., **Chung, E.**, Thompson, R., Brinton, R. Allopregnanolone reverses neurogenic and cognitive deficits in mouse model of Alzheimer's disease. PNAS 107(14): 6498-6503 (2010).
**These authors contributed equally.*

For more information, please visit: http://tirrell.ime.uchicago.edu/People/Eunji_Chung.html

Holistic Multi-scale Approaches for Nanotechnology Convergence Systems

Pil Seung Chung
Department of Chemical Engineering, Carnegie Mellon University, Pittsburgh, PA 15213, USA

During the past decades, along with remarkable scientific and engineering achievements in many areas, the world has faced immense challenges. Increasing need for higher efficiency and reduction in system size motivates the investigation of novel nanoscale systems and materials. Various enhancements have been reported by tuning atomistic properties, chemical structures, and functionalities of advanced materials leading to eye-opening progress. The tremendous improvements in nanotechnology provide the necessary understanding of nanoscale physics to control and design the system properties, while at the same time focusing on the critical issue of integration with macroscopic system scales resulting in converging technology and phenomena at different time and length scales.

My research interest focuses on this challenging problem by introducing a novel holistic multi-scale integration in nanotechnology by hybridizing atomistic / molecular and mesoscale / continuum models, and developing unique methodologies for hierarchical integration of the complete system at different scales, and expanding to include multi-phenomena / physics (Fig.1). Information transfer from one scale to neighboring scales allows one to eventually predict macroscopic behavior or device scale functions from first principles in top-down / bottom-up / middle-out approaches. These approaches along with system optimization based on a clear understanding of nanoscale phenomena will lead to a novel multidisciplinary analysis paradigm.

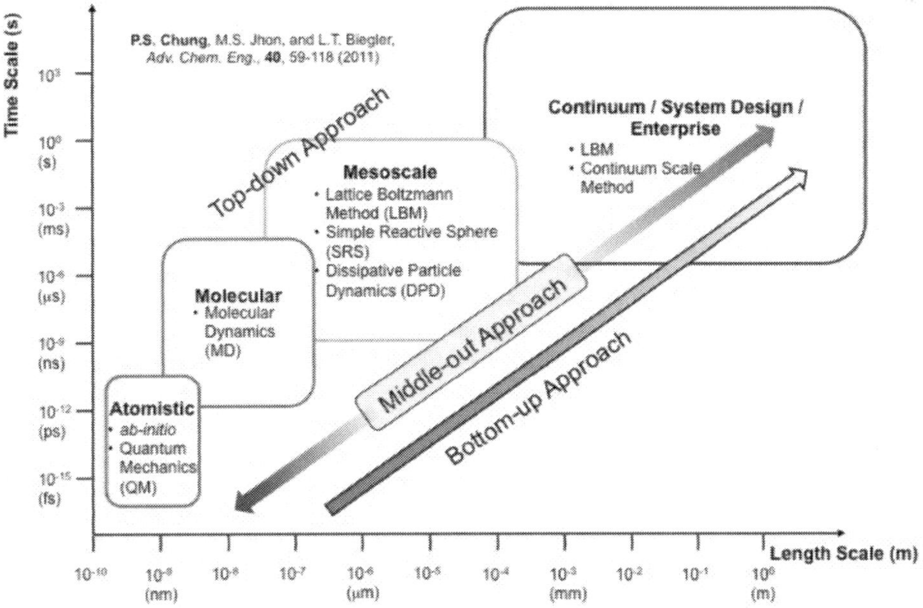

Fig.1. Hierarchy of multi-scale modelings and the integration methodology

In this poster session, I present case studies, which have been examined in the area of information technology and energy technology & sustainability for multi-scale modeling via a hierarchy of scale specific models coupled with optimization for the process system integration. Nano-tribology in the data storage system was chosen as a benchmark example utilizing atomistic / molecular simulations for the nanoscale film of functionalized oligomers / polymers with hybridization of novel graphene technologies and extended to mesoscale / continuum theories for the whole system integration [1,2]. In energy & sustainability area examples, we analyzed polymer electrolyte membrane fuel cell including membrane electrode assembly (polymer electrolyte membrane & catalyst layers) via atomistic / molecular simulations and transport phenomena in the gas diffusion layer via lattice Boltzmann method [3]. In addition, I will introduce and discuss the details of my future research topics and plan in this poster session.

Reference:

[1] P.S. Chung, S.H. Vemuri, S. Park, and M.S. Jhon, "Molecular rheological analysis on binary blends of perfluoropolyether lubricants," J. Appl. Phys., 115, 17B738 (2014).

[2] P.S. Chung, D.S. So, L.T. Biegler, and M.S. Jhon, "Nanotechnology convergence and modeling paradigm of sustainable energy system using polymer electrolyte membrane fuel cell as a benchmark example," J. Nanopart. Res., 14 (8), 853 (2012).

[3] P.S. Chung, R. Smith, S.H. Vemuri, Y.I. Jhon, K. Tak, I. Moon, L.T. Biegler, and M.S. Jhon "Multiscale/ multi-physical modeling in head/disk interface of magnetic data storage," J. Appl. Phys., 111 (7), 07B712 (2012).

386018 A Materials Genome Approach to Metal-Organic Frameworks: From Databases to Advanced Search Algorithms

Abstract:

My current research activity is at the intersection between molecular modeling and data science. In this poster, I will describe my postdoctoral work on metal-organic frameworks (MOFs) in Prof. Randall Snurr's group at Northwestern University and how these activities align with the emerging research area of data science and high-throughput molecular modeling.

In collaboration with researchers from Georgia Tech and Lawrence Berkeley National Laboratory, I have created a database of thousands of computation-ready, experimentally-derived metal-organic frameworks. Construction of the database required removal of impurities and solvents using graph-labeling and cluster-detection algorithms and text-mining. The atomic coordinates of the thousands of structures will soon be made available to the public through the Nanoporous Materials Genome Center. The database is expected to grow as new structures become available in the literature. Following the compilation of thousands of MOFs, we have computationally screened these materials using atomistic grand canonical Monte Carlo simulations to find the top performing MOFs for application in methane storage and hexane isomer separation as an initial demonstration of the utility of the database.

An emerging area in computational chemistry is the high-throughput computational screening of large numbers of structures to identify promising structures and to extract useful structure-property relationship for a given application. MOFs have been a particularly good target for such an approach because of the large number of synthesized structures in the literature. The early high-throughput computational screening of MOFs has been carried out in a brute-force manner, where a number of performance evaluations (e.g., adsorption isotherm at a given pressure) are carried out for every structure in the database. What enables such a high-throughput computational screening approach are several key simplifying assumptions, such as rigid frameworks and the use of generic force field parameters and classical atomistic simulation methods. The new frontiers of high-throughput molecular modeling will involve the removal of such simplifying assumptions and usage of high-level computational chemistry methods (e.g., DFT and QM/MM). These advanced approaches will require large amounts of computational resources, making it impossible to apply them for thousands of structures.

One way to avoid expensive performance evaluations for thousands of structures is to employ advanced search algorithms for metal-organic frameworks. I have developed an evolutionary search algorithm and applied it to a database of hypothetical MOFs to effectively find top MOFs in terms of importance performance criteria such as methane delivery capacity and volumetric surface area. The algorithm that has been developed here will not only be useful in terms of locating the best existing structures in a database but could help to find a new, as-yet synthesized MOFs for gas storage and separation application in the future. The approach could also be extended to optimize search for promising structures in other class of materials, such as metallic glasses that finds a widespread use in electronics and biomedical applications.

My future research direction will focus on the emerging area of high-throughput computation of metal-organic frameworks and other class of materials, with an emphasis on development of a platform for data science approaches and application of molecular modeling to search for best performing materials that could be used in energy and pharmaceutical applications (e.g., gas storage and drug delivery).

385098 Linear and Non-Linear Programming Techniques for System Identification and Green Engineering Applications

Abstract:

My research focuses on linear programming (LP) and non-linear programming (NLP) techniques for system identification and green engineering applications. To this end, I have explored interval analysis-based branch-and-bound techniques for parameter identification, assessment/synthesis of sustainable systems, graphical/analytical techniques for the global solution of the total annualized cost problem for a series of compressors, and the introduction of the novel concept of the attainable region for process networks. Throughout the course of my work, I have experienced first-hand the "curse of dimensionality" prevalent in non-linear optimization algorithms. Thus, I have applied analytical techniques, dimensionality reduction, convexification strategies, and functional analysis to identify global optima and to enhance the efficiency of their computation.

My global optimization expertise has prepared me well to pursue new research avenues in several areas, particularly optimal system identification, initial project feasibility assessment via attainable region concepts, and sustainability assessment and decision making. My strong mathematical and computational background will also allow me to contribute to collaborations that can benefit from this kind of expertise. Furthermore, I will disseminate all developed concepts and lessons learned into course-friendly materials to provide students opportunities to apply basic chemical engineering and optimization concepts to real-world green engineering problems.

385573 Engineered Nanostructures for Imaging, Sensing, Environmental Applications, and Their Implications to Human Health

Abstract:

The primary focus of my research will be on developing nanoscale materials for biomedical applications while also examining the implications through their interactions with biological systems. Light responsive nanomaterials offer a unique advantage in diagnostics and image-guided surgery. Therefore, one of my primary goals is to develop near infra-red light emitting nanomaterials for non-invasive sensative diagnostics tools. In my current postdoctoral research in Professor Ian Kennedy's laboratory at University of California, Davis, I worked on developing an X-ray based optical tomography system and associated contrast agent for deep tissue imaging. I also developed a novel system for early detection of prostate cancer using lanthanides via optical and magnetic resonance imaging (MRI) when I was doing one year of postdoctoral research jointly in the University of Victoria, Canada and the British Columbia Cancer Agency under supervision of Professor Frank van Veggel.

My other major research goals are to predictively determine deposition, translocation of the environmental particulate matters and inhaled drugs in the respiratory track, and to develop efficient materials & techniques for hazardous waste treatment. Very recently, I have developed a magnetic field aided gas phase synthesis technique that resulted in large scale production of extremely efficient magnetic nanochains for arsenic removal from water.

Selected publication (Out of 15):

Sudheendra, L.;* **Das, G. K.**;* Li, C.; Stark, D.; Cena, J.; Cherry, S. R.; Kennedy, I. M. X-ray Luminescence from Europium-doped Lanthanide Fluoride Particles. Chemistry of Materials, **2014,** *26*, 1881–1888. (***Co-First Authors)**

Das, G. K.; Johnson, N. J. J.; Cramen, J.; Blasiak, B.; Latta, P.; Tomanek, B.; van Veggel F. C. J. M. $NaDyF_4$ Nanoparticles as T_2 Contrast Agents for Ultrahigh Field Magnetic Resonance Imaging. The Journal of Physical Chemistry Letters, **2012**, 3, 524–529.

Das, G. K.; Zhang Y.; D'Silva, L.; Loo, J. S. C; Padmanabhan, P.; Heng, B. C.; Loo, J. S. C; Selvan S. T.; Bhakoo, K. K. and Tan T. T. Y. Single-Phase Dy_2O_3:Tb^{3+}Nanocrystals as Dual-Modal Contrast Agent for High Field Magnetic Resonance and Optical Imaging. Chemistry of Materials, **2011**, 23, 2439–2446.

Das, G. K.; Heng, B. C.; Ng, S.-C.; White, T.; Loo, J. S. C; D'Silva, L.; Padmanabhan, P.; Bhakoo, K. K.; Selvan S. T. and Tan T. T. Y. Gadolinium Oxide Ultranarrow Nanorods as Multimodal Contrast Agents for Optical and Magnetic Resonance Imaging. Langmuir, **2010**, 26, 8959 – 8965.

Polymer Membranes for Energy Storage and Delivery

Eric M. Davis

Materials Science and Engineering Division, National Institute of Standards and Technology, Gaithersburg, MD 20899

Developing a fundamental understanding of ion and water transport in ion exchange membranes (IEMs) is vitally important for the design and development of next-generation membrane materials for energy storage and delivery applications. One critical aspect is to understand how film thickness affects transport in these materials, since these materials are often times confined to nanometer size scales in commercial applications. Furthermore, obtaining a deeper understanding of how the addition of nanoparticles in these membranes affects water diffusion and polymer chain dynamics is highly desirable. Therefore, a key part of my future research as a new faculty would be developing and employing *in situ* characterization techniques to elucidate fundamental transport mechanisms in these IEMs in an effort to understand the relationship between material structure and device performance. With this, structural and molecular-level techniques, including small-angle neutron scattering (SANS) and time-resolved infrared spectroscopy, will play a central role in developing an understanding of how changes in material properties affect the mechanics and water/ion transport of IEMs, which ultimately affect fuel cell and battery performance.

My PhD work was primarily focused on the use of infrared spectroscopic techniques to accurately capture and measure water transport in glassy polymers. Additionally, I utilized nonequilibrium thermodynamic modeling to predict water sorption in these nonequilibrium, glassy polymers. My work on transport phenomena, infrared spectroscopy, and nonequilibrium thermodynamics will facilitate my future research endeavors, where I propose to employ these powerful tools to measure the transport of ions and water in IEMs and glassy polymers. During my post-doc, I have gained expertise in use of neutron scattering techniques, where these techniques are currently being utilized to investigate the structure of Nafion nanocomposite membranes. Understanding the location and size of these particles within the membrane is critical in understanding the mechanism by which these membranes exhibit improved performance. Additionally, my post-doc work has provided me with experience in the use of polarization modulation infrared adsorption spectroscopy (PM-IRRAS) and poroelastic relaxation indentation (PRI) to investigate water sorption and transport in polymer thin films. I believe that my PhD work on water transport in glassy polymers, as well as my post-doc work on structure and water/ion transport in IEMs, has provided me with a solid foundation in an area of research that is plagued with unanswered questions, and one with which I believe will be receiving an increasing amount of funding over the next decade.

379195 A Language Learning Focus in Increasing Global Access to Chemical Engineering Education

Abstract:

Both chemical engineering and engineering education are rapidly obtaining international dimension. The U.S. has served as home for leading chemical engineers from all over the world for the last 50 years. The challenges of finding best candidates for chemical engineering career and best chemical engineering education will shape the chemical engineering future in the USA in the 21 Century. In turn, the ability to internationalize best practices in chemical engineering (both at the education and industrial level) is necessary for the successful development of the chemical industry in the global perspective.

Attracting promising students to chemical engineering education system and their transforming into chemical engineering professionals for the US and world chemical industry requires a continuous flow of talents. Successful enrollment of talents by top world universities specialized in chemical engineering is an excellent proof of successful internationalization of chemical industry and education. These universities are, however, facing a rapidly growing challenge of dropping demand for chemical engineering programs both by domestic and international students. Chemical engineering education and engineering education in general are becoming less popular. This fact is confirmed by the statistics of the institute of International Education.

The global demand for world class chemical engineering programs taught in U.S. and other top world chemical engineering universities can be increased if we attract more potential students from Asia, Russia and post-Soviet countries. These countries are proud of talented school graduates and offer government funds programs to sponsor their education in best world chemical education centers. The problem of reduction of the influx of talented school graduates from these countries is that academic mobility in these states is closely linked to the language learning problem (in other words, the ability of involved faculty and students to speak world languages such as English at a level of fluency ensuring their integration into the global community). For example, several percent of Russian secondary school graduates and undergraduate students demonstrate TOEFL results sufficient for enrollment by chemical engineering universities in the USA. These students can be, however, excellent chemists and scientists and be motivated to obtain a good-paid position at world-class chemical industry companies requiring a degree from top universities offering chemical engineering programs.

A new approach is offered to involve more students into the U.S.-offered chemical engineering programs – a language learner focused approach. This strategy is called a "Lifelong Language Learning Environment" (a "Triple-L E") meaning that English learning in the universities from above-mentioned countries should be treated as a truly life-long process with the focus on chemical engineering linguistic aspects. Starting at the secondary school level, engineering English learning continues at a university to offer Bachelors able to pursue Master's degree in a U.S. university or a secondary school graduate with TOEFL scores allowing entering a chemical engineering program. Another component of this "Environment" is that faculty in involved in language learning together with students. An indispensible prerequisite for a successful Lifelong Language Learning Environment is involvement of U.S. universities into its development and functioning at their partnering higher education centers abroad (in other words, it requires strong international partnership that is successfully achieved by all top chemical engineering universities in the USA).

This paper also provides a case study of development and implementing of a "Triple-L E" initiative at one of Russian National Research universities specialized in chemical engineering. A peculiar feature of the Lifelong Language Learning Environment implementation in that university is that mostly technical English training is provided, English for Specific Purposes (ESP). English for Specific Purposes is aimed at developing specific skills of communicating in English on the topics of professional interest, depending on the background and career prospects. English teaching is focused on the engineering field of study with the application of computer aided language learning solutions. The core of the process if formed by professors having both engineering (major) and linguistic (minor) degrees or the linguists with 5-10 years of experience in technical English. Successful involvement of U.S. universities is achieved through the development and implementation of joint short-term and dual degree programs with the focus on the spread of disciplines taught in English from these programs to other university chemical engineering programs. This strategy allowed training of faculty, administrators and students for the internationalized chemical engineering programs and attracting English-speaking students from other countries as well as stimulating outgoing academic mobility.

Process Design Approaches for Seniors

Richard L. Long, Jr., PhD, P.E.
Professor Emeritus, Dept ChE, NMSU
AIChE Annual National Meeting
October 30,2012

Design Course Sequence

Design Course

- Course lab (simulation models- new features):
  - tear streams
  - convergence blocks
  - heat integration
- Class lectures- rules of thumb, equipment sizing, pinch methods, capital cost estimating, etc.
- Group Design Projects

Capstone

- DISTWU/RADFRAC
- Simple reactor models from model library
- Power cycle(students are not familiar with graphical extrapolation methods in Mollier diagrams.
- Special topics (incorporating empirical rate laws into reactor models
- NSDC problem

Group Design Projects

- He recovered from natural gas
- ADN from dichlorobutene
- Acetone recovery from acetone/air binary
- Heat integration

He recovery from natural gas

- The students recognized that they would have to run cold, but they could not figure out how to run such that they were in the two phase region.
- The student ran at a loss of some 30% He from the natural gas.
- The students did not recognize that they needed to use adsorption on charcoal beds with two beds in parallel to account for the breakthrough of the beds and then to switch from one to the other upon breakthrough.
- The students did not have an idea that the price of the natural gas would be only a small increment greater than the price of the raw mixed gas with He in it

ADN from Dichlorobutene

- The students insisted that the instructor provide them with the rate law so they could formulate the design equation for an appropriate Aspen Plus model. The problem was given such that the rate of conversion vs. time was provided as pilot plant data. The data was not conveniently given in compatible input to the Aspen Plus software. Should the instructor provide the modification required to make it compatible with the software?
- The student had no idea what to do to minimize the overall reactor volume. Volume productivity was off by a factor of 10.The students did not realize that the best configuration was a stirred tank followed by a series of small stirred tanks in series to approximate a stirred tank followed by a plug flow reactor.
- This course was from kinetics. Should the instructor in design correct their mistake and allow them to proceed as before?
- How much instruction in using software should be given in design or should the design assume that the background in software is adequate?

Acetone recovery from air/acetone feed

- The students did not realize how to determine the minimum water rate in the absorption column .
- They sent the instructor an email stating that they had "derived" the mass balance. The result was that it was given by the students as 5' height. The instructor responded (correctly) that it was 35' in height. The students stated that Aspen Plus said that they were right. In fact, Aspen Plus did not make a claim either way.
- Question: Should the instructor show them why and how they were wrong? This was a senior design course. What is the right course of action?

Heat Integration

- An oil refiner requires an incremental amount of i-butane for a new alkylation unit which produces gasoline from butylene and i-butane
- The students by day 6 had a flowsheet that eliminated all errors and warnings. They now concluded that they had a working process.
- When they went to optimize it, they could not because they could not properly balance the temperatures.
- So they skipped this required step and went directly to capital cost estimating. And on to economic evaluation.

Curriculum Overview Issues

- The pressure overall in the curriculum is to lower student credit hours for each degree – the result is a lower cost per degree. And the cost is an issue for each state because of taxpayer funding as to state budget costs.
- The incentive in the engineering degree is to lower so-called "wet labs" and to increase the so-called "computer simulation labs" because of the lower cost per credit hour of the "computer simulation labs".
- The skill or art in estimating "reasonable" values of properties is becoming a "lost art".
- The skill or art in performing "wet lab" operations is being lost. It is much more efficient to do "computer simulation labs".
- The "wet labs" are needed when truly unknown results are being determined.
- Question: What is the right balance between "wet labs" and "computer simulation labs"?
- Question: If data is provided in the form of "wet labs", the data may not be compatible with required entry in computer software. What is the appropriate action to take in that circumstance?

Conclusions

The key is that the students are dependent on the understanding from fundamentals as to what the basis of the operations are in simulation. They do not know always how to enter data appropriately into the Aspen Plus simulation, and they require that the instructor show them how to do it.

This creates a conflict in the department.

Questions?

379257 A Language Learner Focus in Internationalizing Demand for Chemical Engineering Education

Abstract:

Chemical engineering education is rapidly becoming internationalized. An important barrier for internationalization is the ability of involved faculty and students to speak global languages such as English at a level of fluency ensuring their efficient integration into the globalized community. English as the world language is the connecting link eliminating borders and distances between continents as well as contributing to the collective intellectual potential of mankind. Integration of non-native English speakers into this global intellectual network faces certain challenges, such as significant efforts needed to form an English-speaking community in countries with no or poor historical linguistic English background. A not so evident challenge, however, is that even English skills proper for day-to-day communication and general understanding needed to form basic contacts and collaborations are insufficient for the real integration of such people into the global league of professionals in industry, education and science. The reason is that excellence needs excellence, while being good but not enough means that you are not competitive.
Attracting promising students to chemical engineering education system and their transforming into chemical engineering professionals for the US and world chemical industry requires a continuous influx of talented students. Successful enrollment of talents by top world universities specialized in chemical engineering is an excellent proof of successful internationalization of chemical industry and education. These universities are, however, facing a rapidly growing challenge of dropping demand for chemical engineering programs both by domestic and international students. Chemical engineering education and engineering education in general are becoming less popular.

The global demand for engineering education can be increased by student enrollment from traditionally not English-speaking countries, such as Asia-Pacific and Russia-CIS countries with large human resources. Another opportunity is to involve best engineers and engineering faculty from these countries into the global engineering education process and well as increase the competitiveness their domestic graduates by inviting best international engineering educators to enrich their academic process. Internationalization is closely linked to the language learning problem for such world regions and, of course, for their internationalization partners in native English-speaking countries.

A new approach is offered to involve more students into the U.S.-offered chemical engineering programs – a language learner focused approach with a lifelong language learning environment meaning that English learning in the universities from above-mentioned countries should be treated as a truly life-long process with the focus on chemical engineering linguistic aspects. Starting at the secondary school level, engineering English learning continues at a university to offer Bachelors able to pursue Master's degree in a U.S. university or a secondary school graduate with TOEFL scores allowing entering a chemical engineering program. Another component of this Environment is that faculty in involved in language learning together with students. An indispensible prerequisite for a successful lifelong language learning environment is involvement of universities implementing programs in English into its development and functioning at their partnering higher education centers abroad.

Rubrics validation to improve teaching-learning process in laboratory subjects of the Chemical Engineering undergraduate degree of the University of Barcelona

M. Iborra[1], E. Ramírez, J. Tejero, R. Bringué, C. Fité, F. Cunill.

University of Barcelona. Faculty of Chemistry. Chemical Engineering Department. Martí i Franqués 1-11, 08028-Barcelona. Spain; [1]Corresponding author: miborra@ub.edu; Tel.: +34934034769; Fax: 34934021291

A rubric is considered as an educational tool that explicitly represents the performance expectations for a specific task or assignment. Basically, a rubric divides the assigned work into component parts and provides clear descriptions of the characteristics of the work associated with each component at varying levels of mastery [1]. To adapt practical subjects of the Chemical Engineering undergraduate degree of the University of Barcelona (Spain) to the European Higher Education Area, rubrics were gradually introduced to practical subjects during the three academic courses (2009-2012) [2]. Tasks and activities ("experimental design and experimental error" written composition, laboratory reports construction by means of wikis, oral presentation, and collaborative work assessment) to be assessed by rubrics were previously chosen and the rubrics construction was done by means of stating the purpose and the learning objectives and developing scoring criteria that address each objective.

Rubrics were introduced in laboratory subjects with the intent that: 1) to communicate to students the specific requirements and acceptable performance standards of the assignments; 2) to provide formative feedback to support and guide ongoing teaching-learning efforts of both instructors and students; and 3) to ensure consistency across multiple graders (other instructors, teaching assistants, etc.) according to an explicit and descriptive set of criteria that was designed to reflect the weighted importance of the objectives of the assignment. In this way, the instructor's grading standards don't change over time because of fatigue, shifting standards based on prior experience, or intrusion of other criteria.

During the 2012-2013 academic course, when the revamping process of teaching-learning methodologies in laboratory subjects of the undergraduate degree was completely implemented for their adjustment to the Bologna Process, instructors and students were queried about the rubrics using offline survey forms to gather evidence that supports the appropriateness of them. That "preliminary" validation was mainly focus on examining content, construct and criterion, and identifying of previous adquisition/development of the same competences in other subjects within undergraduate program in order to adjust the didactic instrument [3]. Validation analysis allowed to infer that the use of rubrics was satisfactory in terms of increasing consistency of scoring between raters, facilitating valid judgment of competencies by identification of students' strengths and learning needs, promoting students' learning and improving instruction. As a counterpart, it was detected that students did not get used to work with rubrics. Despite of the fact of the use of rubrics improved global academic results (general and transferrable competences acquisition/development and students' marks); several students reported not find the qualities in their work even if they know what to look for, since they have a less developed sense of how to interpret criteria. Furthermore, students did not completely engage in the innovation of their own learning process, their participation in validation process was around 68%. Although this percentage seems to be reliable to get conclusions, in some cases gathered information was missing which might made that results of this preliminary rubrics validation analysis were biased. As a consequence, for subsequence academic courses, the methodological changes in the use of rubrics have been focus on the refinement of its content and criteria according to detected strengths and weaknesses in the teaching-learning process, and on the enhancement of the ensuing validation process that will include students and raters.

With this purpose, for the current 2013-2014 academic course, reconstructed rubrics have been used in laboratory subjects. Their validation process has been modified in several ways: 1) validation survey has been reformulated in terms of clarity, wordiness, negative wording, appropriateness of responses listed, balances, etc.; 2) information has been gathered by means of an online survey available via Moodle in the subject virtual environment; and 3) the process includes a student of the laboratory subject as a peer consultant in order to help instructors to formulate the queries to students as well to motivate classmates to engage in validation process. Nowadays, the process of collecting information has just been finished and its analysis, which constituted a second cycle in rubric validation process, is starting up.

References

[1] Eberly Center Teaching Excellence & Educational Innovation, Carnegie Mellon University, Grading and Performance Rúbrics. http://www.cmu.edu/teaching/designteach/rubrics/html. Accessed November 20/2011.

[2] Iborra, M, Ramírez, E, Tejero, J, Bringué, R. Fité, C, Cunill, F. Revamping of teaching-learning methodologies in laboratory subjects of the Chemical Engineering undergraduate degree of the University of Barcelona for their adjustment to the Bologna Process. Education for Chemical Engineers. 2014. doi: http://dx.doi.org/10.1016/j.ece.2014.04.002

[3] Iborra, M, Ramírez, E, Tejero, J, Bringué, R. Fité, C, Cunill, F. Desarrollo y evaluación de competencias transversales (comunicación oral) y validación de las rúbricas de la asignatura de grado experimentación en ingeniería química II. Paper presented at: International Congress on University Teaching and Innovation (CIDUI) 2014; July 2013; Tarragona, Spain.

383201 The Chemical Engineering App for Iphone & Ipad

Abstract:

The Chemical Engineering App for iPad and iPhone were developed to facilitate access to course materials, reference data and calculations for students, educators and professionals in chemical engineering and other STEM disciplines. In just over 2 years, our apps have a user base > 30,000 in over 165 countries. These mobile applications include relevant data and modules for mass & energy balances, thermodynamics, fluids, separations, controls and other courses. The apps are available as free downloads on Apple's App Store, effectively making real chemical engineering course materials available to anyone in the world with an iOS device. Highlights of the Chemical Engineering App will be presented along with our vision for tablets and smartphones in chemical engineering and STEM education.

Lab Training in experimental design applied to product engineering: preparation of a cosmetic cream

Santamaría. E, Maestro.A, Gutiérrez, J.M* and González. C.
Chemical Engineering Department, Faculty of Chemistry, University of Barcelona, Martí i Franqués 1-11, Barcelona 08028, Spain
**e-mail: carme.gonzalez@ub.edu*

As chemical engineering education moves quickly, characterized by competitive global educational market, the role of educators is changing: there is a need to provide modern didactic methods in curriculum delivery. A new Chemical Engineering Master was developed at the University of Barcelona with the objective of approaching the competencies of graduates to the requirements from the chemical industrial sector. Many chemical engineers work in companies, which produce daily use products, and formulated products. By this reason, it is important that students learn by laboratory work the knowledge and the skills involved in product engineering.

A new laboratory work, where students must develop and experimentally study the preparation of a cosmetic cream as a common consumer product, was implemented. To achieve this goal students should know which are the functions and properties that the final product must accomplish, and which composition and preparation variables they could modify in order to obtain the best product.

Students decide the composition and preparation variables to be studied by an experimental design. Results of experimental design could lay down the variables which present a significant influence on the final product properties.

Students studied the influence of the stirring speed (N), addition speed (q) of the dispersed phase (water) and the emulsion dispersed phase ratio. The variables ranges were chosen by them, improving their capabilities of autonomously taking decision with scientific criteria.

The cosmetic cream was obtained as a water-in-oil emulsion composed by an oil phase containing isopropyl miristate, base F-2230 and paraffin. The water phase contained deionized water and glycerin. The oil phase was put into a stirred vessel and homogenized. Dispersed aqueous phase was added by a peristaltic pump.

The properties of the cosmetic cream studied were the stability, droplet size of the emulsion and rheological parameters as yield stress (τ_0) and viscoelastic behavior: viscous modulus (G") and elastic modulus (G'). Response variables were analyzed by an statistical program, so students could observe which composition and preparation variables had a significant importance on the studied parameters.

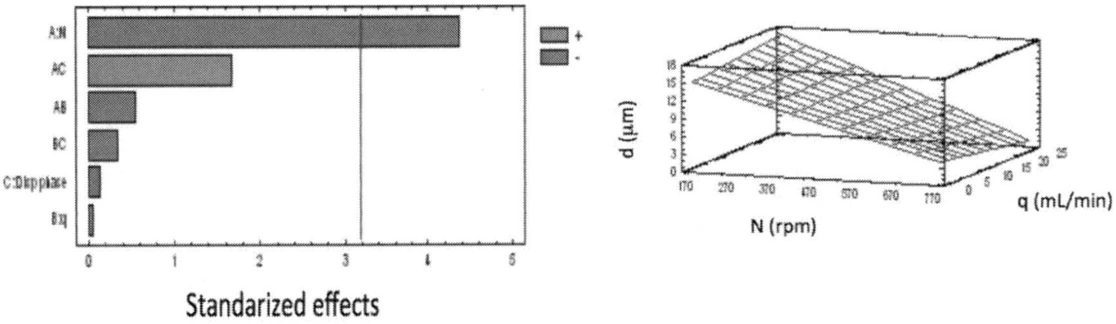

Figure 1. Pareto chart (left) and surface response (right) obtained for droplet diameters.

In the case of the droplet size students obtained the Pareto chart showed in figure 1. The chart indicated that the only variable that had a significant importance in the droplet size was the stirring speed (N). The surface response (Figure 1) showed the droplet size evolution when the stirring speed increases. So, the students conclude that in the studied range the higher stirring speed, the smallest droplet size.

In this laboratory work, students learned to use experimental designs and practiced experimentally the preparation and characterization of common consumer product, acquiring a global view of the parameters to take into account in a process of product engineering. They also worked the three types of transversal skills: (a) instrumental skills (self learning, planning and organization, taking decisions and writing and oral communication skills) (b) social skills (team work, critical reasoning) and (c) systemic skills (creativity, leadership and product care).

Keywords: Product engineering, experimental designs, emulsion, cosmetic cream, lab training.

Financial support from CYCYT CTQ2011-29336-C03-02 is gratefully acknowledged.

384678 The Ball-in-Tube Device: A New Approach to Introductory Fluid Dynamics Education

Abstract:

The ability to apply methods from a course to a real world problem is an essential goal of engineering courses. A few simple experiments are often used to illustrate the theoretical principles behind undergraduate-level fluid dynamics. It is not uncommon for students to be tasked with determining the pressure distribution across a Venturi meter, the flow rate out of a draining tank, or the pressure drop across a fluidized bed. In this work we demonstrate the use of the traditional "ball-in-tube" control experiment as another means to introduce a fluid-dynamics teaching module.

The module is constructed by placing a computer fan below a vertical acrylic pipe and securing an ultrasonic rangefinder in the center of the tube. By controlling the velocity of the air in the tube a ball can be made to float while its distance from the bottom of the pipe is recorded by the rangefinder. This data is then fed into a PID controller which adjusts the velocity of the fan in order to keep the ball at a set height. A differential pressure sensor measures the pressure drop across the tube, and is used in the analysis of the fluid dynamics of the system.

The ball-in-tube experiment has often been used in the classroom to illustrate principles of process control. The traditional goal of this teaching module is to precisely maintain a ball at a height set-point within a tube by controlling the velocity of air flowing past the ball. The ball in tube experiment is an integrating system and experiences significant turbulent noise over small time scales. Despite that behavior, analysis of this system shows that it conforms to traditional undergraduate fluid dynamic theory over extended time intervals, making it amenable to simultaneous illustration of control and fluid dynamic principles.

390693 DIY, Maker Culture, and Chemical Engineering

Abstract:

The semiconductor revolution and the Maker movement have made high technology more accessible than ever before. In this poster, I explore how open-access, open-source resources and a DIY ethic might influence the future of chemical engineering education.

394219 Reverse Electrodialysis: Sustainable Energy from Hydraulic Fracturing Water Recycle

Abstract:

Due to the growing limitations in dependence on fossil fuels, attention has been turned towards searching for alternative means for harvesting energy. One emerging source is energy produced from natural salinity-gradients between freshwater and seawater. Two different processes used to accomplish this are pressure-retarded osmosis and reverse electrodialysis. Significant amounts of research are currently under way regarding pressure-retarded osmosis, however very few studies have been conducted on reverse electrodialysis due to heavy resistance in the diluate chamber of a reverse electrodialysis cell. The focus of this research is to reduce this resistance by using wafers and ion-exchange beads in order to maximize net power gain. Surrogate brackish/saltwater feed streams were used to measure overall power density.

By reducing resistance to levels previously deemed impossible, we are exploring the fundamental limits of separation technology. This research proposes an innovative resolution to high resistance levels through the use of ion exchange wafers to promote ionic movement. Here we show that such wafers have a great impact on the reverse electrodialysis process, and succeeded in achieving a sizeable gain in power density.

Traditional applications include harnessing energy from areas where freshwater bodies meet saltwater, as in an estuary. The future of reverse electrodialysis, however, lies in fracking. Produced water from a fracking well site in Oklahoma was used for power density calculations under realistic conditions. Preliminary results show that our system is capable of producing power densities of up to $10W/m^2$, which are five times higher than the current state-of-the-art. Environmental concerns will be all but eliminated if fracking can evolve to be a sort of cyclic process, where brackish fracking water is converted to energy to, in turn, power the overall operation. This poster presentation will model the power increase of electrode ionization and comment on future potential for improvement, especially in the realm of hydraulic fracturing applications.

394215 Modeling of the Aerobic Cometabolic Transformation of Chlorinated Ethenes By the Mycobacterium Elw-1

Abstract:

ELW1 is a recently isolated aerobic bacterium, in the mycobacterium family, that uses isobutene as a growth substrate. ELW1 was also discovered to have the ability to cometabolically transform substances with chemical structures similar to isobutene, such as chlorinated ethenes, into chloride ion and water. Chlorinated ethenes are notorious carcinogens, and groundwater pollutants, whose remediation has been a worldwide problem for over 40 years. Bioremediation of chlorinated ethenes by the augmentation of contaminated soils with microorganisms such as ELW1 is a proposed method of reducing contaminant concentrations in groundwater. To determine the viability of using ELW1 for groundwater remediation, an initial kinetic analysis of transformation rates must be performed. The focus of this research was to determine the rates of contaminant transformation when this organism was exposed to high concentrations of chlorinated ethenes such as trichloroethylene (TCE), *cis*-dichloroethene (c-DCE), and vinyl chloride (VC).

The ELW1 culture was grown and sustained in both batch culture and steady-state chemostat environments with isobutene as a growth substrate on a minimal growth media. Cells that were used for experiments were concentrated by centrifugation, washed with excess media and placed in a new media solution containing high concentrations of the compound to be tested. Initial rates of transformation were determined using gas chromatography to analyze 100μL samples of headspace gas over a period of one to three days to determine the rate of compound loss. The rates at which ELW1 can utilize isobutene were compared with rates of chlorinated ethene transformation, which overall quantifies the practicality of using this organism bioremediation purposes in the field.

Kinetic parameters of these cometabolic processes were determined using an analytical equation based on a modified Monod equation in order to account for decrease in cell activity over time. Results of this investigation indicate that the ELW-1 culture would be effective for the bioremediation of c-DCE and VC, but would be less effective for treated TCE.

394214 The Enzymatic Hydrolysis of Alfalfa Stalks for Use As a Biofuel Resource

Abstract:

Most biofuel resources, especially plant sources, have two key issues: farming space and slow growth. After overexpressing the sucrose phosphate synthase gene (SPS) in Medicago sativa (alfalfa) plants, large differences in phenotype and growth rate were seen in the plant. While the leaves of the alfalfa plants can be used for animal forage, the stems of the alfalfa can be fractionated into sugars that can then be fermented into bioethanols. This technique could solve the growth problem faced by bioethanol producers.

The purpose of the project was to fractionate and analyze the extractability of fermentable sugars from a control and several transgenic species that show overexpression in the SPS gene. Using a strong acid pre-treatment and enzymatic hydrolysis to break down the cellulose into glucose, the alfalfa stalks were fractionated then analyzed using ultraviolet spectroscopy with anthrone as an indicator for glucose content. Two of the three tested transgenic species were shown to have more glucose content per gram than the control species (2-4% more glucose/gram). A further development of the overexpression technique can lead to faster growth, increased concentration of fermentable carbohydrates, and larger crop yields.

UNIVERSITY OF SOUTH CAROLINA
College of Engineering and Computing

SAGE
Power through Engineering

Synthesis and Characterization of FeAlCr via High Throughput Techniques

Randy Fang

Jonathan Kenneth Bunn and Dr. Jason Hattrick-Simpers

Department of Chemical Engineering

University of South Carolina

AICHE National Conference 2014

11/17/2014

UNIVERSITY OF SOUTH CAROLINA
Honors College

1

Fukushima Daiichi Nuclear Disaster

http://digginginthecIay.files.wordpress.com/2011/03/5525847859_e1934af238_o_d1.jpg

https://coto2.files.wordpress.com/2011/03/2-fuel-pellet-assembly.jpg

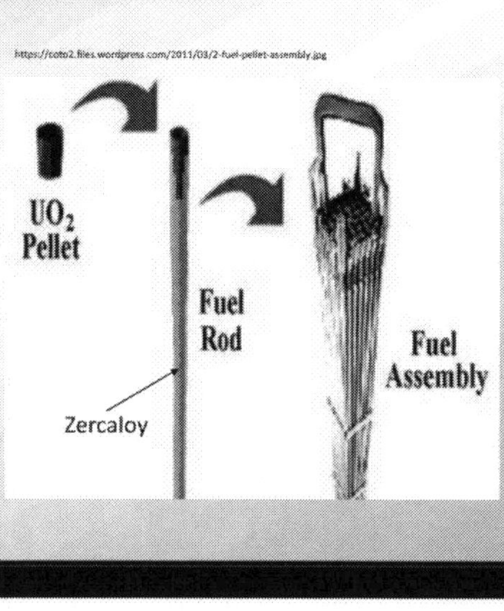

2

FeAlCr Alloy

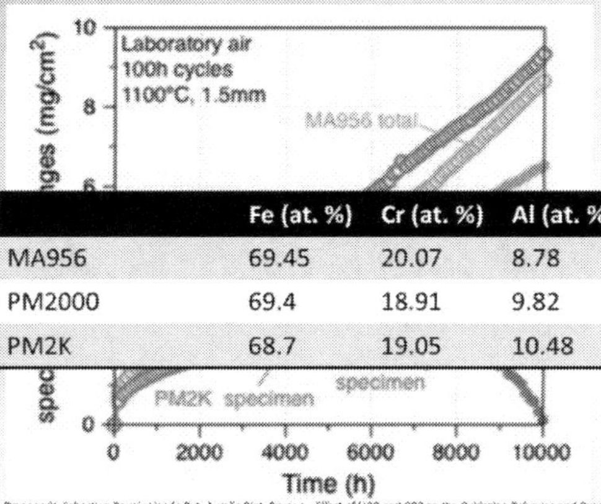

	Fe (at. %)	Cr (at. %)	Al (at. %)
MA956	69.45	20.07	8.78
PM2000	69.4	18.91	9.82
PM2K	68.7	19.05	10.48

Dryepondt, Sebastien Rouaix-Vande Put, Aurelie Pint, Bruce a., *Effect of H2O and CO2 on the Oxidation Behavior and Durability at High Temperature of ODS-FeCrAl*, Oxidation of Metals, Vol 79,2003

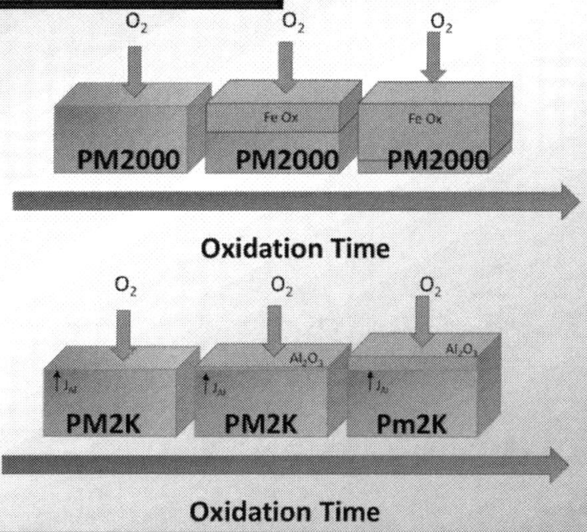

Introduction to High Throughput Techniques

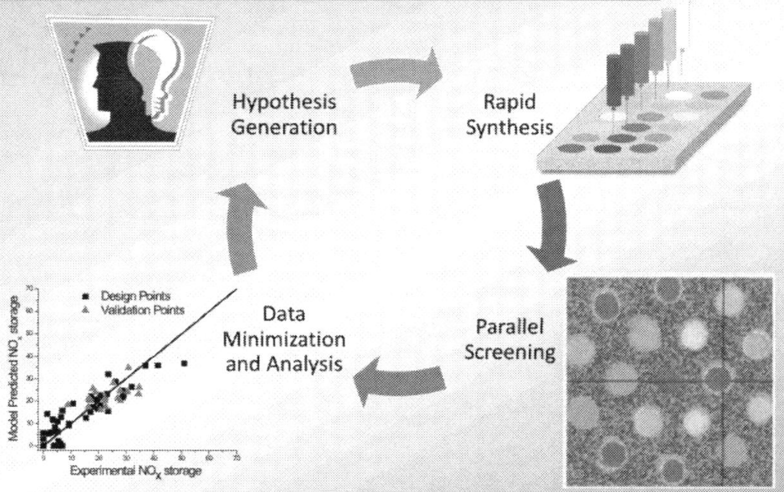

Research Objective

- **Implementing established high-throughput methods for FeAlCr high temperature oxidation resistance screening**
- Non-destructive characterization
- Gain comprehensive understanding of:
  - Time dynamics of oxide formation at high temperature
  - Role of metallurgical phase in promoting growth and continued growth of passivating oxide
- Identification of promising compositions for passivating oxide growth

5

High Throughput Approach to Explore the Ternary Phase Diagram

Qu Materials Science and Engineering A Vol 562 (2013)

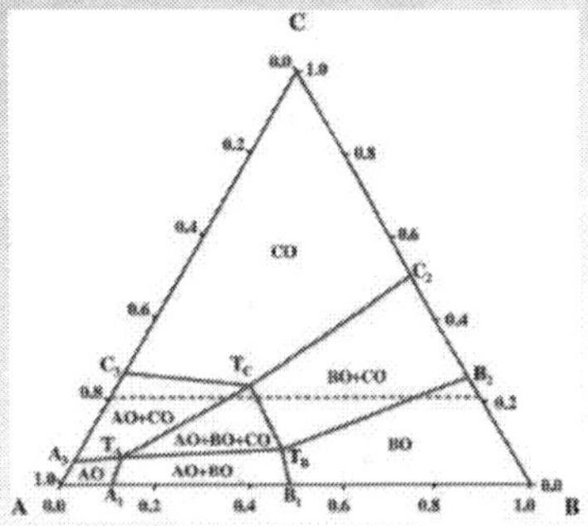

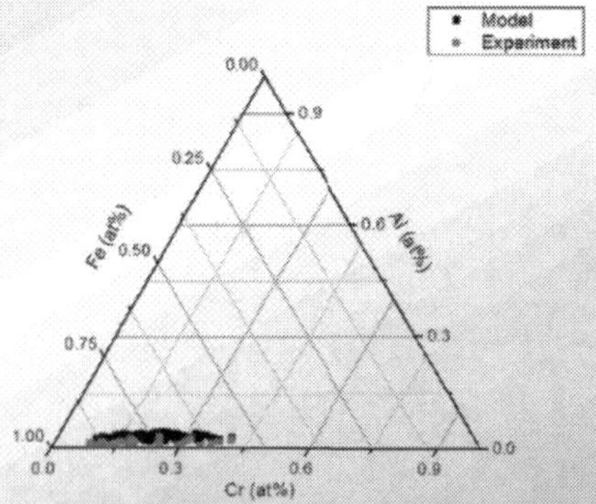

J. K. Bunn, C. J. Metting, and J. Hattrick-Simpers, "A Semi-Empirical Model for Tilted-Gun Planar Magnetron Sputtering Accounting for Chimney Shadowing ", JOM, invited paper, Accepted January 2014 Special Issue

6

Sputtering Geometry and Sample Locations

Ternary Deposition

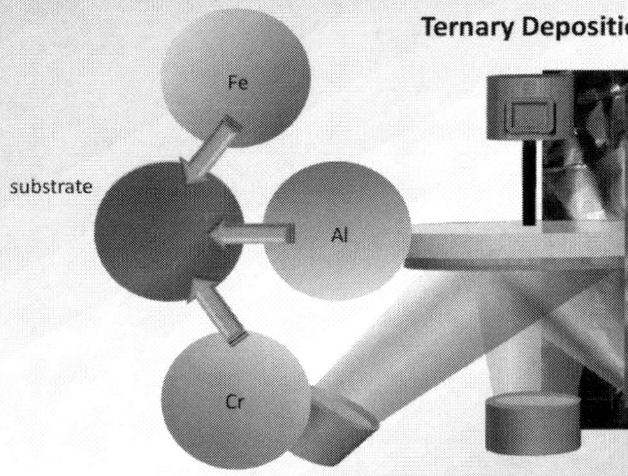

Deposited Sample At 6080CC

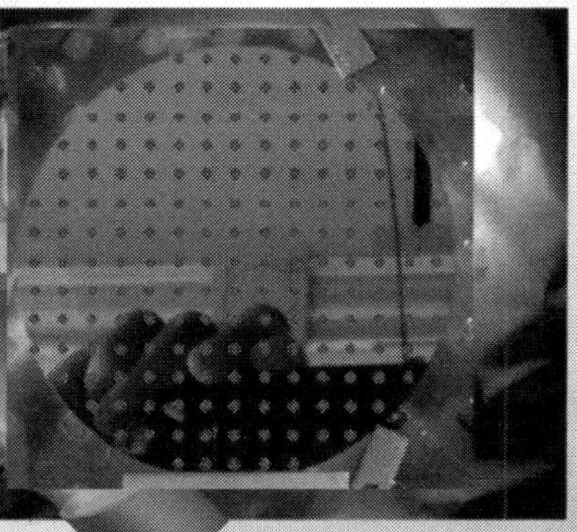

Magnetron Sputtering

- **5 Guns (only used 3)**
- **DC/RF power sources**
- **Ultra high vacuum**
- **Deposition temperature at room temperature**

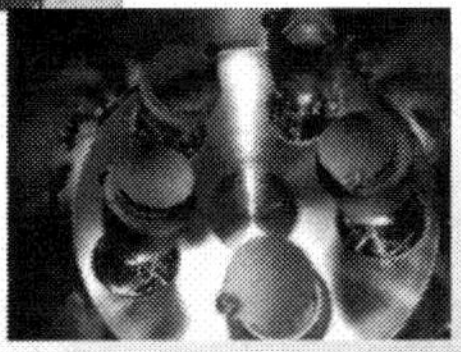

Experimental Design for High Temperature

Sample fabrication
- -12 FeAlCr samples prepared on Al_2O_3 substrate and Ti with adhesion layer
- -1 micron nominal thickness

Heat treatment
- Tube furnace treated
- Treated at 1050°C
- Air quenched

After deposition

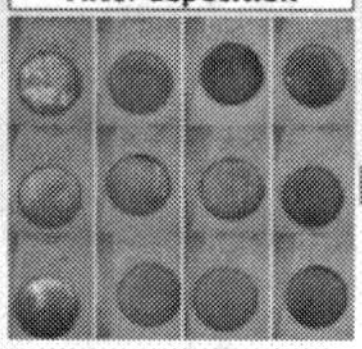

After heat treatments

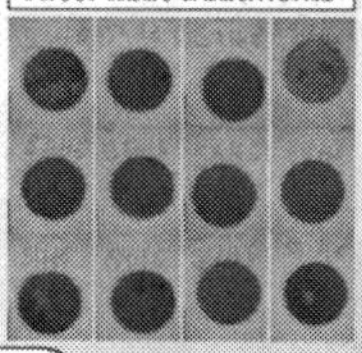

Cycled 7 times for cumulative heat treatments at 1050°C for:

-5 min	-10 min
-20 min	-40 min
-60 min	-120 min
-240 min	-360 min

Characterization
- -SEM/EDS
- -XRD

Characterization Techniques
- **Raman**: Fe_2O_3, Fe_3O_4, and Al_2O_3
- **XRD**: FeCrAl, Surface Oxides, and the substrate
- **Optical Microscopy**

Oxide Formation at 5 minutes, 1050°C

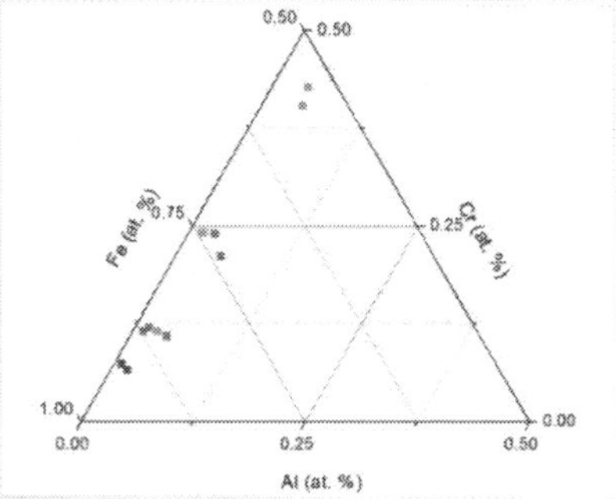

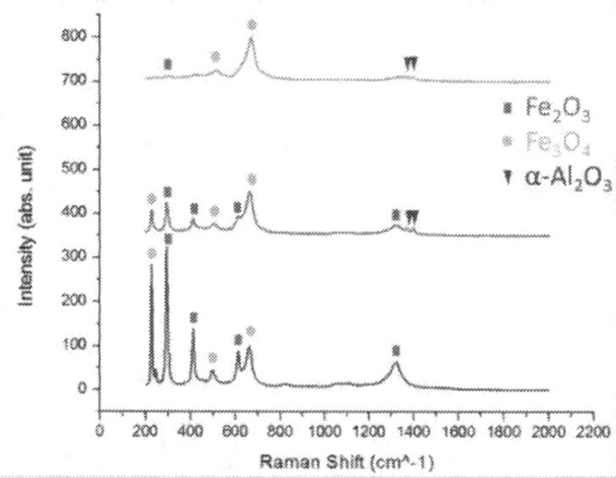

Blue: mostly Fe_2O_3, some Fe_3O_4
Red: mostly Fe_3O_4, some Fe_2O_3, some α-Al_2O_3
Green: very little Fe_2O_3, some α-Al_2O_3, some Fe_3O_4

- No α-Al_2O_3 growth was observed when the composition is Fe heavy

Oxide Formation at 360 minutes, 1050˚C

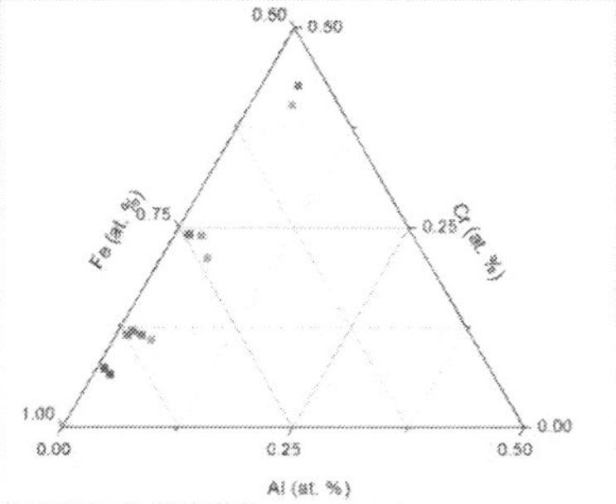

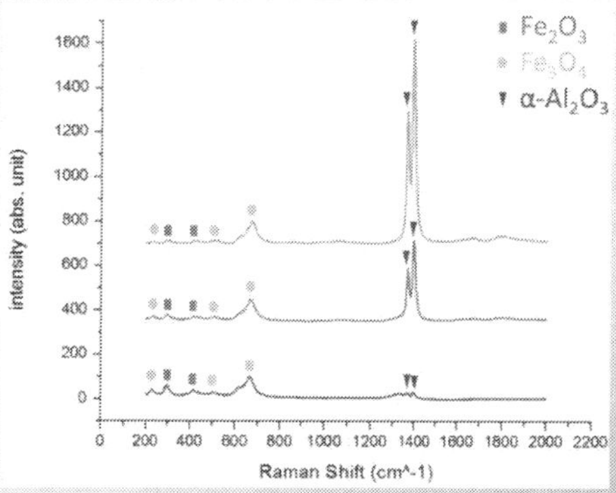

Blue: some Fe_2O_3 and Fe_3O_4, very little α-Al_2O_3
Red: some α-Al_2O_3, some Fe_3O_4
Green: majority α-Al_2O_3, some Fe_3O_4

- Large amount of α-Al_2O_3 growth was observed at the Cr heavy region, indicating that Cr promotes the growth of the alumina oxides

11

Experimental Design at Low Temperature

Sample fabrication
-FeAlCr samples prepared on Si substrate with Ti diffusion barrier
-1 micron nominal thickness

Cycled 6 times for cumulative heat treatments 1 hour each (with the exception of 600°C):

-room temperature	-200°C
-300°C	-400°C
-500°C	-600°C (for 2hr)

Characterization
-*in-situ* GIXRD

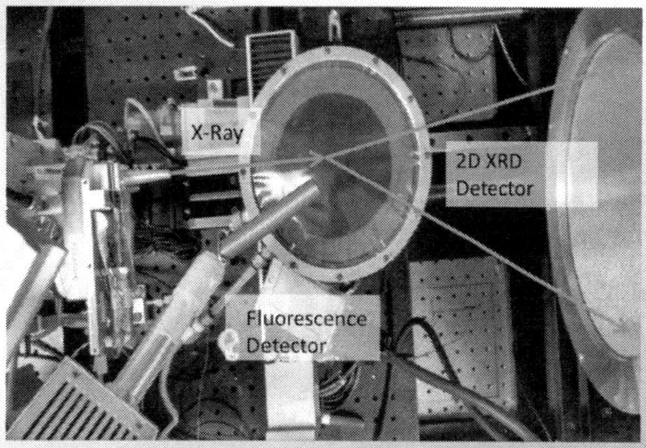

12

Oxide Growth Study through *in-situ* XRD

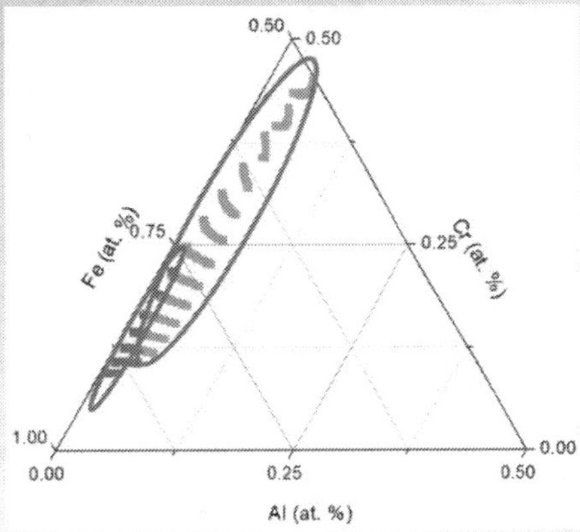

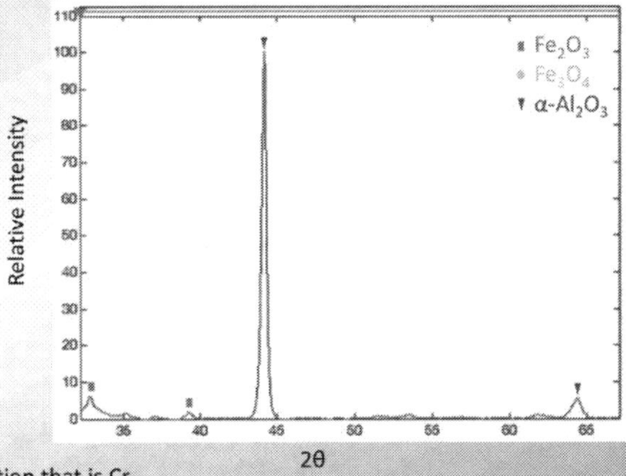

- Iron oxides are suppressed when the composition that is Cr rich (small amount are observed)

Comparison of Trends Between Low and High Temp Oxidation

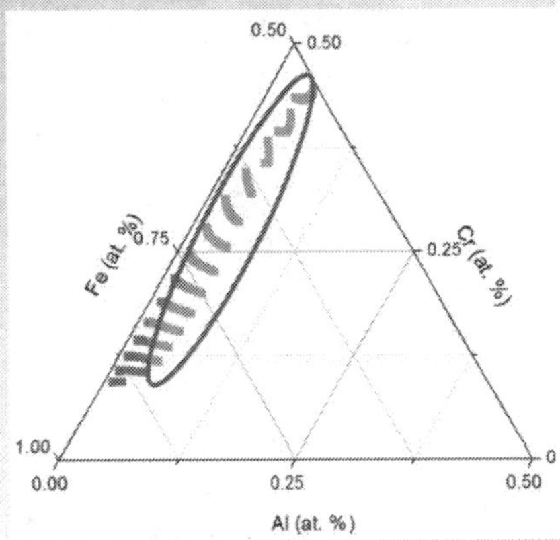

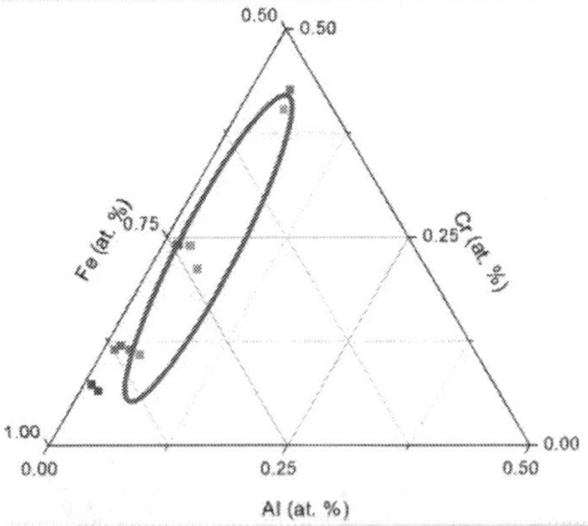

- XRD and Raman matched up where the more Al and Cr rich region are a lot easier to form α-Al_2O_3

Summary

- Probed the temporal effect of temperature on the metallurgic and oxide phase transitions
- A method for the **initial high-throughput and non-destructive screening** of FeCrAl thin film was developed
- **Cr promotes** α-Al_2O_3 growth on the surface of the samples to form a protective layer preventing further development of other oxides
    - This indicates that Cr rich FeCrAl are more favored to operate under high temperature conditions
- Oxidation studies indicate that severe oxide growth was not observed until 400°C
- The methodology can be applied to more complicated systems for high temperature oxidation resistant materials.
- Detailed studies of **oxidation and phase transitions of single compositions at different temperatures** will refine the results

15

Acknowledgement

- **Dr. Jason Hattrick-Simpers**
- **Jonathan Kenneth Bunn**
- **SmartState Center for Strategic Approaches to the Generation of Electricity**
- **USC Honors College (Science Undergraduate Research Fellowship Program)**
- **University of South Carolina**
    - **College of Engineering and Computing Science**

This project was funded in part by the South Carolina Honors College

16

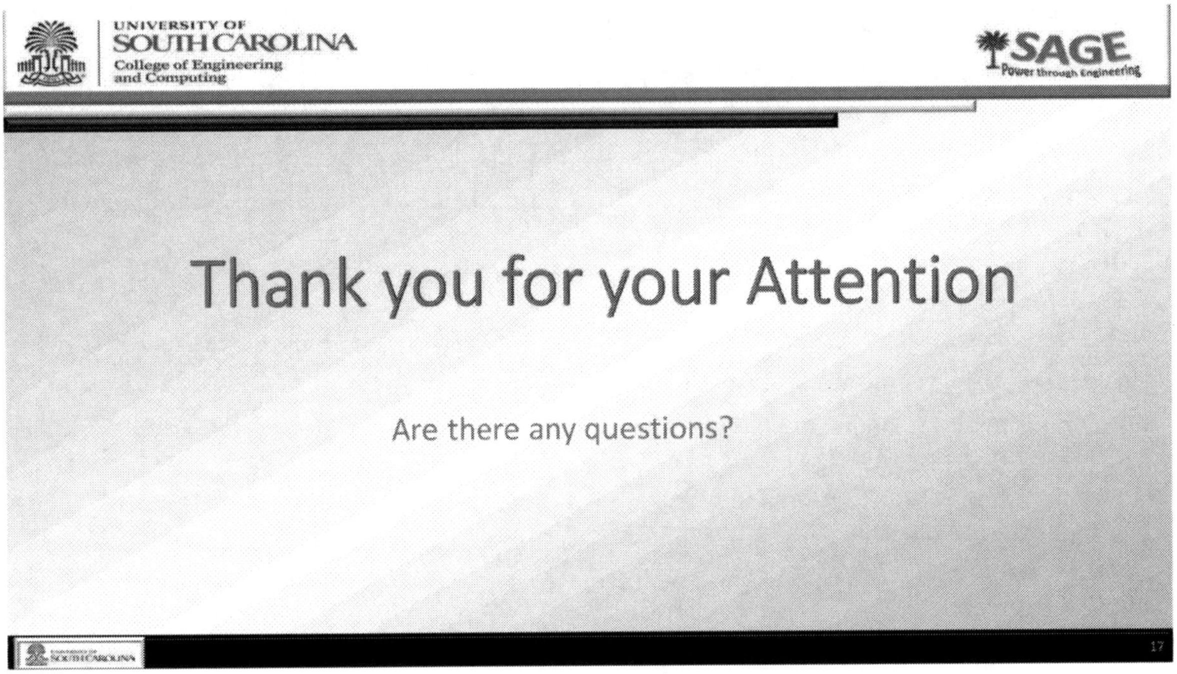
UNIVERSITY OF
SOUTH CAROLINA
College of Engineering
and Computing
SAGE
Power through Engineering
Thank you for your Attention
Are there any questions?
17

Crossing Borders and Communities for Healthy Water

Joshua Gomez, Meng Zhou, Dr. Shuguang Deng

Chemical and Materials Engineering Department, New Mexico State University, Las Cruces, New Mexico, 88003

In Partnership with:

NM Water Resources Research Institute

NMSU Chemical Engineering Department

The New Mexico Alliance for Minority Participation

New Mexico Environment Department

Border Partners (Non-profit)

EPA 2020

National Science Foundation

New Mexico State University NM STATE

Puerto Palomas, Chihuahua, MX

- Located just south of Columbus, NM.
- Population of 4,688.
- Arsenic contamination (0.030-0.1 mg/L).
  - Horrible Carcinogen.
- Fluoride contamination (5-40 mg/L).
  - Bone Fluorosis and neurological damage.
- Concentration depends entirely on well water.
  - Naturally occurring concentrations of arsenic and fluoride exceed the permissible limits of Mexico and the U.S.

New Mexico State University NM STATE

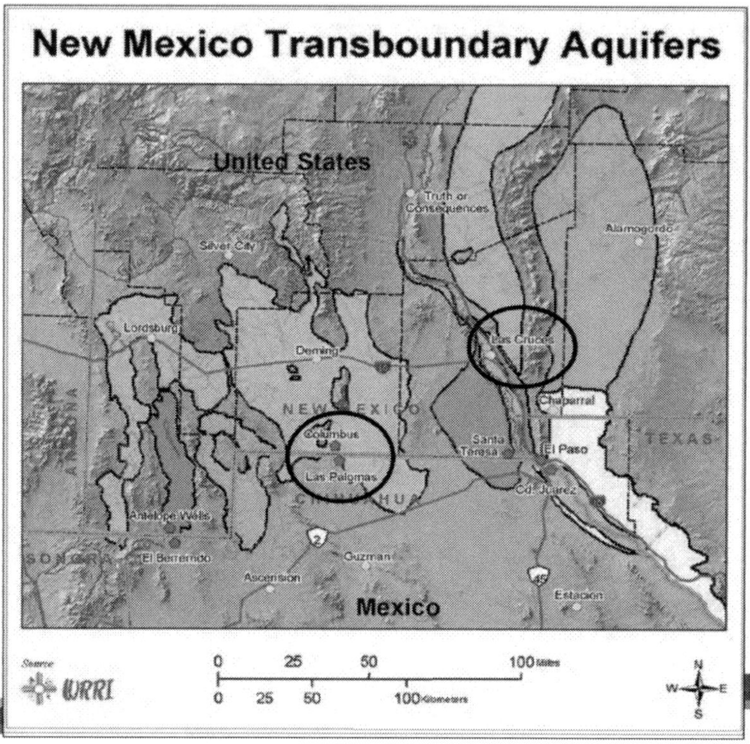

New Mexico State University NM STATE

Proposed Solution

- Homemade water filter
  - Low cost <$15.00
  - Easy to assemble
  - Removes arsenic
  - Removes fluoride
  - Utilizing activated carbon (AC)
  - Utilizing activated alumina(AA)
    - Iron coated
  - Improves the taste

New Mexico State University NM STATE

Filter Designs

- 1 filter of the final design costs $5.00 to make.
- The Activated Alumina costs $1/kg.
- The Activated carbon used cost $35/lb.

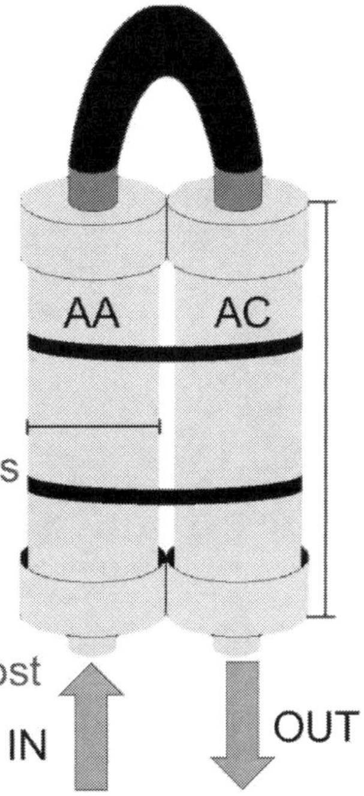

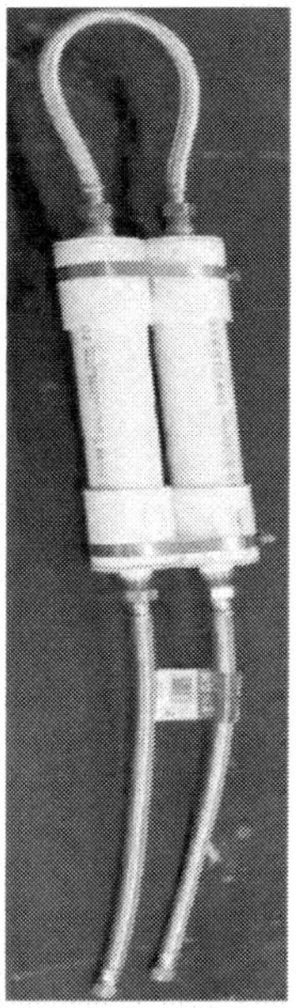

New Mexico State University

Our Approach

NMSU demonstrated to Palomas science teachers how to assemble a simple water filter.

Science teachers Juan Carlos Huerta and Joel Carreon pose with the filter they assembled for installation at the Palomas primary school.

New Mexico State University

What's Working

Hall Environmental Laboratory, Albuquerque

CONTAMINANT	RESULTS	STANDARDS					
		U.S.	Exceeded	%	Mexico	Exceeded	%
Fluoride							
Unfiltered	4.90 mg/L	4.00 mg/L	Y	22.5	1.50 mg/L	Y	226.7
Filtered	0.240 mg/L	4.00 mg/L	N	-	1.50 mg/L	N	-
Arsenic							
Unfiltered	0.030 mg/L	0.01 mg/L	Y	200.0	0.02 mg/L	Y	20.0
Filtered	ND	0.01 mg/L	N	-	0.02 mg/L	N	-

ND = Not Detected

New Mexico State University NM STATE

Optimization of Algae Derived Activated Carbon (AC)

- Low cost alternative to commercial activated carbon.
- More effective adsorbent than commercial carbon.
- Chlorella algae is used as a carbon source.
- Potassium Hydroxide and Iron (II) Chloride
  – Activating agent.
- Carbonization times of 500° C, 600° C, and 700° C.
- Shaker experiments to determine most effective adsorbent for Arsenic(III).

New Mexico State University NM STATE

Carbon Experimental data

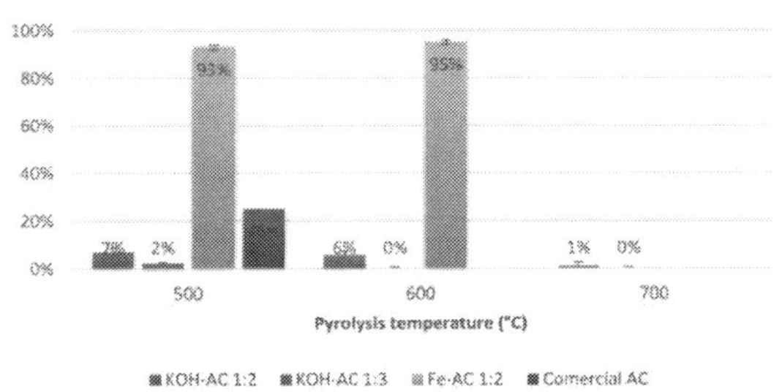

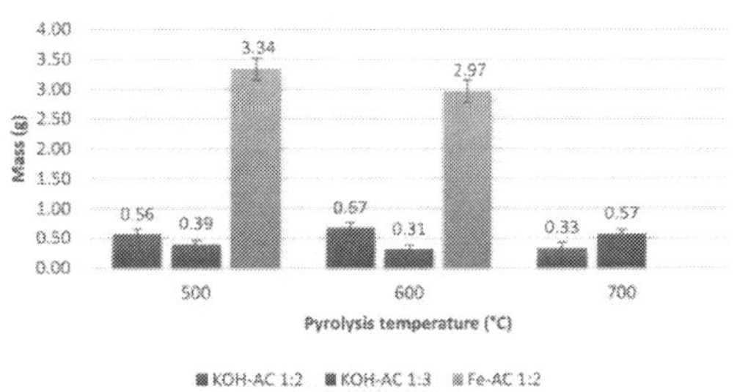

adsorption capacity (mg/g)			
	500°C	600°C	700°C
KOH-AC 1:2	0.082	0.067	0.010
KOH-AC 1:3	0.024	0.000	0.000
Fe-AC 1:2	1.143	1.168	-
Comercial AC	0.275	-	-

$$capacity = \frac{(InitialConcentration - FinalConcentration)Volume}{Mass\ of\ adsorbent}$$

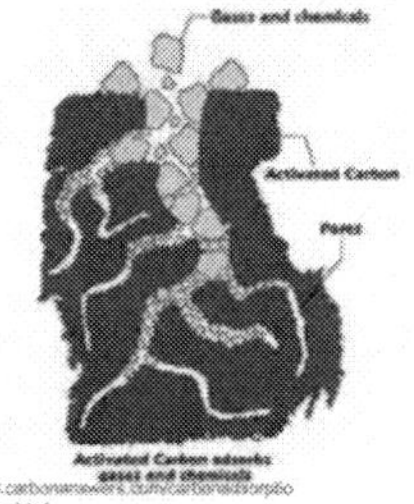

Fasci, M.

New Mexico State University

X-Ray Difraction Pattern

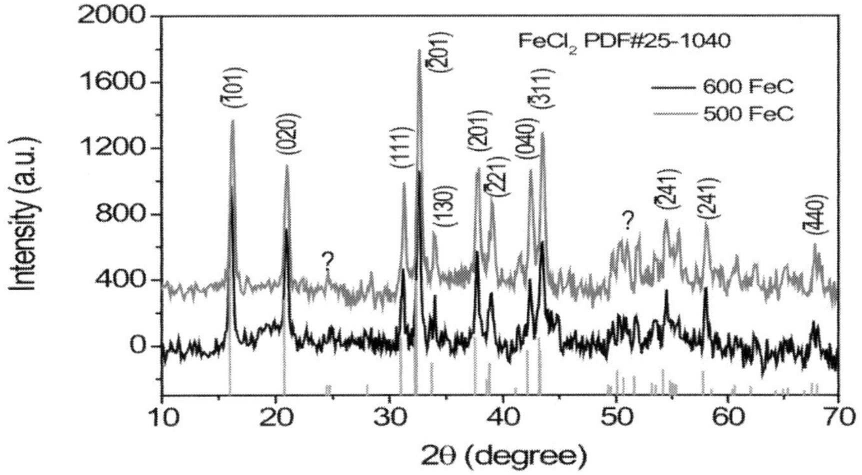

New Mexico State University

Scanning Electron Microscope Imaging

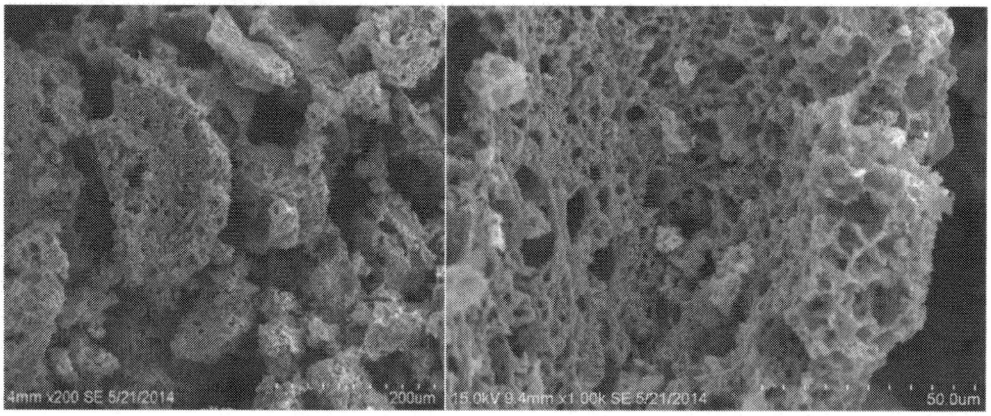

New Mexico State University

Unwashed FeCl(II) Activated Carbon Textural Properties

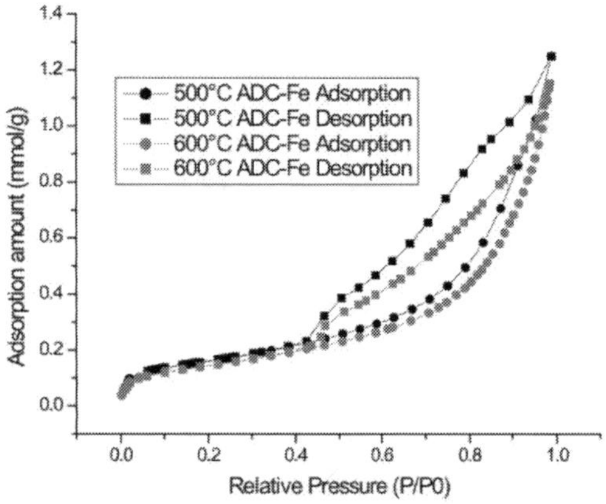

N_2 adsorption and desorption isothermal @ 77K for unwashed samples.

	500 ° C FeCl2/ADC	600 ° C FeCl2/ADC
BET surface area (m^2/g)	12.9893	11.6336
Average pore size (nm)	8.6408	6.5769
Pore volume (cm^3/g)	0.0394	0.0476

New Mexico State University

Washed FeCl(II) Activated Carbon Textural Properties

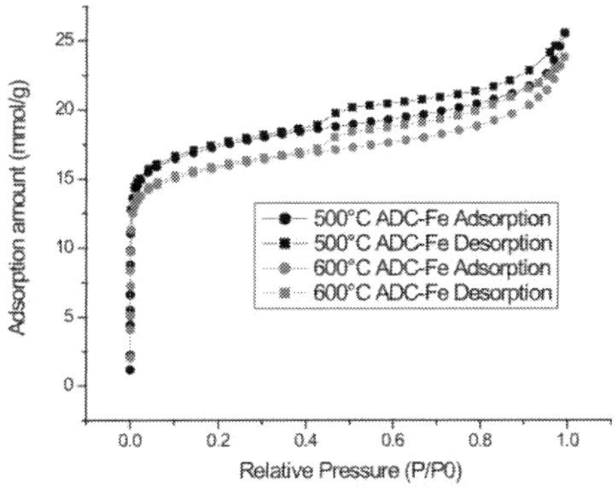

	500 ° C ADC (after washing)	600 ° C ADC (after washing
BET surface area (m^2/g)	1326.0368	1208.1831
Average pore size (nm)	7.8704	7.9340
Pore volume (cm^3/g)	0.3362	0.3305

N2 adsorption and desorption isothermal @ 77K for washed samples.

New Mexico State University

Shaker Experiments

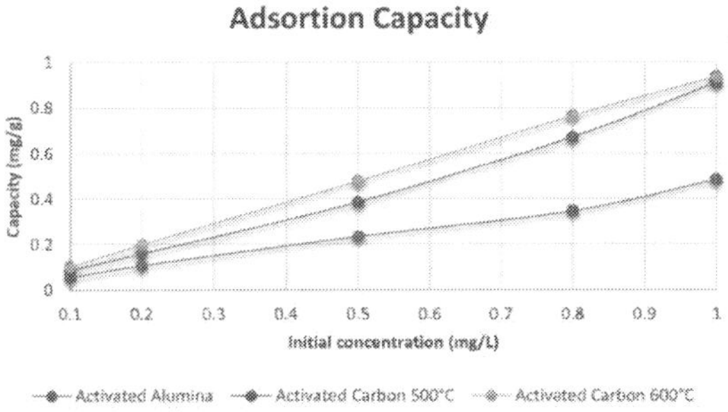

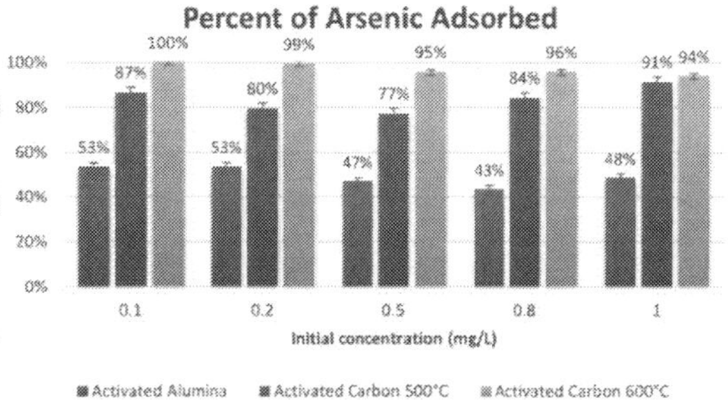

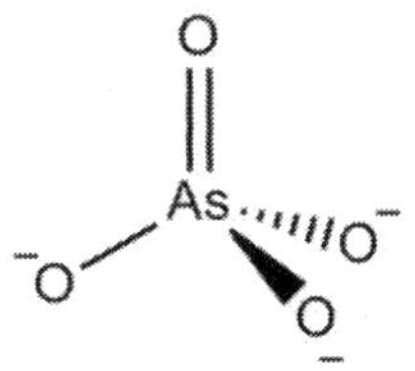

Arsenate

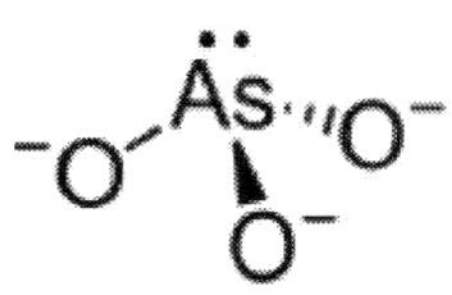

Arsenite

New Mexico State University

Conclusions

- Filters are still being made in Palomas by the teachers today and will continue to be made.
- The results of the shaker experiments and textural property data shows that the carbon activated with Iron (II) Chloride is a viable replacement.

New Mexico State University NM STATE

Next Steps

- Further Development of the filter.
- Reach out to other communities.
- Advance research with optimized carbon.

New Mexico State University NM STATE

Acknowledgments

- Dr. Shuguang Deng, Chemical & Materials Engineering dept.
- Meng Zhou, Chemical & Materials Engineering dept.
- Dr. David A. Rockstraw, Chemical & Materials Engineering dept.
- Erin Ward, Water Resources Research Institute.
- Mark Chidester, Chemical & Materials Engineering dept.
- Peter Edmunds, Border Partners.

NEW MEXICO AMP

NSF HRD # 1305011

Deng's Renewable Energy and Advanced Materials

New Mexico State University

Questions?

Email: jgomez38@nmsu.edu

New Mexico State University

394186 Ferroelectric BTO on Si (001) for High-Efficiency Solar Cell Heterostructures

Abstract:

Ferroelectric oxides such as barium titanate (BTO) have recently become a topic of interest in relation to thin-film solar cell production due to the material's stability at a wide range of temperatures, its spontaneous polarization, and the potentially low-cost of production. Integration of BTO on silicon could enable the creation of highly efficient solar cells, as BTO's ferroelectric properties encourage the detachment of photo excited electrons. We have previously demonstrated successful integration of BTO on silicon carbide (SiC) substrates, using a magnesium oxide (MgO) heteroepitaxial layer to enable single-crystal orientation of the BTO film deposited on SiC.[1] BTO has also been successfully grown on silicon substrates, but with the use of strontium titanate (STO) and MgO buffer layers to resist formation of SiO_2.[2] Integration of BTO on Si with a single MgO interface layer would decrease interfacial losses in a final solar cell device. Based on previous work with SiC, we believe a single interface can be possible by controlling the surface interactions during film growth.

In this research, MgO is grown on Si (001) substrates in an ultrahigh vacuum (UHV) environment using a number of growth methods including molecular beam epitaxy (MBE) and sputtering. In comparing differences in surface structure using reflection high-energy electron diffraction (RHEED) and surface chemistry using X-ray photoelectron spectroscopy (XPS), we can understand surface interactions between Si, Mg, and O. By understanding the relationship between atomic level interactions and ultimate film characteristics, we can engineer the most effective process for BTO integration on Si. The substrates are cleaned using wet chemicals to create a hydrogen-terminated surface that resists contamination from laboratory exposure. XPS and RHEED are used to verify the lack of contamination on the surface and proper crystallographic orientation. Contaminant levels of 7.61 % oxygen (0.15 O/Si ratio) and 18.0% carbon (0.31 C/Si ratio) are the most successful results that have thus far been attained. XPS and RHEED will serve to confirm proper growth of MgO (001) on Si and determine thickness of the MgO film. BTO is then grown on the correctly-oriented MgO film using MBE, as has already been demonstrated on MgO/SiC substrates.[1] This presentation will discuss cleaning and growth conditions for successful production of high quality BTO on Si using the single MgO interface layer.

393873 Natural Gas, the Bridge Fuel

Abstract:

Medium and heavy duty trucks and buses produce almost 18 percent of all carbon emissions from transportation. Alternative energy sources need to be used on a larger scale to reduce these emissions, as well as to reduce dependence on foreign oil. Natural gas is the future of fuel in large vehicles. This paper tackles different issues surrounding the usage of natural gas as the main fuel for natural such as public awareness, public policy, existing infrastructure, consumer choice and controversies surrounding hydraulic fracturing. Current pilot programs, in which public city buses are using natural gas as the fuel of their choice, will be analyzed to determine the best way to implement these programs in other areas of the United States as well.

397811 The Federal Role in Fostering an Innovative U.S. Energy Ecosystem

Abstract:

Although renewable sources of energy – solar, wind, hydro, and geothermal – have been around for over one hundred years, they only generate 12.5 percent of electricity in the United States today. Renewable technology has advanced in cost, reliability, performance, and efficiency, but not enough to allow renewables to be a large contributor to electricity generation and widely used in the U.S. Renewables need continued innovation and progression to become more efficient, more reliable, and more cost competitive with fossil fuels, as well as expanding the possibilities of their use. Increased R&D funding for renewable sources is the key to advancing the technology and giving renewables the tools it needs to provide a larger portion of electricity in the U.S.

With renewables generating a larger portion of electricity, the U.S. can reach environmental sustainability, energy security, and a new economic opportunity. In reference to reaching environmental sustainability, renewables do not emit carbon and greenhouse gases, which is important for a cleaner future and mitigating climate change. Because the "fuel" for renewables includes the sun and wind, electricity generation from these sources does not depend on obtaining fossil fuels inside or outside of our country. They also extend the lifetime of these resources to be used for base load power, providing energy security to the U.S. Finally, if the U.S. is able to quickly and effectively develop competitive renewable technologies and deploy them into the electricity sector, it can provide competitive renewable technologies to the global community, boosting the economy and maintaining world leadership.

Although renewables have the capability to provide many benefits for the U.S., they face obstacles in receiving the funding they need to be widely used. These obstacles include climate change skeptics, the fossil fuel industry, the polarized political environment, and having to cut funding for other areas such as education and defense in order to allocate more money to R&D. The demand for renewables has also been relatively weak because electricity generated by renewables is no different than electricity generated by fossil fuels when the consumer uses it. The main way for renewables to be competitive with fossil fuels is through the cost and since wind is the only renewable that is cost competitive with fossil fuels, a strong customer demand has not been present to spur large private sector investment in renewable technologies.

In order for renewables to accomplish necessary advancements and contribute a larger portion of electricity to the U.S., financing to R&D for the technologies needs to be increased through government funding, capital generated through a carbon tax, public-private partnerships, and incentives for investment in the technology and its R&D. A commitment to innovating renewables in order to reach environmental sustainability, energy security, and a new economic opportunity is necessary from governmental officials to make the tough decisions advocating for widespread use of renewables in the U.S.

397856 A Biowall for Improving Indoor Air Quality

Abstract:

The Biowall is a living plant wall that filters air in the HVAC system of a home or small commercial building. This new technology improves indoor air quality be removing Volatile Organic Compounds and has the potential for energy savings by reducing the need for heating and cooling large amounts of outside air in a building. The presentation provides an overview of biofiltration for buildings and discusses the new research being conducted in this emerging field.

393642 Controlled Release of Type-2 Diabetes Systems

Abstract:

Type-2 diabetes is characterized by insulin-resistance, hyperglycemia, and dysfunctional insulin hormone signaling, and accounts for 90-95% of all total cases of diabetes.[(1)] The total population affected is expected to reach 4.4% worldwide in 2030.[(2)] Current treatment of type-2 diabetes involves a drug regimen, using small molecule pharmaceutics that act on biological pathways to regulate glycemic control. These drugs modulate glucose metabolism, insulin secretion, and/or insulin sensitivity. Best results are observed with optimal patient compliance. Controlled drug delivery could be leveraged to reduce the dependence on patient compliance and enable more autonomous therapies. Specifically, drug delivery from polymeric biodegradable constructs or stimuli-response "smart" gels provide a basis for self-regulated systems.[(3)] In the context of diabetes therapy, glucose is an especially relevant trigger in the design of polymeric drug delivery systems.[(4)] In this work, we have developed several drug-delivery formulations that provide long-term temporal control over drug release. These efforts have resulted in the successful encapsulation of type-2 diabetes drugs, and control over their release kinetics over a period of days. Further efforts to make autonomous therapies will continue to investigate tighter control over glucose-responsive polymeric constructs, with the eventual goal of developing smart glucose-sensing technology for on-demand drug delivery that responds in real time to patient glucose levels. Preliminary efforts have also explored the use of these particles to control blood glucose levels in an animal model of type-2 diabetes. It is our hope that, through this work, the efficacy of type-2 diabetes drug therapy will be improved through smarter design of materials-based therapies.

1. Chronic Disease Prevention and Health Promotion: Diabetes. (2011). Retrieved June 22, 2014, from http://www.cdc.gov/chronicdisease/resources/publications/aag/ddt.htm

2. Wild, S., Roglic, G., Green, A., Sicree, R., King, H. Global Prevalence of Diabetes: Estimates for the year 2000 and projections for 2030. *Diabetes Care* 2004, 27, 1047–1053.

3. Matsumoto, A., Ishii, T., Nishida, J., Matsumoto, H., Kataoka, K., Miyahara, Y. A Synthetic Approach Toward a Self-Regulated Insulin Delivery System. *Angew. Chem. Int. Ed*. 2012, 51, 2124 –2128

4. Kitano, S., Koyama, Y., Kataoka, K., Okano, T., Sakurai, Y. A novel drug delivery system utilizing a glucose responsive polymer complex between poly (vinyl alcohol) and poly (IV-vinyl-2-pyrrolidone) with a phenylboronic acid moiety. *Journal of Controlled Release*. 1992, 19, 162–170.

Performance of Janus Particles in the Self-Assembly Fabrication of Nanoparticle-Decorated Honeycomb-Structured Polymeric Films via the Breath Figures Patterning Technique

Sophie E. Miller[1,2]*, Wei Sun*[2,3]*, and Julia A. Kornfield*[2]

(1) Department of Chemical Engineering, Stanford University, Stanford, CA

(2) Division of Chemistry and Chemical Engineering, California Institute of Technology, Pasadena, CA,

(3) Faculty of Materials Science and Chemical Engineering, Ningbo University, Ningbo, China

Abstract

The breath figures (BFs) patterning technique represents a simple, robust method for producing microstructured honeycomb-patterned porous polymer films. Nanoparticles can self-assemble by spontaneous adsorption onto the interfaces between the templating water droplets and the polymer solvent in order to achieve controlled micrometer and nanometer scale hierarchical structures. Amphiphilic Janus nanoparticles are expected to optimize the films with regards to nanoparticle decorating density within the pores as well as ordering degree of the honeycomb structures given their dual role as Pickering emulsion stabilizing agents and as highly effective solid surfactants, resulting in a significant improvement over what has been achieved using isotropic nanoparticles. Exposed surfaces of silica particles embedded in spherical wax structures (colloidosomes) are chemically modified to render half of the particle hydrophobic. A reliable method for producing optimal colloidosome coverage has been established. Upon successful synthesis of amphiphilic Janus particles, most reliably confirmed by a two-phase interface test, they are employed in the BFs templating method for the first time. Characterization of the resulting films' morphologies will help draw conclusions on the interfacial behavior and self-assembly performance of the Janus particles, as well as the general mechanism of structure formation in the BFs patterning technique.

Introduction

Highly ordered, honeycomb-patterned porous films have potential applications in many areas, including photonic crystals and devices, electronic films, biomimetic materials, picoliter beakers, filtration membranes, cell culturing and immobilization, and catalysis.[1,2] Among the various methods used for fabricating structured hybrid polymer films, the bottom-up, self-assembly breath figures (BFs) templating technique has proven to be among the most simple and reliable.[1,3,4,5]

"Breath figures," studied as early as 1911 by Lord Rayleigh, refer to the organized arrays of water droplets that form when humid air comes in contact with cold solid or liquid surfaces.[2,3] In the BFs patterning technique, water droplets from a flow of humid air condense above the surface of a polymer solution, naturally cooled by the evaporating solvent, in turn providing countless emulsion-like interfaces. When solid nanoparticles are added to the polymer solution, they tend to self-assemble at this interface between the water droplets and the non-polar polymer solution in order to minimize the total energy of the system.[6] This phenomenon, known as a Pickering emulsion, stabilizes the water droplets and prevents them from coalescing.[1,3] The resulting highly-ordered monodisperse array of water droplets leaves an imprint of microscale honeycomb-arranged cavities on the polymer film after the solvent and water evaporate.

In his prior work, Prof. Wei Sun showed that isotropic silica particles employed in the BFs method not only assist in the formation of ordered porous arrays, but also result in hierarchically patterned surfaces with oriented, ordered arrays of silica nanoparticles decorating the microscale honeycombed structure. A trade-off was found, however, between the density of arranged nanoparticles and the degree of order in the honeycomb arrays, with higher concentration of silica particles resulting in altered pore shape and compromised regularity and uniformity of the polymer film.[3]

Janus particles are homogeneous particles with two surfaces of different chemical composition. Their anisotropic surface, combined with their ability to sustain Brownian motion due to their small size, allows them to self-assemble under the right conditions. Janus particles that are hydrophilic on one side and hydrophobic on the other side are the colloid equivalent of a surfactant. When added to the casting polymer solution in the BFs method, they serve to stabilize the water droplets both through the Pickering emulsion effect and by decreasing the surface tension between the droplets and the polymer solution.[7]

Objectives

The goal of this project was to employ, for the first time, novel Janus particles in the BFs method to assist in the fabrication of highly-ordered hybrid films on the micrometer scale and, simultaneously, achieve oriented self-assembly of the Janus particles on the inside walls of the films on the nanometer scale. By modifying the characteristics of the particles employed and investigating the resulting assembled structures, conclusions would be drawn about the interfacial behavior and self-assembly performance of Janus particles in the novel BFs patterning technique. Also, light could be shed on the still-undefined kinetics of structure formation, including how the interfacial stabilization works to prevent the water droplets from coalescing and what kind of processes the templating water droplets undergo during the BFs method.

Given the dual role of Janus particles as Pickering emulsion stabilizing agents and as highly effective solid surfactants, it was expected that the patterned films produced would be optimized with regards to nanoparticle density and ordering degree of the honeycomb structures, resulting in a significant improvement over what has been achieved using isotropic-nanoparticles. Successfully creating uniform, oriented arrangements of Janus particles on hybrid films would have many applications in the design of materials with functional surface properties, such as photonic devices, organic LEDs, electronic films, biosensors and biomimetic membranes.[2, 8]

Results and Discussion

The first part of the project consisted of synthesizing and characterizing the Janus particles. As expected, developing a reliable method for producing optimally covered colloidosomes and confirming the amphiphilicity of the chemically modified nanoparticles proved to be the most challenging step. Once Janus particles were successfully obtained, they were employed in the particle-assisted fabrication of porous polymer films via the BFs templating method. Lastly, the resulting nanoparticle-decorated honeycomb-structured films were characterized.

Synthesis of Wax-Silica Nanoparticle Colloidosomes

Throughout the course of the research, twenty-eight batches of colloidosomes were synthesized by adapting the method employed by Granick (see Methods).[8, 9, 10, 11] This was a challenging step, as the synthesis of wax colloidosomes with adequate coverage and desired contact angle of nanoparticles proved to be highly dependent on the amount of nanoparticles and the methods of emulsification and cooling used. A smooth, even, ordered monolayer of silica particles, with consistent contact angles on the surface of the wax, is required for eventual chemical modification of the Janus particles. Several variables were adjusted in order to achieve optimal coverage of the wax structures.

Nearly all batches were prepared by emulsifying with magnetic stirring. It was hypothesized that sonication could produce smaller emulsion droplets, leading to more regular colloidosomes. However, when sonication was pursued as an alternative to magnetic stirring, the resulting colloidosomes were irregularly shaped with rough, bumpy, cracked surfaces. These particles did not self-assemble onto the interfaces between the wax droplets and the aqueous phase; they were observed throughout the samples, both on the outside and inside of the wax structures. Also, the filtered solution was cloudy, further indicating the presence of unadsorbed silica nanoparticles. Sonication, using the sonicator probe at 30% power, was likely too violent.

Coverage of silica particles on the colloidosome surfaces was found to be highly sensitive to the amount of nanoparticles initially dispersed in the molten wax. The first three batches were prepared with 200 mg 600 nm silica particles. While self-assembly of particles at the wax-aqueous interface was achieved, with particles observed exclusively on the surface of the colloidosomes, SEM images revealed overcrowded surfaces with multilayers of particles instead of the desired monolayer and low contact angles due to jamming. On the other hand, when the amounts of silica particles were reduced to 33.3 mg, 66.7 mg, and 100 mg, the obtained colloidosomes were sparsely covered and the nanoparticles had a wide range of contact angles. It was ultimately concluded that 175 mg silica nanoparticles is optimal for fabricating colloidosomes to make 600 nm Janus particles: resulting surfaces were uniformly covered in highly ordered, well-packed monolayers of silica particles with three-phase contact angles of around 90°.

The most significant improvements in preparing colloidosomes resulted from adjusting the cooling conditions. Several cooling methods and variables, such as stirring rate, were tested.

The greatest challenge encountered was the solidification of silica covered wax structures onto the flask walls. Product had to be scraped off the walls of the flask, resulting in irregular, damaged colloidosomes (Fig.s 1a and 1c). It was hypothesized that slowly cooling the emulsion flask in the oil bath from the "outside-in" was causing the wax to solidify onto the cooler flask walls. When cold water (4°C) was added to the emulsion once it reached 60°C, just above the wax melting point of 53-57°C, small, white particles solidified and floated to the surface. However, the sudden addition of cold water was too intense, and the resulting colloidosomes had irregular nanoparticle coverage and contact angle.

A more gradual inside out cooling was pursued by progressively adding cooler water to the emulsion in two steps after turning off the heating source. First, 50°C water was added to begin steady cooling and to provide a larger volume of liquid in which the emulsion droplets and, eventually, the colloidosomes could be dispersed. When the temperature of the resulting solution reached 60°C, room temperature water was added, causing fine solid particles to rise to the surface. The flask was then promptly removed form the oil bath and the product was filtered. This gradual temperature control during cooling proved to be the most effective method, resulting in fine, powdery wax colloidosomes, with very little sticking to the walls of the flask (Fig. 1b). Colloidosomes with smaller nanoparticles (200 nm and 400 nm instead of 550-600 nm) were also produced; the amounts of nanoparticles used were determined by calculating the total surface area coverage of the particles.

a) 

b) 

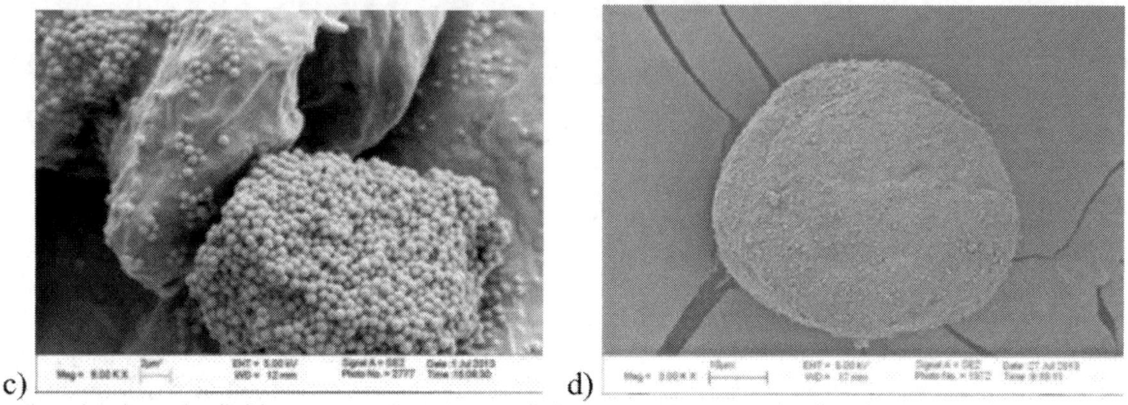

Figure 1. Two batches of colloidosomes produced using different cooling methods. Panels a and c show macroscopic and SEM images, respectively, of colloidosomes obtained mostly by scraping solidified product on the flask walls, after outside-in cooling in an oil bath. Panels b and d show macroscopic and SEM images of improved colloidosomes produced via the two-step, inside-out cooling method.

While the product quality was good macroscopically, SEM images revealed significant crowding on the colloidosome surfaces. Overall, colloidosomes were small and regularly shaped. Most areas were highly packed, with consistent contact angles close to 90°, ideal for Janus particle surface chemical modification, and a few unembedded particles on top of the monolayers (Fig. 2a). However, several showed evidence of wrinkling, a phenomenon which occurs in emulsions in order to maximize the droplets' surface area for nanoparticle adsorption (Fig. 2b).

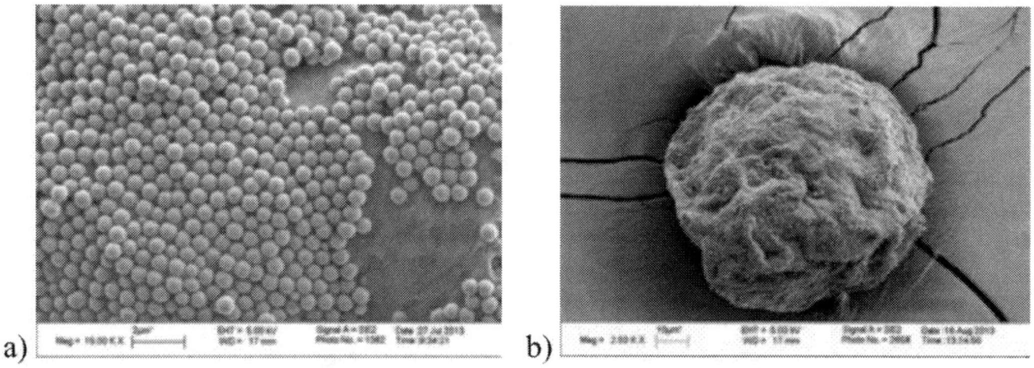

Figure 2. SEM images of colloidosomes. Panel a shows highly packed silica particles with some sitting atop the consistent monolayer. Panel b shows significant surface wrinkling due to nanoparticle overcrowding.

It is counterintuitive that such high crowding of nanoparticles would occur despite the larger number of smaller emulsion droplets produced in this improved method, which maximizes total wax-water interface surface area. In order to improve the reliability of colloidosomes synthesis, the amount of nanoparticles could be fine-tuned to form a consistent, less crowded monolayer.

The effect of DDAB concentration on silica nanoparticle contact angle was tested by reducing surfactant concentration by half. Varying the surfactant concentration changes the particle hydrophobicity and thus the contact angle between the particle surface and the emulsion liquids. At 30 mg/L concentration, the three-phase contact angle was noticeably lowered, with over 80% of the particle exposed instead of the previously observed 50% (Fig. 3a). Also, as was reported by Jiang et al., higher DDAB concentration resulted in closer, more uniform packing, possibly due to screening of electrostatic repulsion between negatively charged silica particles (Fig. 3b).[9]

Figure 3. SEM images of colloidosomes made by emulsifying wax in DDAB solutions with concentrations of 30 mg/L (Panel a) and 60 mg/L (Panel b). Panel a shows 460.2 nm of the 550 nm silica particle fully exposed.

Thus, a successful method for producing fine colloidosomes with regular, packed monolayers of particles was finally achieved by perfecting the emulsion cooling method. However, there remains room for improvement in making the process more reliable and in increasing the yield of spherical colloidosomes with even, hexagonally arranged nanoparticle monolayers.

Janus Particle Fabrication and Characterization

Chemical modification of silica nanoparticles embedded in wax colloidosomes was carried out via the solvent-free method developed by Jiang et al (see Methods).[11] After treatment of the exposed nanoparticles surfaces, the wax was dissolved in chloroform and the nanoparticles collected and washed by centrifugation and redispersion in ethanol.

The first approach to confirm the synthesis of Janus particles involved selectively adsorbing gold nanoparticles onto the chemically modified surfaces of the silica particles and imaging under TEM to observe partly gold-functionalized silica nanoparticles (see Methods). The silane used was (3-aminopropyl)trimethoxysilane (APS), whose positively charged terminal amine group electrostatically attracts negatively charged gold nanoparticles.[12,13,14]

While some TEM images showed possible evidence of Janus character, the technique was overall inconclusive. Particles were definitely modified by the APS, as gold was observed on their surfaces whereas it did not adsorb onto bare silica particles. However, the gold was usually homogenously attached to the entire surface (Fig. 4a).

Possible evidence of Janus particles includes variation in density of gold nanoparticles decorating the surfaces of some chemically modified silica spheres. For example, Figure 4b shows a TEM image of a silica particle whose bottom half appears more densely functionalized by gold. It could be that this half was exposed to the APS vapor treatment and the other half only attracted gold due to residual silane. This density difference was not observed throughout the sample, however, and therefore does not provide definitive confirmation of Janus character.

Several TEM images also showed concentric rings on the silica nanoparticles (Fig.s 4a and 4b). While these may be a result of TEM imaging of the spheres, self-assembly of these particles along the rings, as in Figure 4b, could indicate that the darker central region was unmodified by APS, and gold functionalized the remainder of the particle. Once again, however, this was not observed throughout the samples, and since lack of contrast makes it difficult to see gold nanoparticles in these darker areas, this evidence is unreliable.

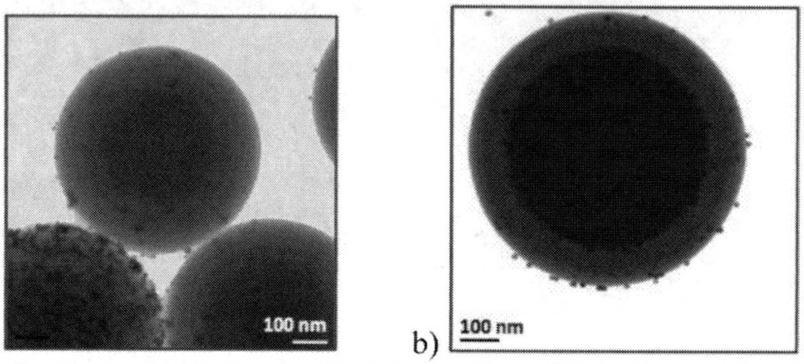

Figure 4. TEM images of gold-nanoparticle-functionalized chemically modified silica particles. Panel a shows varying densities of gold homogenously decorating the silica nanoparticle surfaces. Panel b shows possible evidence of chemical modification on the lower half of the particle due to a greater amount of gold on the surface, as well as self-assembly of gold nanoparticles along the darker, central circular region.

Several problems arose in using gold-functionalization and TEM imaging as a method of Janus particle characterization. First of all, the large number of steps involved in obtaining gold-functionalized particles from bare silica particles introduces several sources of error and result in yield as low as 3-7%. From synthesis and collection of wax colloidosomes, vapor treatment of the exposed silica surfaces, dissolution of wax, and collection of Janus particles to dispersion in colloidal gold, it is difficult to predict where impurities or other issues may have arisen. Secondly, full transmission of the electron beam through the silica particle in TEM images makes it difficult to tell whether the gold nanoparticles observed are on the near or far face of the silica spheres. Finally, early tests of other factors influencing nanoparticle adsorption were carried out, revealing that DDAB residue from colloidosome synthesis may cause gold to adhere to the silica. DDAB remaining on the previously embedded face of collected Janus particles may form a molecular bilayer on the silica surface, with the outward-facing positively charged head of the surfactant molecule attracting gold nanoparticles.

SEM imaging was pursued as an alternative method of characterizing the Janus particles after gold functionalization. Instead of APS, a thiol-terminated silane, 3-mercaptopropyl trimethoxysilane, which chemically reacts with gold, was used to treat the exposed silica surfaces on the wax colloidosomes. These images were similarly ambiguous with regards to selective adhesion of gold to part of the silica surface.

Presence of bare regions (encircled for emphasis in Figure 5) on silica particles otherwise densely covered with gold nanoparticles provides possible evidence of Janus character. However, since the entire particle surfaces could not be imaged, gold decorating density varied throughout the sample, and not all particles showed such distinctly bounded regions. Thus, a more macroscopic approach to confirm Janus particle synthesis was pursued.

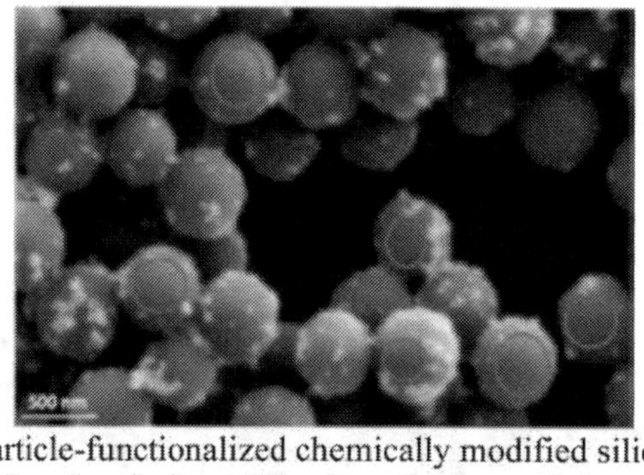

Figure 5. SEM image of gold-nanoparticle-functionalized chemically modified silica particles. Encircled bare regions on some particles indicate possible selective chemical modification of the rest of the particles' surfaces, which are decorated with gold nanoparticles.

Success in fabricating Janus particles was finally confirmed through a two-phase liquid interface test, similar to that performed by Jiang et al (Fig. 6).[11] Three vials were prepared containing a mixture of water (top phase) and chloroform (bottom phase). In one, unmodified, hydrophilic silica particles remained dispersed in the water phase. In another, homogenously hydrophobic silica particles treated with a solution of octadecyltrimethoxysilane (OTS) partitioned into the chloroform phase. If dispersed in either phase alone, the Janus particles collected after selective surface chemical modification quickly settled to the bottom. However, when the particles, half-treated with hydrophobic dichlorodimethylsilane (DCDMS), were placed in the two-phase system, they immediately migrated to the water-chloroform interface. Self-assembly at the liquid interface confirmed successful synthesis of amphiphilic Janus particles. The particles remained at this interface after multiple days, further evidence of these particles' interfacial stability.

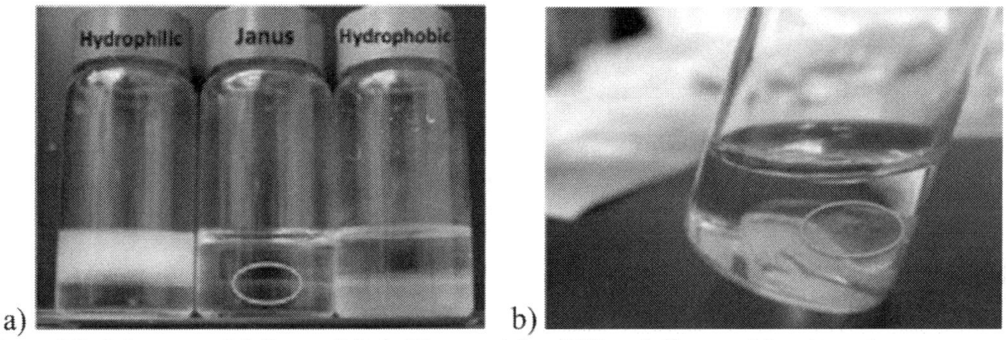

Figure 6. Unmodified, Janus, and fully modified silica particles (550 nm) dispersed in a two-phase system of water and chloroform. Panel a shows unmodified particles dispersed in the water phase (left), homogeneously hydrophobic particles in the chloroform phase (right), and amphiphilic Janus particles adsorbed at the water-chloroform interface with poor dispersion in either phase alone. Panel b shows a more magnified image of Janus particles adsorbed at the fluid interface.

Janus-particle-assisted fabrication of porous polymer films via the BFs method

Upon confirmation of Janus particle synthesis, the amphiphilic particles were employed in the particle-assisted BFs method (see Methods). As was hypothesized, the addition of Janus particles to the polymer casting solution resulted in highly ordered porous structures with high density of nanoparticles decorating the pores, overcoming the previously observed tradeoff between these two qualities (Fig. 7).

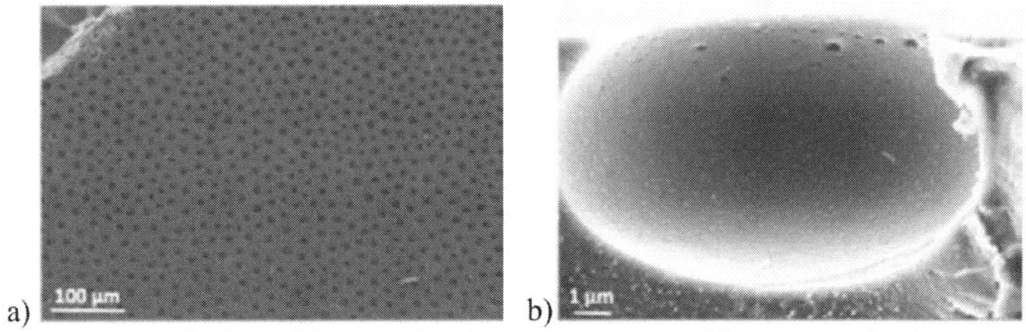

Figure 7. SEM images of polymer BF arrays. Panel a shows a top view of the regular, ordered porous structures, uncompromised by addition of Janus particles. Panel b shows high decorating density of Janus particles (white protrusions) within the pores. Particle diameter is 550 nm.

Several interesting phenomena were observed in the production of Janus-particle-decorated BFs, in addition to achieving higher nanoparticle density within the pores than was previously possible with isotropic silica particles.[3] The Janus particles in the BF arrays were not as fully exposed in the pores as bare silica particles have shown to be. A possible explanation is that, whereas silica particles are negatively charged and thus may immerse further in the templating water droplet during BFs formation, amphiphilic Janus particles are partially attracted to the polymer solvent, causing them to end up further embedded in the polymer matrix. Overall, particles were less than half-exposed in the pores, confirming images of colloidosomes with nanoparticle contact angles below 90° as well as of gold-functionalized nanoparticles indicating that more than half of the particle was treated to be hydrophobic.

Whereas isotropic silica particles were shown to assemble at the openings of the pores near the three-phase water-chloroform-air contact line, Janus particles self-assembled along the entire interface between the water droplets and the polymer solution, evenly decorating the entire pore. In fact, several pores showed distinct boundary lines above which there were few, if any, Janus particles (Fig. 8). This result is consistent with the interfacial stability of amphiphilic Janus particles in Pickering emulsions and as particle surfactants. Whereas bare silica particle were carried to the three phase contact lines via hydrodynamic forces of evaporating polymer solvent, the Janus particles adhered strongly to the water-chloroform interface, avoiding such evaporative influences.

Figure 8. SEM cross-sectional image of BF pore. Janus particles densely and evenly decorate the bottom of the pore. Average protrusion diameter (160 nm) indicates that most of the 550 nm Janus particles are embedded in the polymer matrix.

Breath figures were prepared with varying amounts of Janus particles in the polymer casting solution. At the lowest concentrations, only 5 or 10 µL of a 10 mg/mL particle dispersion were added to 1 mL polystyrene casting solution. Whereas in the case of isotropic silica particles, using such small quantities would have resulted in no particles decorating the pores, amphiphilic Janus particles demonstrate high efficiency of interfacial self-assembly in the BFs method, producing ordered, nanoparticle-decorated porous films (Fig. 9a).

While most pores had uniform nanoparticle density, similar to those imaged in Figures 7b and 8, some contained Janus particle aggregates (Fig. 9b). These formations are particularly interesting because they corroborate literature reporting the assembly of Janus particles into two-dimensional aggregates at the oil-water interface, which suggest dominating attractive interactions between the amphiphilic particles.[15] Therefore, the Janus particles employed in the BFs method display classic Pickering emulsion assembly while also being able to withstand strong hydrodynamic forces.

a) 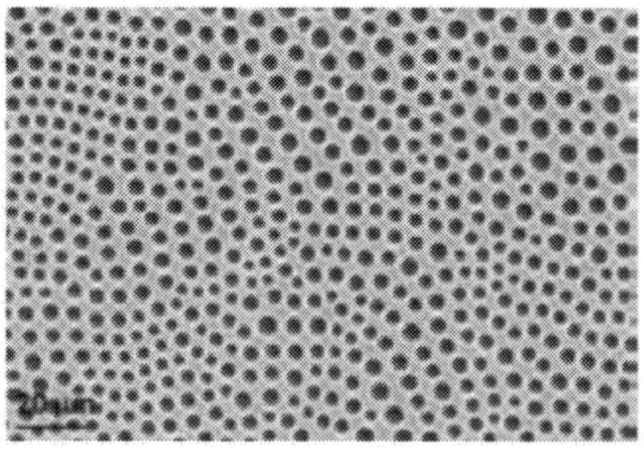 b) 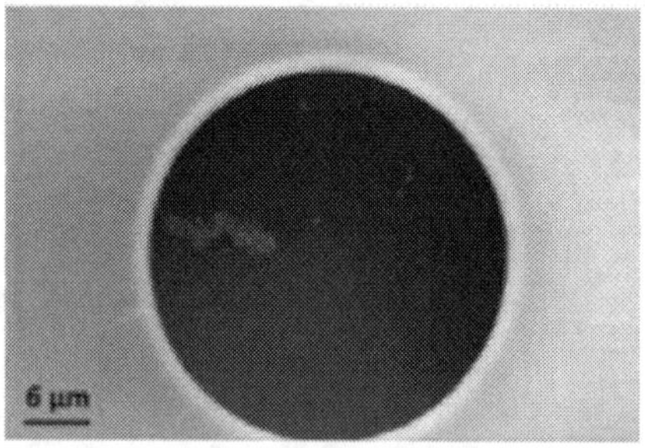

Figure 9. SEM images of breath figures produced with low Janus particle concentration. Panel a shows highly ordered porous structures. Panel b shows an aggregate of Janus particles inside a pore, evidence of self-assembly behavior at the water droplet-polymer solution interface during BFs templating.

At higher Janus particle concentrations (addition ratio 30 μL Janus particle dispersion to 1 mL polystyrene solution), regular, ordered films were still achieved with high decorating density of nanoparticles throughout each of the pores. At similar concentrations, isotropic particles would have resulted in irregular structures and alteration of pore shape. Pore sizes of up to 20 μm were observed throughout the Janus particle decorated BF arrays, significantly larger than typical breath figures with pore sizes of 2-5 μm.[3] This could be a result of increased stabilization of water droplets given the Janus particles' dual role as particle surfactants and as Pickering emulsion stabilizers.

Cross-sectional SEM images shows pores densely and uniformly covered with Janus particles. Once again, most of the particles are over half embedded in the polymer film, with average protrusion radius smaller than Janus particle size. This supports that particles are, on average, more hydrophobic than hydrophilic. Some pores did contain half-exposed particles, indicating desired 50/50 amphiphilic Janus balance.

a) 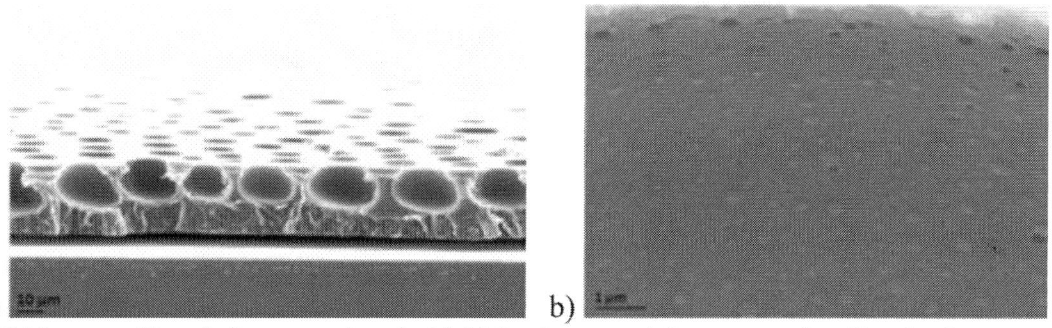 b)

Figure 10. SEM images of breath figures produced with higher Janus particle concentration. Panel a shows preservation of regular porous structures as well as relatively large pore sizes. Panel b shows the inside walls of a pore, densely covered in Janus particles, which are mostly embedded in the polymer film. Average protrusion radius is approximately 350 nm. Janus particle size is 550 nm.

Looking Forward

In the future, much can be done to improve the synthesis of wax colloidosomes to more reliably produce high quality product with even coverage of half-embedded silica nanoparticles. Furthermore, it would be both useful and interesting to be able to confirm Janus character of chemically modified particles via TEM or SEM imaging of gold-functionalized particles. Initial fabrication of Janus-particle decorated honeycomb-structured porous films has laid the foundation for making more breath figure arrays by varying several parameters. While typical Pickering emulsion stabilization decreases with smaller particle size, it is possible that Janus particles' role as particle surfactants could result in high quality BF structures with smaller nanoparticles.[3] Additionally, adjustments could be made to the Janus balance (ratio of hydrophobic to hydrophilic surface area) and to concentration of Janus particles in the polymer casting solution. Eventually, functional hierarchically patterned films can be obtained by, for example, loading the treated Janus particle surfaces with biological molecules for single-cell cultures or catalysts for micro-chemical reactors.

Methods

Synthesis of Wax-Silica Nanoparticle Colloidosomes

Silica particles (600, 550, 400, or 200 nm) are homogeneously dispersed in molten paraffin wax (1.00 g, m.p. 53-57C) at 75°C via magnetic stirring. 10 ml of an aqueous solution of didodecyldimethylammonium bromide (DDAB, 60 mg/L) surfactant, heated to the same temperature, are added to the wax-nanoparticle dispersion and emulsified via magnetic stirring at 1500 rpm for at least 30 minutes. Subsequent cooling to room temperature solidifies the wax, immobilizing the silica particles adsorbed onto the wax/water interface. After filtering the emulsion, washing it 10 times with deionized water, and drying the product under vacuum for 24 hours, the silica-embedded wax colloidosomes are collected.

Silica Nanoparticle (550 nm) Synthesis

550 nm silica nanoparticles are prepared through a variation of the method developed by Stöber et al.[16] A three-neck flask is cleaned with a 1:5 solution of hydrofluoric acid in water and dried. 9.5 mL ammonium hydroxide and 50 mL ethanol are added to the flask. 6 mL tetraethyl orthosilicate (TEOS) and 36 mL ethanol are mixed in a beaker. The TEOS solution is siphoned into a 60 mL syringe equipped with a long metal needle, which is inserted through a rubber stopper into the flask. Using a pump, 35 mL of the TEOS solution are added at a rate of 0.05 mL/min, and the contents of the flasks are continuously stirred at approximately 950 rpm. Once the TEOS is added (over a period of about 12 hours), the contents of the flask are left to stir an additional 12 hours for the reaction to take place. The product is poured into 50 mL conical vials for centrifugation.

Vapor Chemical Treatment of Colloidosomes for Modification of Silica Nanoparticles

Initially, pure silane was used in the vapor chemical treatment method. Eventually, using 5% w/w solutions proved to be equally effective and more economical, as significantly less silane was needed. 5% w/w dichlrodimethylsilane (DCDMS) in toluene is prepared by adding 0.467 mL DCDMS to 9.53 mL toluene in a 20 mL vial. 5% w/w (3-mercaptopropyl)trimethoxysilane (MPS) in toluene is prepared by adding 0.546 mL MPS to 10 mL toluene. 5% w/w (3-aminopropyl)trimethoxysilane (APS) in ethanol is prepared by adding 0.389 mL APS to 9.611 mL ethanol.

Sufficient silane solution is poured into a glass bubbler to submerge the bubbler tip (approximately 10 mL). Nitrogen gas is bubbled at a rate of 2-4 bubbles per second into the silane for 40-60 minutes. The silane vapor is passed into a Buchner funnel containing the colloidosomes on filter paper. The funnel is covered in parafilm punctured with a needle. The particles are rolled in the funnel every 10 minutes to ensure even vapor deposition onto the exposed silica nanoparticle surfaces. Then, for APS-treated particles, dry nitrogen is lightly passed through the sample for 30 minutes to remove any excess silane. When DCDMS or MPS is used for the treatment, dry nitrogen is flowed directly for only 10 minutes, after which it is bubbled through water for 20 minutes to react any remaining silane with water vapor.

Once the vapor treatment is complete, the wax is dissolved in chloroform and the nanoparticles are collected via centrifugation. Particles can then be washed by two cycles of redispersion and centrifugation in ethanol (10,000 rpm for 6-12 minutes). However, this final step significantly reduces yield due to loss of nanoparticles.

Solvent Chemical Modification of Silica Nanoparticles

Adapted from L. Hong et al. And S Jiang et al.[8,9,10] 10 mL APS solution in ethanol (5% w/w) is stirred for several minutes in a 20 mL vial. Fine, powdery colloidosome articles are added to the solution and light stirring continues for 1 hour. Stirring is then stopped, and the colloidosomes are left to continue treatment in the solution for another hour before filtration. The colloidosomes are collected and the wax is dissolved in chloroform before centrifuging and washing as described above.

Bare silica surfaces of Janus particles prepared via APS vapor or solvent treatment could also be reacted with OTS to be rendered hydrophobic. The centrifuged Janus particles are redispersed in 1 mL ethanol via sonication. The dispersion is transferred to a 20 mL vial, to which 4 drops of octadecyltrimethoxysilane (OTS) weighing approximately 0.04 g is added. An additional 1 mL ethanol is added to make a 2% w/w solution, and the particles are left to react overnight with stirring.

Colloidal Gold Synthesis

Prepared using methods described by I. Pastoria-Santos et al. and Y. Takahara et al.[13,14] To synthesize 10-15 nm gold nanoparticles, 95 mL aqueous gold(III) chloride hydrate solution ($HAuCl_4 \cdot xH_2O$, 5.0×10^{-4} M gold ion concentration) is set to boil in a 3-neck round bottom flask equipped with a condenser. The solution is prepared by weighing 16.1 mg gold(III) chloride, which is sensitive to light, in a 20 mL glass vial, and adding some of the 95 mL deionized water. This solution is then transferred to the flask containing the remainder of the water. 5 mL warm sodium citrate solution (1% w/w, 0.05 g sodium citrate dihydrate in 4.95 g water) is added to the stirred, boiling gold(III) chloride solution. Within 5 minutes, the flask contents should turn from yellow to black to a dark violet and finally, once the reaction is finished, to a deep, blood red color.

Adsorption of Gold onto Modified Particles

Gold nanoparticles are adsorbed onto the chemically modified silica particle surfaces by dispersing the dried, collected particles in colloidal gold. Although the exact amounts can vary, I. Pastoria-Santos et al. reports dispersing 3.3 mg silica nanoparticles in 6.2 mL aqueous colloidal gold suspension via sonication. The sol is allowed to equilibrate for at least 5 minutes before removing any non-adsorbed particles by centrifuging and redispersing the functionalized particles three times in water (10,000 rpm for at least 6 minutes).

Breath Figures

Polymer casting solution (15 mg/mL polystyrene in chloroform, MW 192,000 g/mol) is prepared with varying concentrations of Janus particles. A 10 mg/mL dispersion of Janus particles in chloroform is added to the polymer casting solution in ratios of 2.5, 5, 10, 15, and 20 μL dispersion per 1 mL polymer solution.

Dry nitrogen gas is bubbled through water, and the resulting vapor is passed through a glass chamber to create a static environment with 70-90% humidity. Using a syringe, casting solution is added dropwise onto glass or flexible polyethylene terephthalate (PET) substrates. Once the solvent fully evaporates, the films are removed from the chamber.

Acknowledgements

I would like to thank Professor Julia Kornfield for allowing me to work on this project in her lab this summer and providing me with this incredible learning opportunity. I would also like to thank my co-mentor Professor Wei Sun for his supervision, guidance, patience, and support throughout the course of my research. I would like to acknowledge Yuchen Zhou for teaching me several methods crucial in carrying out this project. I would like to extend special thanks to Amy Fu for helping me gain access to equipment in other labs, and to the Tirrell group for allowing me to use their centrifuge. I would like to thank the entire Kornfield group for their encouragement and for making me feel welcome this summer. Finally, thank you to the Caltech Summer Undergraduate Research Fellowship (SURF) Program and Student-Faculty Programs (SFP) Office for providing me the funding for this research and this unforgettable experience.

References

[1] W. Sun, J. Ji, J. Shen, *Langmuir* **2008**, *24*, 11338-11341.
[2] M. Hernández-Guerrero, M. Stenzel, *Polym. Chem.* **2012**, *3*, 563-577.
[3] W. Sun, Z. Shao, J. Ji, *Polymer* **2010**, *51*, 4169-4175.
[4] U. Bunz, *Adv. Mater.* **2006**, *18*, 973-989.
[5] M. Srinivasarao, D. Collings, A. Philips, S. Patel, *Science* **2001**, *292*, 79-83.
[6] B. Binks, *Current Opinion in Colloid & Interface Science* **2002**, *7*, 21-41.
[7] J. Kim, D. Lee, H.C. Shum, D. Weitz, *Adv. Mater.* **2008**, *20*, 3239-3243.
[8] L. Hong, S. Jiang, S. Granick, *Langmuir* **2006**, *22*, 9495-9499.
[9] S. Jiang, Q. Chen, M. Tripathy, E. Luijten, K. Schweizer, S. Granick, *Adv. Mater.* **2010**, *22*, 1060-1071.
[10] S. Jiang, S. Granick, *Langmuir* **2008**, *24*, 2438-2445.
[11] S. Jiang, M. Schultz, Q. Chen, J. Moore, S. Granick, *Langmuir* **2008**, *24*, 10073-10077.
[12] S. Westcott, S. Oldenburg, T. Lee, N. Halas, *Langmuir* **1998**, *14*, 5396-5401.
[13] I. Pastoriza-Santos, D. Gomez, J. Pérez-Juste, L.M. Liz-Marzán, P. Mulvaney, *Phys. Chem. Che. Phys.*, **2004**, *6*, 5056-5060.
[14] Y. K. Takahara, S. Ikeda, S. Ishino, K. Tachi, K. Ikeue, T. Sakata, T. Hasegawa, H. More, M. Matsumura, B. Ohtani, *J. Am. Chem. Soc.* **2005**, *127*, 6271-6275.
[15] B. J. Park, T. Brugarolas, and D. Lee, *Soft Matter*, 2011, **7**, 6413-6417.
[16] W. Stöber, A. Fink, *Journal of Colloid and Interface Science* **1968**, *26*, 62-69.

[17] C. Yu, J. Zhai, X. Gao, M. Wan, L. Jiang, T. Li, *J. Phys. Chem.* **2004**, *108*, 4586-4589.
[18] W. Sun, L. Shen, J. Wang, K. Fu, J. Ji, *Langmuir* **2010**

Supported Fe Fischer-Tropsch Catalysts: Optimizing Preparation Methods Based on a Parametric Study of their Effects

Jonathon Horton[1], Kamyar Keyvanloo[1], Morris D. Argyle[1], William C. Hecker[1]

[1]Department of Chemical Engineering, Brigham Young University, Provo, UT 84602

Background

Fischer Tropsch Synthesis (FTS) is a catalyst-assisted conversion of syngas (CO + H_2) to hydrocarbons (C1–C50). The best FTS catalysts actually select for longer hydrocarbons over shorter hydrocarbons such as methane. Of the two types of metals used for FTS catalysts in industry, iron (Fe) and cobalt (Co), Fe has some advantages over Co including lower cost, lower methane selectivity, and high water-gas shift activity [1]. Commercial Fe catalysts produced for FTS are unsupported because unsupported Fe catalysts have consistently been more active than supported Fe catalysts [2–6]. Therefore, very little work on supported Fe catalysts is reported in the literature [2–4].

Despite their high activity and favorable selectivity, unsupported Fe catalysts are generally too mechanically weak to be used in slurry bubble column reactors (SBCR's), which are the most thermally efficient and economical reactors available [4]. The challenging conditions in SBCR's grind the weaker unsupported catalysts into fine powders, resulting in excessive catalyst loss and plugging the catalyst recovery system [4]. In contrast, supported Fe catalysts are far more attrition resistant [7]. The ideal Fe FTS catalyst would combine the superior activity of unsupported Fe catalysts with the improved strength of supported Fe.

The goal of this study was to gain deeper insights into supported Fe preparation methods as a foundation for future Fe FTS catalyst design. Although many preparation techniques for supported Fe FTS catalysts have been demonstrated [2–4], the reported data lack enough consistency to isolate the effects of specific preparation variables. This study examined five important variables in the preparation of supported Fe catalysts, systematically investigating each individual variable's effects by holding all other variables constant. A series of potassium (K) promoted Fe catalysts supported on alumina (Al_2O_3) were prepared to examine the effect of the following variables: (1) metal deposition method: aqueous incipient wetness deposition or non-aqueous slurry deposition, (2) Fe loading level, (3) K loading level, (4) timing of K and Fe deposition: sequential deposition or co-deposition, and (5) with or without direct K-promotion of the blank support.

Experimental

The six catalysts examined in this study (all Fe/Cu/K/Al_2O_3) are identified as K1–K6. The values of all five preparation variables studied are given for each catalyst in Table 1. Incipient wetness deposition involves using just enough solvent to dissolve and deposit the metal, while slurry deposition involves using excess solvent. Fe was loaded in 10 wt% steps (2 steps for 20 wt%, 4 steps for 40 wt%) with the copper (Cu) followed by drying at 80°C and calcination at 300°C. K was added (4K/100Fe or 8K/100Fe) in one step after all the Fe and Cu for sequential deposition, while for co-deposition, the K was added proportionally with each step of Fe and Cu.

Table 1. Preparation variable values for each of the six catalysts studied in this paper

	Variable 1		Variable 2	Variable 3	Variable 4	Variable 5
Catalyst	**Solvent**	**Deposition Method**	**Fe Loading**[1]	**K Loading**	**Deposition Timing**	**K on blank support**
K1	NA[2]	SD[4]	20	4K/100Fe	Sequential	None
K2	NA	SD	**40**	4K/100Fe	Sequential	None
K3	A[3]	IWD[5]	20	4K/100Fe	Sequential	None
K4	A	IWD	20	**8K/100Fe**	Sequential	None
K5	A	IWD	20	4K/100Fe	**Co-deposition.**	None
K6	A	IWD	20	4K/100Fe[6]	Sequential	**0.2 wt%**

[1]Catalyst wt% Fe. [2]NA = Non-aqueous (50% acetone, 50% isopropanol).
[3]A = Aqueous. [4]SD = Slurry Deposition. [5]IWD = Incipient Wetness Deposition
[6]Plus 0.2 wt% (of the support) K added directly onto the support

Results and Discussion

The effect of each preparation variable was observed by comparing catalyst performance. A few relevant observations for each of the examined preparation variables are given below:

1. Aqueous incipient wetness or non-aqueous slurry: K1 (non-aqueous) has 14% higher activity per gram catalyst and improved stability compared to K3 (aqueous).

2. Fe loading level: K1 (20 wt% Fe) has a 17% higher activity than K2 (40 wt% Fe).

3. K loading level: K3 (4K/100Fe) has a 20% higher activity than K4 (8K/100Fe). In addition, K4 did not show improved selectivity compared to K3.

4. Timing of deposition: K5 (co-deposition) has a 40% higher activity than K3 (sequential deposition).

5. With or without K-promotion of the blank support: K6 (direct K-promotion of the support) deactivates sharply in the first 200 hours on stream, while K3 (no K-promotion of the support) is far more stable.

The performance of the six catalysts of this study was also compared with some of the best catalysts reported in the literature. Two tested preparation methods in particular (non-aqueous slurry deposition and co-deposition) yielded supported Fe catalysts with better catalyst performance than previously thought possible. Figure 1 below shows these two supported iron catalysts (K1 and K5) are superior to the best supported Fe catalysts prior to this work and are even competitive with the best reported unsupported Fe FTS catalysts due to their very high activity (211–267 mmol (CO + H_2)/g_{cat}/MPa/h). K1 and K5 also exhibit high productivity (0.47–0.56 g_{HC}/g_{cat}/h), and reasonable stability. Their superior activity and productivity, coupled with the high strength of a supported catalyst, make these catalysts excellent candidates for use in industry for SBCR applications.

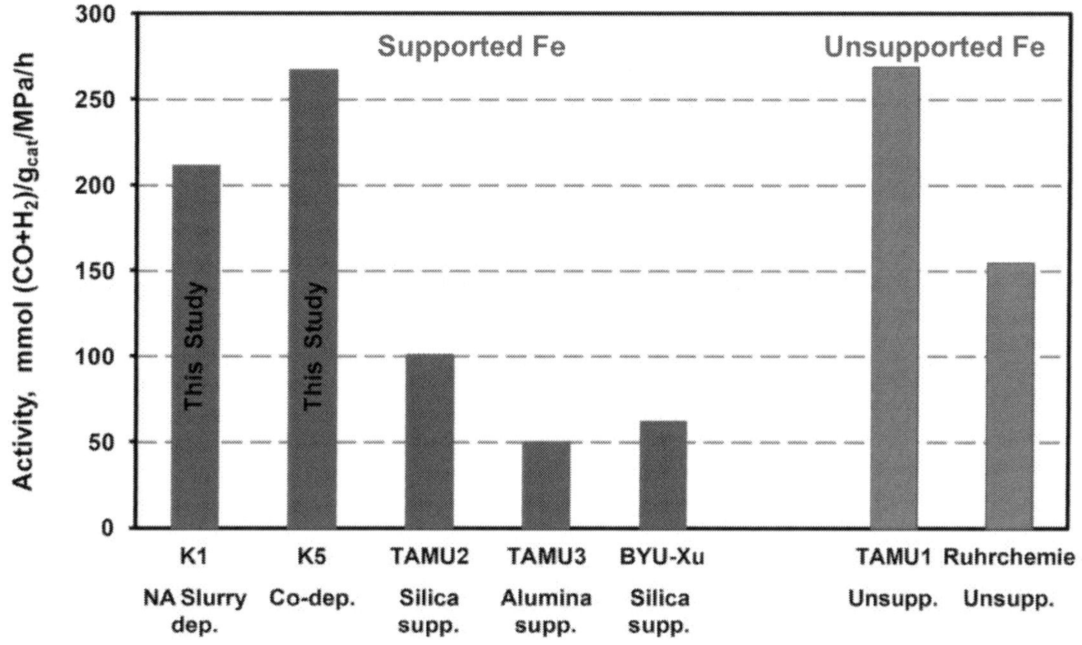

Fig. 1. FTS activity of two catalysts of this study compared with both unsupported and supported Fe catalysts from the literature

References:

1. Bartholomew, C.H. and R.J. Farrauto, *Fundamentals of Inductrial Catalytic processes.* Second ed. 2006: John Wiley & Sons.

2. Bukur, D.B. and C. Sivaraj, *Supported iron catalysts for slurry phase Fischer–Tropsch synthesis.* Applied Catalysis A: General, 2002. **231**(1-2): p. 201-214.

3. O'Brien, R.J., et al., *Activity, selectivity and attrition characteristics of supported iron Fischer–Tropsch catalysts.* Applied Catalysis A: General, 2000. **196**(2): p. 173-178.

4. Xu, J., et al., *Design, synthesis, and catalytic properties of silica-supported, Pt-promoted iron Fischer–Tropsch catalysts.* Topics in Catalysis, 2003. **26**(1-4): p. 55.

5. Bukur, D.B. and X. Lang, *Highly Active and Stable Iron Fischer−Tropsch Catalyst for Synthesis Gas Conversion to Liquid Fuels.* Ind. Eng. Chem. Res., 1999. **38**(9): p. 3270–3275.

6. Keyvanloo, K., et al., *Supported iron Fischer-Tropsch catalyst: superior activity and stability using a thermally stable silica-doped alumina support.* ACS Catalysis, 2014.

7. Anderson, R.B., *The Fischer-Tropsch Synthesis.* 1984, New York: Academic Press.

Comparison of landfill gas treatment options with a novel Gas-to-Liquid process for production of fuels

Ryan Kent, Kirk Jaunich, Tyler Stewart, Zachary Kerbo

Dr. Babu Joseph, Dr. John N. Kuhn

Chemical & Biomedical Engineering, University of South Florida

Abstract:

Approximately 251 MM tons/year of municipal solid waste is produced in United States, with a representative value of Landfill gas (LFG) collection being 2500 ft^3/min per landfill. Landfills are required to collect and dispose of this gas to meet environmental regulations; flaring being the most common option. At the same time landfills spend, on average, around $8,000 per day on diesel fuel for their vehicle fleet. This represents an opportunity if an economic technology can be developed to recover the energy contained in the LFG. A preliminary design and economic analysis was carried out to compare current options (clean and separate the methane in FLG to produce CNG (compressed natural gas), electricity generation, and flaring) for converting LFG with a novel Gas-to-Liquid (GTL) method. The novel process involves two main reactors: a Tri-reforming reactor, which turns carbon dioxide and methane into hydrogen and carbon monoxide (syngas); and a Fischer-Tropsch (FTS) reactor, which converts syngas into long chain hydrocarbons. Using both the Tri-reforming and the FTS reactors in tandem enables conversion of the LFG directly into a tailored cut of middle distillates (diesel and jet fuel) without creation of heavier undesired hydrocarbons. The catalysts, developed in prior research, combat current issues of coke formation and product selectivity, while lowering the cost of synthetic fossil fuel production. Although current options for LFG treatment each have their own advantages, the novel GTL process was the most profitable option with a rate of return at 25%.

Name: Ryan Kent

E-mail: ryankent@mail.usf.edu Cell Phone: 850-324-8499

Department and University: University of South Florida, Chemical & Biomedical Engineering

Advisors: Dr. John N. Kuhn, Dr. Babu Joseph

The authors acknowledge support from:

Trash 2 Cash Energy and its engineers:

- Devin Walker – dmwalker@mail.usf.edu
- Tim Roberge – troberg@mail.usf.edu
- Ali Gardeezi – sgardez2@mail.usf.edu

392757 Recycling Wastewater for Hydraulic Fracturing

Abstract:

Hydraulic fracturing is the use of water to recover natural gas and oil. This process uses approximately 5 million gallons of water per well and is unsustainable. The treatment and recycle of hydraulic fracking water is an engineering best practice not only from an environmental standpoint, but also from an economic standpoint. The current disposal method of deep well injection is a costly and ineffective solution. If frac water is treated and recycled, the savings in water and disposal fees would not only pay for the water treatment program, but also give a return on investment of approximately 30%.

Wastewater treatment options may range from traditional chemical precipitation to evaporation to membrane systems such as Nano filtration, a VSEP system, ultrafiltration, or reverse osmosis. Depending on purity requirements, water treatment methods may be placed in parallel or used independently.

My work has been exploring some of these different membrane treatment options, with a focus in Nano filtration, and doing a full scale design of the economic process. To date I have found that water can be treated for less than $3 / barrel, less than the current cost of disposal. Recycle of hydraulic fracking water can make hydraulic racking more sustainable.

392758 Analyzing RT^2 Profiler PCR Arrays to Profile Cell Matrix Interactions on Smart Biomaterials

Abstract:

Thermoresponsive polymers (TRPs) have the potential to redefine cell culture and medicine in the future. These TRPs change their miscibility at a certain temperature (lower critical solution temperature, LCST), allowing for mechanical, as opposed to biochemical, cell release. This maintains cell-cell and cell-matrix interactions. The gold standard of TRPs is PNIPAM, which is commercially available and has a LCST of 32°C. We have been investigating the biocompatibility of another type of TRPs (copolymers of polymethacrylate and oligoethyleneglycol, PMO), which allows for a tunable LCST between 28 and 90°C. This work investigates the molecular mechanisms of cell adhesion on PMO brushes compared to PNIPAM and tissue culture plastic (TCP) on the basis of gene expression and morphological characterization. Gene expression is studied through RT-PCR with a group of genes coding for various extracellular matrix proteins. Experiments were conducted at both 2 and 24 hour time points to study the changes in gene expression over time as cells adhere to the substrates. Results show an increase in cell adhesion and differential gene expression as a function of both TRP substrate and LCST and highlight a potential opportunity to control cell adhesion and drive cell phenotype through the design of TRP substrates.

Current college students are accustomed to documenting and sharing their experiences through text, photo, and video, thanks to the ready availability of all of these through personal portable devices. The democratization of video production and access has lead to the possibility to both teach and learn with video as never before. This work seeks to capitalize on student expectations and the current technological environment to bring the benefits of both teaching and learning with video into core technical undergraduate engineering courses. Specifically, in this work we ask student teams to create short, targeted, easy to understand videos about concepts in thermodynamics, and then invite them to watch the faculty-vetted library of videos developed by their peers at their own and two collaborating institutions. We are studying changes in students' conceptual learning as a result of participation in this program, and are building a repository of accurate, engaging, videos for thermodynamics learning that will ultimately be shared with other instructors and the public.

Chemical Engineering Technology Lecture & Recitation Material Asynchronous Delivery Mode

presented at the AIChE National Meeting (17 November 2014)

by Chester Little, Ph.D., P.E. and Phillip Hall, M.S., P.E.

Austin Peay State University

College of Science and Mathematics

Convert existing "live" Chemical Engineering Technology program to a fully-online version

this is not a casual decision – conversion manhours are equal to the original development manhours

- adhere to university standards for online programs (substantially-same subject matter, extensive use of standard Course Management System, integrated recitation software)
- recognize rotational schedule and time zone separation (e.g. Afghanistan) of potential student population, transform "live" lectures into an asynchronous format that is easy to navigate
- generate & deliver recorded recitations on a regular schedule using list of problem-solving episodes and questions from students
- lab (pilot plant) classes completed by one extended campus visit
- university core courses already available in online mode

Austin Peay State University 2

College of Science and Mathematics

Course Conversion Methodology

- Starting Point: Well-Developed Live Course
  - book & syllabus
  - full suite of lecture material in electronic format
  - full suite of out-of-class exercise material (homework &/or project assignments)
  - full suite of test material (quizzes and exams)
- Step 1 – Capture live version of course
  - record (audio only) the "normal" lectures during a live delivery of the course
  - these recordings will be used as guides only

designate and list whiteboard, problem-solving episodes (during live version, if possible) – possible recitation material

Austin Peay State University

3

College of Science and Mathematics

Course Conversion Methodology (cont)

- Step 2 – Rearrange course material
  - review entire course (syllabus, slides, audio, video & other support material)
  - develop or modify "Work Breakdown Structure" for course, see next slide
  - rearrange lectures in subject matter groups according to WBS
- Step 3 – Develop a rigorous course plan
  - review live recordings for recitation material – questions & clarification requests from class, periods at whiteboard or document camera, pauses due to confused looks
  - prepare recitation periods by subject matter and adjust syllabus accordingly

4

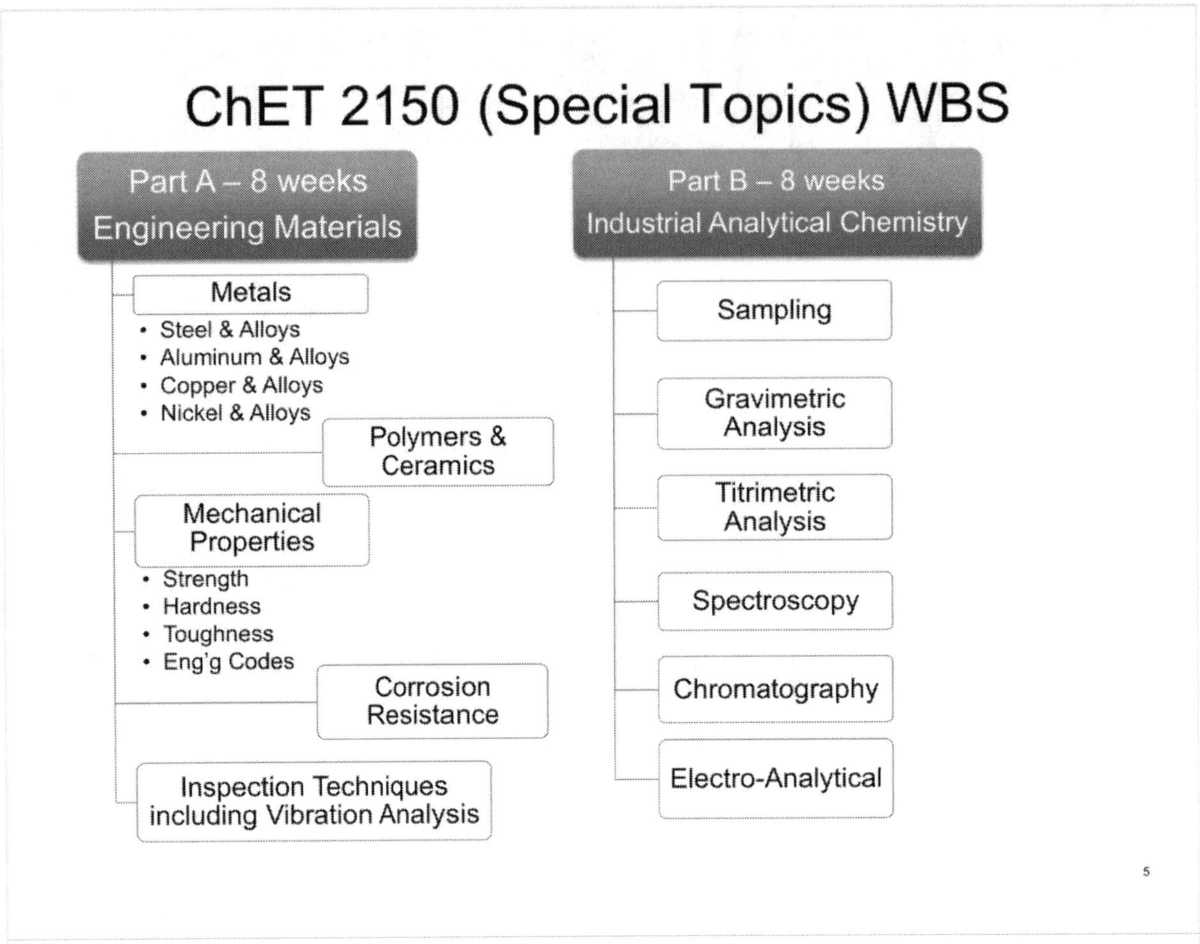
ChET 2150 (Special Topics) WBS
Part A – 8 weeks
Engineering Materials
Metals
• Steel & Alloys
• Aluminum & Alloys
• Copper & Alloys
• Nickel & Alloys
Polymers & Ceramics
Mechanical Properties
• Strength
• Hardness
• Toughness
• Eng'g Codes
Corrosion Resistance
Inspection Techniques including Vibration Analysis
Part B – 8 weeks
Industrial Analytical Chemistry
Sampling
Gravimetric Analysis
Titrimetric Analysis
Spectroscopy
Chromatography
Electro-Analytical
5

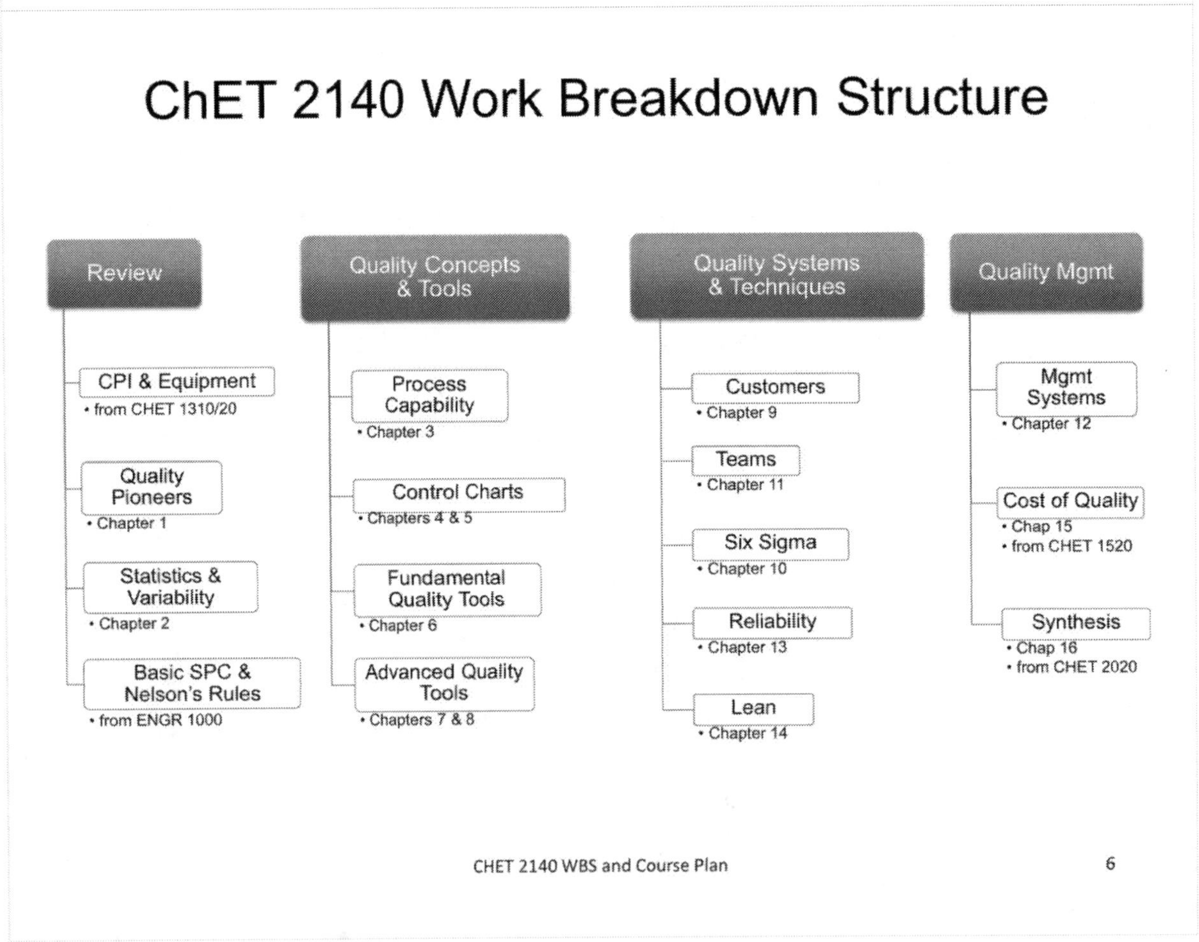
ChET 2140 Work Breakdown Structure
Review
CPI & Equipment
• from CHET 1310/20
Quality Pioneers
• Chapter 1
Statistics & Variability
• Chapter 2
Basic SPC & Nelson's Rules
• from ENGR 1000
Quality Concepts & Tools
Process Capability
• Chapter 3
Control Charts
• Chapters 4 & 5
Fundamental Quality Tools
• Chapter 6
Advanced Quality Tools
• Chapters 7 & 8
Quality Systems & Techniques
Customers
• Chapter 9
Teams
• Chapter 11
Six Sigma
• Chapter 10
Reliability
• Chapter 13
Lean
• Chapter 14
Quality Mgmt
Mgmt Systems
• Chapter 12
Cost of Quality
• Chap 15
• from CHET 1520
Synthesis
• Chap 16
• from CHET 2020
CHET 2140 WBS and Course Plan
6

Course Conversion Methodology (cont)

- Step 4 – Prepare tutorials for CMS
  - many Course Management System providers have video tutorials and training material
  - may be necessary to instruct students how to access and play/pause/rewind lectures and audio
  - same for recitation software (hopefully part of CMS)
- Step 5 – Record "studio" versions of lectures
  - honor the WBS but separate audio tracks by slide (temporarily remove animations)
  - use live versions to assure completeness & accuracy
  - may need a script to assure delivery order

Austin Pea State Universi

7

College of Science and Mathematics

Course Conversion Methodology (cont)

- NOTE: software and formats
  - we currently use the MS Office suite (usually PowerPoint for lecture slides)
  - for audio (only) recording & editing, I use Apple's GarageBand; for video, I use Camtasia
  - we convert all files to MPEG format (.m4a and .mp4)
- Step 6 – Determine appropriate method/vehicle for delivery of recitation (whiteboard) material
  - A. animated lecture slide in a PowerPoint presentation
  - B. movie recorded from display and pasted into PowerPoint presentation as a media file
  - C. stand-alone movie recorded from display, 3 examples: drawing program, Excel spreadsheet and PowerPoint slideshow mode with PEN

8

Animation synchronized with audio (A)

A. animated lecture slide in a PowerPoint presentation

- this method works well with a straightforward example that would normally be done on the whiteboard during a "live" lecture
- don't go overboard with the animation
- pace of delivery is set by the audio not the visual material

• Step 7A – Develop storyboards for the slide

- listen carefully to the slide audio file and note times of key messages (natural groupings of information)
- edit audio file for clarity and emphasis, adjust times
- consider the order and delivery of information

Austin Pea State Universit

9

College of Science and Mathematics

Animation synchronized with audio (A)

• Step 8A – Insert and match audio file

- insert audio file in slide – set to start automatically or with mouse click, include indication when audio is finished
- parse slide elements (bullets, graphs, tables, pic's) and set animation timings according to the storyboard
- test slide for timings and adjust as necessary

• Example (lecture about pipeline wall thickness)

- next slide – original live version
- subsequent slide – the storyboard for the slide
- then – the asynchronous, online version

10

Pipeline Wall Thickness Calculation Example

Dense-Phase CO2 Transmission Pipeline (ASME B31.8)

- 2000 psig, API 5L X52, nominal 8" (8.625" OD) - seamless
- located far (1000 ft) from any inhabited buildings
  see PHMSA website for these (49 CFR 192)

according to ASME B31.8

$$t_c \text{ (calculated wall thickness)} = \frac{P \text{ (max operating pressure) } D \text{ (outside diameter)}}{2\,S\,E\,F\,T}$$

S: SMYS (specified minimum yield strength); this example, $S = 52000\ psi$

E: longitudinal joint factor; this example (seamless), $E = 1$

F: construction type design factor; this example (1000 ft), $F = 0.72$

T: temperature derating factor; this example (ambient), $T = 1$

then $$t_c = \frac{2000\ psi\ (8.625\ in)}{2\ (52000\ psi)\ (1.0)\ (0.72)\ (1.0)} = 0.23\ in$$

t_r (required wall thickness) $= t_c$ + c (corrosion allowance = 0.05") = 0.28"

11

College of Science and Mathematics

P/L Wall Thickness Storyboard

Slide Element #	Slide Element Description	elapsed time	Special Effects
0	Title, Problem Description & Audio File	time zero	launch immediately
1	ASME B31.8, t_c equation	2:42 (162 sec)	dissolve in
2	S - SMYS	3:28 (208 sec)	checker in
3	X52 etc.	3:40 (220 sec)	reveal BOLD & remain
4	E – longitudinal factor	3:46 (226 sec)	dissolve in
5	F – construction factor	3:57 (237 sec)	blinds
6	T – temperature factor	4:22 (262 sec)	appear
7	t_c calculation	4:42 (282 sec)	fly in
8	t_r calculation	4:58 (298 sec)	dissolve in
9	end of audio file	6:17 (317 sec)	title disappearance and indicated in audio – mouse click to advance to next slide

12

Pipeline Wall Thickness Calculation Example

Dense-Phase CO2 Transmission Pipeline (ASME B31.8)

- 2000 psig, API 5L X52, nominal 8" (8.625" OD) - seamless
- located far (1000 ft) from any inhabited buildings
  see PHMSA website for these (49 CFR 192)

according to ASME B31.8

$$t_c \text{ (calculated wall thickness)} = \frac{P \text{ (max operating pressure) } D \text{ (outside diameter)}}{2\,S\,E\,F\,T}$$

S: SMYS (specified minimum yield strength); this example, $S = 52000\ psi$

E: longitudinal joint factor; this example (seamless), $E = 1$

F: construction type design factor; this example (1000 ft), $F = 0.72$

T: temperature derating factor; this example (ambient), $T = 1$

then $$t_c = \frac{2000\ psi\ (8.625\ in)}{2\ (52000\ psi)\ (1.0)\ (0.72)\ (1.0)} = 0.23\ in$$

t_r (required wall thickness) $= t_c + \text{c (corrosion allowance} = 0.05\text{"}) = 0.28\text{"}$

13

College of Science and Mathematics

PowerPoint Slide with Internal Movie (B)

B. movie recorded from display (MS PowerPoint in slideshow mode) and then pasted into another slide

- this method works well with a simple calculation example but requires some simple graphics or animation
- maintains the flow of the lecture with minimal interruption

- Step 7B – Prepare stand-alone slide
  - include all graphics, equations & variable assignments
  - arrange above elements in logical order on the slide
  - record audio file while all elements are visible
  - generate storyboard as discussed previously

14

PowerPoint Slide with Internal Movie (B)

- Step 8B – Insert/match audio file and re-record
  - insert audio file in slide, parse slide elements, set animation timings according to the storyboard and test
  - set video capture to slideshow screen and capture the element animations
  - synchronize audio and video files and convert to a movie
- Step 9B – Insert new movie file and play parameters
  - the movie can a complete slide or in a window as element in a slide
- Example (next slide)

Austin Peay State University

15

College of Science and Mathematics

PowerPoint Slide with Internal Movie (B)

- Movie Example
  - internal to MS PowerPoint slidepack
- Crude Oil Storage Firewall Calculations
  - audience – first year technology students
  - suggest that the movie launch "on click"
  - to allow explanation of the material in this sidebar

16

Stand-Alone Movie (C)

C. separate movie files recorded from display

- ✤ this method is more complicated for the student because they must execute the movie file inside the Course Management System and/or download the movie and execute it using the movie player on their computer
- ✤ of course, the movie format specifications will need to be supplied by your Course Management System administrator

- Scenarios for stand-alone movies
  - ✤ problem solution or explanation requires more detail (or longer time – greater than 10 minutes)
  - ✤ solution involves use of multiple concepts
  - ✤ problem scope is across multiple lectures

Austin Pea
State Universit

17

College of Science and Mathematics

Stand-Alone Movie (C)

- first consideration – which graphical elements to use
- second consideration – which computer input devices (mouse, keyboard commands, pen, drawing tablet)
  - ✤ drawing objects, textboxes, handwriting and sketching – capture the screen of drawing program using mouse & pen
  - ✤ charts, plots, data manipulation – capture the development of an Excel spreadsheet
  - ✤ simple objects with pen highlighting and/or additions – capture a PowerPoint slide in slideshow mode with PEN activation

18

Stand-Alone Movie (C)

- last consideration – Do I have the skill and dexterity to handle all of the graphical elements, input devices while recording a clean audio track?
  - often the answer is "No" - if so, it is recommended to practice in the screen environment to determine the logical progression of graphical elements => storyboard
  - then record the audio track (only) while manipulating the screen environment without concern for exact placement of the graphical elements nor the precision of handwriting and sketching => storyboard timings (from audio track)
  - then re-do and capture the screen manipulations honoring the storyboard (and listening to the audio)

Austin Peay State University

19

College of Science and Mathematics

Stand-Alone Movie (C)

Examples have been provided for each of the above scenarios

- screen capture of drawing program using mouse & pen – Basic Engineering exercise for Units and Conversions
- screen capture of an Excel spreadsheet development – Basic Statistics and Generation of a Zone Chart
- screen capture of a PowerPoint slideshow mode with PEN activation – Descriptive Statistics (Calculating by Hand)

20

Course Conversion – Other Considerations

- keep in mind that asynchronous lectures are not limited to preset time durations
- however lectures could be limited to subject matter associated with chapters in the textbook or elements (sub-elements) from the course WBS
- realistically asynchronous lectures are often limited by the maximum file size allowed by your CMS or the mental saturation of your students
- rule of thumb – regarding elapsed time, an asynchronous lecture is slightly more than half a normal, live lecture
- lectures – target 30 to 45 minutes
- recitations – target 5 to 15 minutes

AP Austin Pea State Universit

21

College of Science and Mathematics

Thank You

Questions?

22

Improving Student Interaction with Chemical Engineering Learning Tools: Screencasts and Simulations

Chemical & biological engineering faculty have developed over 1,050 screencasts covering topics in chemical engineering courses. Screencasts are short videos (typically less than 10 minutes) with narration and are made by digital capture of a tablet PC screen. Screencasts can introduce a topic, solve an example problem, explain a concept, explain a diagram and process, demonstrate software use, review for an exam, or present a mini-lecture. They can be used in combination with textbooks, online reading quizzes, homework assignments, and office hours. This personalized method of learning empowers students by giving them control over the rate of information delivery and when they receive information. As of May 2014, these videos had been watched/downloaded over 4 million times and have an overwhelmingly positive response from students in our classes and as seen through YouTube comments.

Although many screencasts demonstrate problem solving skills and suggest students attempt to solve the problems on their own before watching the step-by-step solutions, they are unidirectional in their information delivery. Without student comments, we are unable to determine student misconceptions and issues with the materials. This became the motivation to create interactive screencasts. Interactive screencasts start by posing a conceptual question that is followed by embedded video links to video responses based on the answers chosen (see figure on next page). If a student clicks on a wrong answer, they are led to a video explaining why their answer is wrong and then asked to choose another answer. This continues until the correct answer is chosen and the video solution is revealed. These screencasts provide a significant advantage over textbook examples since they require the student to answer the question without being able to look at the solution. Students have been tremendously excited about these videos and have used them to "test themselves" after classes and prior to exams. Analytics also enable us to track how answers are being chosen, thus aiding our efforts to identify confusing concepts.

Another effort to improve student interaction involves hands on computer simulations. We are using *Mathematica* based simulations (see figure on next page) to enhance student learning and better connect conceptual mastery with physical modeling of systems. These simulations allows users to manually control variables and almost instantly visualize the effects on the system behavior. This provides a useful resource for promoting student interaction during assignments and supporting in class questioning where students are asked to predict system outcomes. There is a growing library of chemical engineering simulations at the Wolfram Demonstrations Project. We have also begun developing our own simulations and screencasts that explain their use.

We will present on these resources and how to use them within a flipped classroom approach to improve student interaction and active learning methods within chemical engineering classes.

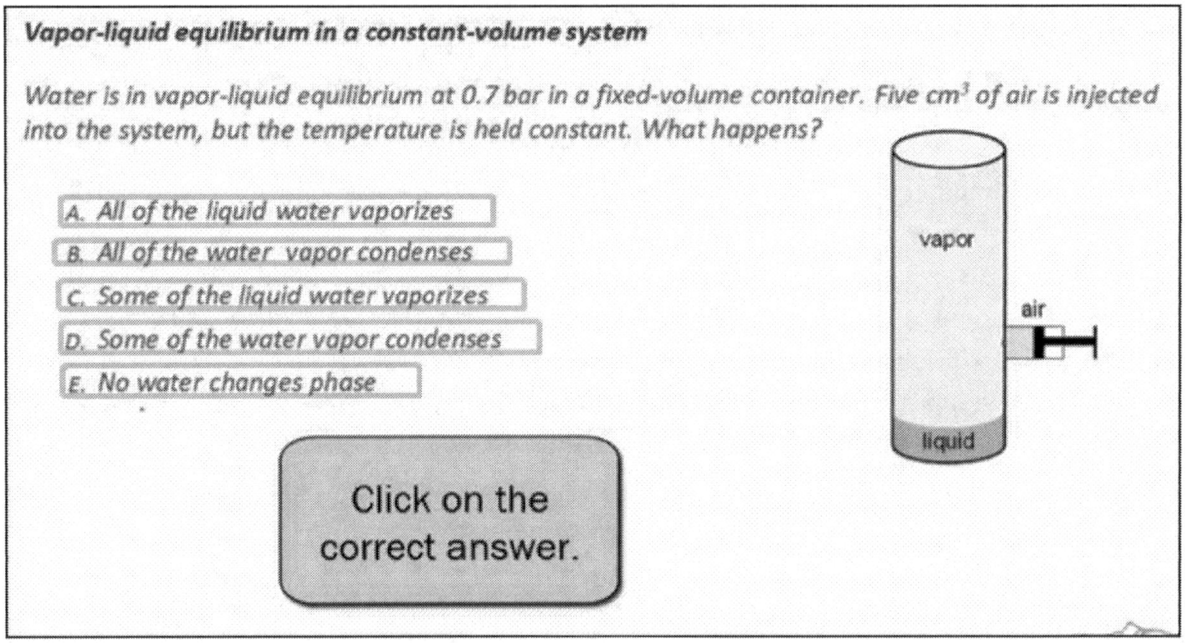

At the end of a short presentation of the question, the video pops up answers that have embedded links. The boxes around the answers indicate "hotspots" that are clickable. Each link opens up a separate video.

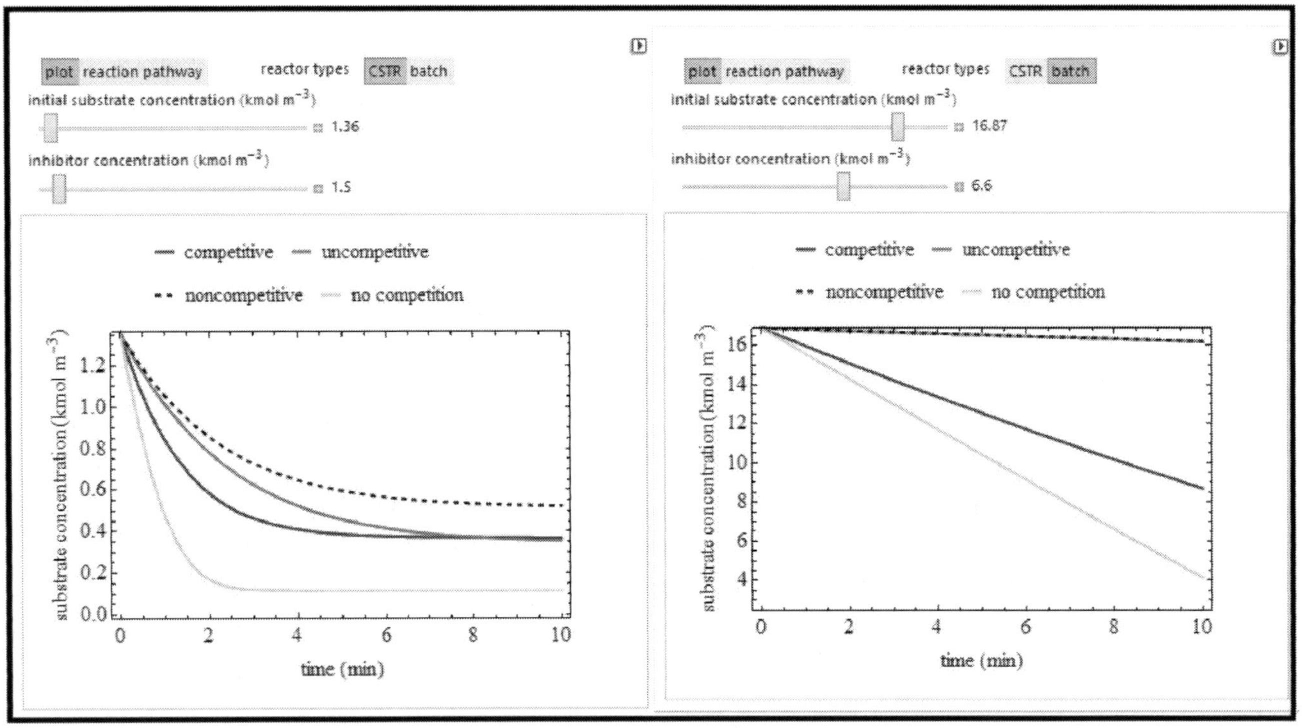

Wolfram Demonstrations Project Simulation: Inhibition of Enzyme Reactions in Continuous Stirred-Tank Reactor and Batch Reactor. Sliders and output buttons control how resulting analysis is presented.

378714 NSF Overview

Abstract:

The National Science Foundation (NSF) is an independent federal agency created by Congress in 1950 "to promote the progress of science; to advance the national health, prosperity, and welfare; to secure the national defense…" With an annual budget of $7.2 billion (FY 2014), we are the funding source for approximately 24 percent of all federally supported basic research conducted by America's colleges and universities. In many fields such as mathematics, computer science and the social sciences, NSF is the major source of federal backing.

394222 Slow Growth Increases Myo-Inositol Availability for Glucaric Acid Production in S. Cerevisiae

Abstract:

Glucaric acid has been classified as a top value-added chemical from biomass for potential uses ranging from cancer therapies to sustainable nylons by the Department of Energy. A biosynthetic pathway was constructed in *E. coli* to convert glucose to glucaric acid but activity of the *Mus musculus* myo-inositol oxygenase (MIOX) enzyme was rate-limiting. New expression constructs enabled a one-step knock-in of the four pathway genes to *S. cerevisiae.* Protein sequence similarity mapping paired with simple structure-function hypotheses enabled efficient sampling of over 300 MIOX enzymes. Substituting a plant MIOX in the pathway increased glucaric acid titers detected in the YPD culture broth at 72hrs. Supplementation with 60mM myo-inositol tripled glucaric acid titers for yeast expressing either the plant MIOX (1.5 g/L) or the mouse MIOX (0.94g/L). However, myo-inositol supplementation did not affect glucaric acid titers in yeast expressing the mouse MIOX during slower growth on non-fermentable carbon sources such as glycerol or ethanol. Myo-inositol is a component of yeast membrane phospholipids and carbon-limited chemostat experiments demonstrated that slow growth rates lead to increased myo-inositol availability and glucaric acid titers. We pursued a fed-batch strategy in YPD with yeast expressing the plant MIOX and achieved final titers of 1.2g/L.

Development of Tissue Phantoms for a New Breast Cancer Detection Technique

Matthew A. Conrad
Caitlin B. Douglas
Adam J. Nolte, PhD

Rose-Hulman Institute of Technology | Department of Chemical Engineering

Background Information

- Healthy breast tissue has a Young's modulus of ~3 kPa and tumorous tissue has a Young's modulus of ~20 kPa

⟹ Gelatin hydrogels seem to be acceptably soft

- 250 Bloom B Gelatin was utilized

Research Goals

- To prepare gelatin tissue phantoms of a Young's modulus of healthy tissue with tumors of various sizes and geometries
- To characterize the mechanical properties such as the effects of aging, multiple tests, strain rate, and sodium alginate as an additive

Rose-Hulman Institute of Technology | Department of Chemical Engineering

Tissue Phantom Preparation

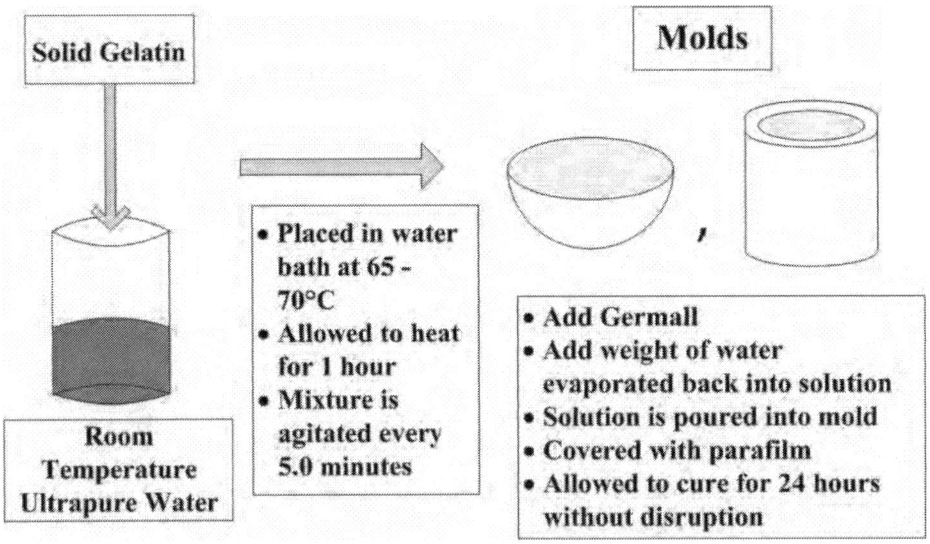

Rose-Hulman Institute of Technology | Department of Chemical Engineering

Tissue Phantom Preparation

Rose-Hulman Institute of Technology | Department of Chemical Engineering

Tumor Inclusion

Two proposed methods:

Liquid Injection via syringe

- 12.0% w/w gelatin liquid solution syringe-injected into congealing 4.0% w/w gelatin solution

⟹ • Not effective

Pre-made inclusions

- 12.0% w/w solid gelatin inclusion placed in congealing 4.0% w/w gelatin solution (T ~ 23.5°C)

⟹ • More effective approach

Rose-Hulman Institute of Technology | Department of Chemical Engineering

Mechanical Testing

Mechanical properties measured:

Young's Modulus
Fracture Strength
Toughness

- Samples tested in compression with an Instron with a 10 kN LC
  - Young's Modulus from strains less than 10%
  - Some samples tested to failure

Rose-Hulman Institute of Technology | Department of Chemical Engineering

Concentration Dependence

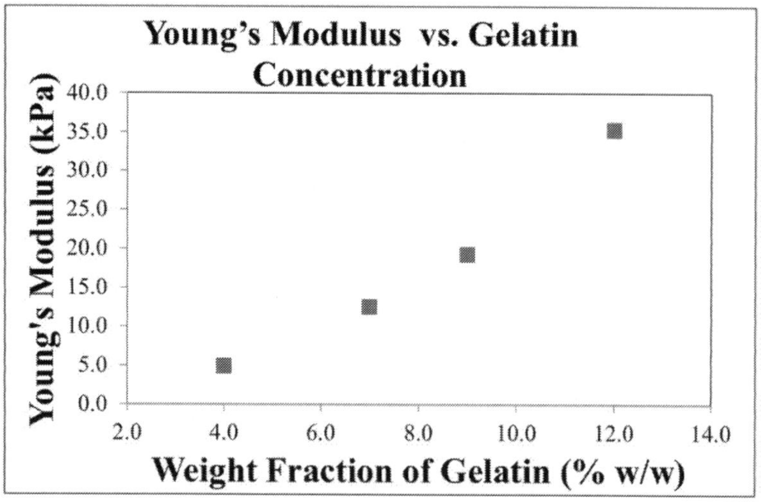

Rose-Hulman Institute of Technology | Department of Chemical Engineering

Aging Effects

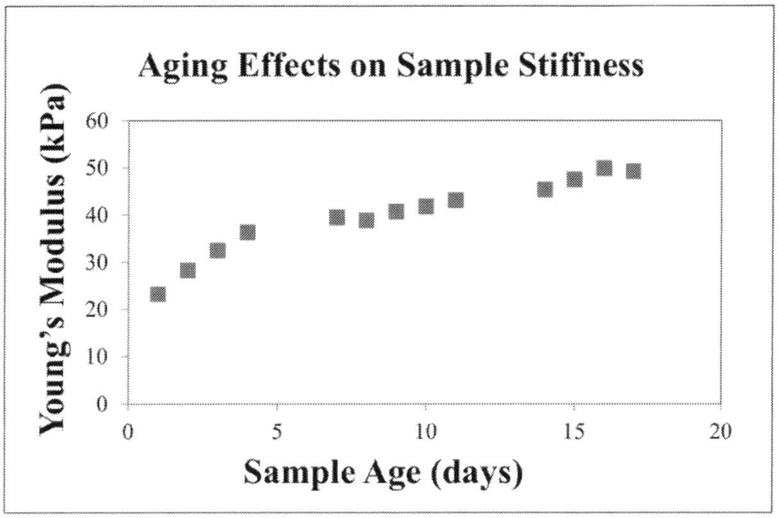

Rose-Hulman Institute of Technology | Department of Chemical Engineering

Testing Speed

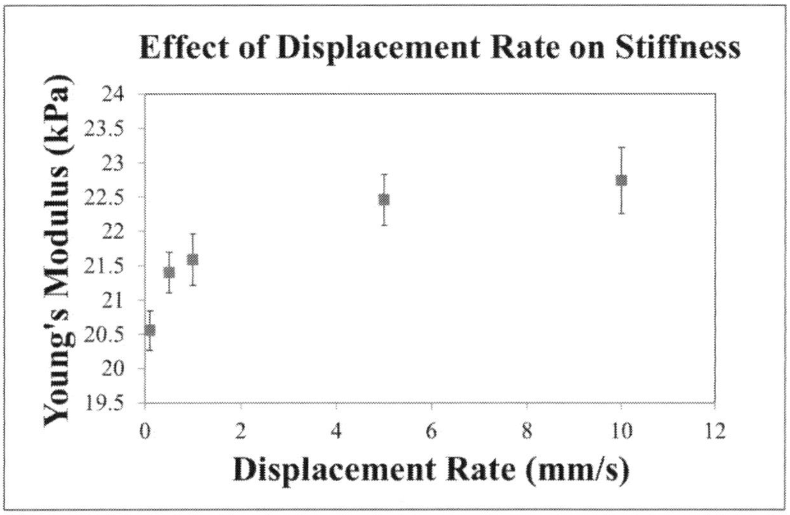

Rose-Hulman Institute of Technology | Department of Chemical Engineering

Multiple Test Effect on Stiffness

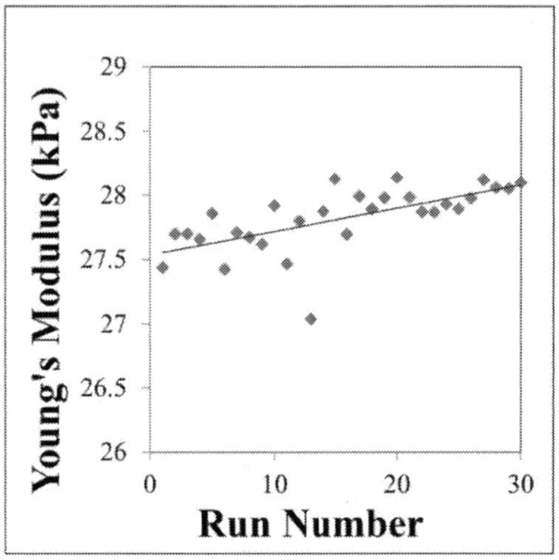

Uncovered during testing

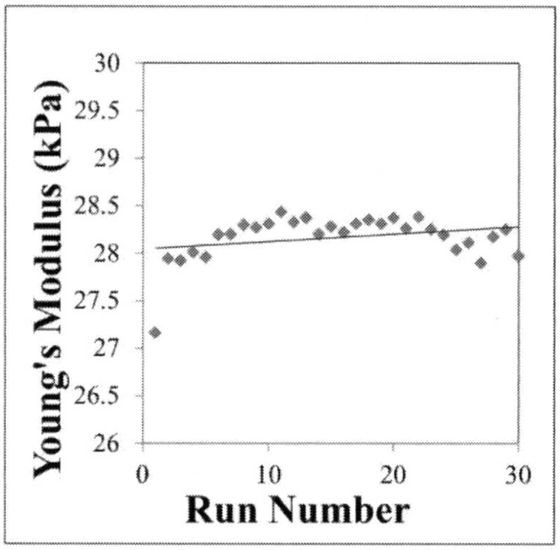

Covered during testing

Rose-Hulman Institute of Technology | Department of Chemical Engineering

Additive Effects: On Stiffness

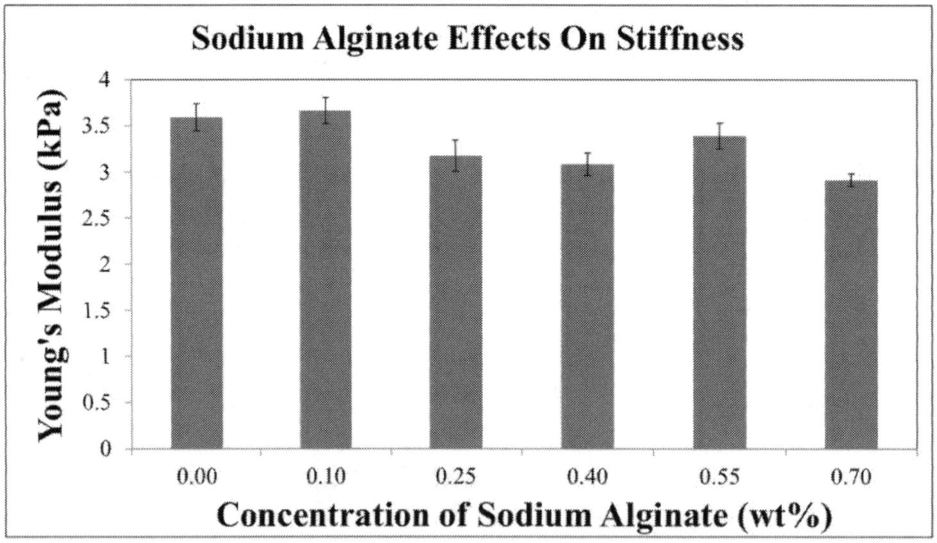

Rose-Hulman Institute of Technology | Department of Chemical Engineering

Additive Effects: On Fracture Strength

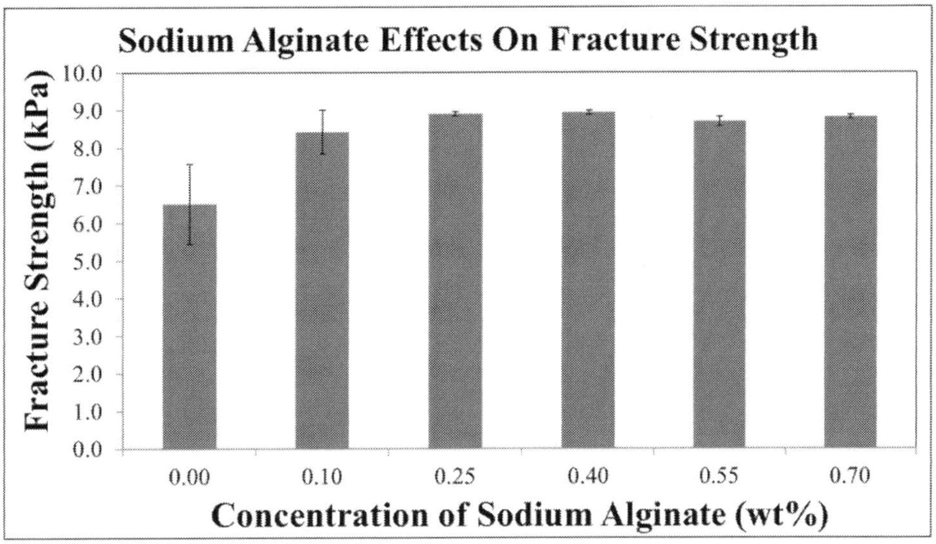

Rose-Hulman Institute of Technology | Department of Chemical Engineering

Additive Effects: On Toughness

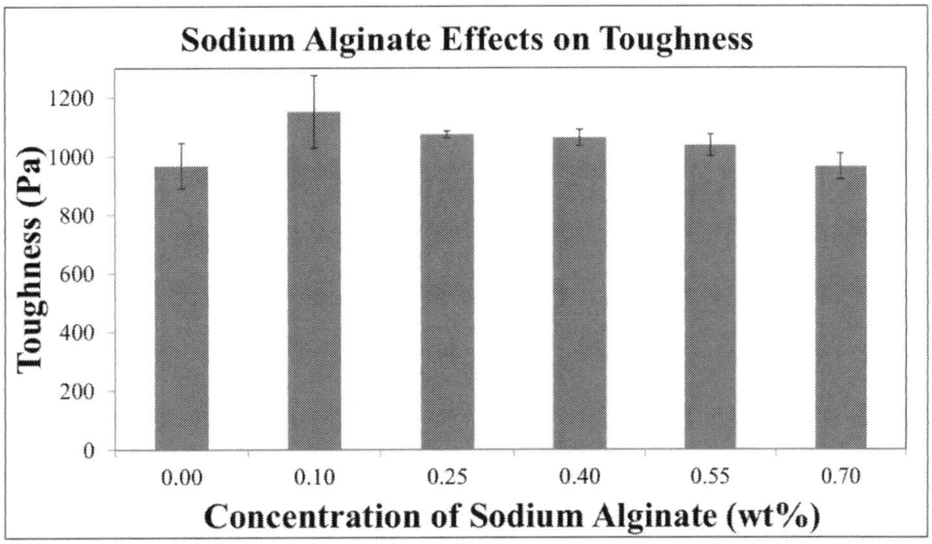

Rose-Hulman Institute of Technology | Department of Chemical Engineering

Conclusion

- Gelatin hemispheres can be made to mimic healthy breast tissue with tumor inclusions by inserting a pre-made tumor into a congealing gelatin solution
- Increasing the sodium alginate concentration appears to improve the fracture strength of gelatin hydrogels while decreasing the stiffness

Rose-Hulman Institute of Technology | Department of Chemical Engineering

Future

- Confirmation of results with texture analyzer
- More accurate relaxation tests
- Optimizing concentrations of additives

Rose-Hulman Institute of Technology | Department of Chemical Engineering

Acknowledgements

- *This research was funded by the National Science Foundation under Grant No. ECSS-1306808*
- *Professor Michael Insana and research group members (UIUC)*
- *Professors Olson and Throne (RHIT)*

Rose-Hulman Institute of Technology | Department of Chemical Engineering

Relaxation Effects

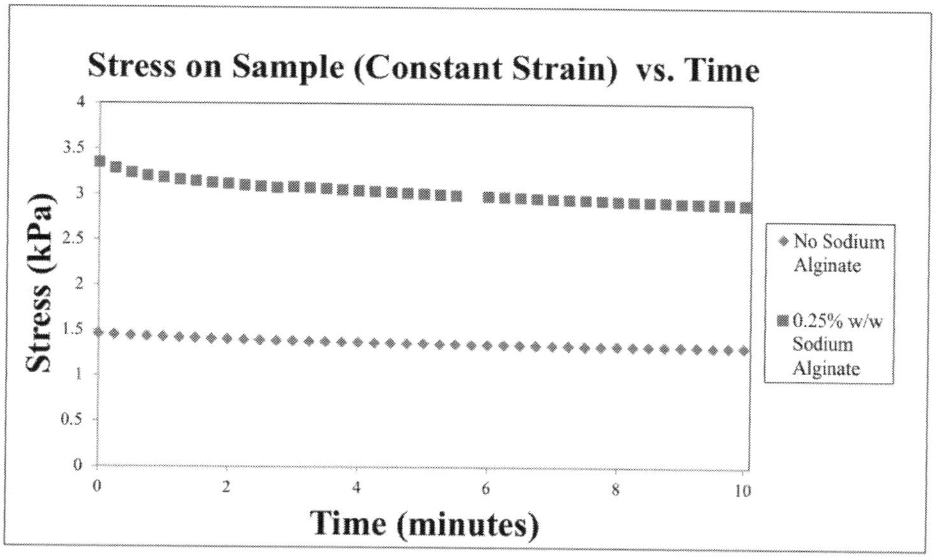

Rose-Hulman Institute of Technology | Department of Chemical Engineering

394192 Effective Antisense Design Using Ensemble of Energetically Sub-Optimal Secondary mRNA Structures

Abstract:

There is a deficit of effective therapies in the current research against bacterial infection. Many therapeutic strategies seek to use small molecules to target the infectious pathogen. One approach to disease inhibition involves direct manipulation of the pathogen at the RNA level. Part of the central dogma, messenger RNA (mRNA) is a transcript of genetic information that encodes the fundamental instruction for protein production. Inhibiting translation of mRNA effectively prevents synthesis of proteins.

In order to effectively block the RNA's message from being read, it must be physically accessed by the therapeutic agent. A field of study has arisen in which a complementary nucleic acid binding strand, called antisense[1], is generated to impede protein synthesis. One of the issues that arises in creating effective antisense RNA is the ability to find appropriate target sites on the mRNA to be inhibited. This problem is in part due to fluctuating secondary structure sequestering target sites. In fact, the kinetics of physical accession of the antisense to the target strand are suggested to be the rate-limiting factor[2] and thus the cause of inefficiency of this method. However, the fluctuations among suboptimal conformations are energetically inherent to the nucleic acid sequence and can be predicted. A software package, GenAVERT, has been developed to thermodynamically determine effective target sites by identifying the ensemble of the most probable energetically suboptimal states and determine which regions of mRNA are most accessible[3]. Once properly identified, a successful and effective antisense can, in theory, be synthesized.

In this work, a proof of principle will be demonstrated in *E. coli* with the downregulation of a short-lived variant of the green fluorescent protein through construction of an inducible antisense molecule. Statistical reduction in the fluorescence upon antisense induction will affirm the efficacy of this model.
Supported by fundamental energetic principles, this method of producing an antisense sequence complement for the most accessible mRNA target region has the potential to greatly reduce the strife of pathogenic infection.

[1] Lee, L. and Roth, C. (2003). Antisense technology in molecular and cellular bioengineering. *Current Opinion in Biotechnology*, [online] 14(5), pp.505-511.

[2] Patzel, V. and Sczakiel, G. (2000). In vitro selection supports the view of a kinetic control of antisense RNA-mediated inhibition of gene expression in mammalian cells. *Nucleic Acids Research*, [online] 28(13), pp.2462-2466.

[3] Johnson, E. and Srivastava, R. (2013). Volatility in mRNA secondary structure as a design principle for antisense. *Nucleic Acids Research*, [online] 41(3), pp.e43-e43.

Nanotechnology REU Summer 2014

Raghav Malik

AIChE National Conference: Undergraduate Research Forum II: Bioengineering, November 17th

Supervising Graduate Student: Giulia Suarato

Principal Investigator: Dr. Yizhi Meng

Department of Materials Science and Engineering

Stony Brook University

2

Proanthrocyanidin (PAC)

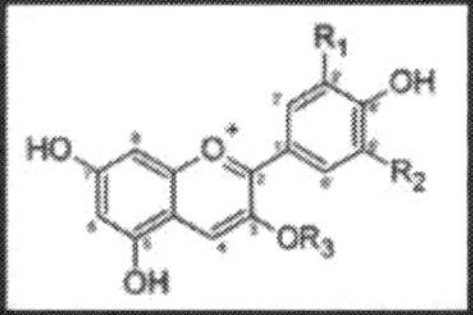

- Organically derived (from fruits and wines)
- Digests in saliva
- Promising chemotherapeutic effects, novel cancer cytotoxicity
- Encapsulation to improve bioavailability

Stony Brook University

Kom Kamonpatana, et. al. Susceptibility of anthocyanins to ex vivo degradation in human saliva, Food Chemistry. 2012: Vol. 135, 738-747

Chemotherapeutic Effects of PAC

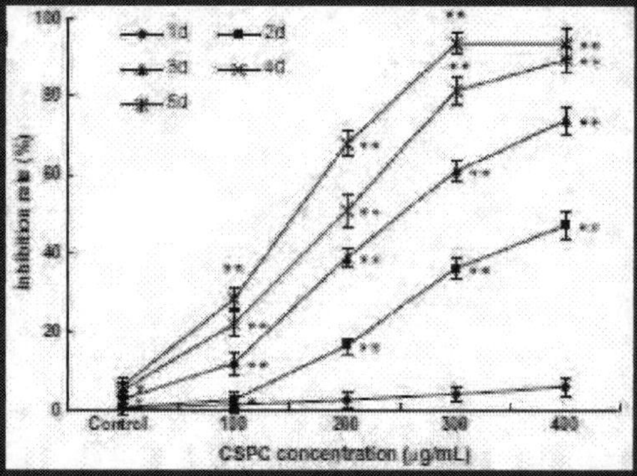

- Human Hepatoma G2 Cells
- Inhibition Rate: (Control – Treated / Control)*100
- Inhibition Rates Dose-Dependent
- Inhibition Time Dependent

H. Zhang, J. Ke, T. Shao, J. Li, Y. Duan, Y. He, C. Zhang, G. Chen, G. Sun, X. Sun. Food and Chemical Toxicology. 2014: Vol. 64, 166-176

Objective

Encapsulate PAC using chitosan and characterize the resulting nanoparticle as well as study the effect on MG-63 Osteosarcoma cells *in vitro*.

5

Why Chitosan?

- High Bioavailability
- Organic, derived from Chitin
- Biodegradable
- Hydrophilic

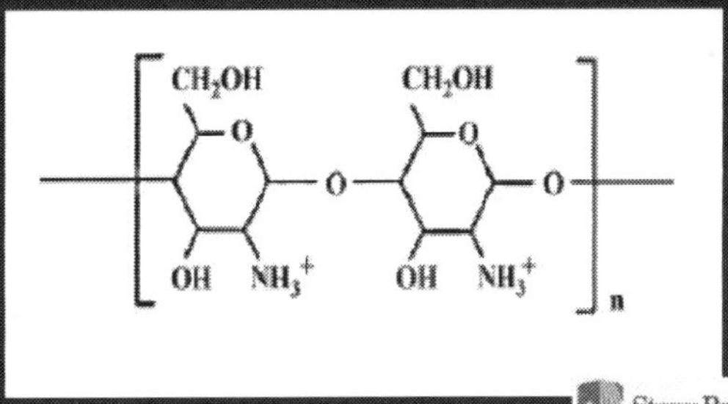

Chitosan

Stony Brook University

6

Encapsulation Potential

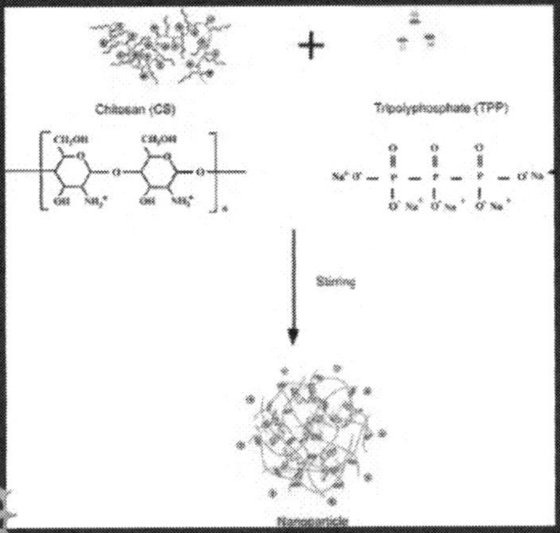

- Ionic Gelation Method
  - Pluronic F68 (detergent)
  - GCS:TPP 8.7:1 and 2.9:1
- Sizes are typically 200-300 nm

Stony Brook University

Adriana Trapani, Johannes Sitterberg, Udo Bakowsky, Thomas Kissel. International Journal of Pharmaceutics, 2009: Vol. 375, 97-106

Structure of Glycol Chitosan

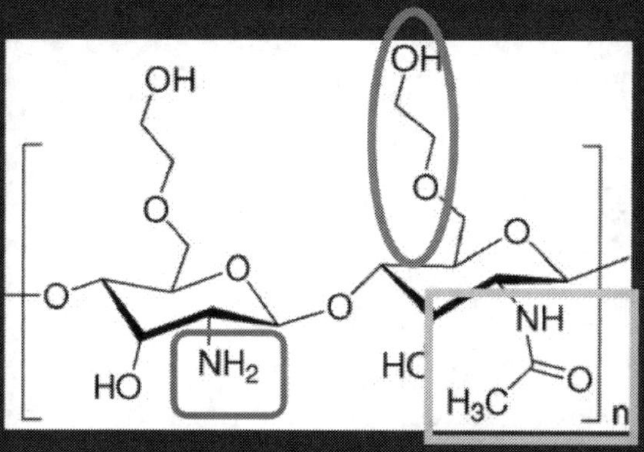

Approach

- Synthesis
- Characterization
- Delivery *in vitro*

9

Fabrication and Characterization of NPs

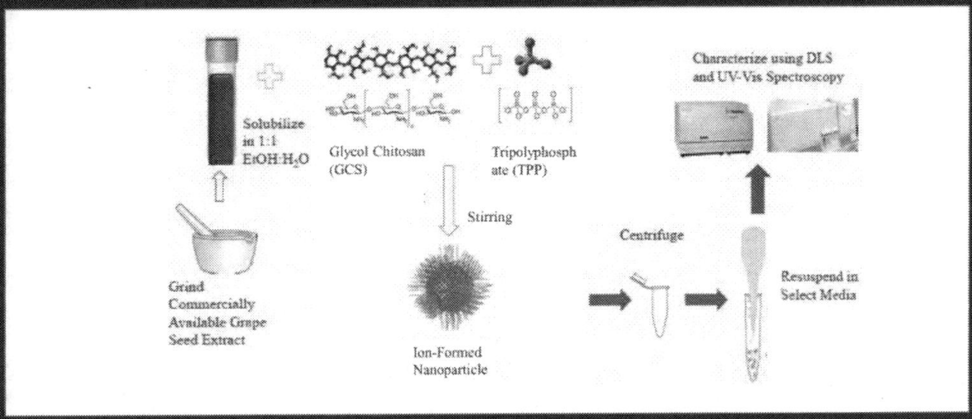

10

Nanoparticle Characterization

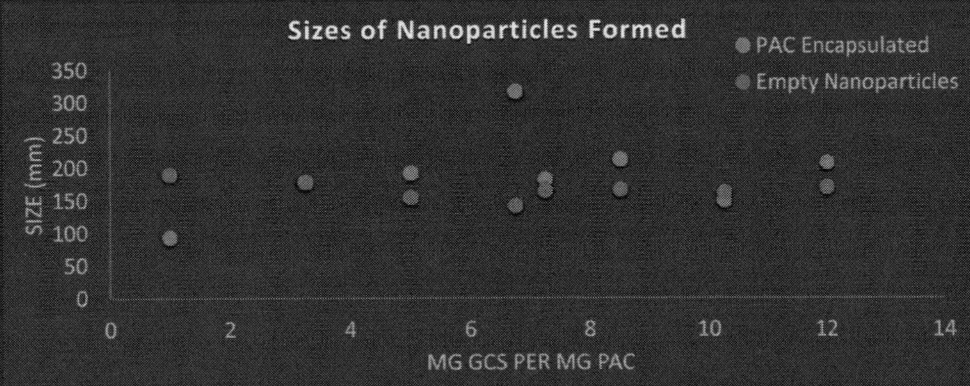

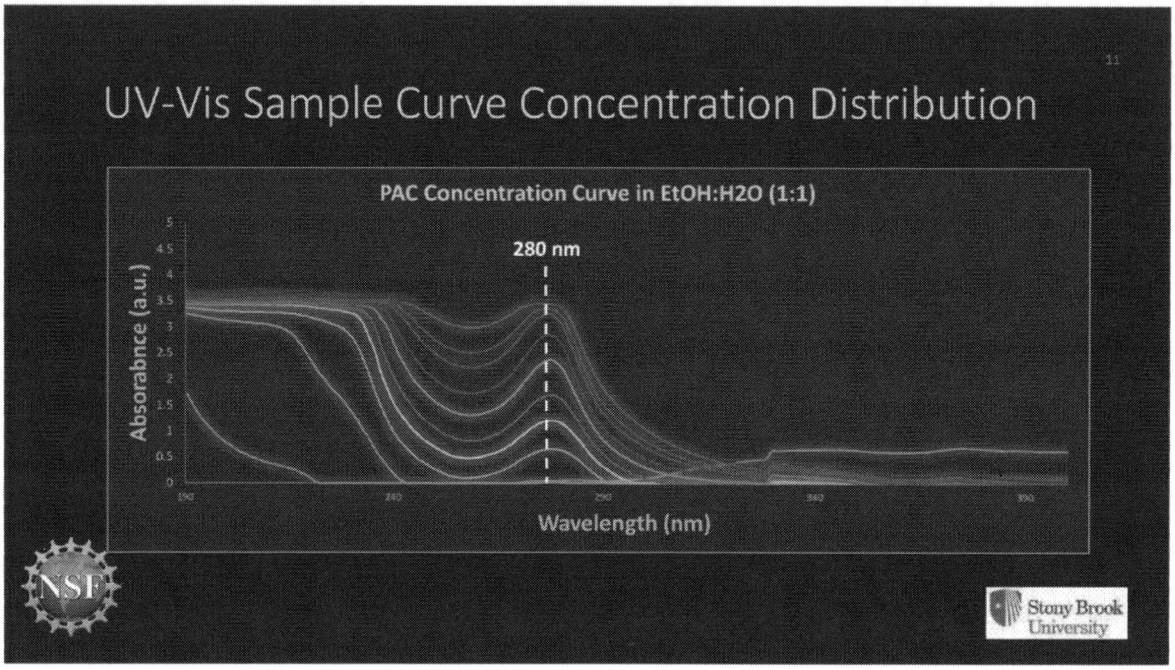
UV-Vis Sample Curve Concentration Distribution
PAC Concentration Curve in EtOH:H2O (1:1)
280 nm
Absorabnce (a.u.)
Wavelength (nm)
NSF
Stony Brook University

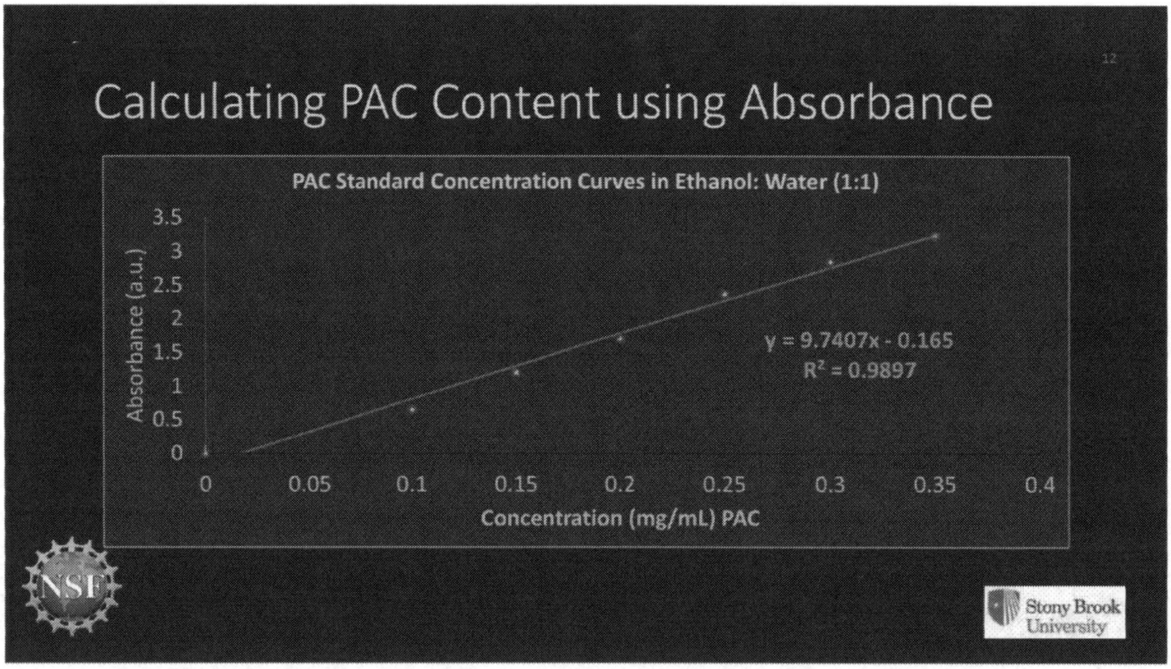
Calculating PAC Content using Absorbance
PAC Standard Concentration Curves in Ethanol: Water (1:1)
y = 9.7407x - 0.165
R² = 0.9897
Absorbance (a.u.)
0 0.5 1 1.5 2 2.5 3 3.5
0 0.05 0.1 0.15 0.2 0.25 0.3 0.35 0.4
Concentration (mg/mL) PAC
NSF
Stony Brook University

13

Encapsulation Efficiencies of GCS NPs

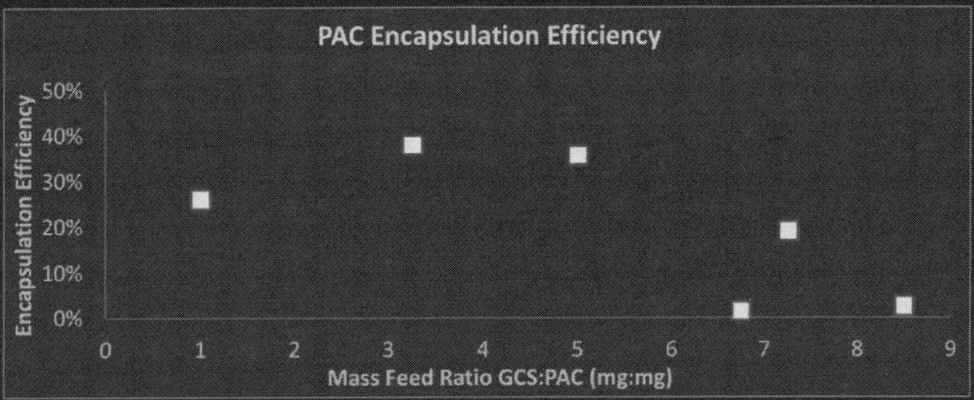

NSF

Stony Brook University

14

Delivery

- Cells used: MG-63 Human Osteosarcoma Cells
- Morphological study of cells after application
  - Comparing free and encapsulated PAC
- Cell Viability Assay done using MTS Assay
  - Comparing free and encapsulated PAC

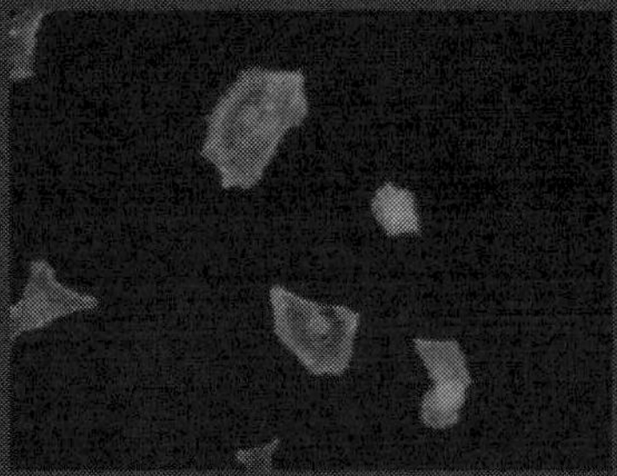

NSF

Stony Brook University

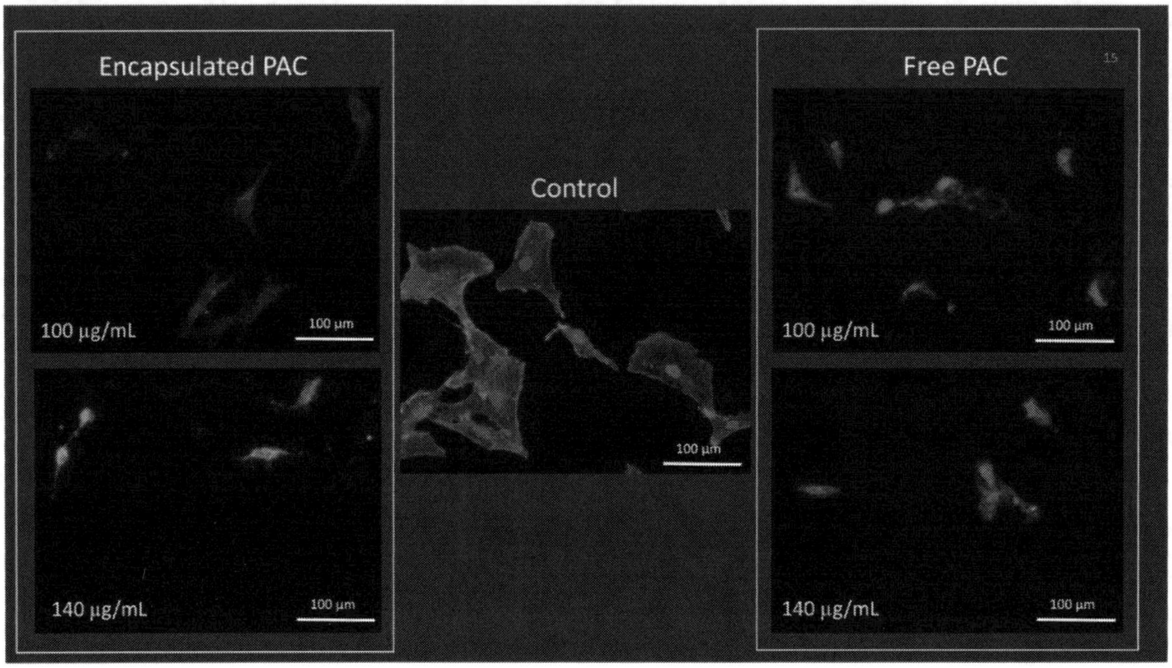

Cell Viability after 72 Hours

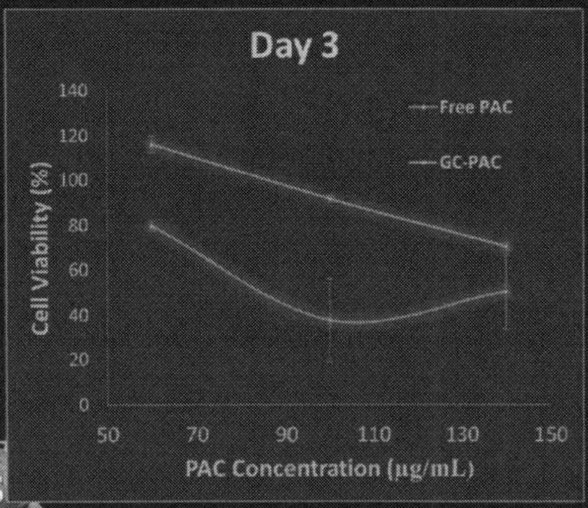

- MG-63 Osteosarcoma cells doubling time <72 hours
- Cell Viability is dose dependent
  - Can be adjusted for desired effect

Stony Brook University

17

Percent Cell Viability

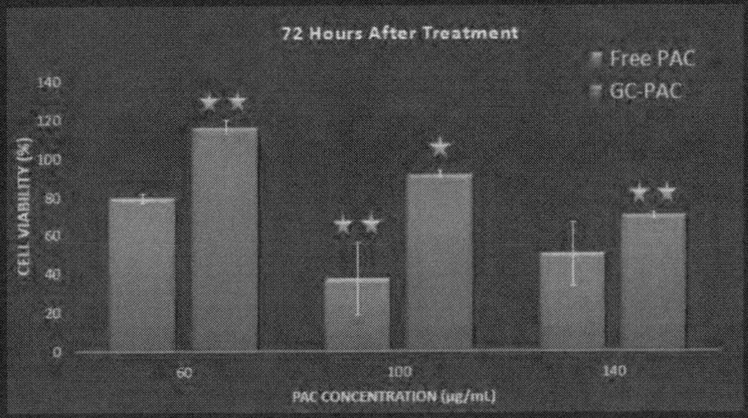

★ = P < 0.05
★★ = P < 0.01

NSF

Stony Brook University

18

Conclusions

- Most Efficient Mass Feed Ratios: 5 mg : 1 mg GCS:PAC
- Morphological analysis shows similar effects by both free and encapsulated PAC
- MTS Analysis indicates encapsulated PAC is advantageous over free PAC in its dose-dependency as well as effectiveness with greater amounts of PAC

Future Studies

- Release Kinetics of GCS-PAC NPs
- Amount of GCS in formed NPs

NSF

Stony Brook University

Acknowledgements

I would like to thank the National Science Foundation, Giulia Suarato, Dr. Yizhi Meng, the Center for Inclusive Education, and all other students and mentors involved in the REU for making this research possible by providing the resources necessary

394191 Single Cell Isolation Via Microfluidic-Based Droplet Delivery

Abstract:

Commercial technologies such as fluorescent-activated cell sorting (FACS) for isolating and delivering single cell samples are often inappropriate for smaller research facilities due to their size and cost. Researcher labs conducting experiments on single cells require an inexpensive and accessible method for creating samples. We demonstrate the practicability of using a gravity-driven microfluidic device under a bench-top optical trap to generate droplets containing isolated single cells for delivery onto a 96 well plate. The microfluidic device dispenses droplets of constant volume at fixed intervals. We establish functionality by delivering droplets containing glutaraldehyde fixed red blood cells (RBCs) onto glass slides. A fluorescent confocal microscope capable of imaging the glutaraldehyde-induced fluorescence of the RBCs is employed to count the number of cells contained per droplet. We attained 64% fidelity of single-cell samples, indicating that delivery from our device is not stochastic. Biomedical researchers often expect droplet delivery methods to be compatible with a 96 well plate standard; to this end we achieved automated delivery into each well of a 96 well plate using a Teensy 3.0 encoded motorized stage. Additionally, we implemented a side and forward scatter detection scheme around the optical trap to show the feasibility of fully automating the cell isolation and droplet delivery process. The entire apparatus is compact and portable enough to fit within a sterile environment to prevent contamination of samples. We propose that our setup is an inexpensive and compact alternative to flow cytometers that is accessible to most labs and can reliably sort single cells into droplets for sterile delivery.

394190 Self-Assembling Zwitterionic Nanogels As Immune Isolating Coatings for Stem Cell Derived Pancreatic Islet Transplantation

Abstract:

The concept of pancreatic islet transplantation to patients with diabetes to restore normoglycemia is a viable clinical strategy. However, widespread clinical application of islet transplantation remains limited because of the harmful side effects of the necessary immunosuppressive therapy to prevent host rejection of transplanted cells. Here we report the fabrication of a superbiocompatible nanogel and a novel coating strategy for protection of stem cell derived pancreatic islets against host rejection in vivo. Zwitterionic nanogels (100-300nm in size) were fabricated using an inverse emulsion polymerization synthesis scheme and further surface modified with CLICK reactive functional groups. Stem cell derived pancreatic islet cells were engineered to express azide functionalities on their cell membranes. Confocal fluorescence microscopy was used to verify covalent reaction and assembly of dual-layered nanogel coatings on engineered islet cell clusters. Cytocompatibility of coating and islet cell functionality were examined using a combination of in vitro glucose tolerance tests, live/dead staining, and in vivo studies on mice treated with streptozocin, a small molecule drug that destroys the beta cells. Our results indicate that the developed nanogel formulation and encapsulation strategy provides a viable strategy for prolonged protection and isolation of transplanted islets against host rejection

394187 Modeling of Selenium Nanoparticle Deposition for Optimized Production of Antibacterial Surfaces

Abstract:

The Webster Nanomedicine lab designs and explores applications of nanomaterials to inhibit bacterial growth and cancer cell responses. Due to its beneficial effects on the body, and known chemopreventive and chemotherapeutic properties[1], selenium has been used in its nano-form to impart nano-roughness and chemical effects in these experiments. To ensure precise and consistent selenium nanoparticle (SeNP) coverage, we aim to tabulate several relevant aspects of the particle. By adjusting the parameters of the reaction producing SeNP, a variety of data can be collected and analyzed. Characteristics that are pertinent include the nanoparticle size, coverage, surface energy, and roughness. Eventually, a cubic response surface model will be developed using a central composite design (CCD), which will be able to predict the interaction of the particles with bacteria and cells.

SeNP are produced through a reaction of glutathione (GSH) with sodium selenite (Na_2SeO_3). The reactants are introduced in a 4:1 molar ratio, respectively. Thin discs of poly(l-lactic acid) (PLLA) are added to the reaction, providing a surface for the cells to grow. In order to precipitate out the SeNP, 2M NaOH is added to the mixture. The reaction is then halted by the addition of DiH_2O. The experimental design used four factors: time in GSH/ Na_2SeO_3, time in NaOH, reaction volume, and the number of samples in the volume. Three results are compared: nanoparticle size, nanoparticle coverage, and surface energy of each sample. Both the size and the coverage of the produced nanoparticles can be measured using scanning electron microscopy. The surface roughness is portrayed through atomic force microscopy. Surface energy can be calculated utilizing the contact angle, which is defined using a Goniometer.

Previous studies show longer time increments (in both cases) increase the deposition of SeNP on the substrate. However, as the increments reach larger values of time (30-60 seconds), there is less of a change in coverage. This suggests the existence of a saturation point. On the PLLA, there is a finite number of binding sites for the nanoparticles. The size of the nanoparticle, with time being the only deviation, remains consistent. It is predicted that development time and reaction volume directly correlate. Recent observations of previous studies have shown that increased coverage correlates with development time in both steps, and therefore, roughness.

Through characterization of the SeNP product, a cubic response surface will be calculated relating the input parameters to expected results, optimizing the coverage of nanoparticles on substrates. These efforts conserve both time and resources because reactions can be designed with focused specifications.

394223 Exploiting Polymer-Nanoparticle Interactions to Create a Hydrogel with Biomedical Applications

Abstract:

Injectable hydrogels have recently become a relevant material platform in applications such as in vivo drug delivery or as cell carriers because they can be implanted in a minimally invasive manner and closely mimic biological tissue. Attractive properties for such injectable hydrogels are that they easily flow under applied stress (*shear-thinning*) and rapidly heal upon relaxation of the applied stress (*self-healing*). Previously, we developed a shear-thinning and self-healing injectable hydrogel driven by non-covalent interactions between hydrophobically modified cellulose and PEG-PLA nanoparticles. [1] The nature of this gel enables facile loading with a variety of therapeutics as well as tunable release profiles of multiple therapeutics. Specifically, the PEG-PLA nanoparticles allow for loading with hydrophobic therapeutics while a hydrophilic therapeutic can be entrapped within the aqueous phase of the gel.[2] In this work, we have quantified the factors that affect the mechanical properties and release profiles of the gels, demonstrating the range of their tunability. Additionally, we explored other nanoparticles and biopolymers for the fabrication of similar materials. Lastly, we studied the use of this injectable hydrogel as a cell carrier with osteogenic properties. Human mesenchymal stem cells (hMSCs), which are capable of differentiating into bone, cartilage, and fat cells, were encapsulated in the gel and survived injection through a 28 gauge needle into three-dimensional collagen hydrogels. Additionally, PEG-PLA nanoparticles were loaded with dexamethasone to stimulate osteogenic differentiation of the encapsulated hMSCs. We are currently exploring the use of this drug-loaded cell carrier as an injectable system for bone healing. Overall, we present an injectable hydrogel with tunable properties for the controlled delivery of therapeutics with facile synthesis and minimally invasive application *in vivo* that is capable of serving as a stimulatory cell carrier.

---

[1] Appel, E.A., Tibbitt, M.W., Langer, R., Nature Communications., **2014**, *in revision*.

[2] Greer, J.M., Appel, E.A., Tibbitt, M.W., Langer, R., *manuscript in preparation*.

Validation of High Throughput Electrochemical Gas Sensing Screening System

Zixuan Wang

November 17, 2014

Background

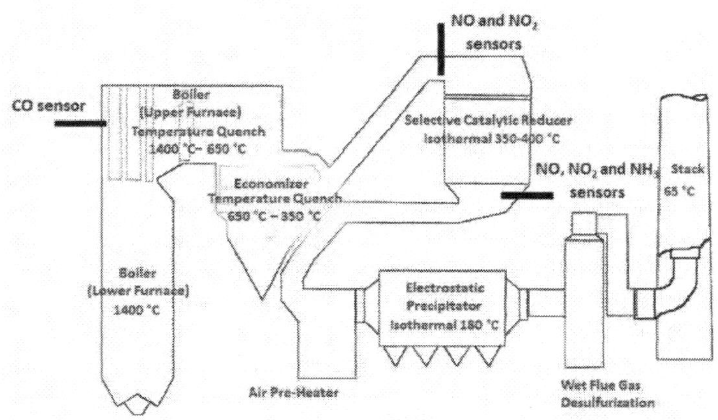

Harsh combustion environment - 800 ° C
Accurate and selective to target gas

UNIVERSITY OF SOUTH CAROLINA

High Throughput Methodology

Subatomic difference affect large scale

↓

Not feasible to individually test material

↓

High throughput synthesis and characterization methods

High Throughput Methodology

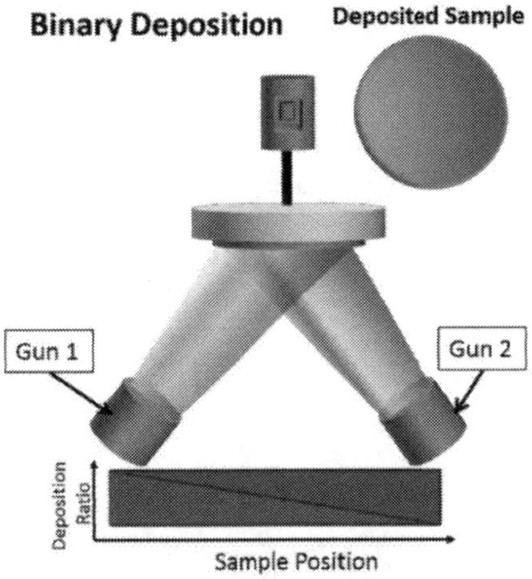

- Normalize affect of distance on the potential

Development of Uniform Standard Sample

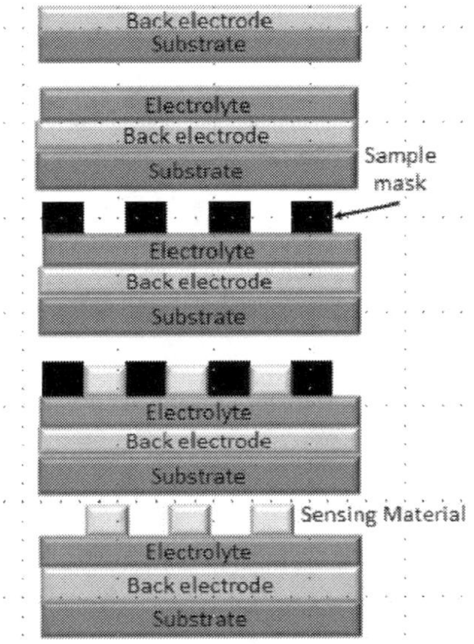

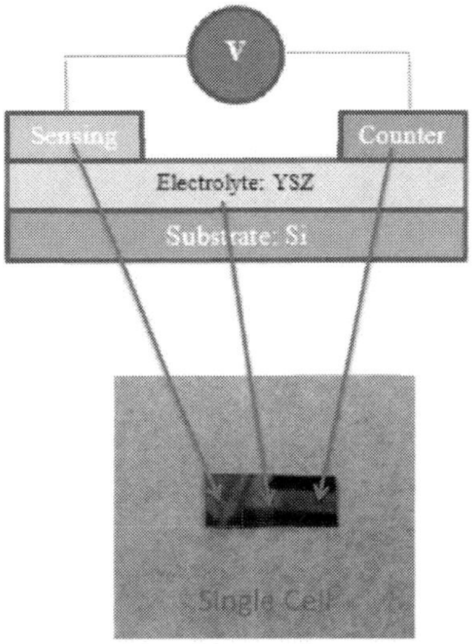

UNIVERSITY OF SOUTH CAROLINA

Surface Mechanisms

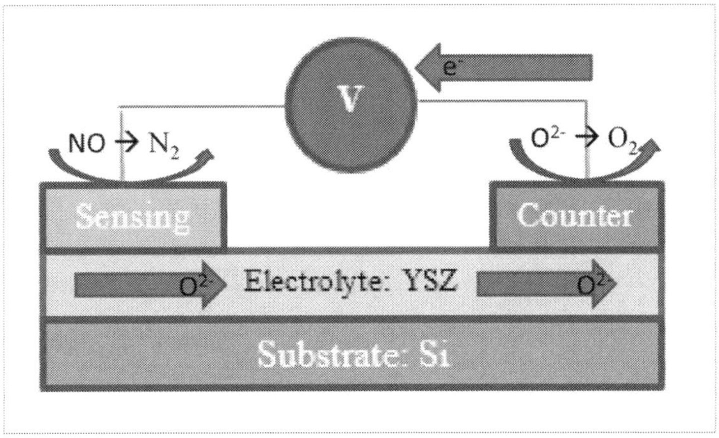

UNIVERSITY OF SOUTH CAROLINA

Sensors and Actuators B 121 (2007) 652-663

Reflectometry Results - YSZ

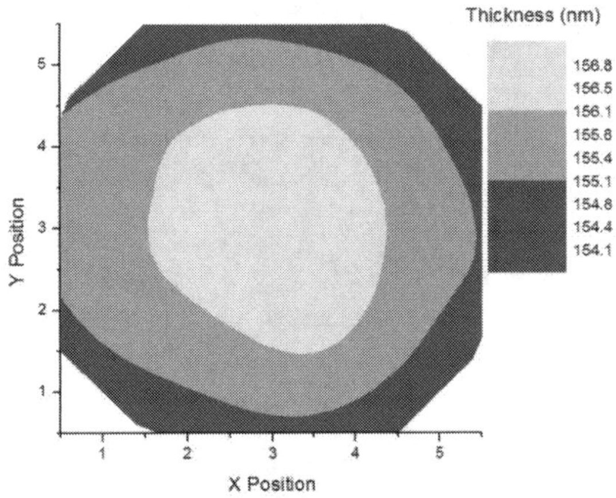

- Results confirm relative uniformity of sample (1.8% difference)

UNIVERSITY OF SOUTH CAROLINA

XRD Results - YSZ

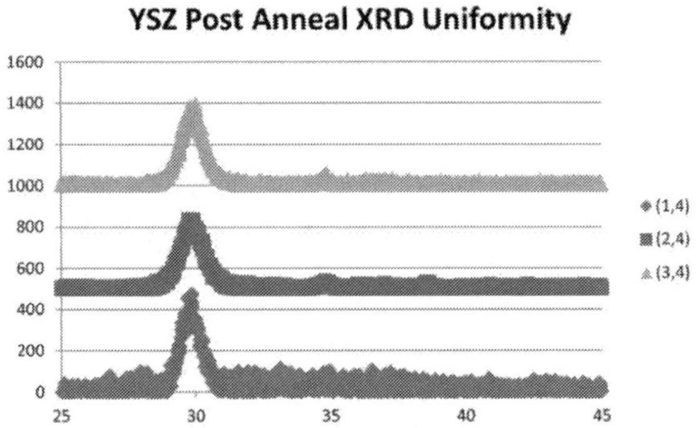

(1,4)	29.745	
(2,4)	29.836	34.848
(3,4)	29.876	34.859

- XRD analysis using Jade
- YSZ Cubic:
  - (111): 29.675°
  - (200): 34.398°
- Confirm identical phases

UNIVERSITY OF SOUTH CAROLINA

Reflectometry Results – WO_3

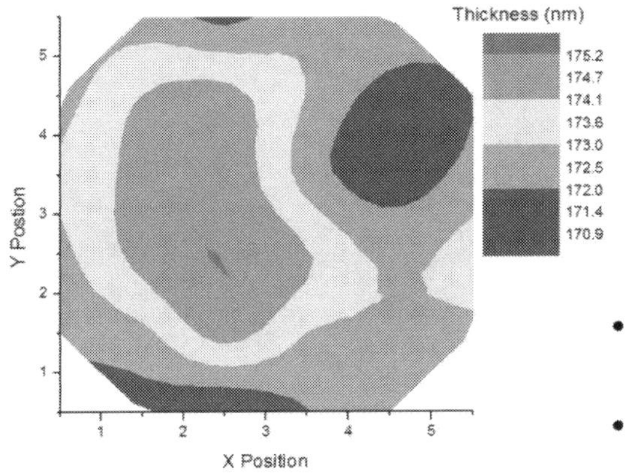

- Confirms relative uniformity (2.5% difference)
- Evidence some parts of the plate that are slightly thicker or thinner

UNIVERSITY OF SOUTH CAROLINA

XRD Results – Post Annealed WO_3

(1,4)	23.143	23.653	24.362
(2,4)	23.140	23.671	24.397
(3,4)	23.127	23.641	24.377

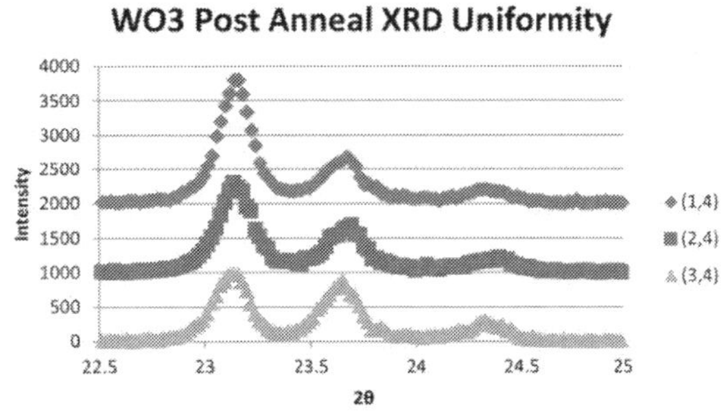

- XRD analysis using Jade
- WO_3 monoclinic:
  - (002): 23.118°
  - (020): 23.582°
  - (200): 24.365°
- Confirm identical phases with phase shifts present

UNIVERSITY OF SOUTH CAROLINA

Reflectometry Results - WO_3-YSZ

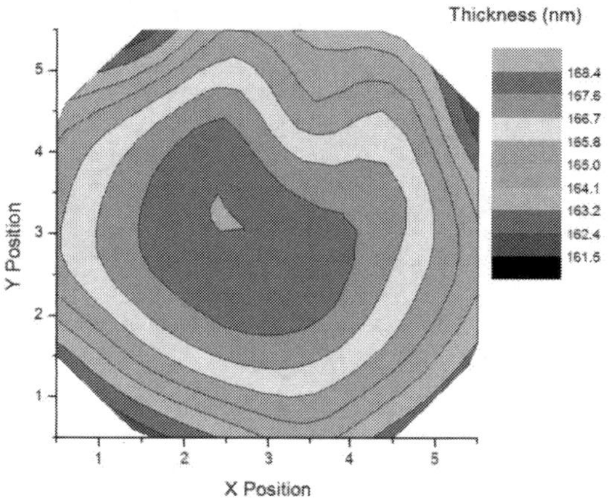

- Sputtering conditions: same
- YSZ held constant
- Confirms relative uniformity (4.3% difference)

UNIVERSITY OF SOUTH CAROLINA

XRD Results – WO_3-YSZ Post-annealed

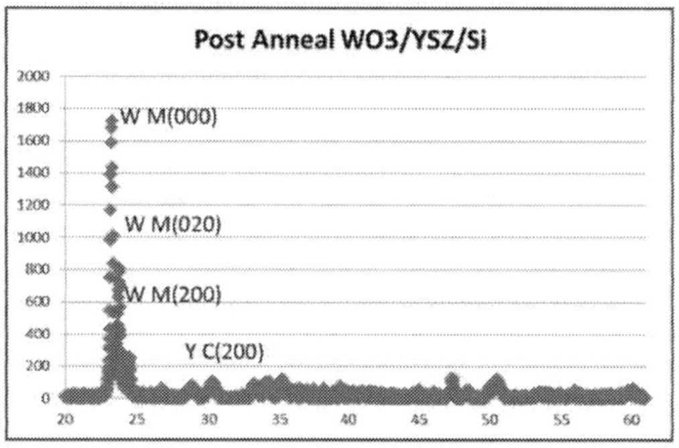

WO_3 monoclinic:

(002): 23.118°

(020): 23.582°

(200): 24.365°

YSZ Cubic:

(111): 29.675°

(200): 34.398°

UNIVERSITY OF SOUTH CAROLINA

Electrochemical Cell Setup

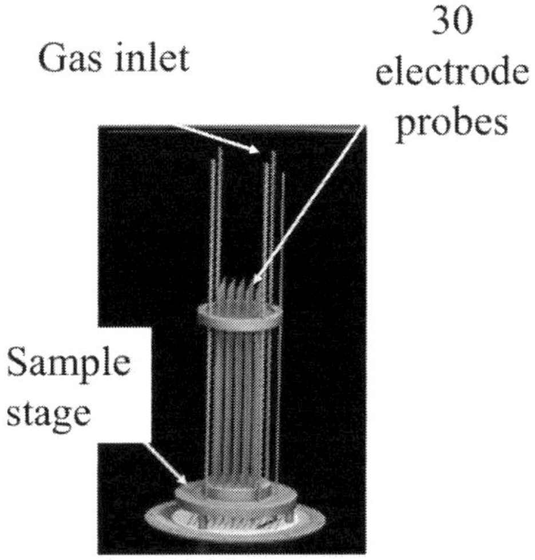

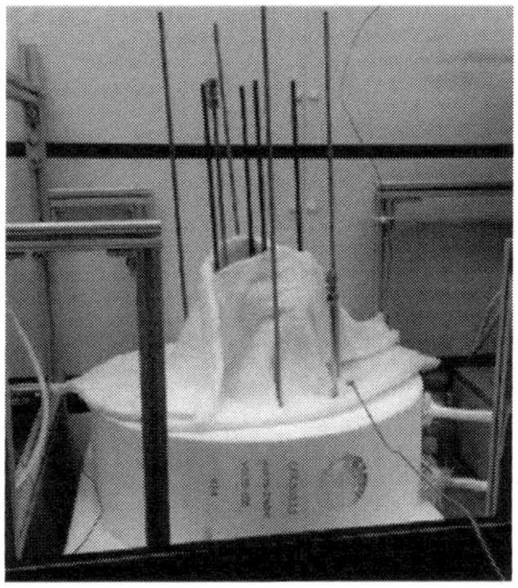

Single Electrochemical Cell Tests

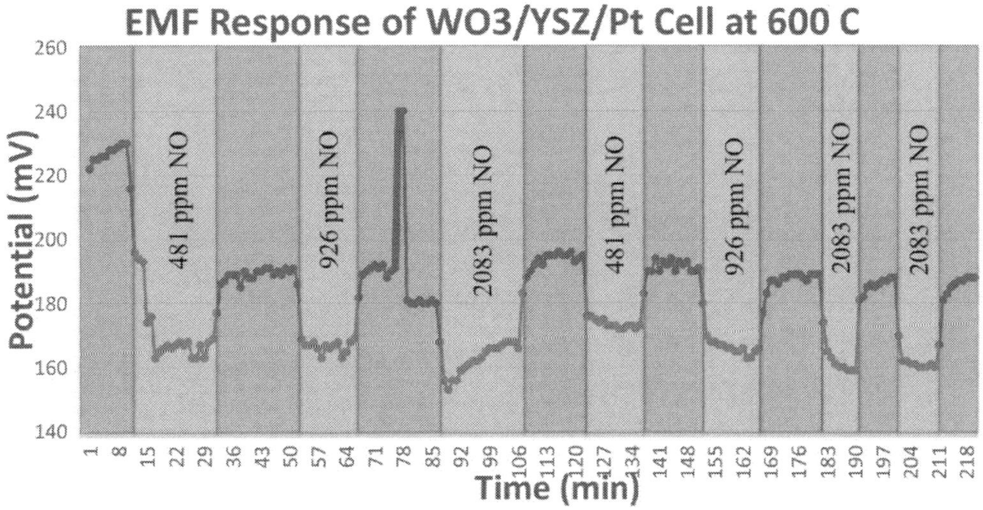

UNIVERSITY OF
SOUTH CAROLINA

Electrochemical Cell Tests

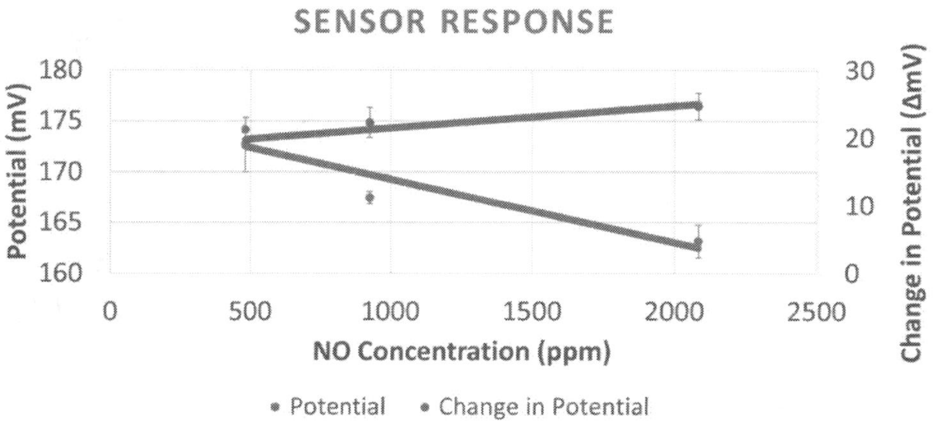

Multielectrode Cell Tests

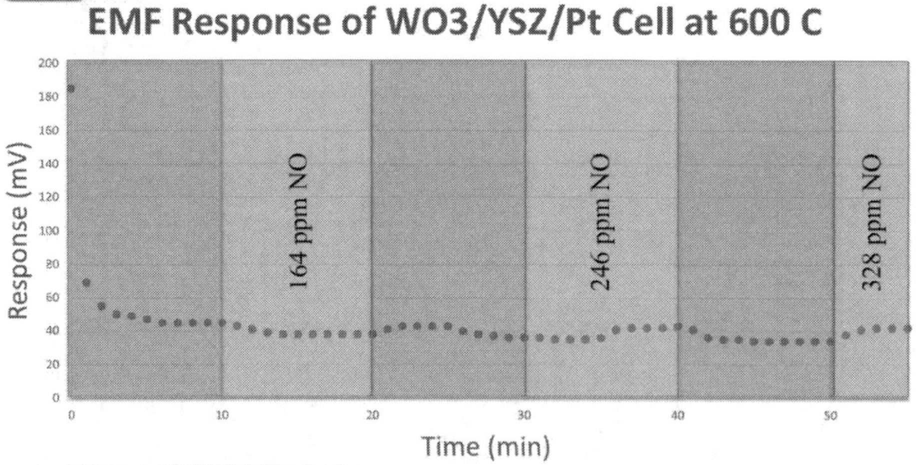

Conclusions

- Designed planar sensor array
  - Use for combinatorial screening of electrochemical gas sensors
- Confirm deposition uniformity
  - Reflectometry: thickness
  - XRD: phases
- Cell potential
  - Changes valid use of planar sensor design

Future Work

- Create multielectrode with increased surface area/perimeter of the electrodes
- Normalize potential to distance of electrodes
- Create multielectrode with different composition gradient

Acknowledgements

- National Science Foundation (NSF)
- University of South Carolina – College of Engineering and Computing
- Strategic Approaches to the Generation of Electricity (SAGE)
- Principal investigators: Dr. Jochen Lauterbach and Dr. Jason Hattrick-Simpers
- Graduate mentor: Ben Ruiz-Yi

UNIVERSITY OF SOUTH CAROLINA

394216 Long-Range Correlations in Liquid Water

Abstract:

The purpose of this research project was to observe water molecules' environments to determine whether or not the system forms identifiable clusters of similar environments. The different clusters aid in proving the hypothesis that different water patches at differing high and low densities values form in a size of 1nm in a system at room and lower temperatures. Molecular dynamics computer simulations were done using the dl_poly classic program with the water model potential TIP4P/2005. The simulations were run at various configurations for density, temperature, and pressure. The results were read by a program, which took the file from the simulations and determined the coordinate positions of the water molecules. Different clusters were found based on the results. These results are indicative of distinctive structures formed in supercooled water that may lead to the water patches of varying densities.

394155 Frictional Study of Polyethylene Glycol Monolayers on Silica Substrate

Abstract:

Lubricating micro- and nano-electromechanical systems (MEMS and NEMS) has become increasingly challenging as devices are continuously miniaturized. As the dimensions of these systems decrease, their surface area to volume ratios increase significantly and frictional forces become dominant where moving parts exist. Solid-state lubrication via self-assembled monolayers presents a promising alternative to traditional macro-scale lubricants which are too bulky to use in such small systems.[1,2] Previous studies have shown that self-assembled monolayers containing alkylsilanes and mixed alkylsilane and perfluoroalkylsilane chains reduce friction coefficients compared to equal length systems. Although, hydrocarbon-based monolayers have been widely studied and shown to improve frictional properties and device life, mixed monolayers impart the beneficial properties of perfluoroalkylsilanes (low critical surface tension and high thermal and mechanical stability) to the film. Furthermore, a mixed length configuration with longer alkylsilane chains relative to the perfluoroalkylsilane chains leads to a liquid-like layer formed by the taller chains that fold over and align with the shearing direction.[3]

In this study, we investigate polyethylene glycol (PEG) monolayers using molecular dynamics simulations as a potential solid-state lubricant for NEMS. The presence of oxygen atoms within the chains render these chains more flexible than their alkyl- and perfluoroalkyl-counterparts. Thus, even sparsely functionalizing a surface leads to a uniform film. We have investigated the tribological properties of such films across a range of surface coverages and chain lengths. Preliminary results show the effects of surface coverage on friction coefficients is diminished as chain length is increased. This suggests that equivalent friction coefficients could be achieved while reducing separation distances by using longer chains on sparser surfaces.

394001 A Machine-Learning Model to Predict Activation Energies of Hydrogenation Reactions

Abstract:

The goal of this research is to develop a machine-learning model to predict activation energies associated to the hydrogenation of organic compounds in the presence of specific catalytic substrates. The current method of predicting activation energies involves a linear relationship with the change in Gibbs free energy of the reaction, which can be very inaccurate. Other methods include experimentation and simulation, both of which have significant cost and/or time drawbacks. This research seeks to improve on these methods to develop a means to predict activation energies quickly, accurately, and with minimal cost.

393764 Microscopic Modeling of the Self Assembly of Poly(ethylene oxide)-Poly(propylene oxide)- Poly(ethylene oxide) (PEO-PPO-PEO) Block Copolymers: Critical Micelle Concentrations

Abstract:

Single chain mean field theory (SCMFT) is applied as a coarse-grain simulation model for Pluronics, a trademarked branch of triblock copolymers. The microscopic model is used to quantify the micellization process of the nonionic class of surfactants at 37 degrees Celsius for drug delivery applications. The simple model with only three interaction parameters is shown to be able to reproduce with high accuracy experimental data for the critical micelle concentrations of different chain lengths of the polymers. These parameters are physically significant and are close to the experimental values from Flory models. In addition, aggregation number of the surfactants and volume fraction profiles are explored. The effects of the chain lengths of hydrophobic tails and hydrophilic heads on CMC values are discussed. SCMFT is transferable in surfactants of different ABA block conformations and other more complex polymer structures where previous simulations have not excelled.

Recommendations for Legislative Actions to Reduce Carbon Emissions in the Electricity Production Sector

Sam White

American Institute of Chemical Engineers
University of Michigan
Washington Internships for Students in Engineering

August 2014

Executive Summary

There is currently a movement in America pushing for environmental effects to be taken into consideration when assessing new and existing business ventures. As emphasized by President Obama"s 2014 State of the Union Address, a major area for this movement to make progress is within the energy sector [1]. Electricity production currently produces the most greenhouse gas emissions of any energy sector in the US, making it the energy sector where the largest environmental impact can be made with proper policy action [2].

Luckily, currently-available, low-emission renewable energy technologies are most well suited for electricity production over any other energy sector application. As a result of recent government regulations combined with high coal prices, coal plants, which currently dominate the electricity production landscape, are being retired at an accelerated rate [8]. This, combined with increased electricity demand, has opened up the door for new electricity production technology to be implemented. However, due to the low price of natural gas resulting mainly from the onset of fracking, efforts to implement new renewable-sourced electricity production facilities have been stalled in favor of natural gas power plants. While natural gas plants produce only half the carbon emissions of coal, and do not produce harmful byproducts such as sulfur dioxide or mercury, they are not as environmentally friendly as the other available alternatives, such as nuclear, solar, or wind [11] [12] [13] [14].

The fact that natural gas dominates the shift away from coal is not surprising, however, given the current economic landscape. The figure below shows the levelized cost of electricity (LCOE) for select available electricity production technologies, giving a $/Mwh cost of production by factoring in all applicable costs (construction, fuel, maintenance, operation, etc.) [23].

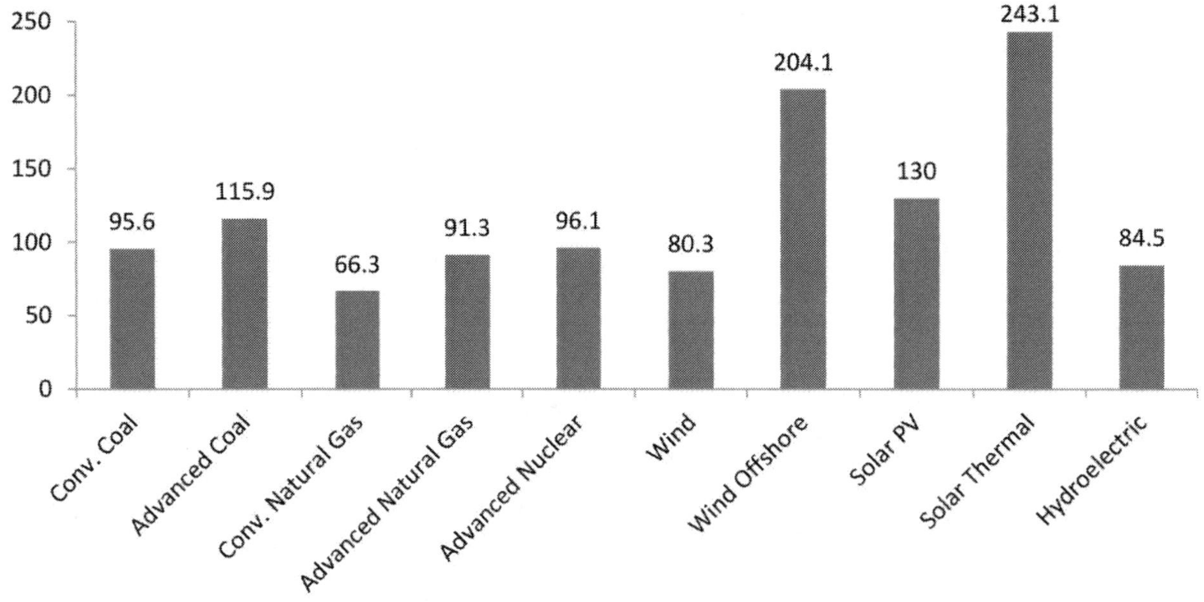

US Average LCOE for Plants Entering Service in 2019, in 2012 $/Mwh [23 EIA LCOE]

As seen in figure above, conventional natural gas currently offers the cheapest form of electricity production. It is important to note, though, that some alternatives like nuclear or wind are already competitive with other conventional electricity production methods like coal. Other alternatives, like solar or offshore wind, are still too expensive to be competitive at the moment, mainly due to the fact that electricity production from renewable sources is still too variable. With the advent of advanced electricity storage technology this issue would be eliminated, but such technology is currently unavailable or too expensive to be implemented economically [19].

Because of their positive environmental impacts, alternative electricity production methods are given some assistance by the federal government in order to make them more economically competitive. These economic aids mainly come in two forms: tax credits and loan programs. On the tax side, an installment tax credit allows solar production facilities a credit equal to 30% of the cost of implementation, while a production tax credit allows wind, hydro, and geothermal plants to receive a .23 cents/kwh credit. However, the installment tax credit is only available until the end of 2016, and the production tax credit already expired at the beginning of 2014 [26]. On the loan program side, the government has extended loans to help both nuclear and renewable power plants, though the renewable plant loan program expired in 2011 [27] [28].

On top of aid being given to alternative electricity production methods, the federal government can also level the playing field by putting a burden on conventional electricity sources. This mainly comes in one of two forms, a carbon tax or a cap-and-trade program, both of which seek to charge electricity producers for the amount of carbon emissions they produce. On top of producing revenue for the federal government, these programs help incentivize electricity producers to innovate new ways to reduce their carbon emissions. Though no such program has ever been implemented, a cap-and-trade program was able to at least pass the House in 2009 [24] [25].

Even small actions by the federal government can have noticeable effects in shifting the production methods used in the electricity production sector. Just a relatively small tax such as $10/metric ton of carbon emissions would cause significant carbon emission reductions compared a scenario in which no government actions are taken [33]. Additional analysis specific to natural gas and renewables shows that any carbon tax would actually increase the amount of natural gas-based electricity production for the next couple of decades, and increase the amount of renewable-based electricity production indefinitely. Additionally, if current favorable tax policies for renewables are renewed long-term, the use of renewables is significantly higher than those projected given current tax policy deadlines, especially after 2030 [34] [35].

As a result, it is recommended that a two-pronged approach be taken to limit the amount of economic disturbance while ensuring the electricity production sector makes a greater shift towards renewables given the reduction of coal use. First, a carbon tax equivalent of $10/metric ton of carbon emissions should be enacted to reflect the social cost of emissions. Each year this tax would increase 5% per year after inflation in order to push companies to innovate new carbon reduction technologies. Second, the production tax credit should be extended for all renewable electricity production methods using the current credit of .23 cents/kwh. This credit would decrease 5% per year after inflation to mirror the carbon tax in order to account for the increased rate of improvement in renewable technologies and the relative advantage given by the increasing carbon tax.

It is important that such a program be long-term in order to allow industry to accurately model the economic landscape their projects will enter. Given the current gridlock in Washington and the long-term nature of implementing new electricity production facilities, a long-term tax plan is the only way to ensure economic stability. Factoring in only revenue from taxes on natural gas use and costs from credits extended to renewables use, such a program would pull in an average surplus of $5.5 billion per year between 2015 and 2040; factoring in coal would increase this figure significantly, particularly in the near future. Feeding at least part of this surplus into loan programs for nuclear plants would ensure that an increased amount of low carbon electricity production facilities are able to come to market.

Table of Contents

Table of Figures

Table of Tables

1.0 Introduction

In his 2014 State of the Union address, President Obama stated that "[O]ne of the biggest factors in bringing more jobs back is our commitment to American energy"[1]. He stressed the importance of "a cleaner energy economy" and acknowledged that "[it] won't happen overnight, and it will require tough choices along the way." Though he touted the success of his "all-of-the-above energy strategy", and highlighted the advantages of natural gas and solar, the President laid out no definitive plans for how the US would approach energy production in the future. However, with the release of recent carbon emission regulations, the executive branch seems to be taking at least some action to direct the country towards a cleaner energy economy. However, if a cleaner energy economy is going to happen in America, the legislative branch needs to get more involved by developing and implementing a long-term energy policy plan.

Of all the areas of energy production in the America, the most promising area for change is the electricity production sector, since a vast majority of alternative energy sources are most viable for electricity production instead of applications like transportation or residential heating. Luckily, this is also the energy sector in which the most impactful changes can be made: in 2012, electricity production was the largest single contributor to greenhouse gas emissions in the US, accounting for 32% of greenhouse gas emissions as seen in Figure 1 below [2].

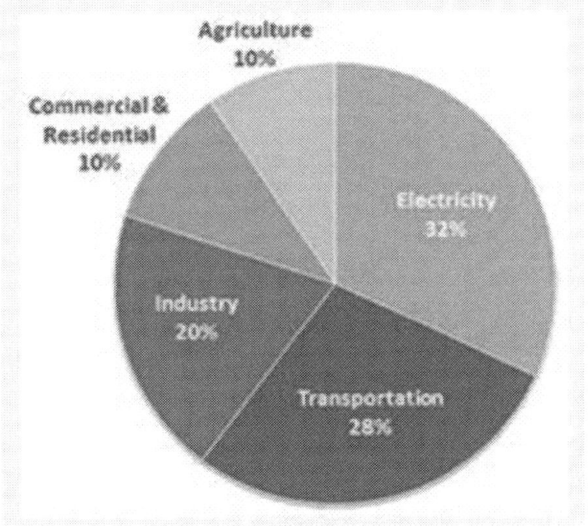

Figure 1. Total Greenhouse Gas Emissions by Economic Sector in 2012 [2]

The implementation of a cleaner energy economy has shown to be a popular policy among voters from both parties. According to a Washington Post – ABC News poll from June 2014, 70 percent of voters say that the federal government should require limits to greenhouse gases from existing power plants. 57 percent of Republicans, 76 percent of Independents, and 79 percent of Democrats support state-level limits on greenhouse gas emissions. Most significantly, voters are willing to pay extra for energy if it means a reduction in greenhouse gas emissions. "Asked whether Washington should still go forward with limits [on energy-sector greenhouse gas emissions] if ‚they significantly lowered greenhouse gases but raised your monthly energy bill by 20 dollars a month," 63 percent of respondents say yes, including 51 percent of

Republicans, 64 percent of Independents and 71 percent of Democrats" [3]. These results suggest that Americans on both sides of the aisle are in support of implementing a cleaner energy economy.

In addition to the popularity of a cleaner energy economy, it also makes sense from an environmental and health stand point. On top of the fact that the reduction of greenhouse gas emissions reduces the rate of climate change, it brings along health benefits as well. With the release of its Clean Power Plan in June 2014, which seeks to cut carbon pollution from the power sector by 30% by 2030 by imposing stricter emission limits on existing plants, the EPA projects that "annual public health benefits will total somewhere between $55 billion and $93 billion by avoiding up to 100,000 asthma attacks and 2,100 heart attacks each year." By contrast, they predict compliance cost of $7.3 to $8.8 billion per year, significantly lower than the expected public health benefits [4]

However, despite the ongoing executive branch actions regarding energy policy, the legislative branch has taken relatively little action regarding energy policy lately, mainly due to party-politics-driven gridlock. At the beginning of 2014, the Production Tax Credit for renewables was allowed to expire for the first time since 2004, and only the fourth time since its enactment in 1992 [5]. In May of 2014, the Senate allowed a bipartisan energy efficiency bill to die due to disagreement over other energy related issues, such as the fate of the Keystone XL pipeline and possible legislative response to President Obama"s climate change regulations. In fact, it has been 7 years since Congress passed a major energy bill, which means there has been no major legislative action in response to the changing energy landscape as a result of the shale fracking boom [6].

A plan for implementing a cleaner energy economy becomes even more important as energy demands increase, particularly when it comes to electricity production. As electricity demand increases, more electricity production facilities are built, and how these facilities produce electricity greatly affects their environmental impact. According to the EIA"s Annual Energy Outlook 2014 report, US electricity demand will increase from anywhere between 14% and 41% between 2012 and 2040, though their best estimates suggest that electricity demand will increase by 29% over this time period. Additionally, the growth of electricity demand is projected to increase sometime around 2025, marking the beginning of an upwards slope in electricity demand growth for the first time in 75 years [7].

In order to meet this demand and fill the gap left behind by plant retirements, new electricity production facilities need to be built. As a result of recent emission regulations like the Clean Power Plan, coal power plants are being retired at an accelerated rate. However, due to the low price of natural gas in recent years thanks to fracking, much of this gap has been filled by natural gas fired plants instead of alternatives, like nuclear or renewables. This is mainly due to economics: even the cheapest alternative energy sources, which are on the verge of being if not already cost competitive to coal, simply cannot compete with the low price of natural gas. While natural gas is a lower-emissions alternative to coal, it has emissions nowhere near as low as alternative electricity sources.

The onset of natural gas is not a bad thing, however: using natural gas as a bridge fuel to reduce emissions in the immediate future while hoping that renewables become more cost competitive in the future is not unreasonable given the rate of development and investment in renewable technologies. Natural gas also offers the unique advantage of "spinning reserve", a source that can quickly be powered up to meet rapid fluctuations in demand. This means natural gas can offer a flexibility that base load electricity sources like coal or nuclear cannot, while acting as a buffer against the variability of renewables like solar or wind.

Embracing natural gas in the here and now is also economically responsible, as it keeps electricity prices low by offering a cheap alternative to coal and helps support the growing American natural gas production market. However, without government action, this shift towards natural gas will dwarf any efforts to make renewables a larger share of our electricity production sector. If the government is serious about using natural gas as a bridge fuel to a cleaner energy economy in the electricity sector, a long-term policy needs to be enacted to slowly guide the market in the right direction.

2.0 Overview of Electricity Production Methods and Their Environmental Impacts

In order to discuss energy policy, it is first important to establish the electricity needs of the country, as well as the current capabilities of various electricity production methods and their effects of the environment. As seen in Figure 2 below, coal and natural gas currently dominate the electricity production sector, with nuclear and renewables making up the bulk of the remaining production.

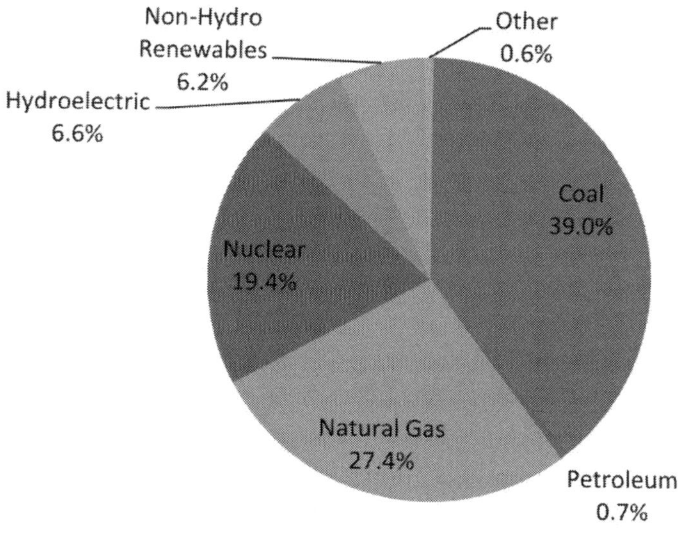

Figure 2. Share of Total US Electricity Production by Source [Adapted from 8]

2.1 Coal and Natural Gas

The two main fossil fuel based / conventional energy producing sources, coal and natural gas power plants are currently the largest source of electricity in America, accounting for nearly two-thirds of the country"s total electricity production [8]. Coal plants are relatively easy to keep in operation over a long time scale, but it is difficult for them to make relatively quick changes to production capacity in response to the changing needs of the grid, particularly during peak-usage hours. As a result, coal plants are very useful in providing a baseline power supply somewhere between peak and off-peak requirements.

Natural gas, on the other hand, allows for a very flexible energy production method. Gas flow rates into a furnace can be controlled with relative precision, unlike solid input streams found in coal plants. As a result, natural gas plants can be very responsive to the changing needs of the energy grid. In fact, natural gas plants are typically used to augment other energy production facilities for this very reason.

From an environmental perspective, the EIA considers conventional electricity sources like coal and natural gas as being the only contributors to CO_2 emissions in the electricity production sector. In 2010, the EIA reported that 1969 million metric tons of CO_2 were released by coal plants nation-wide, while 1285 million metric tons of CO_2 were release by natural gas plants [9].

However, coal tends to have other unique environmental impacts. First, burning coal tends to release nitrogen oxides, which tend to have larger warming potentials than CO_2. For example, N_2O has a SAR 100 year warming potential of 310, meaning that one ton of N_2O has 310 times the impact on global warming compared to one ton of CO_2 over the course of 100 years [10]. From a life cycle assessment standpoint, coal plants produce 974 metric tons of CO_2 equivalent per gigawatt-hour, which accounts for this increased warming potential as well as CO_2 emissions due to non-production factors, such as plant construction or fuel transportation [11]. Coal also tends to produce SO_2, which, along with nitrogen oxides, is the major man-made cause of acid rain. According to the EPA, "In the United States, roughly two-thirds of all SO_2 and one-quarter of all nitrogen oxides come from electric power generation that relies on burning fossil fuels, like coal" [12]. Additionally, coal plants are a major source of mercury, a highly toxic element. According to the EPA, "Coal-burning power plants are the largest human-caused source of mercury emissions to the air in the United States, accounting for over 50 percent of all domestic human-caused mercury emissions" [13].

In the case of natural gas, environmental impacts are lower than that of coal. According to the EPA, "… the burning of natural gas produces nitrogen oxides and carbon dioxides, but in lower quantities than burning coal". In fact, from a life cycle assessment standpoint, natural gas plants produce 469 metric tons of CO_2 equivalent per gigawatt-hour, less than half that of coal [11]. Additionally, the EPA states that "[e]missions of SO_2 and mercury compounds from burning natural gas are negligible". [14]. While the EPA also mentions that unburned natural gas being released either from a power plant or during transportation is a concern, though in 2009 the

EIA stated that "[i]ncreases in energy-related methane emissions [were] largely from underground coal mining", suggesting that methane releases are more prevalent in the procurement of coal than in the use and transport of natural gas [14] [15].

2.2 Nuclear

In the US, nuclear energy makes up nearly one-fifth of all electricity production [8]. According to the World Nuclear Association, "it [nuclear power] is especially suitable for large-scale, continuous electricity demands, which requires reliability (i.e. base load)" [16]. In this way, nuclear power plants are analogous to coal plants in the sense that they are very useful in meeting baseline power needs, but have trouble making quick changes in response to fluctuating peak-hour demand.

From an environmental standpoint, "nuclear power plants do not emit CO_2, SO_2, or nitrogen oxides as part of the power generation process" according to the EPA [17]. However, as with all power production facilities, the construction of nuclear power plants leads to non-negligible greenhouse gas emissions. Based off a compilation of life-cycle emissions studies gathered by the Nuclear Energy Institute (NEI), nuclear energy"s life-cycle emissions are significantly lower than those of coal or natural gas plants, and are in fact comparable to those of renewable like solar, wind, or hydro [18]. From a life cycle assessment standpoint, nuclear produces 15 metric tons of CO_2 equivalent per gigawatt-hour, which is the second-lowest behind wind, and only about 1.6% that of coal plants [11]. Much of these emissions come from the construction of plants, which are significantly larger than other electricity production facilities, and the transportation of fuel.

2.3 Solar and Wind

As shown in Figure 2, only about 6% of electricity production in the US comes from non-hydro renewable sources. Of this 6%, a majority comes from solar (in the form of both photovoltaic cells and solar thermal) and wind [8]. Of all of the available technologies, solar and wind are the most challenging to integrate into the grid due to the fact that they are considered variable sources: if the sun doesn"t shine or the wind doesn"t blow, these technologies are not actively putting energy into the grid, meaning demand has to be met by some other means [19] [20]. In order to offset this, large-scale production facilities are looking into using some form of energy storage system. This would allow excess energy collected during times of sunshine or wind gusts to be used when the sun doesn"t shine or the wind doesn"t blow [19]. However, the storage capacity of current electricity storage technology is limited. As a result, any useful amount of electricity storage is currently too expensive, which limits the viability of renewables.

From an environmental standpoint, neither solar nor wind produce carbon emissions during their operating lifetime, since their energy is derived from the sun and not from burning fossil fuels [19] [20]. However, like nuclear, solar and wind do have some emission impacts due to the processes used to manufacture materials and construct plants: wind is responsible for 14 metrics tons of CO_2-equivalent per gigawatt-hour, while solar photovoltaics are responsible for

39 metric tons of CO_2 equivalent per gigawatt-hour [11]. As with nuclear plants, many of the emissions related to renewables come from construction of production facilities. In the case of solar photovoltaics, higher life-cycle emissions are seen as a result of the emissions that result from the manufacturing process of the solar cells themselves.

2.4 Other Renewables

Solar and wind are not the only renewable sources of energy available to the US. As seen in Figure 2, nearly 6.6% of electricity production comes from hydro plants. The main issue with hydro is that most of the available sites for dams in the US have already been dammed, meaning there is little room for expansion. As a result, the Obama administration has identified a collection of dams that currently do not produce electricity, but outfitting these dams with turbines would only increase the current US hydro capacity by roughly 15% [21]. Another alternative energy source is geothermal, which utilizes naturally-occurring underground sources of heat to power turbines. While this energy is clean, it is only available on a regional basis: Hawaii and the western US have a large geothermal production potential, while areas like New England do not [22]. Since hydropower does not have a lot of potential for growth, and geothermal is regionally specific, they will not be explored as in-depth in this paper.

2.5 Environmental Impact Summary Table

Table 1 below summarizes the life-cycle environmental impact per gigawatt-hour of the main electricity production methods highlighted in this paper.

Technology	Metric Tons of CO_2 released / GWh
Coal	974
Natural Gas	469
Nuclear	15
Solar	39
Wind	14

Table 1. Life-Cycle Environmental Impact Summary of Electricity Production Technologies [Adapted from 11]

3.0 Economics in the Electricity Production Sector

On top of technological limitations and environmental concerns, economics plays a major role in determining which technologies get used for energy production. On the production side, economic factors tend to play a major role in determining which electricity production technologies are utilized by producers. On the consumer side, the electricity production technology used by electricity providers plays a role in the price of electricity sold to consumers.

3.1 Comparing Costs of Electricity Production Technologies

Due the wide range of variables present in energy production, it can be difficult to compare the cost of one electricity production technology to another. However, a useful metric, levelized cost of electricity (LCOE), lays out the cost of given technologies on a per-MWh basis, factoring in such costs as construction, fuel, operating costs, maintenance, and financing costs [23]. Figure 3 below shows the US average LCOE for plants entering service in 2019, in 2012 dollars/MWh, and does not factor in any public financial support (tax credits, subsidies, etc.).

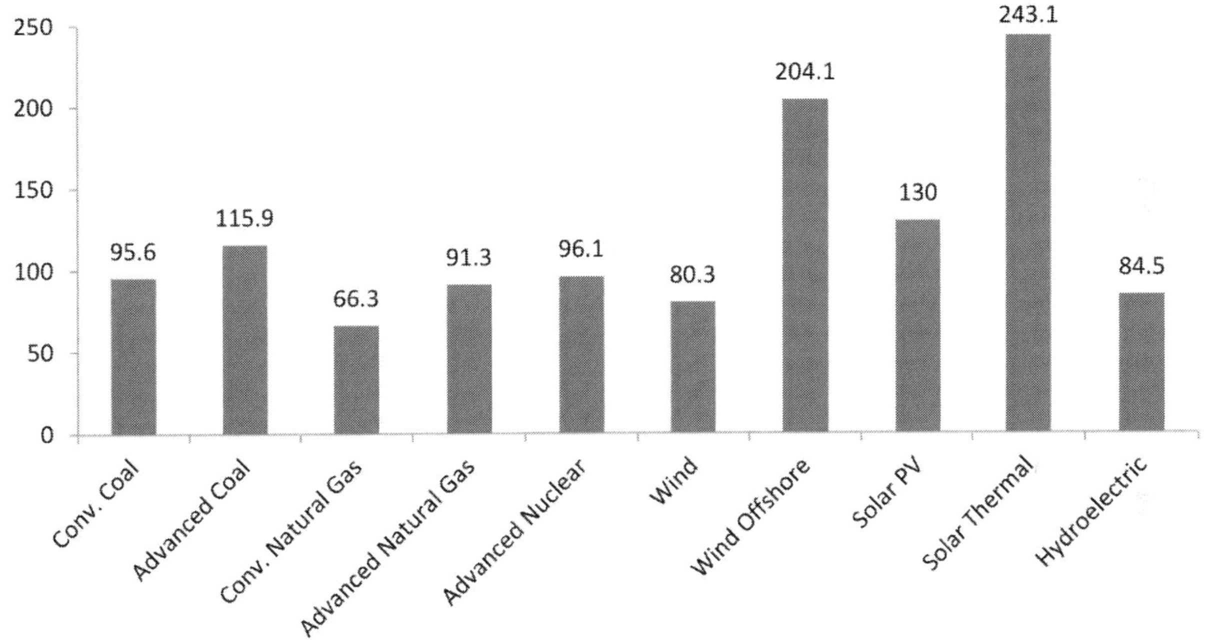

Figure 3. US Average LCOE for Plants Enetering Service in 2019, in 2012$/MWh [23]

As Figure 3 shows, the conventional energy production sources, coal and natural gas, are currently the cheapest available electricity production technologies, which explains why they currently account for such a large portion of US electricity production. Much of this can be contributed to the low price of fuel. Natural gas in particular is benefitting greatly from low fuel prices as a result of the recent American shale fracking boom. However, advanced coal and natural gas facilities, which incorporate carbon capture and sequestration technologies, are significantly more expensive than their conventional counterparts.

While coal and natural gas remain the cheapest electricity production technologies, it is important to note that there are alternatives that are already competitive. For example, the LCOE of nuclear energy is nearly the same as conventional coal and advanced natural gas, and is in fact less expensive than advanced coal. This is particularly noteworthy since nuclear fills the same base load electricity production role as coal, and produces no greenhouse gas emissions during its operational lifetime. It is also important to note that, while many of the renewable energy sources are significantly more expensive than conventional sources or nuclear, wind built on land has the second lowest LCOE, higher than only conventional natural gas.

3.2 Capacity Factor

One factor contributing to the relative expense of renewable sources, such as offshore wind or solar, is their variable nature. Since they are not able to produce electricity at all times of the day, renewable sources suffer from prolonged periods of not operating at peak capacity. The easiest way of quantifying this is through capacity factors, the ratio of actual output to the theoretical output if the plant could operate at full capacity indefinitely. A comparison of capacity factors for each electricity production technology can be seen in Figure 4 below.

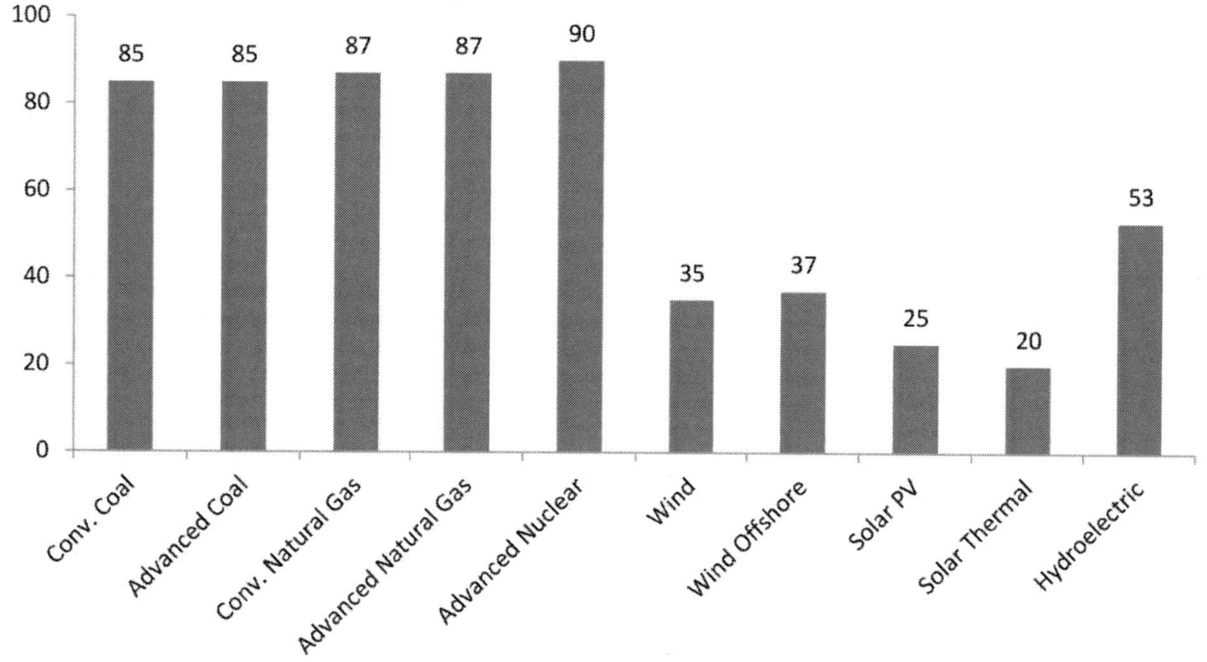

Figure 4. Capacity Factor (%) of Various Electricity Production Technologies [23]

It is important to note that there is no electricity production technology with a capacity factor of 100%: plants do not always run at maximum capacity, and require downtime for things like routine maintenance or system upgrades. However, it is easy to see that renewable sources have significantly lower capacity factors, mainly due to the fact that they cannot produce electricity at all times of the day. Because of this, capital and operating costs of renewable energy plants factor significantly more into their LCOE.

However, with advancements in technology, particularly in the energy storage field, the capacity of renewable sources could significantly increase, and possibly even become comparable with conventional sources. This would cut LCOE by anywhere from 40-70% (depending on current capacity factor). If this electricity storage is cheap enough, even the most currently expensive renewable energy technologies would be competitive with conventional sources.

4.0 Recent Policy Strategies

In recent years, legislators have used a wide variety of strategies to help shape the electricity production sector, most notably by offering some form of economic incentive to desirable technologies. These economic incentives typically come in the form of tax credits, though they may also come in the form of subsidies or guaranteed loan programs. Congress has also made efforts in the past to enact a carbon tax program, but such a program has never been able to make it out of Congress [24] [25].

4.1 Tax Credits for the Electricity Production Sector

The major economic incentive congress has implemented is by extending tax credits to electricity producers, which reduces the amount of taxes they have to pay to the federal government and thus improving their bottom line[1]. Recently, these tax credits have been targeted at renewable electricity resources, most notably solar and wind, though to a lesser extent hydroelectric and geothermal as well [26]. These tax credits have taken on two different forms: an investment tax credit and a production tax credit.

The investment tax credit allows qualifying electricity producers to receive a tax credit equal to a certain percentage of their initial cost of implementation. Solar energy installments are eligible for a 30% credit, while geothermal plants are eligible for a 10% credit. These credits have no limit. As of right now, these tax credits extend to any qualifying facility put in service before January 1, 2017. Wind power facilities put in service between 2009 and 2012 were also eligible for a 30% investment credit, but that has not been renewed [26].

The production tax credit, on the other hand, allows qualifying electricity producers to receive a tax credit based off the amount of electricity produced. As of the end of 2013, this tax credit was 2.3 cents / kWh, and extended to wind, geothermal, and hydropower facilities [5] [26]. If a facility qualifies for both the investment and production tax credit, they are only permitted to claim one of them, at their choosing. The production tax credit extended to any qualifying facility under construction by January 1, 2014, and is available for qualifying facilities for the first 10 years of operation. However, since January 1, 2014, the production tax credit for renewable electricity production facilities has not been extended [26].

4.2 Loan Programs

The federal government also offers loan programs for electricity producers. These loan programs differ from tax cuts mainly in the fact that producers are required to pay back their loan; while there is a cost to the federal government, it is only in the risk of borrowers defaulting,

---

[1] It is important to note that non-profits, such as municipal utilities and utility cooperatives, do not pay taxes, meaning that federal tax cuts do not have an effect on their bottom line. While alternative programs to incentivize non-profits do exist, they are not covered in the scope of this paper.

which has historically been low. Under Title XVII of the Energy Policy Act of 2005, two loan programs were outlined that are still used today: section 1703 and section 1705 [27] [28].

Section 1703 provides loans for "a new or significantly improved technology that is NOT a commercial technology" [27]. Qualifying facilities must also be located in the US. In the current state of the electricity production sector, the only facilities that really meet these requirements are related to nuclear energy; the most recent loans are an $8.3 billion loan for two new nuclear power plants in Georgia and a $2 billion loan for a uranium enrichment plant in Idaho [27]. This program is essential to nuclear power plants, as the most prohibitive cost to entering the market for nuclear plants is upfront capital costs. Due to relatively low operating costs once the plant is built, loan repayment is usually not an issue.

Section 1705, on the other hand, focuses on loan programs for renewable energy production facilities. Unlike section 1703 rules, section 1705 has a time limit, and was only available to projects that had begun construction before September 30, 2011. Since then, this project has not been renewed. Section 1705 loans tend to be lower than their section 1703 counterparts: one of the more expensive section 1705 loans was for $1.3 billion to an Oregon-based wind farm [28].

The major significant difference between section 1703 and 1705 loans is how the Credit Subsidy Cost (CSC) is handled. By law, the CSC, which is "the expected long-term liability to the Federal Government by issuing the loan guarantee", must be paid either by budget appropriations or by the borrower directly. For section 1703, which tend to have more expensive loans and thus larger CSCs, the CSC is paid by the borrower. For section 1705, the CSC was paid for by appropriations [27] [28]. In the American Recovery and Reinvestment Act of 2009, $6 billion was appropriated towards section 1705 CSCs. [29]

4.3 Carbon Tax and Cap-and-Trade Program

On top of offering economic benefits to desirable electricity production methods, Congress can also place economic burdens on undesirable electricity production methods. In the electricity production sector, this can be practically done in one of two ways: either a carbon tax or a cap-trade-program.

A carbon tax works very simply: electricity producers pay an additional tax for each unit of carbon produced during electricity production. This means that carbon-heavy electricity production facilities like coal (and to a lesser extent, natural gas) power plants would pay an additional amount for their electricity production, thus increasing their LCOE. In the same way that tax credits and subsidies seek to level the economic playing field by making renewables cheaper, a carbon tax would seek to level the playing field by increasing the cost of conventional electricity production methods. However, there has never been a carbon tax approved by either chamber of Congress.

An alternative method to a carbon tax is a cap-and-trade program, which seeks to place an artificial limit on the amount of carbon that can be emitted. Once this limit has been determined, electricity producers can buy permits to emit certain amounts of carbon dioxide. The government only sells as many permits as it takes to reach the cap, and then monitors the emissions of each producer to make sure they do not exceed their permitted limit. Producers are able to buy or sell permits to or from other producers as needed.

Cap-and-trade programs create economic incentive to reduce carbon emissions in two ways: first, reducing carbon emissions means that electricity producers need to buy fewer permits, saving them money. Second, if producers reach their permitted cap, their competitors determine the cost of additional permits (since each company now control its own excess permits, not the government). This means exceeding the permitted carbon emission limit can be disproportionately costly, either through the purchase of competitor permits or paying government fines.

In order to ensure long-term carbon emission reduction, cap-and-trade programs also incorporate a permit allowance reduction system. This means that, each year, the amount of permits allowed is reduced, effectively lowering the acceptable amount of carbon dioxide that can be emitted by electricity producers. While a cap-and-trade program did make it out of the House in 2009, the bill was never heard by the Senate, and died when that session of Congress ended [24] [25].

5.0 Projected Electricity Sector Growth: Electricity Sector Capacity Additions for 2013

In order to establish a reference case, it is useful to look at how the electricity production sector has changed recently. Figure 5 below shows the EIA projections for electricity sector capacity additions from 2008.

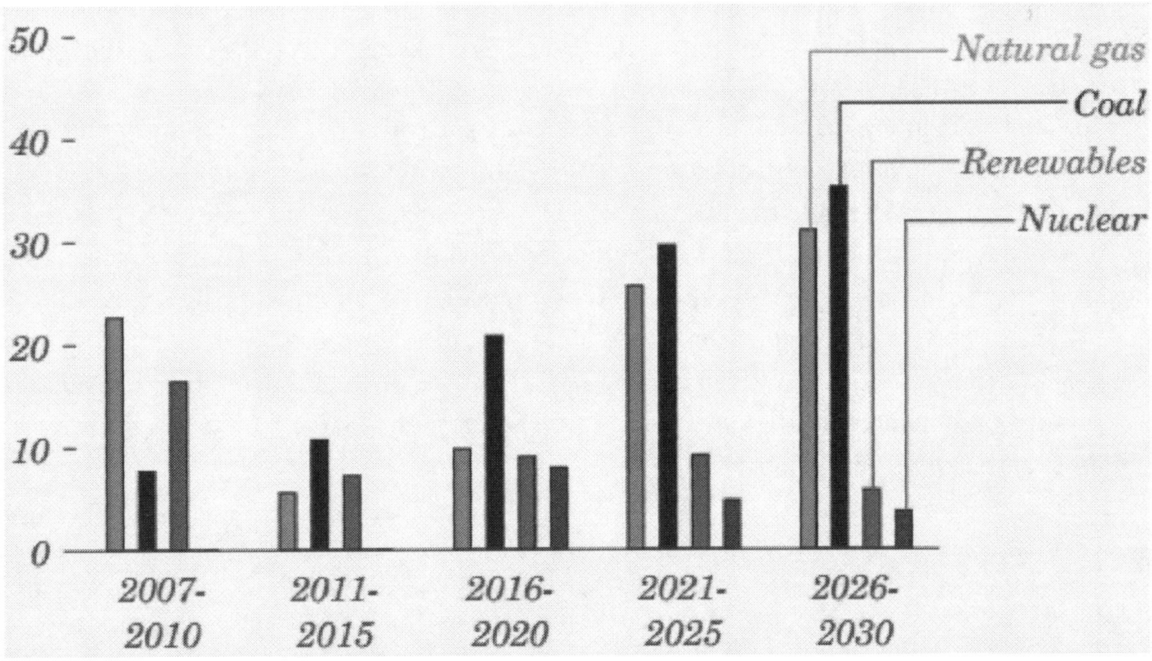

Figure 5. Electricity Generation Capacity Addition Projections by Fuel Type, 2008 (Gw) [30]

As Figure 5 shows, in 2008, it was projected that coal would begin to dominate electricity production additions starting in 2011. Much of this can be attributed to the increasing cost of natural gas, as shale fracking was still in its infancy in 2008. Additionally, renewables, which see a large spike in 2007-2010, were projected to fall by the wayside after 2010 as they begin to be dwarfed by both natural gas and coal.

However, as a result of the shale fracking boom, the reduced cost of renewables, and increased government pressure to reduce carbon emissions, these 2008 projections do not reflect what has actually happened in recent years. Figure 6 below shows the electricity generation capacity additions for 2013.

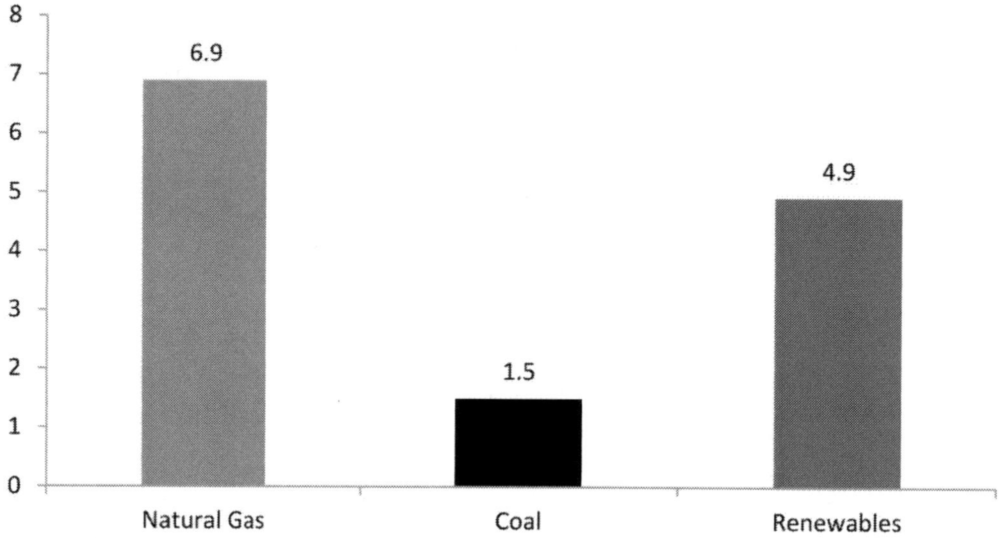

Figure 6. Electricity Generation Capacity Addition by Fuel Type 2013 (GW) [Adapted from 31]

As seen in Figure 6, natural gas dominated new capacity additions, accounting for more than half of the electricity capacity additions in 2013. Renewables made up roughly one-third of new additions, which is attributed to the completion of a number of large scale solar thermal plants in Arizona and California as well as "falling [photovoltaic] costs, aggressive state renewable portfolio standards, and continued federal investment tax credits" [31]. Coal makes up a meager 11% of new capacity additions, in stark contrast to the projections shown in Figure 5.

Long term projections have also changed to reflect the changing economic and political landscape surrounding electricity production. Figure 7 below shows the EIA projections for electricity sector capacity additions from 2014.

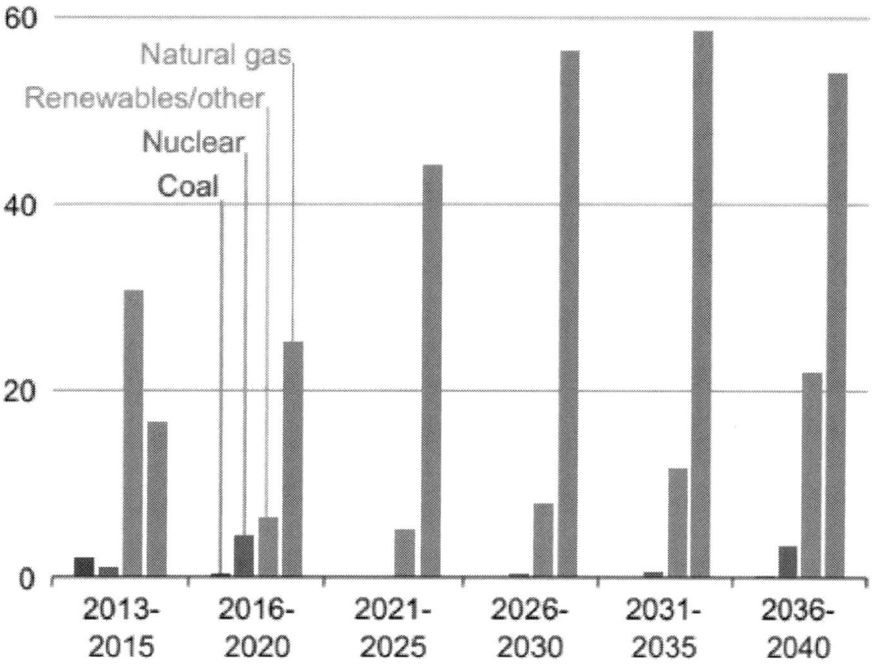

Figure 7. Electricity Generation Addition Projections by Fuel Type 2014 (GW) [32]

As Figure 7 shows, natural gas dominates new additions through 2040, mainly as a result of low gas prices. Coal additions fall virtually to 0 as a result of increased emissions regulations. Most alarmingly, the addition of renewables is actually lower than those projected in 2008 for the period between 2016 and 2035. This is mainly due to the extremely low cost of natural gas relative to any other fuel source, making it take up a larger share of the new additions in the 2014 projections than conventional sources did in the 2008 projections. It is important to note, however, that these projections only show a reference case where no additional government actions are taken.

5.1 Policy Action Effects: Overall Carbon Emissions

Despite the recent transition from coal to natural gas, as well as the increased onset of recently-cheap renewables, carbon emissions from the electricity production sector are projected to increase over the next several decades if current laws and regulations remain in effect. This is mainly attributed to the shift towards natural gas-based electricity production and the shift away from renewables and nuclear as a result of cheap natural gas for the foreseeable future [33]. However, if some form of carbon tax is implemented, carbon emissions could decrease instead increase, as seen in Figure 8 below.

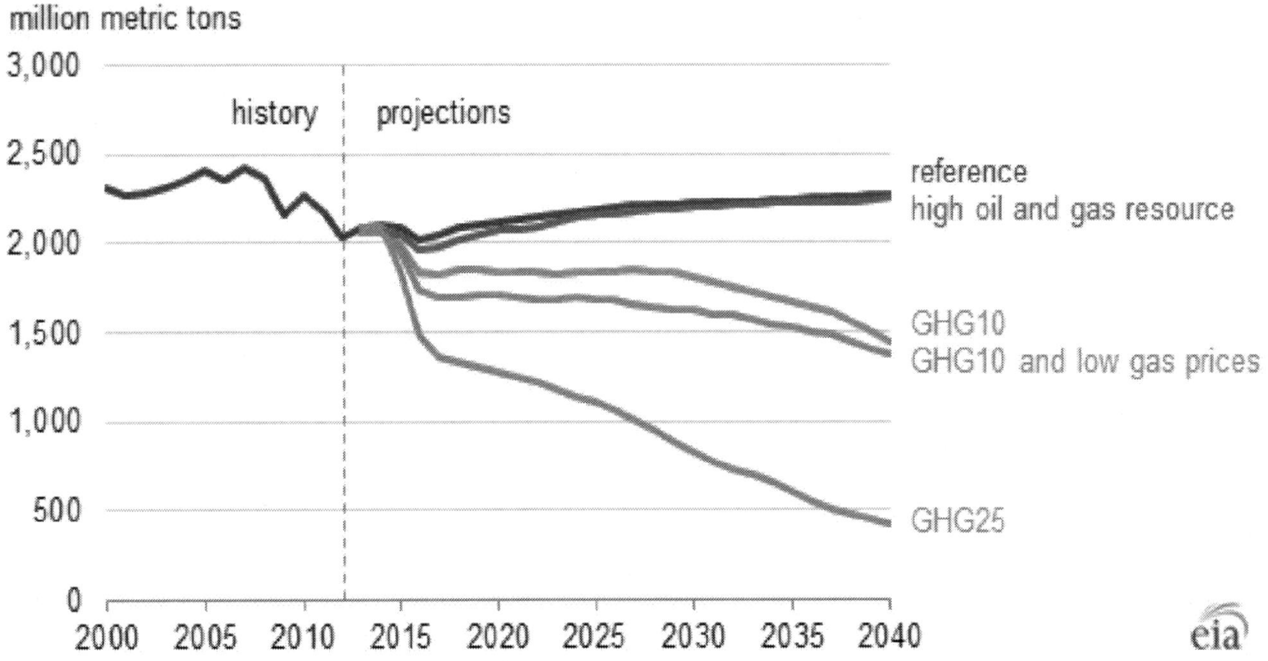

Figure 8. Electricity-Related CO_2 Emissions in Five Cases, 2000-40 [33]

Figure 8 takes into account five different potential cases. The reference case reflects no change in laws or regulations surrounding electricity production and assumes relatively constant economic factors relative to those in 2014. As seen in Figure 8, the reference case shows a 12% increase in emissions up through 2040 [33].

The GHG 10 and GHG 25 cases both refer to the implementation of a carbon tax (or equivalent cap-and-trade permit cost) in 2015 of \$10 / metric ton of CO_2 or \$25 / metric ton of CO_2, respectively. The GHG 10 case shows a 16% lower CO_2 emission rate in 2025 and a 36% lower CO_2 emission rate in 2040 relative to the reference case. The GHG 25 case shows an even greater decrease in CO_2 emission rates, with a 49% reduction in 2025 and an 82% reduction in 2040. In both cases, there is a large dip in emission rates before 2020, partially due to an accelerated shift away from coal and partially due to an increased implementation of renewables [33].

The High Oil and Gas Resource case assumes the same conditions as the reference case, except that oil and gas resources become more abundant than currently projected. In this case, CO_2 emissions decrease ever so slightly more than in the reference case as a result of increased shift towards natural gas, but by 2040 emission rates are roughly the same as the reference case as a result of producers expressing and increased favor towards natural gas relative to renewables [33].

Finally, the GHG 10 and Low Gas Prices case effectively combines the GHG 10 and High Oil and Gas Resources cases. In this case, reduction of CO_2 emissions before 2020 is greater than those seen in the GHG 10 case, mainly due to an even greater shift from coal to natural gas. However, because a greater share of this shift away from coal is given to natural gas and not renewables, CO_2 emissions do not decrease as drastically after 2030 compared to the

GHG 10 case; it can be extrapolated that CO_2 emissions would actually begin to be higher than the GHG 10 case sometime between 2040 and 2050 [33].

Overall, these cases suggest that some form of action needs to be taken in order to produce a marked reduction in carbon emissions from the electricity production sector. If nothing changes, CO_2 emissions will increase slightly over the next 40 years, in contrast to the significant emission reductions we‟ve seen since 2005. Emissions get even worse with an increased gas resource and lower gas prices, which may result with the advancement of fracking and other extraction technologies. Based off of this projection, the only way to continue a trend of CO_2 emission reduction is to accelerate the shift away from coal and ensure that an increased share of that shift is made up of renewables or nuclear.

5.2 Policy Action Effects: Conventional Fuels

When assessing the viability of the cases listed in the previous section, it is important to understand the effects of each action on each specific electricity production method. As discussed previously, there is already a shift towards natural gas and away from coal in response to lower gas prices and increased carbon emission regulations. Figure 9 below shows how possible policy actions may augment that shift by showing the projected natural gas electricity generation in the five previous cases.

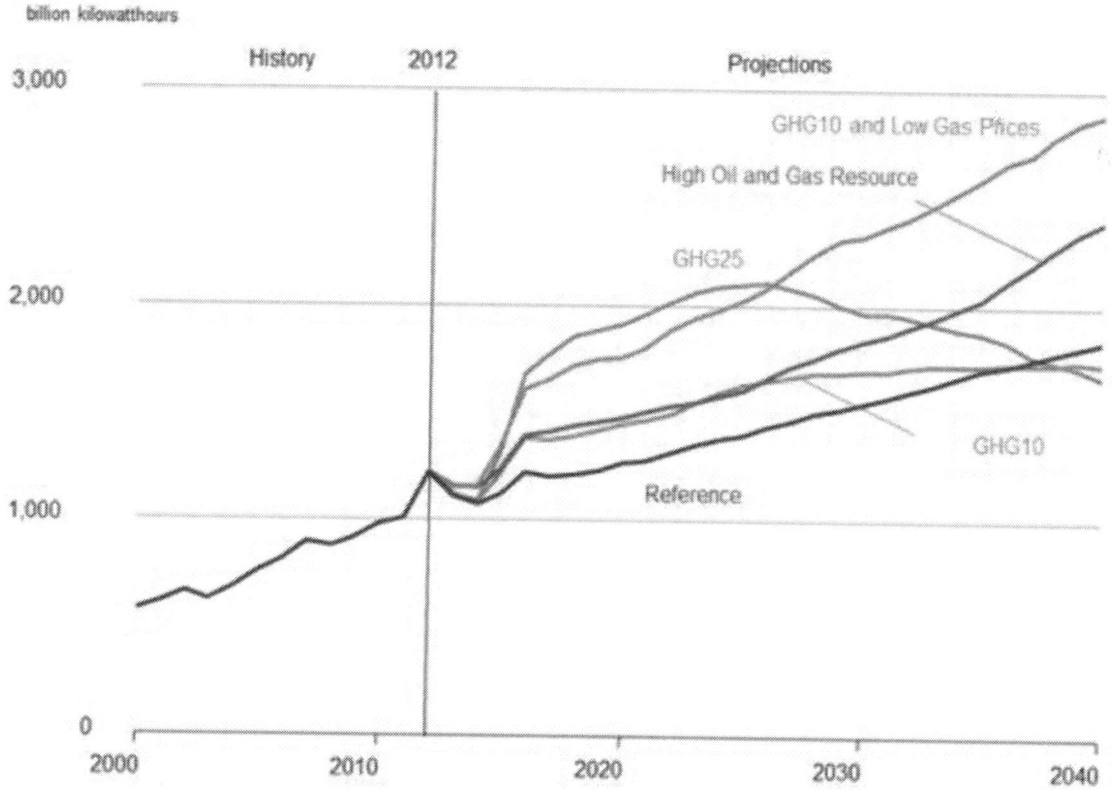

Figure 9. Natural Gas Electricity Generation in Five Cases, 2000-40 [34]

As seen in Figure 9 above, electricity production via natural gas is projected to increase over the next several decades in all cases. In the reference case, we see a relatively stable and constant increase of natural gas fired plants as they replace retiring coal plants; in 2012, 37% of electricity generation comes from coal, and in 2040 that share is 32%, marking a relatively slow shift [34]. With a high gas resource, this shift is more dramatic in the near future, and the long term rate of increase in natural gas plants is also slightly higher than the reference case.

The most interesting part of this projection, however, is the effects of a carbon tax. In all GHG cases, the shift towards natural gas in the near future is increased relative to the reference case. In the GHG 10 case, the share of electricity from coal is 28% in 2025 and 19% in 2040. For the GHG 25 case, the shift is even more drastic, with the coal share being 10% in 2025 and 1% in 2040 [34]. In both cases, the increase in natural gas production is curbed by the increased implementation of nuclear of renewables: in the GHG 10 case, natural gas electricity production levels off around 2030, while in the GHG 25 case, it actually decreases starting in 2025. However, if gas prices remain low, natural gas electricity production continues to increase through 2040, reaching the highest level of any of the five test cases.

5.3 Policy Action Effects: Renewables

Despite the fact that natural gas dominates new electricity production additions, renewables are also seeing an increase in implementation as the electricity production shifts away from coal. However, without some form of government action, the rate of renewable additions will be dwarfed by those of natural gas, leading to the increased carbon emissions seen in Figure 8. Figure 10 below shows the effects of some possible government actions by plotting the projected renewable electricity generation in four cases.

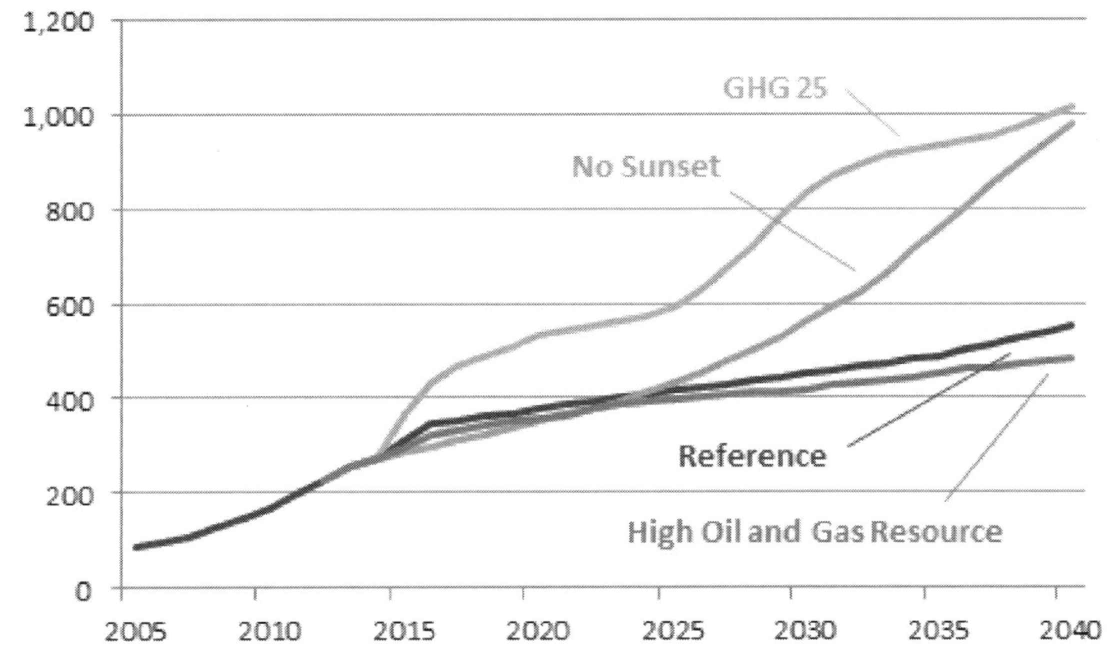

Figure 10. Renewable Electricity Generation in Four Cases, 2000-40 (billion kWh) [Adapted from 35]

In all cases, electricity generation from renewables increases over the next few decades. However, for the reference and high gas resource cases, the rate of increase in renewables is less than that seen between 2000 and 2014, mainly due to the increased implementation of natural gas power plants. The no sunset case shows the effects of extending renewable tax credits (i.e. the installment tax credit and the production tax credit) indefinitely in their current forms. In this case, the growth seen between 2000 and 2014 remains relatively constant through 2025, at which point the rate of implementation of renewables begins to increase steadily. In the GHG 25 case, a large increase is seen just after 2015 in response to the implementation of the carbon tax, after which the rate of implementation of renewables begins to level off. However, around 2025, electricity generation from renewables begins to increase dramatically once again, as more projects come into completion in response to the increased cost of natural gas power plants, whose production begins to decline as seen in Figure 9.

Compared to the reference case, both the no sunset and GHG 25 cases see disproportionate increases in solar and wind relative to other renewables like hydro or geothermal. When it comes to solar, electricity production in both cases is triple that of the reference case in 2040; for wind, it"s double. In both cases, geothermal actually sees slightly lower electricity production in 2040 compared to the reference case, and hydro sees relatively little growth between 2014 and 2040 in all cases [35].

6.0 Policy Recommendations

The electricity production sector in the US is rapidly changing as a result of the shale fracking boom. While the large shift from coal to natural gas will improve carbon emissions on a per kwh basis, the low price of natural gas is also reducing the amount of renewable sources entering the market. If no government action is taken, the growth of electricity demand will still mean an increase in carbon emissions from the electricity production sector over the next 30 years due to the low prices of natural gas. If the government truly wishes to use natural gas as a bridge fuel to a low-carbon electricity sector, a carbon tax needs to be implemented and tax credits for renewables need to be extended.

A carbon tax is necessary to appropriately reflect the societal cost of carbon-emissions-heavy technology. However, a large carbon tax is not necessary, and would cause unnecessary instability as the market acted swiftly to adjust. Instead, a low carbon tax that progressively increases is the most responsible way to implement such a change. Based off the projections in Figures 8 and 9, a carbon tax of $10 / metric ton of CO_2 would lower carbon emissions in the electricity production sector without drastically changing the economic viability of natural gas facilities. This tax would increase at a rate of 5% per year in order to incentivize investment in further carbon emission reduction technology. For a cap-and-trade program equivalent, this means the acceptable level of carbon emissions each year would be reduced, effectively increasing the cost of emissions a proportional amount.

On the other side of things, renewable technology still needs some assistance to be competitive with conventional fuels, as seen in Figure 3. However, the current split between the investment tax credit and the production tax credit makes the development of new and improved renewable technologies difficult, as it is more difficult to predict the economic environment these technologies will emerge into. A single, long-term tax strategy would offer economic stability and predictability that promotes private investment that has been yet unseen. In this way, a single, long-term tax strategy should be used to incentivize the use of all types of renewables in the electricity sector.

Of the two programs currently available, fully embracing the production tax credit makes the most sense, as it incentivizes the implementation of energy efficient and energy dense technologies[2]. It also incentivizes the development of improvements that reduce the upfront cost of these technologies, which help open them up to small scale residential and commercial consumers. The current rate of 2.3 cents per kwh is an appropriate start-point for such a tax program, but with the implementation of a carbon tax, indirect incentive towards renewables increases proportionately. The increasing nature of the carbon tax combined with the increased rate of improvement for renewable technologies actually means that the production tax credit can be slowly reduced. As such, a 5% reduction per year to mirror the carbon tax rate increase would be appropriate.

In both the carbon tax and the production tax credit, both programs need to be comprehensive and long term in order to allow the electricity producers to accurately model the economic environment that they will be entering with new projects. Due to the long-nature of construction of electricity production facilities, as well as the long operating life-time of these

[2] Again, this credit would not incentivize non-profit utility providers. Alternative programs, not covered in the scope of this paper, could be used to incentivize these utility providers.

facilities, stability in the tax structure from the government is paramount. The unexpected expiration of the production tax credit in 2013 mainly to political gridlock is unacceptable, and should be avoided in the future if at all possible.

All in all, if such a two pronged approach of a $10 / metric ton carbon tax combined with an extension of the production tax credit is used, the government will turn a net profit as a result. Based on the projections in Figures 9 and 10, there will be an average profit of $5.5 billion per year between 2015 and 2040 under such a program, with yearly profits ranging between $4.5 billion and $6.1 billion [adapted from 40 and 41]. Keep in mind that these projections only factor in the revenue from natural gas facilities; factoring in the revenue of coal should increase revenue further, despite the increase in cost from the increased use of renewables.

It is further recommended that the money made from this program be reinvested in the loan programs for electricity producers. While the reintroduction of the section 1705 program may be beneficial, it seems more beneficial to use this money to offset or eliminate the CSC cost burden on borrowers of section 1703 loans. With the accelerated retirement of coal, reduced or eliminated CSC costs would incentivize increased use of nuclear power, which acts as a more practical and carbon neutral base load power supply alternative compared to natural gas.

7.0 References

[1] Obama, B. (2014, January 28). *President Barack Obama's State of the Union Address.* Speech presented at the House Chamber, Capital Building, Washington, DC. Retrieved from http://www.whitehouse.gov

[2] EPA. (2014, June 2). *Learn About Carbon Pollution From Power Plants*. Retrieved from http:// www2.epa.gov

[3] Clement, S. (2014, June 2). A huge majority of American support regulating carbon from power plants. And they're even willing to pay for it. *The Washington Post*. Retrieved from http://www.washingtonpost.com

[4] Koch, W. (2014, June 3). EPA carbon emission limits may shift energy mix. *USA Today*. Retrieved from http://www.usatoday.com

[5] Union of Concerned Scientists. (2014, January 31). Production Tax Credit for Renewable Energy. Retrieved from http://www.ucsusa.org

[6] Davenport, Coral. (2014, May 12). Amid Pipeline and Climate Debate, Energy-Efficiency Bill Is Derailed. *New York Times*. Retrieved from http://www.nytimes.com

[7] EIA. (2014, May 7). *Annual Energy Outlook 2014*. "Market Trends: Electricity demand". Retrieved from http://www.eia.gov

[8] EIA. (2014, May 21). *Electric Power Monthly*. "Table 1.1. Net Generation by Energy Source: Total (All Sectors), 2004 – March 2014". Retrieved from http://www.eia.gov

[9] EIA. (2013, May). *State-Level Energy-Related Carbon Dioxide Emissions, 2000-2010.* "Table 2. 2010 state energy-related carbon dioxide emissions by fuel". Retrieved from http://www.eia.gov

[10] EIA. (2011, February). *Emissions of Greenhouse Gases in the United States 2009*. "Table 5. Greenhouse gases and 100-year net global warming potentials". Retrieved from http://www.eia.gov

[11] Meier, P.J. (2002, August). *Life-Cycle Assessment of Electricity Generation Systems and Applications for Climate Change Policy Analysis*. pg 70. Madison, Wisconsin: University of Wisconsin, Fusion Technology Institute. Retrieved from fti.neep.wisc.edu/pdf/fdm1181.pdf

[12] EPA. (2012, December 4). *What is Acid Rain?*. Retrieved from http://www.epa.gov

[13] EPA. (2014, March 10). *Basic Information | Mercury*. Retrieved from http://www.epa.gov

[14] EPA. (2013, September 25). *Natural Gas | Clean Energy*. Retrieved from http://www.epa.gov

[15] EIA. (2011, March 31). *Emissions of Greenhouse Gases in the US*. Retrieved from http://www.eia.gov

[16] World Nuclear Association. (2014, May). *World Energy Needs and Nuclear Power*. Retrieved from http://www.world-nuclear.org

[17] EPA. (2013, October 22). *Nuclear Energy | Clean Energy*. Retrieved from http://www.epa.gov

[18] Nuclear Energy Institute. (2014). *Life Cycle Emissions Analyses*. Retrieved from http://www.nei.org

[19] EPA. (2012, February 28). *Solar Energy | Energy and Global Climate Change in New England*. Retrieved from http://www.epa.gov

[20] EPA. (2014, May 1). *Wind Energy | Energy and Global Climate Change in New England*. Retrieved from http://www.epa.gov

[21] DOE. (2012, April 17). *Energy Department Report Finds Major Potential to Increase Clean Hydroelectric Power*. Retrieved from http://energy.gov

[22] EPA. (2014, May 1). *Geothermal Energy | Energy and Global Climate Change in New England*. Retrieved from http://www.epa.gov

[23] EIA. (2014, April). *Levelized Cost and Levelized Avoided Cost of New Generation Resources in the Annual Energy Outlook 2014*. Retrieved from http://www.eia.gov

[24] American Clean Energy and Security Act of 2009. H.R. 2454 of the 11th Congress. Retrieved from http://www.opencongress.org/bill/111-h2454/actions_votes

[25] Hulse, Carl. (2010, July 22). Democrats Call Off Climate Effort Bill. *New York Times.* Retrieved from http://www.nytimes.com

[26] 26 U.S.C. §§46, 48

[27] DOE. *Section 1703 Loan Program*. Retrieved from http://energy.gov

[28] DOE. *Section 1705 Loan Program*. Retrieved from http://energy.gov

[29] Wilson Sonsini Goodrich & Rosati. *Status Update – Department of Energy Title XVII Loan Guarantee Program.* Retrieved from http://www.wsgr.com

[30] EIA. (2008, June). *Annual Energy Outlook 2008*. "Early Capacity Additions Use Natural Gas, Coal plants Are Added Later". Retrieved from http://www.eia.gov

[31] EIA. (2014, April 8). *Half of power plant capacity additions in 2013 came from natural gas.* Retrieved from http://www.eia.gov

[32] EIA. (2014, May 7). *Annual Energy Outlook 2014*. "Most new capacity uses natural gas and renewables". Retrieved from http://www.eia.gov

[33] EIA. (2014, June 9). *Power sector CO_2 emissions sensitive to policy changes and natural gas supply.* Retrieved from http://www.eia.gov

[34] EIA. (2014, May 7). *Market Trends: Emissions*. "Energy-related carbon dioxide emissions are sensitive to potential policy changes". Retrieved from http://www.eia.gov

[35] EIA. (2014, April 29). *Renewable electricity projections show growth under alternative assumptions in AEO2014*. Retrieved from http:// www.eia.gov

396486 Reactive Gelation: a new Method for Producing Polymeric Porous Materials

Abstract:

Porous materials are in general devised to increase the surface area exhibited by the unit mass of a given material. This is needed for a number of applications in different areas, each of which requires a certain pore size which determines the actual accessibility to the surface. Accordingly, it is important to be able to realize the accessible surface area and the corresponding pore size distribution best suited for a specific application.

Current technologies for the production of polymeric porous materials involve some kind of suspension polymerization in the presence of suitable porogenic species. In this presentation a new technology is discussed based on the coagulation/coalescence of spent polymer latexes previously prepared in emulsion polymerization without the addition of porogenic species. This methodology allows a close control of the pore size distribution particularly in the range of tens of nanometres to tens of microns. It can be used for the preparation of monolyths as well as particles of various sizes.

Various applications, starting with the support for stationary chromatographic phases for protein adsorption, are discussed.

396484 Chemical Reaction Engineering at the Molecular Scale

Abstract:

Chemical reaction engineering (CRE) has traditionally addressed the impact of heat and mass transfer occurring in catalyst pellets and catalytic reactors. Advances in the characterization of catalysts and in the ability to described the thermodynamics of adsorption and the kinetics of elementary processes at the atomic and molecular levels have opened the way for extending the length scale of CRE from the scale of meters to microns down to the level of Ångstroms. This talk will illustrate what can be learned about the performance of catalysts at the atomic and molecular scale through the application of experimental techniques. Illustrations will be drawn from recent studies devoted to the reactions of hydrocarbons over zeolites, the condensation of aldehydes and ketones at acid-base pairs supported on silica, and the electrochemical splitting of water. Some comments on what can be expected for CRE in the future will also be offered.

396487 Applied Mathematics and Reaction Engineering: Sustainable Elements of the Chemical Engineering Discipline

Abstract:

The chemical engineering discipline continues to evolve and expand its reach, influenced by societal needs and advances in the basic sciences. Throughout this evolution two core elements of the discipline remain critical in chemical engineering practice, applied mathematics and reaction engineering. These have been the main focus areas of the graduate education and research activities of Professor Arvind Varma throughout his influential career. In this talk several salient examples from chemical engineering practice and technology will be described which underscore the need to sustain education and research in applied math and reaction engineering.

396488 Design principles of bimetallic core-shell catalysts

Abstract:

Despite the unique properties of core-shell bimetallic catalysts, the fundamental reasons for their high activity remain elusive. We will review a framework we developed to predict novel core-shell bimetallics. We will show how conventional models fail to quantitatively predict their activity and demonstrate that these materials exhibit bifunctional behavior that is key to understanding their performance.

396489 Dealkylation and Disproportionation of Long Chain Alkylaromatics over Immobilized Ionic Liquid Catalysts

Abstract:

Long chain alkylaromatics, such as n-dodecylbenzene, are important starting points for the production of surfactants and form the basis for the production of many biodegradeable detergents. It this presentation we use n-dodecylbenzene disproportionation as a probe for acid activity and selectivity in ionic liquid catalyzed dealkylation and disporportionation of long chain alkylaromatics. The ionic liquid catalysts used in this study comprise ionic liquid – functionalized high surface area silica materials including MCM-41. The study showed that $AlCl_3$-based ionic liquids supported on silicas and MCM-41 are active catalytic materials for alkylbenzene (DDB) transformation (mainly disproportionation and dealkylation) at low reaction temperatures. We also concluded that the catalytic dealkylation of n-dodecylbenzene requires a high density of very strong acid sites.

357976 What Can Textbook Authors Do to Make Solutions Manuals More Secure?

Abstract:

It is in the best interests of textbook authors to have more secure solution manuals. Developing problems and a solution manual require a significant amount of time. Professors want problems that are new and do not have solutions that are readily available to students on the internet. One effect of internet pirates' illegal posting of copyrighted solution manuals without permission is more frequent new editions, which increases the cost of textbooks to students. The following possible partial responses to the proliferation of solution manuals will be explored and audience feedback will be solicited:

1. Initially, create a partial solution manual that shows solutions to only a portion of the problems. Make available every year a manual that has an increased number of solutions. A password protected web site or method to have professors e-mail the author with requests would be required.
2. Initially, include a limited number of new problems plus a number of older problems in the new edition. The new problems would be identified in the solution manual. The solution manual would have solutions to all of the problems. Solutions to the old problems are in solution manuals for previous editions; thus, students probably have access to solutions. Author makes available every year additional problems on the publisher's public web site. The solution manual with the additional solutions would be available by subscription.
3. Author prepares a series of similar problems with a spread sheet. In the solution manual the author shows the solution to only one of the problems in a non-spreadsheet format. The solution manual could contain answers to the other problems but no solutions. This technique could be tried either with the spreadsheet in the solution manual and without the spreadsheet.
4. Prepare solution manual, but with mistakes in some solutions. Inform professors privately of the problems with mistakes. Students copying these problems could be handled in a fashion considered appropriate by the professor.
5. Have a few problems in the book that do not have solutions in the manual. These would need to be solved in advance to be absolutely positive that they are solvable.
6. Give problems in the textbook (e.g., design problems) that do not have a single solution. Perhaps present one solution and request the students find alternate solutions.
7. In the "if you can't beat them, join them" approach old problems and solutions are bundled into the equivalent of a Schaum's Outline Series, which is sold to students. Let students have all solutions, but tests are closed book.
8. Complain vociferously about the pirate sites. Have your university block these sites to make it more difficult for students to find solution manuals.
9. Make internet pirates "walk the plank."

A 21st Century Textbook for Material and Energy Balances

Matthew Liberatore

COLORADO SCHOOL OF MINES
engineering the way

When you renew your AIChE membership, Please join the Education Division of AIChE!!!!

Disclaimer: I own Read Swipe Solve LLC and will financially benefit from sales of FMEB

Start With A Literature Review

"The students receiving the multimedia learning modules performed significantly better on both tests than the students experiencing the text-based presentations" T. Stelzer, et al., Am. J. Phys. 2009, *77*, 184–190.

"Students also reported limitations of the textbook, contrasting it to the diversity of resources available via the Internet" C. S. Lee, et al., J. Eng. Ed. 2013, 102, 269–288.

"Participants with pre-lesson quiz scores in the lower-quartile improved 64% more (p-value < 0.001) with the interactive web-native content than with the static web content." Edgcomb &Vahid, ASEE Conf. Proc. 2014. (zyante.com)

3

Textbook = 20th Century Technology

OLD Technology | 20th Century Technologies | 2014 Nobel Prize

https://www.inkling.com/blog/2014/10/prototype-fidelity-digital-content/ accessed 10/2014.

4

21st Century Textbook = Software

5

Why Software Trumps 'Print'

Type of Textbook	Cost	Interactivity	Updates
Fundamentals of Material and Energy Balances	$29.99 for web, iOS, Android	Yes	Frequent
Traditional Textbook	$71.50 for Kindle	No	Once per decade

Learn more at www.readswipesolve.com

MEB = The comprehension course

The 1st chemical engineering course

A gateway to the rest of the curriculum

Concepts covered:

- Mass/mole/atom balances
- Reacting systems
- Vapor-liquid equilibrium
- Energy balances
- Transient systems

Building from the bottom up

Goal: Develop problem solving skills and conceptual understanding

How:

Quick hitting concepts

Scaffolded, tiered examples

Interactivity and decision making

Bring Active Learning To The Textbook

8

Scaffolded Tool for Active Reading©

Minimum Text

Non linear

Interactive Feedback

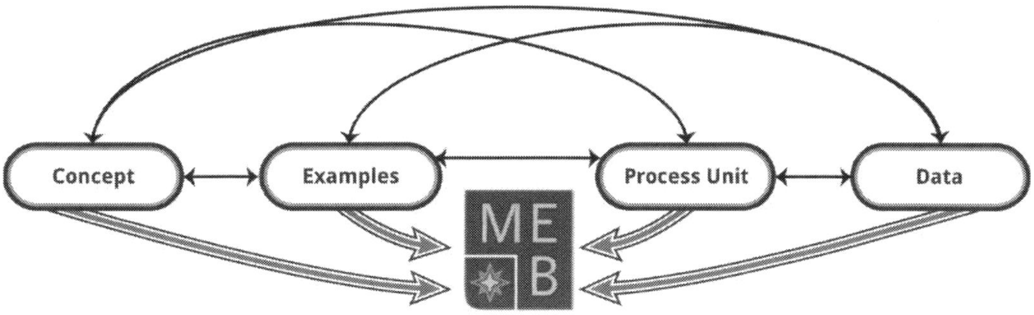

9

Quick Hitting Concepts

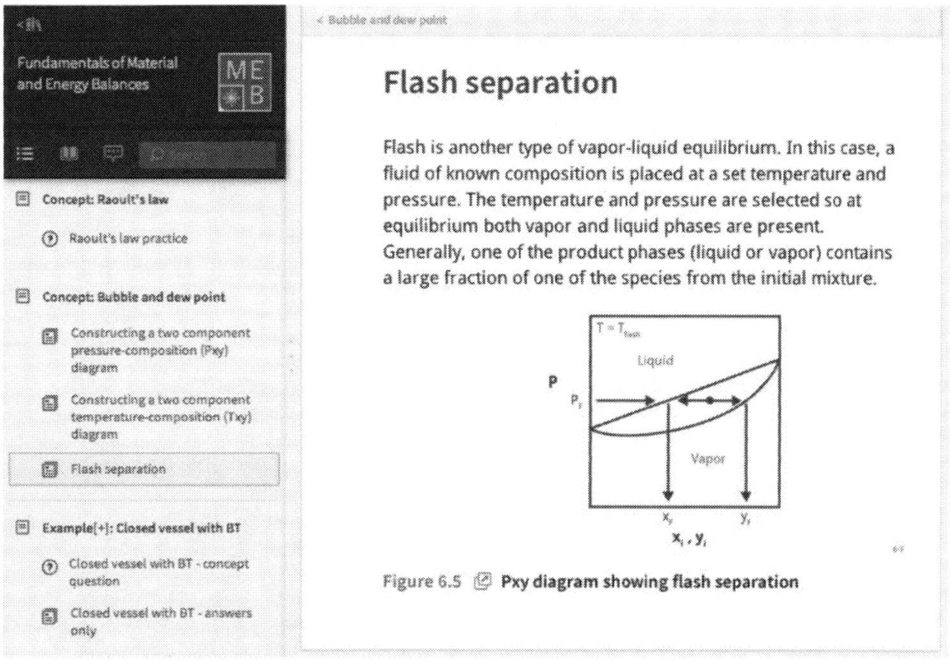

10

Flashcards = interactive definitions

Concept: Properties of water

Notes 0

Properties of water

Water is a very well studied substance, and a wealth of tabulated water properties are available. Reading the "steam tables" is an important skill to determining properties from density to enthalpy for liquid and vapor water. Some general vocabulary and simple phase diagrams are introduced and then some rules of thumb on determining the phase and properties from given information concludes this concept. While steam tables are provided in the Appendix, the concepts for the properties of water translate to most single species.

Six basic terms are defined here. The terms bubble point and dew point are covered in Chapter 6.

PHASE	PHASE BOUNDARY	SATURATED
SATURATION T	SATURATION P	PHASE DIAGRAM

Next, the basic phases of water and pure substances are defined.

11

Flashcards = interactive definitions

Concept: Properties of water

Notes 0

Properties of water

Water is a very well studied substance, and a wealth of tabulated water properties are available. Reading the "steam tables" is an important skill to determining properties from density to enthalpy for liquid and vapor water. Some general vocabulary and simple phase diagrams are introduced and then some rules of thumb on determining the phase and properties from given information concludes this concept. While steam tables are provided in the Appendix, the concepts for the properties of water translate to most single species.

Six basic terms are defined here. The terms bubble point and dew point are covered in Chapter 6.

Different forms that a pure substance can exist. Solid, liquid, and gas are the most common	PHASE BOUNDARY	Two phases existing in equilibrium
SATURATION T	P where vaporization starts at a given T	PHASE DIAGRAM

Next, the basic phases of water and pure substances are defined.

12

Choose Your Example/Adventure

C6 flash

A liquid mixture containing 79 mol% cyclohexane and the balance n-pentane enters a flash tank. The tank is maintained at 98°C and 29 psia. The flow rate of the entering stream is 96 mol/min.
Part a. Draw and label a process flow diagram.
Part b. Find the molar flow rates (mol/min) and mole fractions of the exiting stream(s).
Concept question. The entering flow rate increases by 17%, the composition of the pentane in the exiting liquid stream will increase, decrease or stay the same.

Jump to different levels using links:

Problem statement
Numerical answers
Diagram only
Equations only
Full Solution
Concept question

13

Example: Numerical Answers

Numerical answers

$$x_{2,C} = \boxed{0.914}$$

$$y_{3,C} = \boxed{0.757}$$

$$x_{2,P} = \boxed{0.086}$$

$$y_{3,P} = \boxed{0.243}$$

$$\dot{n}_2 = \boxed{20.4 \frac{mol}{min}}$$

$$\dot{n}_3 = \boxed{75.6 \frac{mol}{min}}$$

14

Example: Diagram

Diagram

Figure 6.6a C6 flash PFD

15

Example: Equations

Equations

$$\dot{n}_1 = \dot{n}_2 + \dot{n}_3$$
$$y_{1,C} \cdot \dot{n}_1 = x_{2,C} \cdot \dot{n}_2 + y_{3,C} \cdot \dot{n}_3$$
$$y_{3,C} \cdot P = x_{2,C} \cdot P_C^{sat}$$
$$y_{3,P} \cdot P = x_{2,P} \cdot P_P^{sat}$$
$$1 = x_{2,P} + x_{2,C}$$
$$1 = y_{3,P} + y_{3,C}$$

16

Example: Full Solution

[Step 2] Stream 1

[Step 3] Only 1 system to write balances.

[Steps 4, 5, 6] Mole balances

Overall: $\dot{n}_1 = \dot{n}_2 + \dot{n}_3$

Compenent C: $y_{1,C} \cdot \dot{n}_1 = x_{2,C} \cdot \dot{n}_2 + y_{3,C} \cdot \dot{n}_3$

[Step 7] Two Raoult's laws, Antoine equation, and two sum of mole fractions.

Raoult's Law C: $y_{3,C} \cdot P = x_{2,C} \cdot P_C^{sat}$

Raoult's Law P: $y_{3,P} \cdot P = x_{2,P} \cdot P_P^{sat}$

Sum mole fractions: $1 = x_{2,P} + x_{2,C}$

Sum mole fractions: $1 = y_{3,P} + y_{3,C}$

Antoine equation: $\log_{10} P^{sat} = A - \frac{B}{T + C}$

[Steps 8, 9, 10, 11] 6 unknowns: 2 balances, 2 Raoult's laws, 2 sum mole fractions. Solve.

Use given T and Antoine equation to find P^{sat} values.

$P_C^{sat}(T = 98°C) = 1242 \text{ mmHg} = 165600 \text{ Pa}$

$P_P^{sat}(T = 98°C) = 4224 \text{ mmHg} = 563100 \text{ Pa}$

Many ways to solve this system of equations. One way is:

Plug sum mole fractions in Raoult's Law for P. Solve Raoult's law C for $y_{3,C}$ and plug $y_{3,C}$ into Raoult's Law P. Solve for $x_{2,C}$.

$$y_{3,C} = \frac{x_{2,C} \cdot P_C^{sat}}{P} \text{ and } (1 - y_{3,C}) \cdot P = (1 - x_{2,C}) \cdot P_P^{sat} \rightarrow$$

$$x_{2,C} = \frac{P - P_P^{sat}}{P_C^{sat} - P_P^{sat}} \rightarrow x_{2,C} = \boxed{0.914}$$

$$y_{3,C} = \frac{x_{2,C} \cdot P_C^{sat}}{P} \rightarrow y_{3,C} = \boxed{0.757} \text{ and } x_{2,P} = \boxed{0.086} \; y_{3,P} = \boxed{0.243}$$

Plug overall balance into component balance to find flow rates.

$$y_{1,C} \cdot \dot{n}_1 = x_{2,C} \cdot \dot{n}_2 + y_{3,C} \cdot (\dot{n}_1 - \dot{n}_2) \rightarrow \dot{n}_2 = \boxed{20.4 \frac{mol}{min}} \text{ and } \dot{n}_3 = \boxed{75.6 \frac{mol}{min}}$$

[Step 12] Check using component P balance or any atom balance.

12 steps problem solving method

17

Example: Full Solution

[Step 2] Stream 1

[Step 3] Only 1 system to write balances.

[Steps 4, 5, 6] Mole balances

Overall: $\dot{n}_1 = \dot{n}_2 + \dot{n}_3$

Compenent C: $y_{1,C} \cdot \dot{n}_1 = x_{2,C} \cdot \dot{n}_2 + y_{3,C} \cdot \dot{n}_3$

[Step 7] Two Raoult's laws, Antoine equation, and two sum of mole fractions.

Raoult's Law C: $y_{3,C} \cdot P = x_{2,C} \cdot P_C^{sat}$

Raoult's Law P: $y_{3,P} \cdot P = x_{2,P} \cdot P_P^{sat}$

Sum mole fractions: $1 = x_{2,P} + x_{2,C}$

Sum mole fractions: $1 = y_{3,P} + y_{3,C}$

Antoine equation: $\log_{10} P^{sat} = A - \frac{B}{T + C}$

[Steps 8, 9, 10, 11] 6 unknowns: 2 balances, 2 Raoult's laws, 2 sum mole fractions. Solve.

12 steps problem solving method

18

Example: Full Solution

Steps 8, 9, 10, 11 6 unknowns: 2 balances, 2 Raoult's laws, 2 sum mole fractions. Solve.

Use given T and Antoine equation to find P^{sat} values.

$P_C^{sat}(T = 98°C) = 1242$ mmHg $= 165600$ Pa

$P_P^{sat}(T = 98°C) = 4224$ mmHg $= 563100$ Pa

Many ways to solve this system of equations. One way is:

Plug sum mole fractions in Raoult's Law for P. Solve Raoult's law C for $y_{3,C}$ and plug $y_{3,C}$ into Raoult's Law P. Solve for $x_{2,C}$.

$$y_{3,C} = \frac{x_{2,C} \cdot P_C^{sat}}{P} \text{ and } (1 - y_{3,C}) \cdot P = (1 - x_{2,C}) \cdot P_P^{sat} \rightarrow$$

$$x_{2,C} = \frac{P - P_P^{sat}}{P_C^{sat} - P_P^{sat}} \rightarrow x_{2,C} = \boxed{0.914}$$

$$y_{3,C} = \frac{x_{2,C} \cdot P_C^{sat}}{P} \rightarrow y_{3,C} = \boxed{0.757} \text{ and } x_{2,P} = \boxed{0.086} \; y_{3,P} = \boxed{0.243}$$

Plug overall balance into component balance to find flow rates.

$$y_{1,C} \cdot \dot{n}_1 = x_{2,C} \cdot \dot{n}_2 + y_{3,C} \cdot (\dot{n}_1 - \dot{n}_2) \rightarrow \dot{n}_2 = \boxed{20.4 \frac{mol}{min}} \text{ and } \dot{n}_3 = \boxed{75.6 \frac{mol}{min}}$$

Step 12 Check using component P balance or any atom balance.

12 steps problem solving method

19

Example: Concept Question

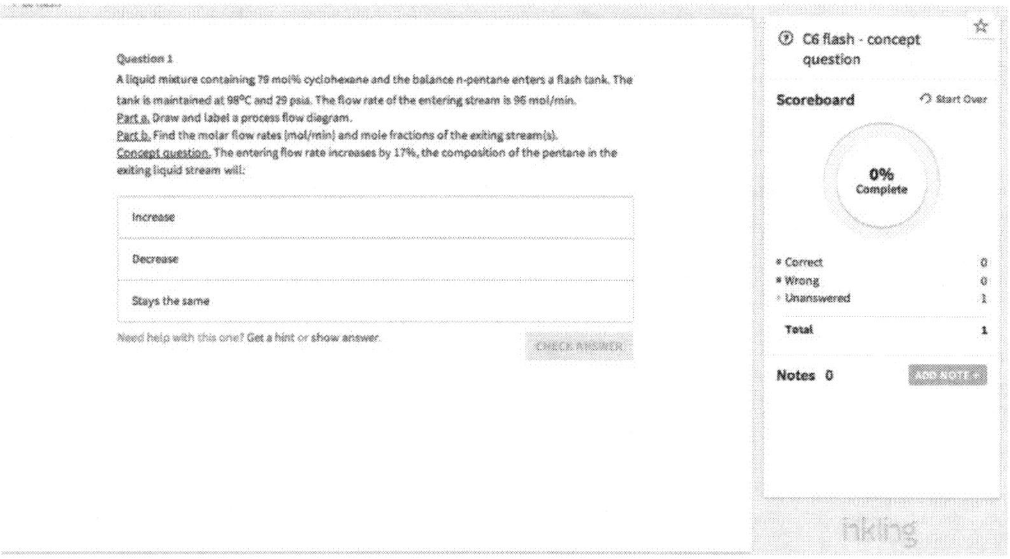

20

Example: Concept Question

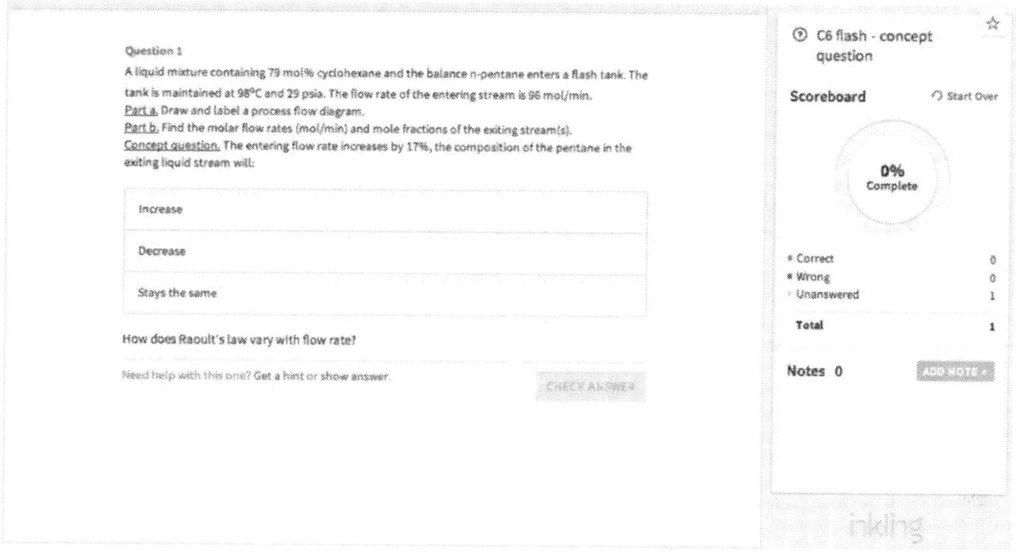

21

Example: Concept Question

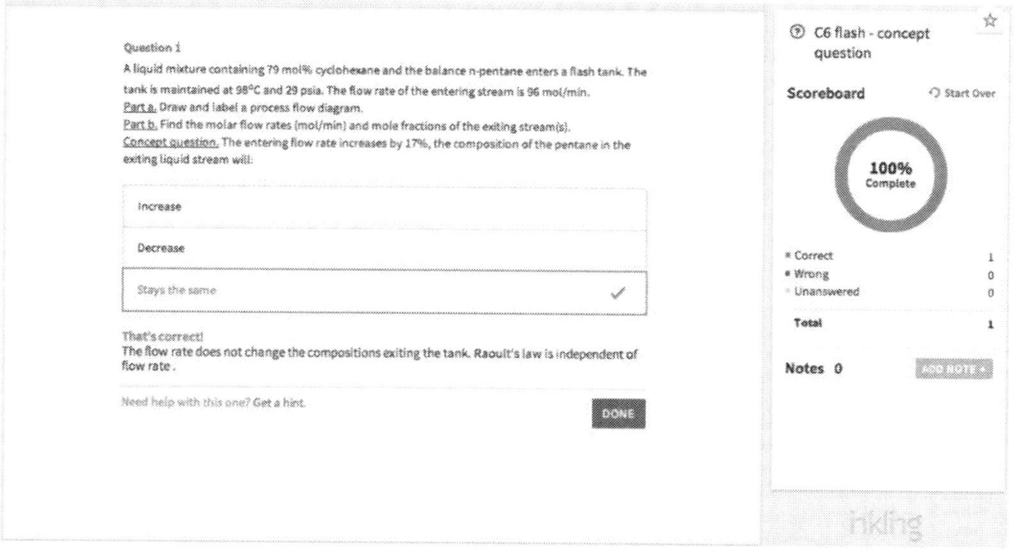

22

Process Unit

Flash tank

Flash separation is a simple process unit where T, P are altered to create separation of one or more components. The exiting vapor and liquid streams are in equilibrium providing one extra equation for each component in equilibrium. The process may be called flash evaporation and the equipment is commonly known as a flash drum. Additional information on flash can be found elsewhere.

Figure 6-13 **Flash tank.**

Process Unit: External Resources

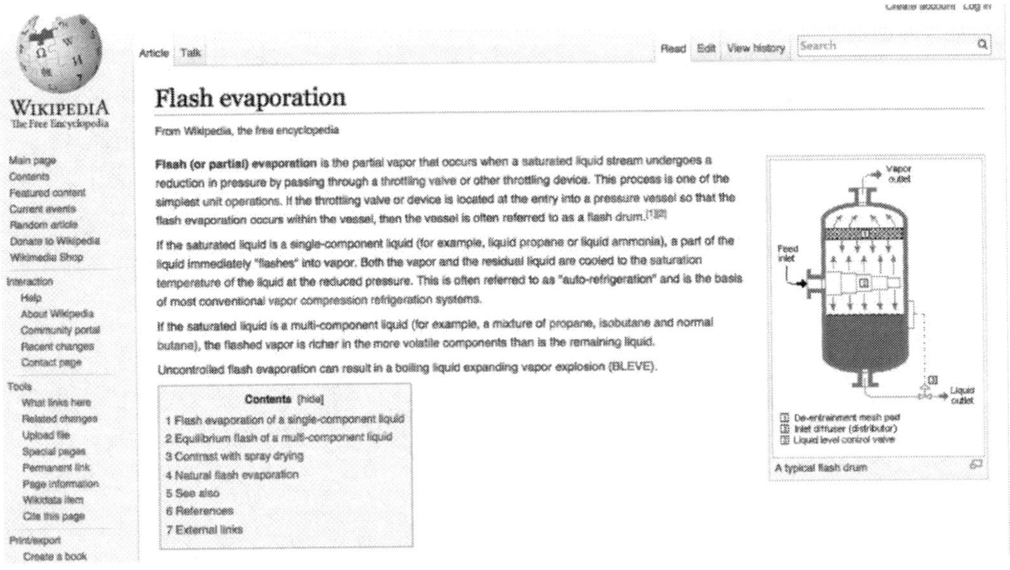

Create account Log in

Article Talk Read Edit View history Search

WIKIPEDIA
The Free Encyclopedia

Main page
Contents
Featured content
Current events
Random article
Donate to Wikipedia
Wikimedia Shop

Interaction
Help
About Wikipedia
Community portal
Recent changes
Contact page

Tools
What links here
Related changes
Upload file
Special pages
Permanent link
Page information
Wikidata item
Cite this page

Print/export
Create a book

Flash evaporation

From Wikipedia, the free encyclopedia

Flash (or partial) evaporation is the partial vapor that occurs when a saturated liquid stream undergoes a reduction in pressure by passing through a throttling valve or other throttling device. This process is one of the simplest unit operations. If the throttling valve or device is located at the entry into a pressure vessel so that the flash evaporation occurs within the vessel, then the vessel is often referred to as a flash drum.[1][2]

If the saturated liquid is a single-component liquid (for example, liquid propane or liquid ammonia), a part of the liquid immediately "flashes" into vapor. Both the vapor and the residual liquid are cooled to the saturation temperature of the liquid at the reduced pressure. This is often referred to as "auto-refrigeration" and is the basis of most conventional vapor compression refrigeration systems.

If the saturated liquid is a multi-component liquid (for example, a mixture of propane, isobutane and normal butane), the flashed vapor is richer in the more volatile components than is the remaining liquid.

Uncontrolled flash evaporation can result in a boiling liquid expanding vapor explosion (BLEVE).

Contents [hide]

Vapor outlet
Feed inlet
Liquid outlet
[1] De-entrainment mesh pad
[2] Inlet diffuser (distributor)
[3] Liquid level control valve

A typical flash drum

Data

Antoine equation constants for determining vapor pressure

Vapor Pressure (or Saturation Pressure) is discussed in Chapter 5. The Antoine equation is a very common way to determine vapor pressure as a function of temperature. Three constants are needed to calculate vapor pressures from the Antoine equations. These constants are tabulated here for many common compounds.

$$\log_{10} P^{sat} = A - \frac{B}{T+C}$$

where A, B, C are tabulated here. P^{sat} has units of mmHg and T has units of °C. These tables can easily be copied to a spreadsheet for simplifying vapor pressure calculations.

Compound	Formula	A	B	C	Range (Low T)	Range (High T)
Chloroform	$CHCl_3$	6.90328	1163.03	227.4	-30	150
Cyclohexane	C_6H_{12}	6.84941	1206.001	223.148	19.9	81.6
Cyclohexanol	$C_6H_{12}O$	6.2553	912.866	109.126	93.7	160.7
n-Decane	$n\text{-}C_{10}H_{22}$	6.95707	1503.568	194.738	94.5	175.1
1-Decene	$C_{10}H_{20}$	6.95433	1497.527	197.056	86.8	171.6

25

Affordable, Customizable Cost

Full book = $29.99

Individual chapters (9+Appendix) = $4.99/each

Buy chapters as you go = max $29.99

26

Surveying the Students

Beta Test: Spring 2014 with 153 students

- 80% of class purchased FMEB
- 66% of class found FMEB useful

Most common complaints: Fix the typos, include all of the major course concepts, and data tables

27

Surveying the Students

Beta Test: Spring 2014 with 153 students

- 80% of class purchased FMEB
- 66% of class found FMEB useful

Most common complai[illegible]ypos, include all of th[illegible]urse concepts and data ta[illegible]

Completed Summer 2014

28

What's next? – Research Questions

Usage analytics: How well are students engaged, how long students reading the STAR format versus a traditional paper textbook, which levels of the questions are students using, which difficulty of scaffolded examples are the students using.

Can the 12 step problem solving technique be observed in a time restricted exam setting.

Can this technology be fully integrated with online homework (e.g., Sapling Learning) to provide instructors real time assessment of the students progress in problem solving and conceptual knowledge.

29

Why Software Trumps 'Print'

Type of Textbook	Cost	Interactivity	Updates
Fundamentals of Material and Energy Balances	$29.99 for web, iOS, Android	Yes	Frequent
Traditional Textbook	$71.50 for Kindle	No	Once per decade

Learn more at www.readswipesolve.com

azengineering

Online Course Materials and Do Students Use Them in Ways that We Expect Rational Beings Would?

Paul Blowers and Chris Jabczynski
Plus Virgil Mayberry and Ali Khaghani
Chemical and Environmental Engineering
The University of Arizona

1

THE UNIVERSITY OF ARIZONA

azengineering

Faculty Assumption…

Why would we as faculty ever assume that a student knows how to study?

Partner up with someone else and come up with some ideas…

2

THE UNIVERSITY OF ARIZONA

azengineering

In the Literature…

Active learning is crucial

8410–8415 | PNAS | June 10, 2014 | vol. 111 | no. 23

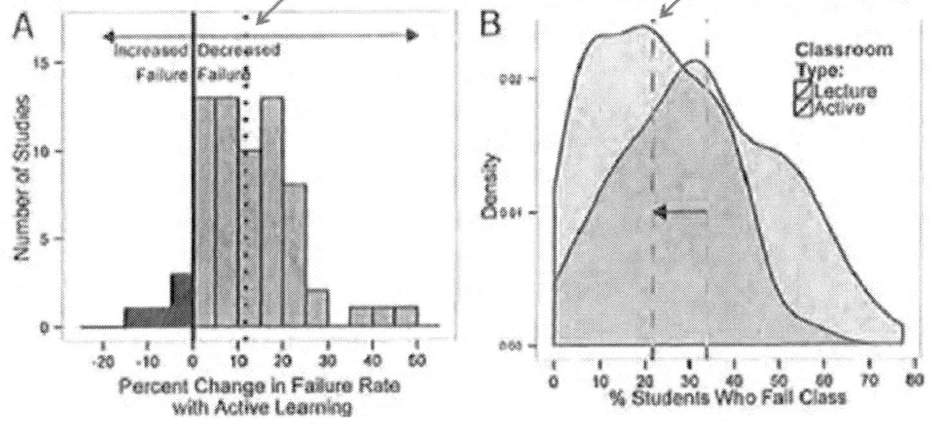

Active learning increases student performance in science, engineering, and mathematics

Scott Freeman[a,1], Sarah L. Eddy[a], Miles McDonough[a], Michelle K. Smith[b], Nnadozie Okoroafor[a], Hannah Jordt[a], and Mary Pat Wenderoth[a]

THE UNIVERSITY OF ARIZONA®

azengineering

More Literature Informing the Work

Course structure matters

Increased Structure and Active Learning Reduce the Achievement Gap in Introductory Biology

David C. Haak,[1]* Janneke HilleRisLambers,[1] Emile Pitre,[2] Scott Freeman[1]†

Active Learning and Peer Instruction

Peer Instruction: Ten years of experience and results

Catherine H. Crouch and Eric Mazur[a)]
Department of Physics, Harvard University, Cambridge, Massachusetts 02138

Am. J. Phys. 69 (9), September 2001

Innovation in large lectures--teaching for active learning

Diane Ebert-May; Carol Brewer; Sylvester Allred

Bioscience; Oct 1997; 47, 9; Research Library Core
pg. 601

Transforming the lecture-hall environment: The fully interactive physics lecture

David E. Meltzer[a)]
Department of Physics and Astronomy, Iowa State University, Ames, Iowa 50011
Kandiah Manivannan
Department of Physics, Astronomy, and Materials Science, Southwest Missouri State University, 901 South National Avenue, Springfield, Missouri 65804

Am. J. Phys. 70 (6), June 2002

4

THE UNIVERSITY OF ARIZONA®

azengineering

What we Assume Students are Doing…

Week	Date	Day	Lec #	Reading Assigned	Homework Due	Topic
1	8/25	M	1	Preface, Chap 1, 2.1-2.4	Recent Picture and 2-3 personal facts	introduction to course conversion factors
	8/27	W	2	3.1-3.3		process variables
	8/29	F	3	3.3-3.5	Behavioral Survey	definitions and concepts
2	9/1	M		Labor Day - No Class	Concept Inventory Tool on D2L	
	9/3	W	4		Writing Assignment HW 1	material balances
	9/5	F	5	4.0-4.3c		
3	9/8	M	6	4.3e-4.4		
	9/10	W	7	4.5, Chapter Summaries 1-4	HW2	
	9/12	F	8	Life Skills Practice 1		
4	9/15	M	9	**Test # 1**		

What really happens, especially in lower level courses…

They show up unprepared and the time is less likely to be productive for them.

5

THE UNIVERSITY OF ARIZONA.

azengineering

Explicit Syllabus and Course Structure

Students were given explicit pre-class activities that involved a recording, reading, and an ungraded pre-quiz. Students were then quizzed on the material in class with a graded quiz each day that cumulatively makes up part of their grade. Required weekly homework assignments contribute another 10%, while three midterm exams contribute 10% each and the final makes up the remainder.

Prior lecture ended	Pre-recording that is 3-10 min long	Reading assignment that is 10-25 min long	CMS – Prequiz – ungraded, but have answers	Lecture with in Class Graded Quiz

40 % of their total class grade!!!

6

THE UNIVERSITY OF ARIZONA.

azengineering

Course Structure II

Final exam: (20 % of grade).
Comprehensive final on December 18th, 1 pm to 3:00 pm. A comprehensive final will be given during the scheduled period during finals.

In class quizzes on homework and pre-class activities (40% of grade)
Quizzes will be given regularly at the beginning of class on the homework and pre-class activities/recordings. These quizzes will be created to make sure students are personally responsible for completing materials when they are due. At this point, students are less than one year away from being a fully functioning professional in the workforce and the expectation then will be that you meet your timelines and get your work done as required. This is practice for those scenarios of the future.

Exam Policy on Underperforming:
All students are eligible to earn some points back they miss on an exam by going to office hours and working with the preceptor, TA, or faculty member to have their questions answered and then to be evaluated on that understanding. Students who receive less than a 60 % on an exam will be required to go to office hours twice in the following week and will be required to turn in a copy of the Exam Office Hour Form to the instructor to demonstrat they met in office hours as they attempted to earn points back. Students may go to any two office hours or may also schedule additional time with the instructor outside those office hours. In the first office hour, students who received less than 60 % are expected to ask questions that help them master the content from the exam, which does not mean that they just watch the problems being solved. During the second office hour meeting, or the first for those who scored higher than 60%, the student will be given a make-up question where they will either solve the problem, or explain the steps that would reach the solution. Students who successfully do this can earn up to 5 points back on that particular exam, unless their score on the exam goes above 100. A total of 20 points may be earned back on the midterms by any student if they are successful in the make-up sessions in demonstrating the percentage of knowledge they have mastered.

THE UNIVERSITY OF ARIZONA

azengineering

Assessment

IRB approval was obtained and students were consented in the first week of class

Pre-class concept inventory and Pre-class Surveys on student behavior were conducted

Question 3 (1 point)

Which of the following statements is true?

Question 5

In a general science/engineering course, how often do you do each of the following?

#	Statement	Never	Rarely	Occasionally	Regularly	Always
1	Read the textbook prior to a lecture on a given topic?	○	○	○	○	○
2	Organize information you have come across into a framework of what you already know?	○	○	○	○	○
3	Attempt to solve problems that are not assigned?	○	○	○	○	○
4	Review content from a prior class meeting?	○	○	○	○	○

THE UNIVERSITY OF ARIZONA

azengineering

Assessment II

After each exam, students were asked to reflect

Question 4 (1 point)

How many days before Test 1 did you begin to study actively for it?

- ○ 1
- ○ 2
- ○ 3
- ○ 4
- ○ 5
- ○ 6

How important was reading the textbook to helping you earn the grade you hope to earn?

- ○ Not applicable
- ○ Unimportant
- ○ Neutral
- ○ Important
- ○ Very important

9

THE UNIVERSITY OF ARIZONA.

azengineering

D2L and Course Management Software

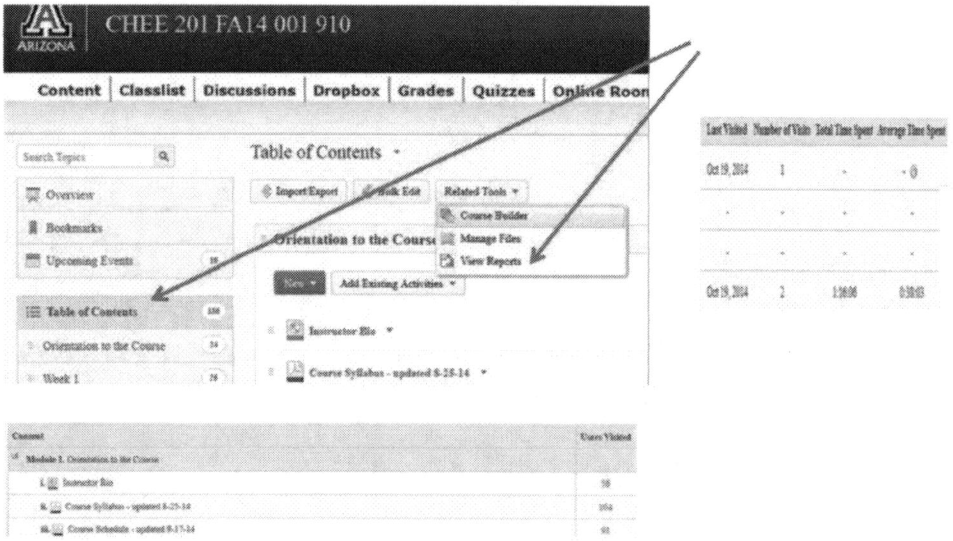

10

THE UNIVERSITY OF ARIZONA.

Data Pulling and Analysis

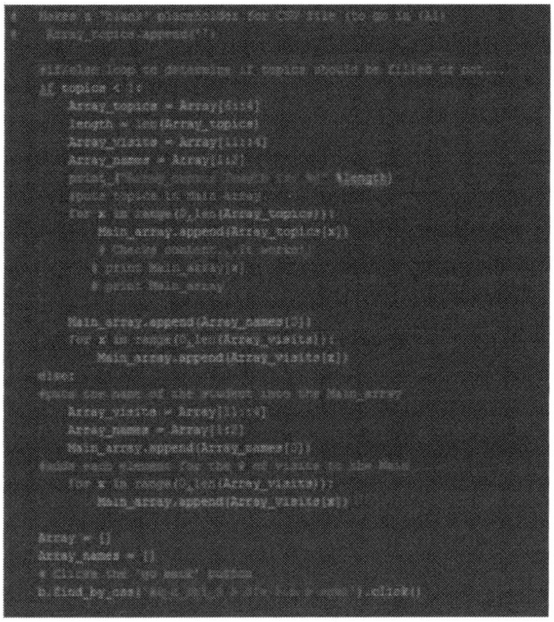

Custom Program

Used to pull data nightly, and create a historical record.

Data Integration and Sorting

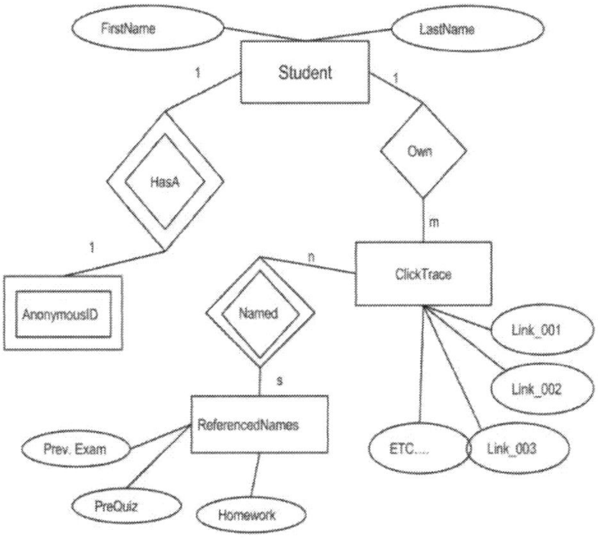

This is a simplified view of the schema of the underlying database that is used to more efficiently work with the data.

Tools used:
SQLite3
Python3.4
MatPlotLib
Hosted on Linux servers

azengineering

Data Analysis and Trends

The group of A students elevated quiz performance substantially

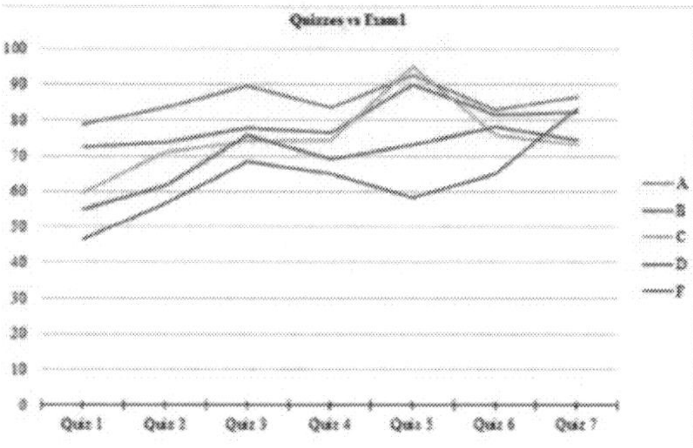

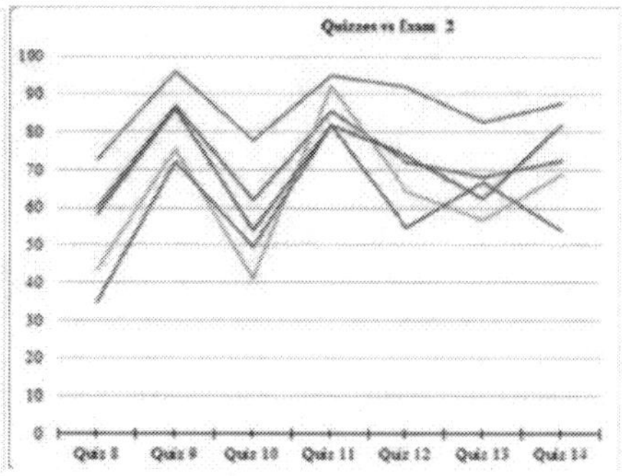

13

THE UNIVERSITY OF ARIZONA®

azengineering

Data and Results I

Some students are realizing they need to transition now. Others have transitioned

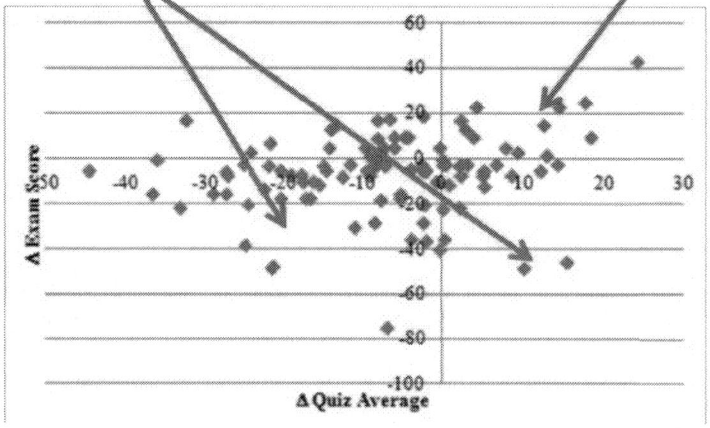

14

THE UNIVERSITY OF ARIZONA®

azengineering

Homework Assignments and Student Behavior

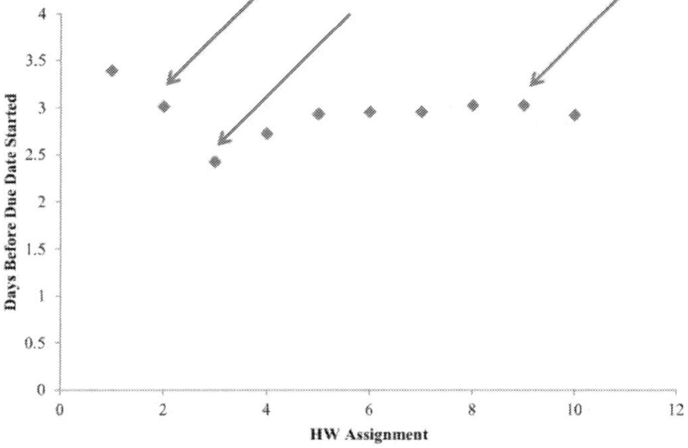

THE UNIVERSITY OF ARIZONA

azengineering

Data and Results II

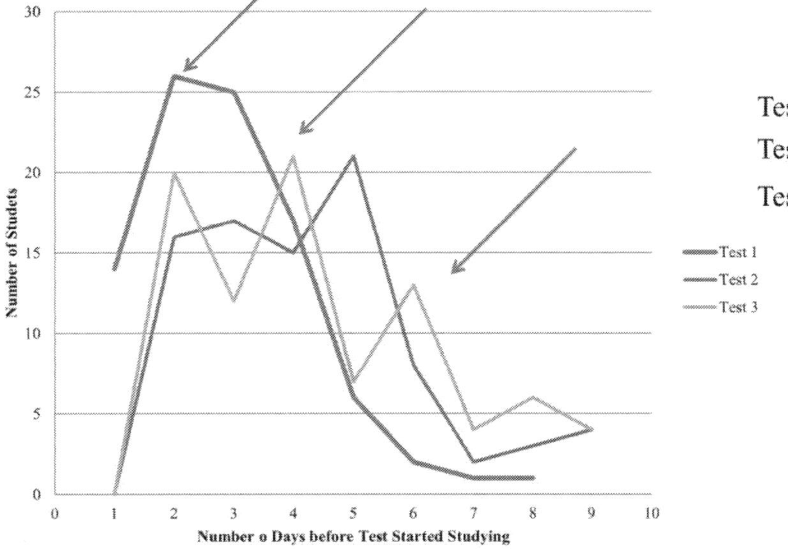

Test 1 2.89 days
Test 2 3.30 days
Test 3 3.43 days

16

THE UNIVERSITY OF ARIZONA

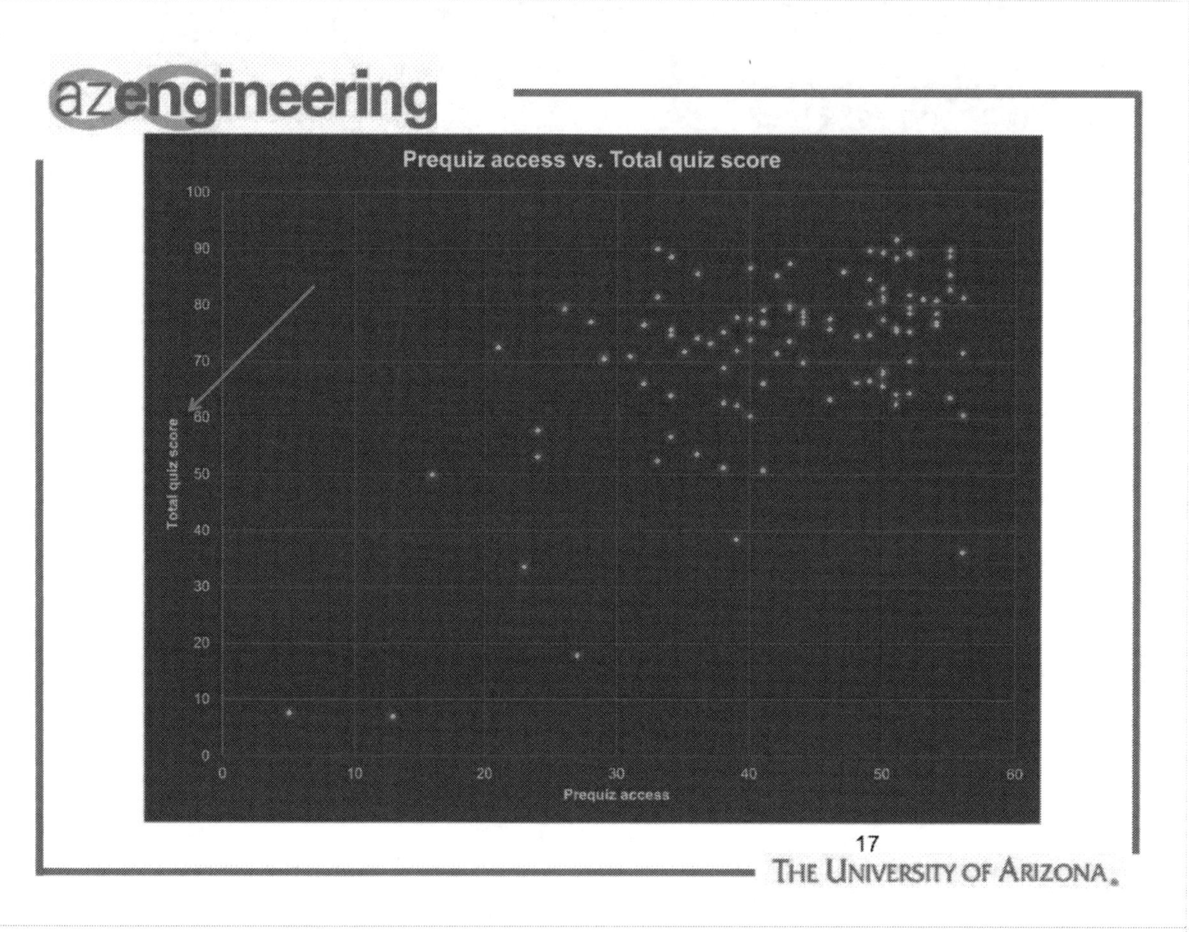
azengineering
Prequiz access vs. Total quiz score
Total quiz score
Prequiz access
17
THE UNIVERSITY OF ARIZONA

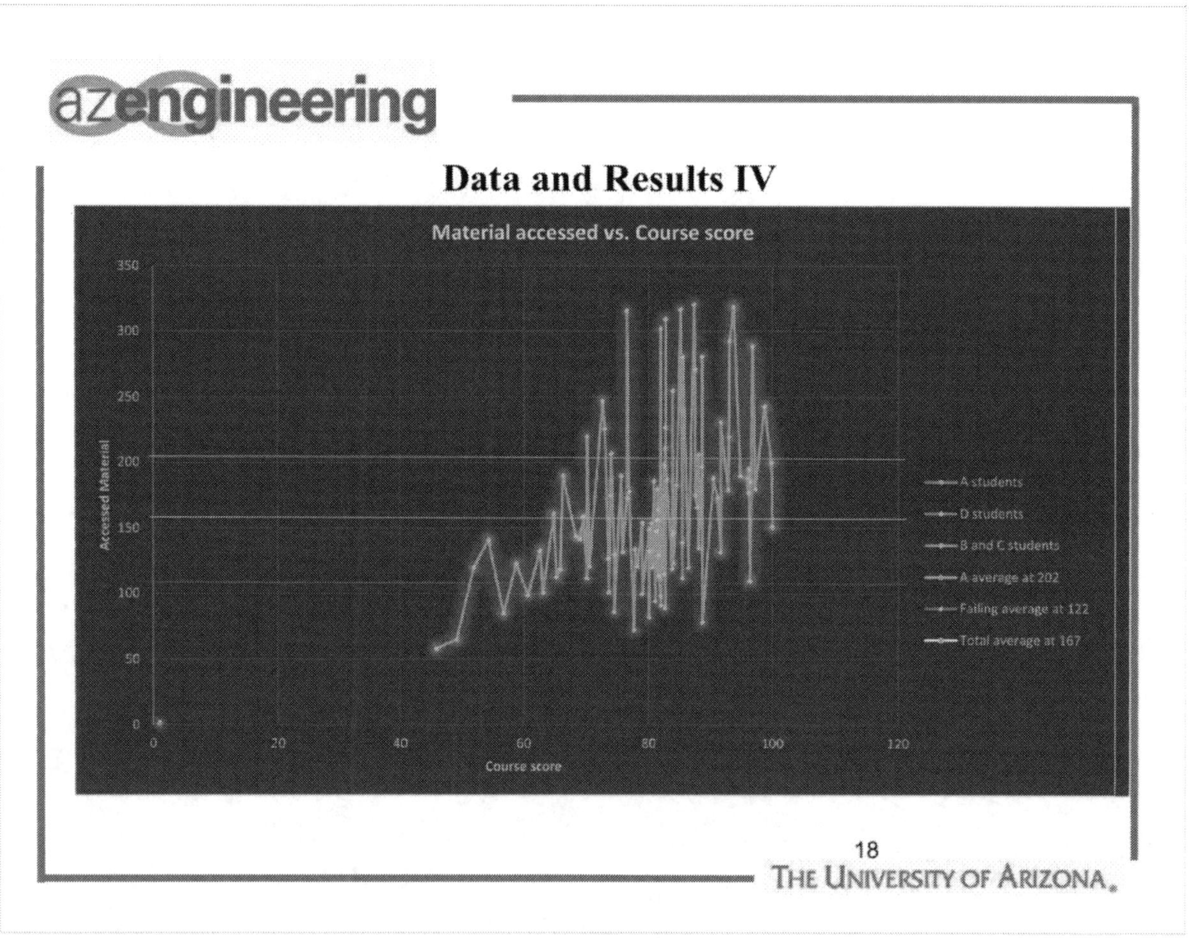
azengineering
Data and Results IV
Material accessed vs. Course score
Accessed Material
Course score
A students
D students
B and C students
A average at 202
Failing average at 122
Total average at 167
18
THE UNIVERSITY OF ARIZONA

azengineering

Data and Results IV:
Test Average Versus In Class Quiz Average

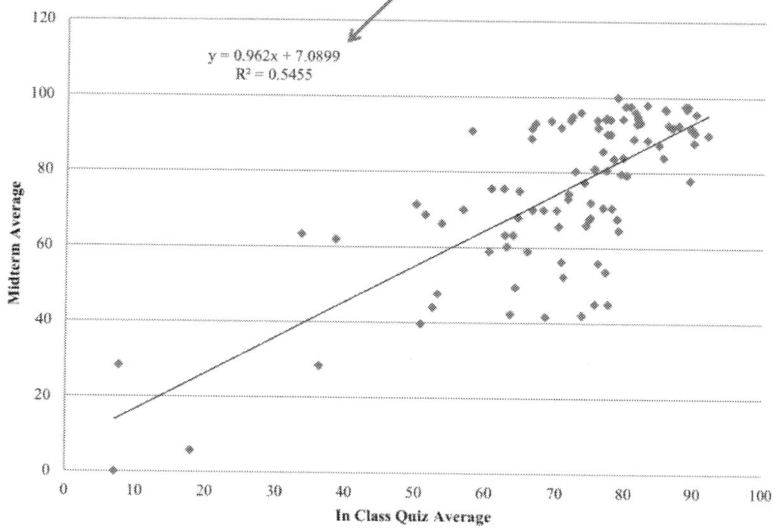

19

THE UNIVERSITY OF ARIZONA®

azengineering

Data and Results: Comments from a Student

The key to my success, I guess you could say, was understanding. I did not pull out many different problems that would have me run through similar motions, instead, I picked one great problem and set my mind to understanding what the heck was going on. Usually the great problems came from lectures, most of them being the ones that had been printed and handed out by Dr. Blowers. I judged a good problem based off of its ability to mix key ideas from previous lectures into a worded problem that I had to keep rereading in order to decipher what was being asked. Then, I began to recite in my head what I already knew, and I what I didn't know as I went about solving the question. As a boost of self-confidence, I would redo the problem a couple of times later on so that the flow chart that we have been working on in class sort of became second nature. Before this, I used to perform lots of problems about the same topic, but the different numbers can be confusing and ARE NOT IMPORTANT. What is way more important is the steps that I took to solving that problem. That's the problem with new problems, they take too much effort to solve the meticulous math and it strays one's focus from the BIG PICTURE. Starting a problem that one has already completed gets their mind away from getting the correct number, and the second, third, even fourth time they solve it, that person begins to fully understand what was being asked in the first place, and the flowchart in the back of their mind begins to solidify.

Exam Average Went Up By 20 % Over the Three Exams!

20

THE UNIVERSITY OF ARIZONA®

azengineering

Do Students Change Behavior after Getting Feedback?

Question 11 — **Difficulty: 1 Average Grade: 0 / 1 (0 %)**

What elements of the course/campus activities/life/work are impeding your ability to do as well as you might like to be doing in this course, if anything?

- Working 30 hours of week, and coaching two soccer teams. This exam was a wake up call though, ready to buckle down, but feel like I need more assistance and I cant make office hours.
- Time management, and having other exams/homework.
- Clubs
- Balancing work schedule
- the time put doing assigned exercises from the book.
- personal issues. not enough time to work through extra problems.
- transfering the question into mathematical equations
- Getting better with time management but still difficult to work, train and get everything done for all classes.
- Deadlines overlapping with other classes. But I'm getting better at getting things done ahead of time.

Notice the large number of "other" issues discussed openly.
These students are trying to work hard but many feel overwhelmed.
Increasing course structure is acknowledged, but not always as a good end point.

azengineering

What are the Essential Elements of a Course to Stress to Sophomores?

	0 is unimportant: 4 is very important		
Summary:	Test 1	Test 2	Test 3
Reading the textbook?	2.78	2.56	2.49
Organizing	3.05	3.15	3.12
Attempting problems not assigned	2.60	2.34	2.20
Reviewing information from prior classes	3.28	3.23	3.34
Summarizing notes	2.37	2.19	2.35
Office hour for HW	1.88	1.89	2.16
Office hours for other issues	1.65	1.42	1.23
Reviewing prior exams	3.32	3.24	2.88
Reviewing exam solutions before next exam	3.37	3.45	3.36
Class pre or post recordings	3.04	2.82	3.04

azengineering

Future Questions

- Finish constructing database management tool for extracting and compiling data
- Identify trends after exam 4 and how student access is evolving regarding grades.
- Re-evaluate exam averages as students drop for fair comparison
- Work with students to help in skill development (time management skills)

23

THE UNIVERSITY OF ARIZONA.

azengineering

Acknowledgments and Questions

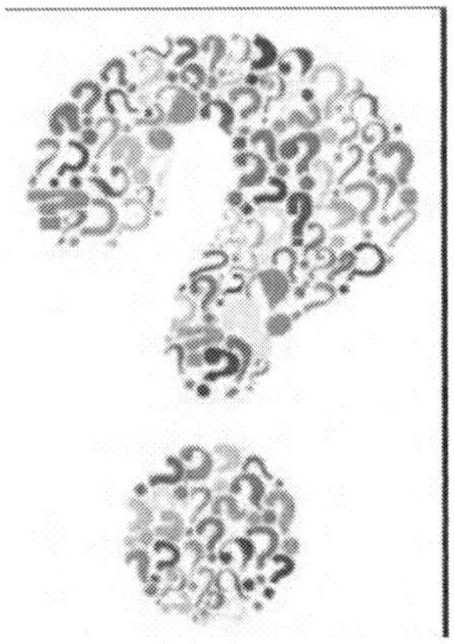

Undergraduate STEM Education Initiative

Association of American Universities

24

THE UNIVERSITY OF ARIZONA.

Title: ASSESMENT OF TEXTBOOK-FREE COURSES IN THE BIOCHEMICAL ENGINEERING FIELD AS VEHICLES FOR LIFELONG LEARNING.

Author: Daniel Forcinti
Chemical and Biochemical Engineering Department
Missouri University of Science and Technology
Rolla MO 65409
Email: forcinit@mst.edu

Abstract

The purpose of this project was to investigate the effect of using alternative ways to deliver content in place of textbooks on the ability of the students to become lifelong learners. The results of this study will be used to design, in a follow up project, a route map for those who desire replacing the textbook by alternative materials.

The author has been teaching a bioseparations course without a textbook for over ten years. The author's motivation for not using a textbook in bioseparations (and currently also in a bioreactors class) is that textbooks in biochemical engineering are of little use because of the dynamics of the field. For example, the hybridoma technique to produce monoclonal antibodies was developed 40 years ago and today the production of monoclonal antibodies is the most rapidly growing pharmaceutical sector. Moreover, monoclonal antibodies or antibodies fragment are produced today mostly using disposable technology, non-existent 20 years ago. There are no indications, in the students' comments, that the lack of a textbook is a major hurdle in this bioseparations class. There were, however, some indications that some students feel "fragile" without a textbook. The effect of the absence of a textbook on lifelong learning and the inclusion of specific activities aimed at facilitating lifelong learning have not been explored. This is addressed in this paper.

Short term results were obtained by comparing the results of a survey administered to students who took the sequence bioseparations/bioseparations lab/bioreactors. Students' opinions about (1) the effect of the absence of a textbook on their development as lifelong learners and (2) the significance of the initiatives introduced in the classes as lifelong learning incentives were requested as part of the end of the semester teaching effectiveness survey.

Introduction

I have been teaching a bioseparations course without a textbook for more than ten years and a textbook free bioreactors class for a couple of years. Bioseparations is a second semester junior class whereas bioreactors is a second semester senior one. The material for the bioseparations class has been given to the students through Blackboard first and for the last few years in a CD containing: (1) A number of scientific papers and vendors' material for each of the main core topics, (2) A series of power point slides that I use to support my lectures (the slides are not used as the main teaching method in this class), and (3) Reading guidelines. Table 1 shows an example of reading guidelines used in the first week of the class. The material for the bioreactors class has been given to the students in Blackboard. In both classes monographs covering the main topics in the class are put in the reserve section of the library.

Table 1. Reading guidelines for the introductory chapter of Bioseparations

Manuscript	Comments
Graslund, S. et al. (more than twenty authors).. Protein production and purification. Nature Methods, 5: 135-147 (2008).	**Good review (quite detailed) of purification of recombinant proteins**
Ladisch, M. 2004. The role of bioprocess engineering in biotechnology. The Bridge. A publication of the National Academy of Engineering. 34(3).	**Opinion by one of the leaders of the field. (#2)**
Asenjo, J.A. and Andrews, B.A. Mini-review—Challenges and trends in bioseparations. J. of Chemical Technology and Biotechnology, 83:117-120 (2008).	**Broad review by an engineer. The modeling part is to be taken with care.**
Gottschalk, U., The renaissance of protein purification. BioPharm International, 2006.	**Informative**
Dolnik, V. Capillary electrophoresis of proteins 2005-2007. Electrophoresis, 29: 143-156 (2008).	**Good review about capillary electrophoresis of proteins. It has a good description of the technique. We will focus on the modeling part, which is not covered by this article.**
Genentech, MetMAb Fact sheet. Genentech, Pertuzumab Genentech, History.	**About one of the leading companies**

Shen, S-H. Multiple joined genes prevent product degradation in Escherichia coli., Proc. Natl. Acad. Sci, USA. 81:4627-4631 (1984).	
Wilken, L.R. and Nikolov, Z.L. Recovery and purification of plant-made recombinant proteins. Biotechnology Advances, 30: 419-433 (2012).	**Good review about the challenges of the recovery of recombinant proteins expressed in plants**
Ponchon, L. and Dardel, F. Large scale expression and purification of recombinant RNA in Escherichia coli. Methods, 54: 267-273 (2011).	**A good review about RNA production and purification.**
Deshmukh, R.R., Warner, T.N., Hutchison, F., Murphy, M., Leitch II, W.E., De Leon, P., Srivatsa, G.S., Cole, D.L., Sanghvi, Y.S. Large-scale purification of antisense oligonucleotides by high-performance membrane adsorber chromatography. J. of Chromatography A, 890:179-192 (2000).	**Specific example of RNA purification. For the avid reader.**
Josefberg, J.O. and Buickland, B. Vaccine Process Technology. Biotechnology and Bioengineering, 109:1443-1460 (2012).	**A bit out of topic but it shows how broad is the bioseparation field**
Aggarwal, S. What's fueling the biotech engine –2011 to 2012. Nature Biotechnology, 30: 1191-1198 (2012).	**A very good compilation of economic data. This is a "must" read.**
Biopharma report on infectious diseases.	**You should read this article.**

Only recently the author has asked the question whether or not the lack of a textbook in this classes encourage the students to become lifelong learners. In the following paragraphs I discussed the use of textbooks in higher education and the identification and assessment of lifelong learners.

Results and Analysis

There are few studies about textbook free classes (an E-Book is still a book). For the most part the studies are in Biology and Physics courses but not in engineering. The editor of CBE Life Science Education, G. Reines, recognized that very few professors (In America) would consider teaching a course without using a textbook. Science and engineering books have become more colorful, more encyclopedic, and accompanied by more ancillary materials such as CD-ROMs, study guides, and websites. He continued arguing that most (American) instructors ask which textbook to use, not whether to use a textbook. What is not clear in his view is whether or not textbooks help students learn better.

Podolefsky and Finkelstein (2006) did a relatively large study about textbook use in four physics classes in the U.S. The study affected 800 students. Their main findings were: 1) Only

37% of students regularly read, 2) Less than 13% read often and before the lecture, 3) There is no correlation between reading habits and course grades and 4) Students identify textbook with homework and lecture with exams. The latest point can be used as a justification as to why there is no correlation between reading habits and course grades. Still, one may argue that if there is a connection between homework and tests then there is an indirect connection between the textbook and the tests.

Carpenter et al. (2006) did a study upon a request from British publishers. They found that students consider textbooks important but they have difficulties in explaining why. The authors argue that the something similar may happen to instructors. In Great Britain some instructors still provide the students a reading list from which the students can pick up one or more books. An equal number ask students to pick one of several alternative main course texts. It is worth comparing this British experience in which only 15% of lecturers 'adopt' a required text with the American practice in which the almost universal practice is for the professor to choose one book for his or her course, which is then bought new or used by the vast majority of students. A greater anxiety is that some lecturers may not be aware of how critical they are in the decision to purchase and may not appreciate that a less than strong endorsement for a book means is it less likely to be bought.

A few thoughts from Klymkowsky (2007) summarize the most relevant points relative to the complex topic of the use of textbooks in higher education. First, we must recognize that the issue of whether to use a textbook is complex, and it is dependent upon course and curricular goals. Students (and colleagues) expect a textbook; yet often, the textbook is not used, except as a reference. Then, "Why don't we offer the students a list of reference books?", asked Klymkowsky. Moreover, most textbooks are not written with current evidence about best teaching and learning practices in mind, so they may be difficult to integrate into the design and presentation of a course that is based on this evidence.

Figure 1puts lifelong learning in ABET's language. The learning outcome is for the students to become lifelong learners. Self-directed, self-regulated, self-motivated and reflective learning skills define a lifelong learner. A few performance criteria are summarized in the last block. The students must prove that they can plan their own learning, they should be able to monitor and assess their learning and they should be able to independently find and use technical information. A textbook free course facilitates to develop these skills and makes possible their monitoring.

Paulsen and Feldman (1999) reflected about the role of college teachers in the development of epistemological beliefs by the students. They claim that faculty, in their roles as college teachers and designers of learning environments, should assume a greater responsibility for promoting motivationally and educationally productive epistemological beliefs among their students. They concluded that first year college students have naïve epistemological beliefs whereas the instructors have sophisticated epistemological belief. This is supported by the study of Hoskin et al. (2011), who surveyed epistemological beliefs (Hoskins et al. 2011) among a group of students as part of a larger study aimed at monitoring and quantifying the ability of the students to read primary sources. Table 2 reproduces portions of the survey. The development of sophisticated epistemological beliefs is a feature of lifelong learners. On the contrary, naïve beliefs become a hurdle that the students need to overcome to become lifelong learners. This survey will be administered to the bioseparations and bioreactors class starting in the Spring of 2015.

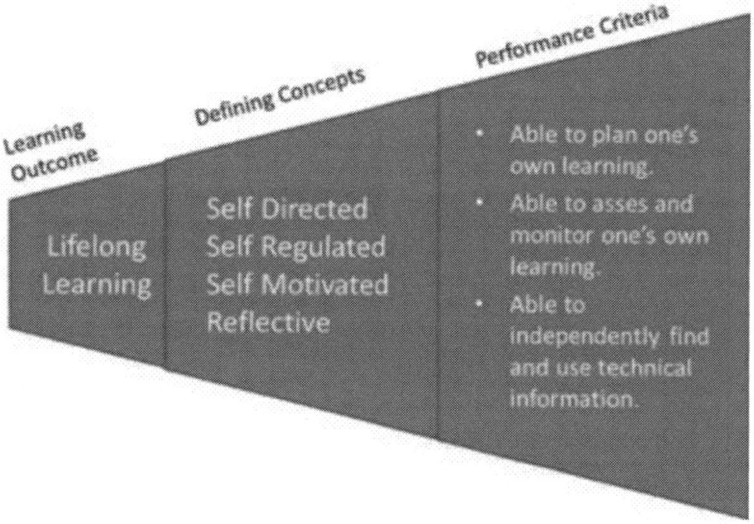

Figure 1. Schematic of lifelong learning from an ABET perspective (adapted from http://depts.washington.edu/celtweb/

Table 2 . Items form the C.R.E.A.T.E. survey that measure epistemological beliefs. (Hoskins et al. 2011)

Knowledge is certain.
If two different groups of scientists study the same question, they will come to similar conclusions. (R)
The data from a scientific experiment can only be interpreted in one way. (R
Because scientific papers have been critically reviewed before being published, it is unlikely that there will be flaws in scientific papers. (R)
Because all scientific papers are reviewed by other scientists before they are published, the information in the papers must be true. (R)
Sometimes published papers must be reinterpreted when new data emerge years later.
Results that do not fit into the established theory are probably wrong. (R)
Ability is innate.
I think professionals carrying out scientific research were probably straight-A students as undergrads. (R)
You must have a special talent in order to do scientific research. (R)
Attitude toward science
Science is a creative endeavor
I have a good sense of what research scientists are like as people
I do not have a good sense of what motivates people to go into research. (R)
Scientists usually know what the outcome of their experiments will be. (R)
Collaboration is an important aspect of scientific experimentation.
Items followed by a (R) are reverse scored.

One practical motivation not use textbooks in biochemical engineering classes is the fast dynamics of the field. For example, the hybridoma technique to produce monoclonal antibodies was developed 40 years ago and today the production of monoclonal antibodies is the most rapidly growing pharmaceutical sector. Moreover, monoclonal antibodies or antibodies fragment are produced today using disposable technology, non-existent 20 years ago.

The material for both classes is updated yearly and that creates a burden that the instructor tries to overcome over the winter break (both classes are offered once a year in the

Spring semester). We limit our searches to a single search engine (Scopus). New reviews on a particular topic have the highest priority. Part of the time is also spent glancing through new topics. Both classes are heavily focused on protein pharmaceuticals (with even more emphasis on monoclonal antibodies and fragments). Manuscripts are assigned to the students as reading assignments (the students in groups prepare a short report and present the report to all of us).

There is no indication in long term (10 years) students' evaluations, in the students' comments, that the lack of a textbook is a major hurdle in the bioseparations class. There are, however, some indications that some students feel "fragile" without a textbook. A survey to alumni for the last 20 years is being prepared.

Table 3 shows the results of a recent survey administered to the students last academic year. Fifteen responses were collected in the senior year class and ten in the junior one.

Table 3. Exit survey for the Bioseparations and Bioreactors classes

Question	Senior Class	Junior Class
Do you think this class has helped you to: (a) know how knowledge is organized, (b) how to find information (c) how to use information in such a way that others can learn from it (d) None of the above (e) All of the above	100% (e)	Mix (b) and (c)
Lifelong learners are those who are: (1) better prepared to plan their own learning, (2) able to assess and monitor their own learning, and (3) able to independently find and use technical information Do you think that a textbook free class like this one has helped you to become a lifelong learner? (1)Strongly agree (2) Agree (3) Disagree (4) Strongly disagree.	50% 1, 50%2	90% (2) and 10% (3)
The best material for a class is: 1. A single textbook 2. A combination of scientific papers and vendors literature 3. A combination of scientific papers, vendors literature and textbooks.	Mixed results 2 and 3	50% (3), 40% (2) 10% (1)

The data so far seems to indicate that the less mature students (those taken bioseparations) have more problems handling the class in the absence of a textbook. The students in the advanced class seem to have receive the teaching strategy quite well. A survey about their epistemological beliefs was not administrated but it will be this Spring to the new batch of students.

During this year, surveys will be sent out to alumni who have taken the bioseparations and bioreactors sequence with and without textbook. The author has recently collected the names and location of the students that have graduated from the Biochemical Engineering Program since 1993. Approximately half of them have used a textbook in the bioseparations class and half haven't. All the students have used a textbook in the bioreactors class with the exception of this year graduating seniors. We are developing a survey that will reach the alumni before the end of this calendar year. The data collected will be evaluated to decide whether the information collected was sufficient to verify our hypothesis and justify continuing teaching textbook free and/or redesigning the classes further.

References

Carpenter, P., Bullock, A., Potter, J. Textbooks in teaching and learning. Brookes eJornal of Learning and Teaching, April 2014, 11 pp.

Hoskins, S.G., Lopatto, D., Stevens, L.M. The C.R.E.A.T.E. approach to primary literature shifts undergraduates' self-assessed ability to read and analyze journal articles, attitudes about science , and epistemological beliefs. CBE Life Science Education, 10: 368 (2011).

Klymkowsky, M.W. Teaching without a Textbook: Strategies to focus learning on fundamental concepts and scientific process. CBE –Life Sciences Education, 6:190-193 (2007).

Paulsen, M.B. and Feldman, K.A. Student motivation and epistemological beliefs. New Directions for Teaching and Learning, no 78: 17 (1999).

Podolefsky, N. and Finkelstein, N. The perceived value of college physics textbooks: students and instructors may not see eye to eye. The Physics Teacher, 44: 338 (2006).

Reines, G. Note from the editor introducing Klymkowsky manuscript (2007).

387408 Investigating Creativity in an Open-Ended Project

Abstract:

Creativity, the ability to develop something innovative and useful from pre-existing knowledge and experience (Liu & Schonwetter, 2004), is a significant component of engineering; engineers must be creative in order to develop original processes and products and solve problems in novel ways. However, a prominent aspect of a ChE curriculum is to teach students common problem-solving techniques or algorithms to reach a solution; as a result, students may lose interest in creativity and instead focus on following a demonstrated path and getting the "right" answer. Open-ended problems encourage creativity since they have multiple viable solutions and allow students to create their own solution path. Through examining student work products from an open-ended project, we seek to develop an evidence-based understanding of how creativity is embodied. Specifically, we investigate three aspects of creativity in engineering: fluency, or the number of ideas; flexibility, the variety of ideas; and originality, the novelty of the idea.

We consider the following research questions: how do these aspects of creativity compare in student teams working on an open-ended virtual laboratory project versus more constrained physical laboratory projects? How do these aspects compare for student teams and experts, i.e., to what extent does domain expertise affect creativity? Does creativity relate to project performance (the team's final grade on the project) and if so, how?

The open-ended project, the Industrially-Situated Virtual Laboratory Project, was used as one of three laboratory projects in the senior laboratory course at Oregon State University; during the term, students also completed two physical laboratory projects, including a heat exchange experiment and an ion exchange experiment. As a part of each project, student teams created several work products, including memos, reports, and laboratory notebooks, which provide evidence of the team's solution path and document the elements that may contribute to the team's understanding and progress. These elements are identified, arranged in chronological order, and presented in a graphical form called a Model Map. For the student teams, Model Maps have been generated for all three laboratory projects while the expert teams only completed the virtual laboratory project. The Model Maps provide a platform for analyzing the three previously mentioned aspects of creativity: fluency (the number of models on a team's model map), flexibility (the different types of models developed), and originality (the uniqueness of a model compared to the other teams). By looking at a subset of these Model Maps, we found that students tend to be more fluent and flexible in the open-ended virtual laboratory project than in the physical laboratory projects, utilizing a wider range of elements in their solution path. Surprisingly, the experts scored lower on these measures of creativity. We speculate that this result is due to the richer domain knowledge of the experts, but it warrants further study.

Liu, Z., & Schonwetter, D. (2004). Teaching creativity in engineering. *International Journal of Engineering Education*, *20*(5).

386339 When Students Develop e-Learning Material for Their Peers

Abstract:

The Department of Chemical Engineering have long had a desire to develop e-learning material for their undergraduate modules. Unfortunately, given our very high student/staff ratio, there is never enough time to do this properly. With funding from the Faculty, an ExxonMobil teaching award and industry, ten undergraduate students were hired for eight weeks during the Summer 2013 and Summer 2014 to develop e-learning material based on Moodle for most of the department's first year modules. The posts were advertised to all undergraduates and students applied by submitting a one page application detailing what they wanted to develop, and why they were right for the job. The selected students were mainly from Year 2, but there were also a few Year 1 students and one and two Year 3 students, repectively, in 2013 and 2014.

The experience far surpassed the expectations. The students, who only knew each other vaguely beforehand, came together as a team from Day 1. The Year 3 students assumed a clear leadership position and led the team brilliantly. The material the teams developed was imaginative, comprehensive, technically sound, and with just the right dash of humour. The material was much better than what a member of academic staff would have developed, in particular, in terms of the feedback included for multiple choice questions etc. The students also developed videos to support some of the learning.

The poster will show examples of all the material produced, and will discuss the experience of letting students take the lead in developing e-learning material for their friends.

The Impact of Effective Staff Involvement on the Successful Realization of ABET Requirements

Amer Hamouz*, Ayham Jaaron, Farouq F.I. Halawa, Basheer W.F. Shaheen, and Maryam Hmouda

Faculty of Engineering and Information Technology

An-Najah National University- Palestine.

***Corresponding Author: Professor of Chemical Engineering,**

E mail: elhamouz@najah,edu

One of the major endeavors of the faculty of engineering at An-Najah National University over the past few years was the "ABET project" which aimed at boosting the level of quality of education in the faculty. The "ABET project" started with seven engineering programs, Civil, Electrical, Telecommunication, Chemical, Industrial, Mechanical and Mechatronics, at once.

Although several obstacles were faced at the beginning of the project such as the resistance of change by both students and instructors, and the high number of evaluated programs, the seven engineering programs were evaluated by the ABET program evaluators during their visit which took place in October 2103, due process is underway, and the final results will be determined at the ABET Commission meeting in July 2014. The success of the "ABET project" was accentuated by several elements and innovative ideas.

Innovation was exhibited when a committee on the faculty's level was set up that encompassed faculty members from diverse engineering backgrounds who were familiar with the quality standards of education, and had outstanding experience in the academic environment. The committee's role was to lead the ABET process and provide the necessary guidance and advice.

Moreover, the faculty of engineering had realized that creating a successful assessment and evaluation processes require a dedicated executive team and a unit to coordinate, facilitate, implement and follow up the achievement of the quality requirements. That's why a special center was established called "ABET Center - Quality and Accreditation unit".

The center and its dedicated staff who have engineering backgrounds played a vital role in the success of the project by acting as a link between ABET committee and the engineering staff. This was followed by creating a computerized assessment system administered by the ABET center. The system, which was user-friendly, made it easy for the instructors to fill in marks and to monitor the students' attainment in Student Outcomes through courses. It also made it easy for the evaluators to conduct their review smoothly.

Furthermore, several committees have been formed in the engineering departments with different roles and activities. For example, the Program Educational Objectives (PEOs) maintenance committees were appointed in all departments to review and maintain the PEOs. In addition, Quality Assurance Committees (QAC) were assigned in all departments to set and follow up the continuous improvement action plans.

It is important to note that ABET assessment process requires a considerable amount of data collection and paperwork. The environmental aspect, however, has been a priority to the faculty of engineering. This was reflected in the green policy implemented in the ABET assessment process by relying on electronic data (copies) instead of hard-copies.

All the above mentioned elements not only enabled the faculty of engineering to make it successfully to the ABET on-site review in a relatively short preparation period of time, but also immensely improved the academic process in the engineering programs. For example, the existence of a computerized system for assessing, monitoring and evaluating the Student Outcomes helped the engineering programs track and improve students' weaknesses effectively. In fact, over the past two years the engineering programs were able to detect some weaknesses in students' soft-skills, particularly in communication, working in teams and reporting. As a corrective measure, all the engineering departments have increased the additional criteria in the core courses that reached 88% by end of Fall semester 2013-2014, the fact that has improved students' attainment in these outcomes. Moreover, the "ABET project"t contributed in strengthening the communication with the local market. This was translated in forming the Industrial Advisory Boards (IAB) in departments and increasing the number of graduation projects implemented jointly with the local market which has exceeded the 20 % in most of the engineering programs. Furthermore, several developments were made in the laboratories and facilities in the engineering departments. These developments comprised safety precautions, adding new equipment, and increasing the number of the Teaching Assistants that made the students experience much more effective and enjoyable.

In this paper a case success study from the chemical engineering program will be presented.

Finally, as for any other quality systems, success can't be achieved without real staff involvement, sound top management support, and existence of motivated employees who are eager to continuously improve the overall performance.

392297 Hazards Associated with Chemical, Physical and/or Biological Processes in the ChE Curriculum

Abstract:

The new chemical engineering program criteria state: *The curriculum must provide a thorough grounding in the basic sciences, including chemistry, physics, and/or biology, with some content at an advanced level, as appropriate to the objectives of the program. The curriculum must include the engineering application of these sciences to the design, analysis, and control of chemical, physical, and/or biological processes, including the hazards associated with these processes*. Solicited information from program evaluators and department chairs regarding this topic will be analyzed and discussed.

383643 The Fundamentals of Engineering (FE) Exam Converts to Computer Base Testing (CBT) and the Impact of the FE-CBT As an ABET Assessment Tool

Abstract:

Traditionally, the Fundamentals of Engineering (FE) exam involved an 8 hour test using pen and paper. In January of 2014 the pen and paper examination process was terminated and converted to computer base testing (CBT) exam. Historically, this exam has been used as part of the academic ABET assessment process, being one of the only direct measure of a student mastery of engineer skills. While the FE exam still serves as an excellent assessment tool, the chemical engineering academic community must be informed of the exam changes in order to take full advantage of the FE as an assessment tool.

This presentation will present the new CBT examination process for the FE. In addition, the presentation will present the impacts of these changes on the usefulness of the FE, CBT as and assessment tool, from the prospective of an educator attempting to use the FE-CBT for assessment, a *National Council of Examiners for Engineering and Surveying* (NCEES) examination preparer, and an ABET- Program Evaluator (PEV).

386396 Accreditation at Rose-Hulman: Lessons from a Recent ABET Visit

Abstract:

This presentation will discuss best-practice lessons gleaned from various resources on the occassion of our program's recent ABET accreditation visit. In our case, there was a change in department head between the data collection/self-study preparation and the visit years; I will highlight the importance our data collection and continuous improvement procedures played in ensuring a smooth transition and a successful visit.

The occasion of the visit also afforded our department an opportunity to revise our program educational objectives (PEOs), and to demonstrate the manner in which our department was satisfying the recent program criteria revision requiring inclusion of hazards and process safety. I will discuss the procedures we used to review and modify our PEOs, and I will also give an overview of how we integrate education in process safety and hazard analysis throughout our curriculum, with tips on how to ease the job of program evaluators attempting to document evidence for the new hazard requirement in the chemical engineering program criteria.

386615 Enhancing Hands-on Problem-Solving Across the Chemical Engineering Curriculum

Abstract:

Our chemical engineering curriculum contains at its "core" a rigid sequence of required courses that need to be taken in a precise order. Traditionally, this sequence is delivered as a set of lecture-based courses. This means, especially in the junior-year sequence, there is little to no opportunity for hands-on laboratory work to supplement the core engineering, mathematical, and scientific theory. By introducing a deliberately designed series of hands-on activities that increase in complexity as students progress through the curriculum, we seek to improve student ability to make connections between interrelated fundamental theories. The primary objective of this project is to investigate the effects on student learning and retention of knowledge and skills as a result of threading activities through a sequence of required courses.

At our university, one obstacle to successful supplementation of lecture with hands-on activity is a lack of proper equipment. The material used in a sophomore-level experimental methods course is deliberately comprised mostly of household items, so that emphasis is on experimental design. The equipment used in the senior capstone laboratory has a focus on effective writing about complicated theory, and the laboratory space consists of a set of disconnected pieces of equipment to examine one specific aspect of theory from sophomore and junior courses.

We have designed and constructed a customizable, multi-function laboratory apparatus, deliberately designed to allow for the integration or segregation of modern chemical engineering equipment, including pumps and piping systems, heat exchanger, and process control hardware and software. The equipment is used directly in the required chemical engineering laboratory courses, plus as a new component to traditionally lecture-only courses in the sophomore and junior sequences. Modifications are made to each of five courses directly involved in the use of this equipment to make more explicit the connections between important chemical engineering concepts.

We outline here the timeline for implementation of this project, the collection of attitude and assessment data from students in our program, and preliminary results.

387201 Engagement, Teamwork, and Transfer: A Comparison of Physical Laboratories and Virtual Laboratories

Abstract:

Laboratory courses are an important part of engineering curricula, allowing students to act as practicing engineers by engaging in a hands-on experiment on a physical reactor, working as a team and applying the theory and concepts they have gained in other classrooms. With the advent of technology, virtual laboratories have become a complement to physical laboratories. Virtual laboratories allow students to work with complex, large-scale processes that may not be available in the classroom physically, due to space, time and monetary constraints. They can also be used as a scaffolding tool, reducing the students' focus on making the physical equipment work and allowing them to focus more on their experimental strategy, conceptual understanding and data analysis. This work describes survey results from a cohort of students participating in laboratory course that includes both physical laboratories and a virtual laboratory in order to investigate how the students' perceptions differ between the two. For example, how does student engagement differ in physical and virtual laboratories? Do students approach collaboration and cooperation similarly in the two settings? Is it easier to apply or even recognize the need for previous knowledge in a physical or in a virtual laboratory?

Throughout the course, 133 students worked in teams to complete three different projects. Two projects are based on physical laboratories, including heat exchange and ion exchange experiments, and one is based the Industrially-Situated Virtual Laboratory, using either a chemical vapor deposition reactor or a bioreactor. Through a set of five surveys, we investigated the overall change in students' views and compared the students' perceptions of each of the three laboratory projects. The survey results were analyzed using principal component analysis to determine the significance of and correlation between factors such as engagement, teamwork, agency and transfer in both the virtual and physical laboratories. By analyzing the pre- and post-course surveys, we consider if the laboratory course as a whole affected the students' interest in engineering and their view of success. Analysis of the post-laboratory surveys allows for a comparison of student perceptions of physical and virtual laboratory experiences. We found that when students work in virtual laboratories, they are more engaged, find it easier to work in teams and are able to better transfer past knowledge, thus demonstrating that virtual laboratories are an engaging, realistic component of the engineering laboratory curriculum.

388630 Developing a New Undergraduate Laboratory Curriculum

Abstract:

The paper will describe the implementation of a new practice based curriculum which has been developed during the past 3 years at Imperial College in London for our chemical engineering undergraduates.

The programme centres on two major new teaching facilities which we have developed.

The first was the design and installation of a 4 storey high industrial quality pilot plant in the heart of our Department. This allows innovative and industrially relevant training to be delivered to our undergraduates in an academically rigorous environment.

Our Pilot Plant facility includes a number of unique teaching innovations :

- Pilot Plant can be operated by iPad's by the students or lead learners, in the plant space or indeed anywhere in the world.

- Pilot plant control software has student configurable process control options, allowing a range of control strategies to be evaluated during teaching by the students.

- Plant has 250 sensor/instruments, approximately 3 times the normal number used. These extra instruments allow us to deliver a number of unique teaching and learning experiences. For example, we have a number of process lines with multiple flow measurement technologies deployed in series allowing students to assess the real differences in alternative flow measurement technologies.

- An interactive e-learning module for student training on the Pilot Plant control system has been developed in conjunction with ABB.

- Students work in an ergometrically designed control room ideontical to what they would find in best industrial practice, with optimised human-machine interfaces.

- The Pilot Plant contains all the safety features you would expect to find on a real industrial plant included SIL rated instrumentation, permit to work system, 3 phase UPS system, independent dual processor safety controller, SMS messaging, multi-lingual emergency paging, alarm management system, integrated klaxon warning system and remote video visualisation of all pilot plant locations.

The second major new facility was a large dedicated UG teaching laboratory which can house over 60 students at any one time using any of the 40 experimental systems deployed in the laboratory. Supported by a new cohort of professional teaching fellows, the new laboratories provide a rich learning experience for our 1st year, 2nd year and 3rd year laboratories. These UG laboratories are themed:

- 1st Year Foundation Laboratories

-2nd Year Knowledge Laboratories

-3rd Year Discovery Laboratories

Together these two facilities are our Discovery Teaching and Learning Space which I descibe in detail this presentation and how they form a core part of our UG curriculum.

391672 Introducing Microfluidics and Chemical Processes in Micro-Devices to Chemical Engineering Undergraduates

Abstract:

Microfluidics is a new field of study with exciting opportunities for chemical and biomedical engineers. Introducing microfluidics at the undergraduate level can not only ignite the interest of young engineers to the field but also can help introduce fundamental concepts of fluid flow, mass transfer, reaction engineering and separation techniques. The present study focuses on introducing microfluidics experimental modules into chemical engineering undergraduate classes. The fluid flow module focuses on laminar flow within a simple Y shaped channel where the students can observe the flow and calculate the corresponding Reynolds number in order to understand the relationship between flow regime and device and fluid parameters. The Mass transfer module introduces the concepts of diffusion and mixing within a micro device, and the students are able to calculate the diffusivity based on their observations of the phenomena in the device. Reaction kinetics and the combined effects of diffusion and reaction are discussed in the reaction engineering module, where the students can visually monitor the reaction via changes in coloration of the reactants. The devices used for the study are made using simplified soft lithography without the need for clean room facilities and complex processing so that the modules can be easily developed and taught in the chemical engineering curriculum.

A Visual Demonstration of Process Control Principles

AIChE National Meeting

Thomas O. Spicer, III
Professor

Chase Swaffar
Research Assistant

Ralph E. Martin Department of Chemical Engineering
18 November 2014

Background

- Students often struggle with basic concepts in process control.
- Objectives:
  - Demonstrate the connection between controller (software) and valves/sensors (hardware).
  - Use heuristics to tune a control system.
  - Show the destabilizing effects of dead time.
  - Demonstrate how the stability of a 2x2 system changes depending on operating conditions.
  - Reach the objectives without saying the words "Laplace transform."

Background (2)

- System design:
  - Develop a visual system that goes beyond plots of variables as a function of time and illustrates basic control principles.
  - Focus on smaller scale apparatus that allows quick experimental response (a few minutes).
  - Develop the system so it does not seem trivial to students.
  - Provide support to reach the objectives without requiring mathematical transient analysis or turning the experiment into a trivial game.
  - Trade off: industrial hardware/software v. small scale implementation
  - Spencer (2009) CEE

Mixing System

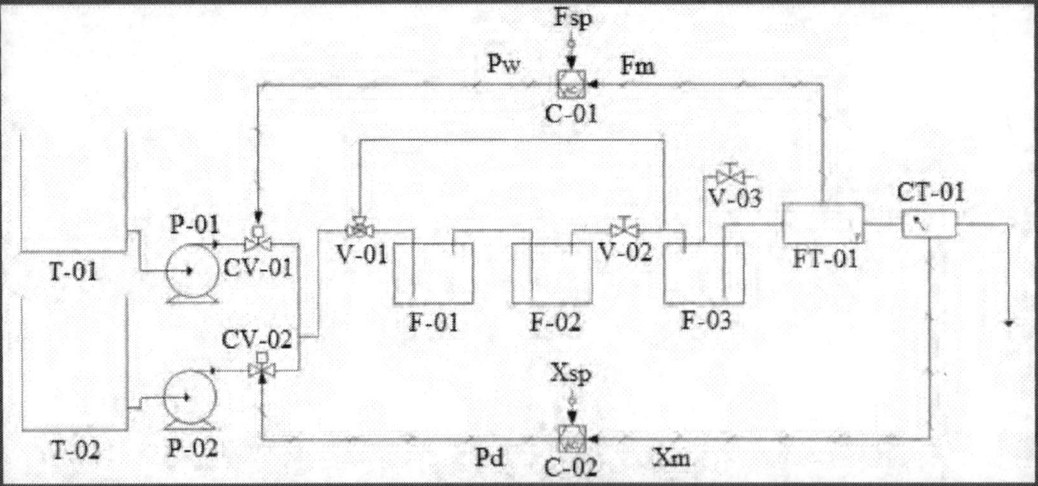

Equipment list

- 20 liter Nalgene Polyethylene Carboys
- Shurflo AC Magnetic Drive Centrifugal Pumps; Model #: 8020-503-250;45 psi internal bypass; 1.4 gpm
- 290 mL Erlenmeyer flasks
- Kelly Pneumatics High Flow Mini Proportional Valve and Driver Board; Model #: KPIH-TPW-20-90-50 (2.9 L/min); Model #: KPI-TPW-20-60-50 (0.45 L/min)
- Flow Technologies Omniflo Turbine Flow Meter with Linear Link Model #: FTO-4NINWBLHC-1
- Unico 1100 Spectrophotometer (methylene-blue dye at 640 nm)
- Data Acquisition (DAQ) module (NI USB- 6009)
- LabView

Before Experiments

- Safety overview
- Steady state analysis of the system
- Basic process control terminology:
  - Controlled variables (flow rate or concentration); manipulated variables (valve positions)
  - Set point (controlled variables)
  - Concept of gain (output/input)
  - Concept of dead time (pipe flow, sensor response)
  - PID control
  - Disturbances/set point changes (overshoot)
  - Gain scheduling (depending on operating conditions)
  - Adequacy of control

Preliminaries

- Dye solution preparation (in advance)
- Calibration of flow meters (flow rate/valve voltage)
- Calibration of spectrophotometer (%A or %T/spec voltage; dye concentration/%A or %T)
- Calibration included in lab to emphasize the importance of the process in a real setting.

Tuning Heuristics (ControlSoft, Inc.)

Make a small change in set point (~5%). Monitor CV (flow rate, concentration) and MV (valve voltages)

- If no visible instantaneous change in CV or no overshoot (loop is overdamped), increase controller gain by 50%. Repeat.
- If the MV is unstable or oscillates with overshoot greater than 25%, reduce controller gain by 50% and increate integral reset rate by 50%.
- If MV continues to oscillate but tolerable overshoot, decrease gain by 20% and reduce integral reset rate by 50%.
- If 3 or more peaks after set point change, reduce integral reset rate by 30% and increase derivative gain by 50%
- If MV is flat and offset for a long time, increase integral reset rate by 100%.

Experiments/Exercises

- Tuning First Order system (single flask)
  - Proportional control only
  - Set point change will generate offset
  - PI control – tune by heuristics; offset eliminated (demonstrate with set point change)
  - Disturbance rejection (switch off power to a pump)
  - PID control – improvement over PI?

Experiments/Exercises (2)

- Destabilizing effect of dead time (FOPDT)
- Turn valve so flow is through all three tanks in series (dead time in first two tanks)
- Unstable system
- Bring back to control and use heuristics to tune FOPDT system.
- (Resulting settings can be compared with estimate methods such as Direct Synthesis)

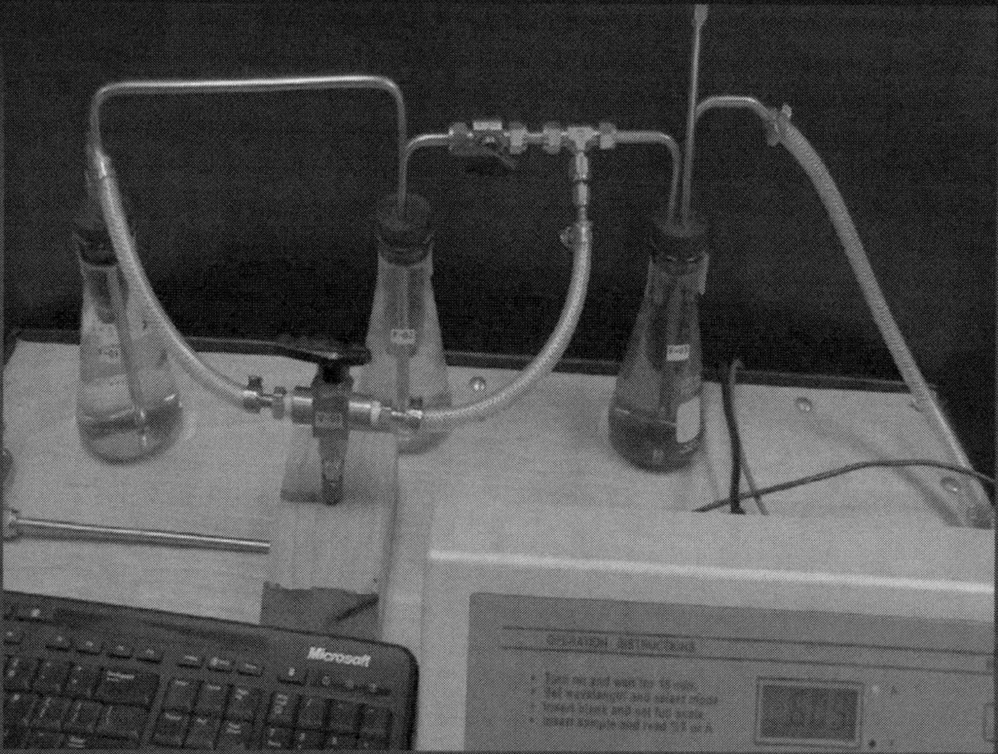

MIMO – 2x2 Pairing

- Relative Gain Array can be derived for this system from steady state mass balances.
- For the dilute concentration range, preferred pairings are:
  - Total flow – higher flow rate (water)
  - Dye concentration – lower flow rate (dye)
- For the concentrated solutions, preferred pairings are:
  - Total flow – higher flow rate (dye)
  - Dye concentration – lower flow rate (water)
- Demonstrate that if pairings are incorrect, system impossible to control.

Conclusions

- A system has been developed to visually demonstrate:
  - How to tune a control system using heuristics.
  - Destabilizing effect of dead time.
  - Importance of pairing inputs/outputs in a simple system depending on operating conditions.
- Total cost of the system $3500 (80% of the cost in the flow meter and the spectrophotometer)

Questions?

Tom Spicer, PhD, PE
Professor and Barker Chair in Chemical Engineering
Ralph E. Martin Department of Chemical Engineering
3202 BEC/CHEG
University of Arkansas
Fayetteville, AR 72701

479-575-6516
479-575-7926 fax
http://www.cheg.uark.edu

Engaging High School Students in Chemical Engineering through Hands-On Laboratories

In many rural schools, there is a lack of opportunities for high school students to participate in advanced science programs, especially those that expose them to the fields of engineering. However, this is a critical time period for students, as they formulate ideas about career choices and make curricular decisions that may affect future college paths, major choices, and ultimately the number of students pursuing engineering careers. Given the location of Cornell in central upstate New York, we have an opportunity to impact students in rural communities and encourage the next generation of engineers. To this end, for the past five year the graduate students of Cornell's School of Chemical and Biomolecular Engineering have organized and hosted a day-long program for rural tenth grade girls and their parents that has grown from a small pilot program to a large event involving 40-50 families and nearly thirty volunteers. The program, WOMEN (Women's Outreach in Materials, Energy, and Nanobiotechnology), is aimed at fostering the girls' interest in STEM fields through a series of short, interactive laboratory modules led by graduate students and focused on different subject areas: Materials, Process Engineering, Biology/Biomolecular Engineering, and Fluid Mechanics. WOMEN also encourages the students to begin considering college and career paths and provides their parents with information about college admissions and financial aid while their daughters participate in the laboratory activities. Based on survey feedback, the event has been highly successful in broadening the participating girls' view of engineering and increasing their reported interest in STEM careers. The students respond especially favorably to hands-on activities and explanations of scientific ideas that tie back to their own experiences. In this presentation I will provide an overview of the entire program, highlight some of the experiments run by the students, and discuss feedback on the program and how suggested changes have been incorporated to improve the program over the years.

Use of Combination Carbon-Dioxide/Oxygen Off-Gas Sensor to Monitor Aerobic Fermentation

A BlueSens "Blue-In-One" combination CO_2/O_2 sensor was used to monitor aerobic *Saccharomyces cerevisiae* fermentations for the purpose of undergraduate laboratory development. Batch fermentations with different aeration rates and substrate concentrations were conducted to determine optimal conditions for cell yield or ethanol production. Off-gasses were measured over the course of each fermentation, which allowed for real-time metabolic characterization of the culture. At a constant airflow rate, it was possible to calculate the carbon dioxide evolution rate (CER) as well as the oxygen uptake rate (OUR), and dividing the CER by the OUR gives the respiratory quotient (RQ). This information can be used to gauge the extent to which glucose is utilized aerobically, as high substrate concentrations can lead to glucose being fermented to ethanol even under aerobic conditions (known as the Crabtree effect). Therefore, industrial yeast production is almost always conducted in fed-batch mode to keep the substrate concentration low, lessening the Crabtree effect and increasing cell yield as opposed to producing ethanol. The RQ indicates the Crabtree effect when greater than unity, and varies between 0.7 and 1 depending on the oxidation level of the substrate being utilized by the culture. In addition, the total CO_2 evolved can be used in conjunction with ethanol and cell yields to complete a carbon balance for each batch. The impacts on student learning objectives, as well as the application of these tools to commercial fermentations in industry, will be discussed.

SIMULATING FAST CHEMICAL REACTIONS AND MIXING WITH RANDOM COALESCENCE-DISPERSION MODELING

Gary K. Patterson
Missouri University of Science and Technology
Rolla, Missouri

Motivation for Research

- Need for simple method to simulate yields for very complex reactions in a mixed vessel – tubular, fed batch, continuous flow stirred vessel
- Alternative to closure hypotheses for chemical reactions used in CFD, which can be questionable for more than one reaction
- Could possibly be extended to two-phase reactions with mass transfer issues

Macro/Meso/Micro Mixing

C-D simulates Mesomixing, which is the predominant mechanism affecting chemical reactions in a stirred vessel

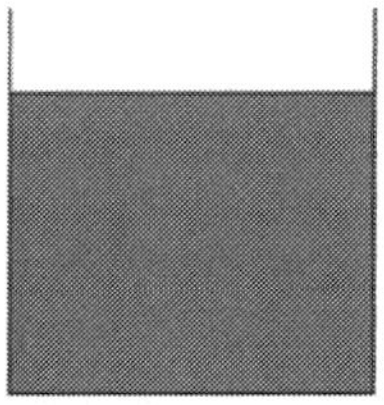

Macromixing:
-scale of the tank
-blend time

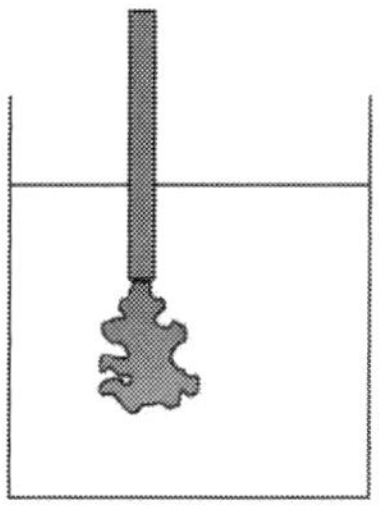

Mesomixing:
-scale of the feedzone
-Corrsin mix time

Micromixing:
-smallest scales of turbulence
-diffusion time

Hierarchy of Scales for Mixing and Chemical Reaction

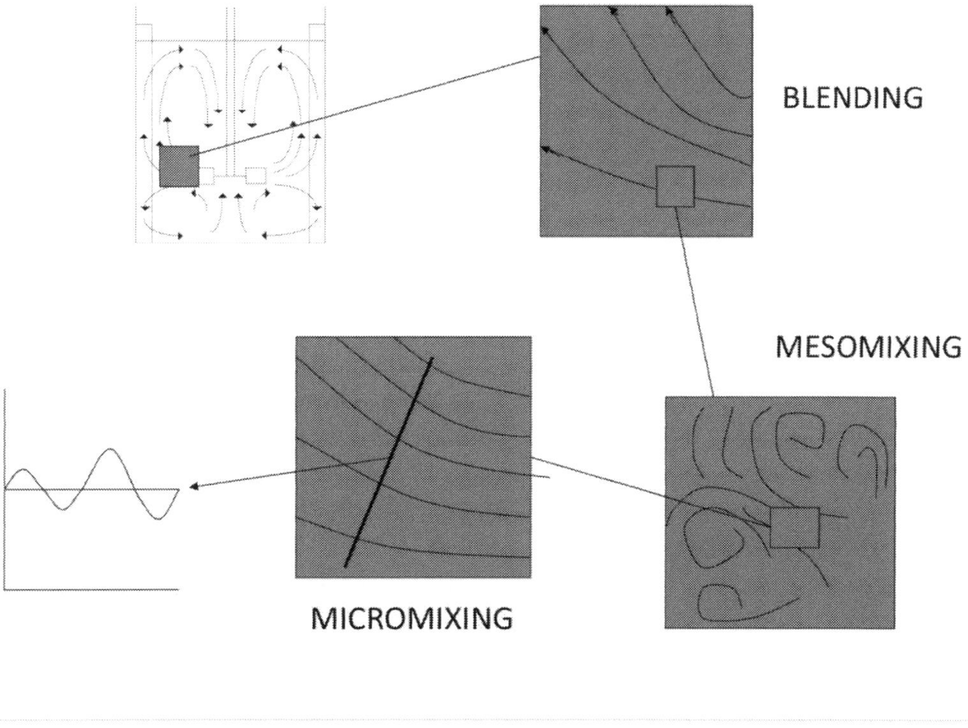

RANDOM COALESCENCE AND DISPERSION IN A MIXING VESSEL

- Let the fluid be represented by small discrete fluid elements which can move and mix with one another
- The fluid elements are moved within vessel segments and from segment to segment in accordance with probable flow rates and patterns (based on CFD or experimental evidence) at the start of a time increment
- Feed is simulated by new fluid elements moved into a selected segment
- Mixing is simulated by randomly choosing pairs of fluid elements at a computed rate for each segment (pairs chosen=[coalescences/time] X Inc. Time), then averaging their concentrations (coalescence) – then they are separated (dispersion)
- Chemical reactions are computed in each fluid element as if it is a batch reactor for the computed time increment
- Fluid elements are again moved based on flow rates

RANDOM COALESCENCE-DISPERSION MIXING MODEL

$I_{CD} = 0.1\ (\varepsilon/L_S^2)^{1/3} \equiv 0.186\ (\varepsilon/k)$; C-Ds/time/#elements

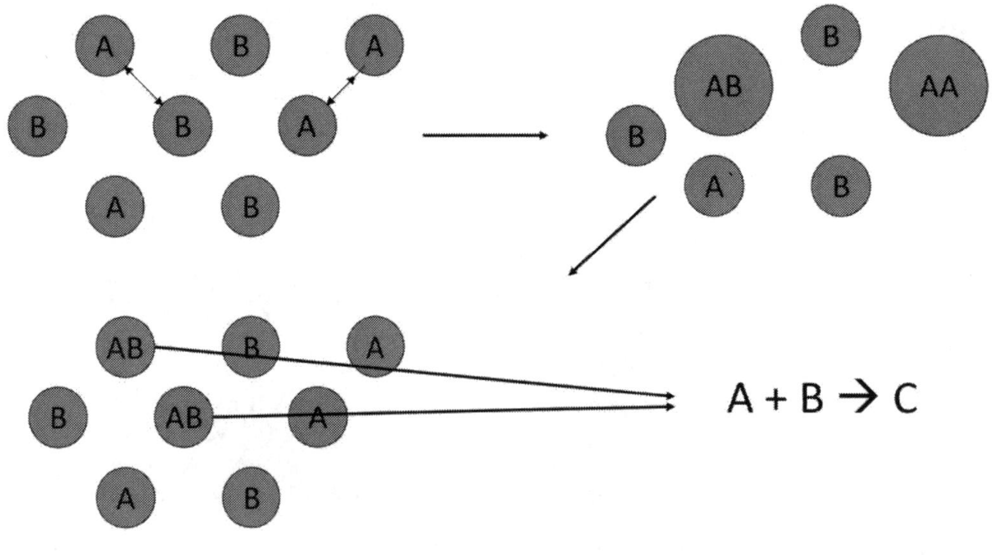

C-D Rate is Related to Mixing Rate and Fluid Dynamics

- The C-D Parameter I ~ 0.1 $(\varepsilon/L_s^2)^{1/3}$, in units of coalescences/time increment/fluid element; where $(\varepsilon/L_s^2)^{1/3}$ is approx. $-4ds_i/s_i dt$ based on Corrsin mixing theory [$ds_i/dt = -s_i/(4\,(L_s^2/\varepsilon)^{1/3})$ for meso-mixing]

- Based on comparison of simulation results with extent of mixing measured by use of a fast reaction bv Vassilatos and Toor

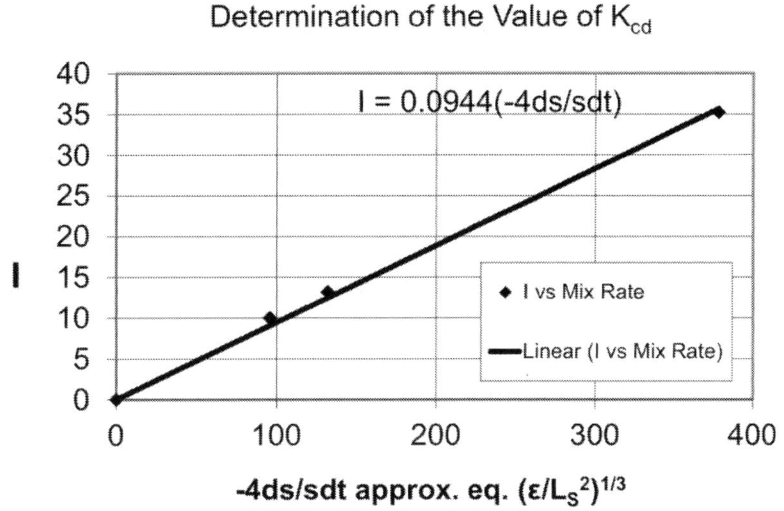

SIMULATION IN A TURBULENT FLOW TUBULAR REACTOR

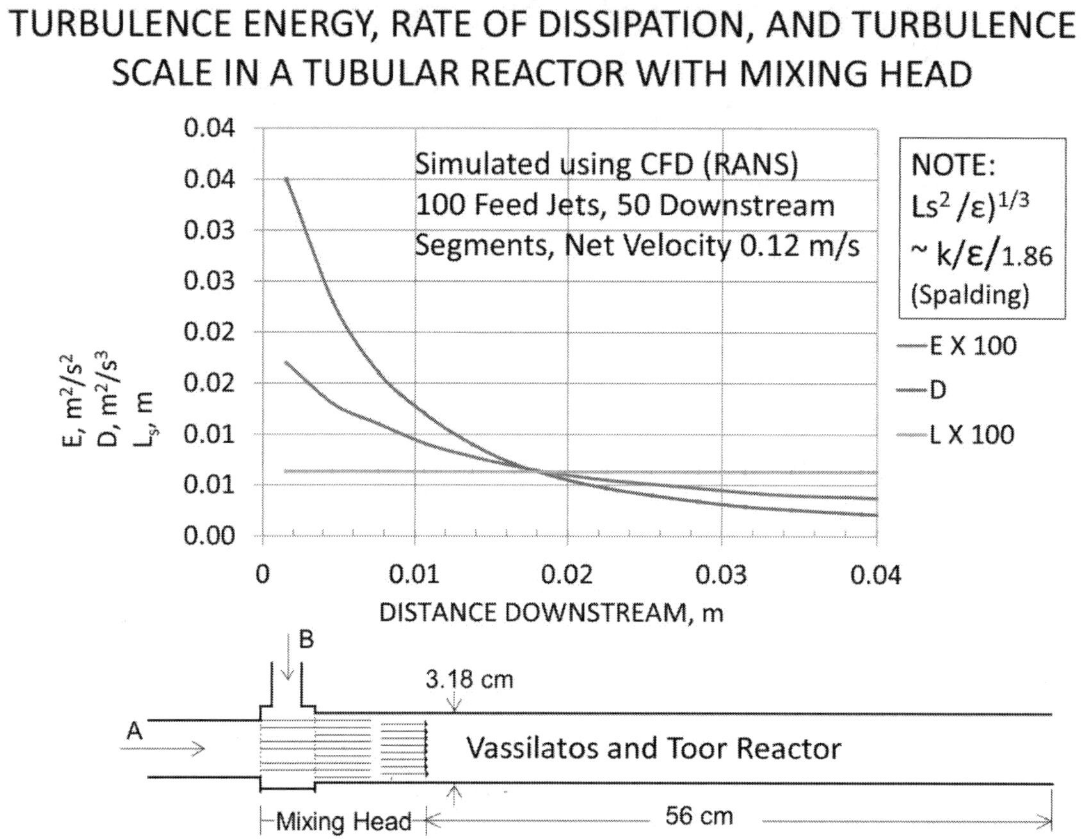

Fluid Element Flows Applied to a Tubular Reactor Using C-D

No. Elements Flowed = Vel*Cross Area*(Elem/Seg)*Delt/Seg Vol

Elements are Identified by Seg. No. and Elem. No. within the Seg.

Each Element has Properties, such as Comp.Conc., Density, Temp.

Elements Move According to First In – First Out

When Moved, Elements Change Name According to Segment Entered And Position in That Segment

COMPARISON OF EXP. DATA (V-T) WITH C-D SIM. FOR TUBULAR REACTOR WITH MIXING HEAD BASED ON C-D PARAMETER – I; 100 Elements/ Segment; 5000 Segments; U=0.12m/s; F + O→ C; K_1=12,600 L/M S

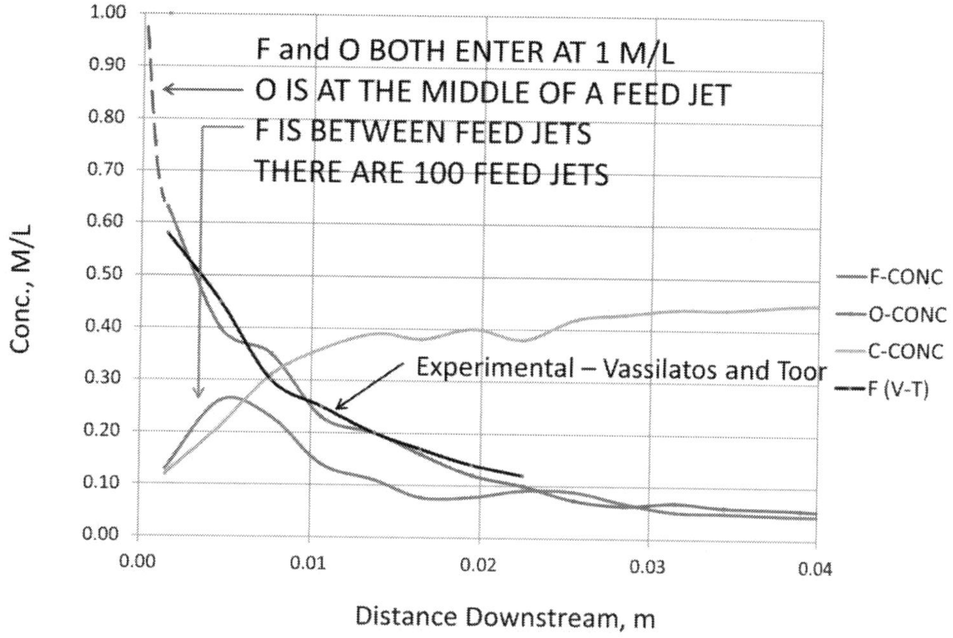

C-D SIMULATION USING EXPERIMENTAL FLOWS AND TURBULENCE IN A STIRRED VESSEL

Application of the C-D Method to a Stirred Vessel Using Exp. Flow Rates and Turbulence

- Divide the reactor into volume segments – 6 and 15 in these examples

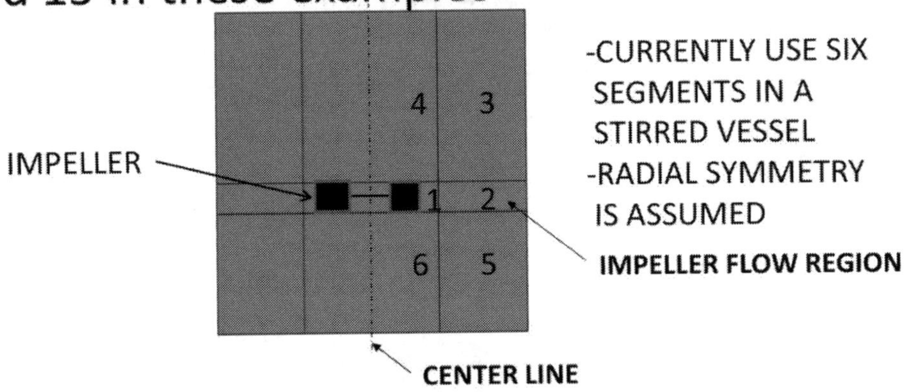

Assign a number of fluid elements to each segment, typically 100 to 500

APPLICATION OF C-D TO A STIRRED VESSEL WITH VALIDATION DATA FROM PAUL AND TREYBAL (1971)

CONCENTRATIONS OR YIELD OF C IN PAUL-TREYBAL REACTOR;
A+B-->C; C+B-->D; K_1=34 L/M S; K_2=3.8 L/M S; TOP FEED; 300 RPM; D/T=1/2

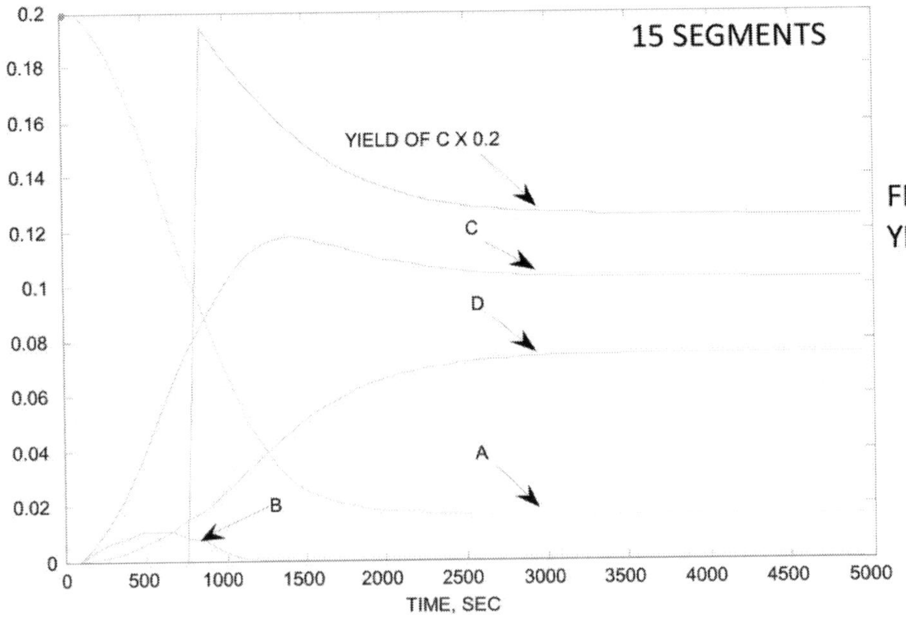

FINAL YIELD=0.62

SUMMARY OF CD-SIMULATION RESULTS FOR THE PAUL-TREYBAL REACTOR (TALL 36 L TANK) COMPARED WITH EXPERIMENTAL RESULTS

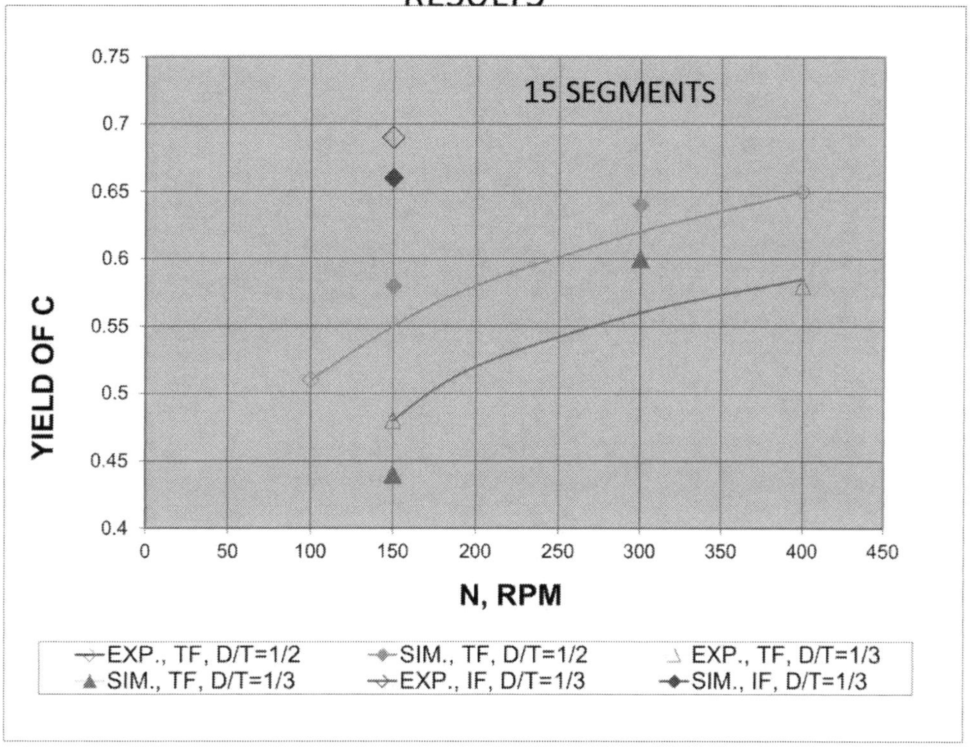

EXAMPLE SIMULATION WITH LARGE SEGMENTS AND USE OF EXPERIMENTAL FLOW DATA (EXP. DATA MIDDLETON et al. 1983)

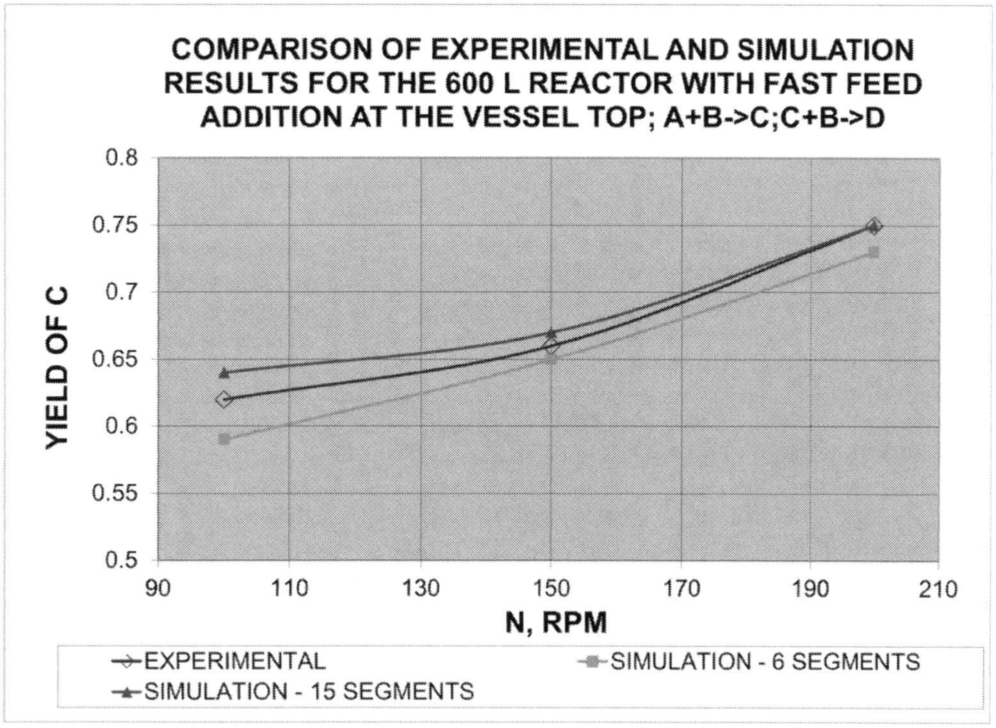

C-D SIMULATIONS USING CFD FOR THE FLOWS AND TURBULENCE

FLOW PATTERN IN STIRRED VESSEL WITH DISK TURBINE; FEED OF F NEAR IMPELLER TIP SHOWN IN LIGHT GREEN; C-D USED FOR CONCENTRATIONS

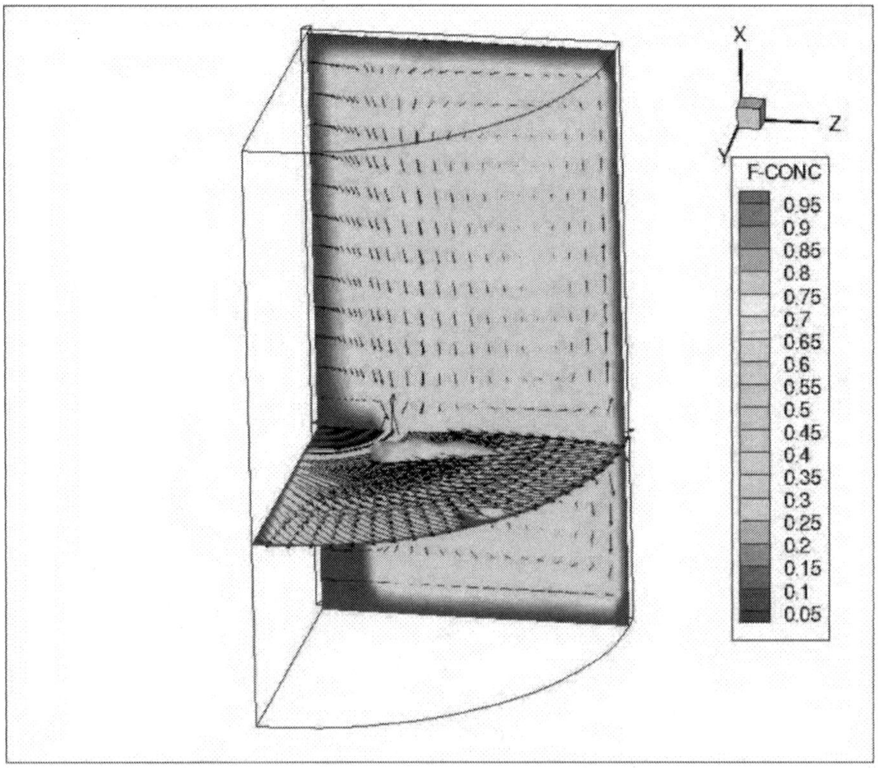

Flow Pattern and Concentrations of Feed into a Fed-Batch Vessel: Impeller Feed

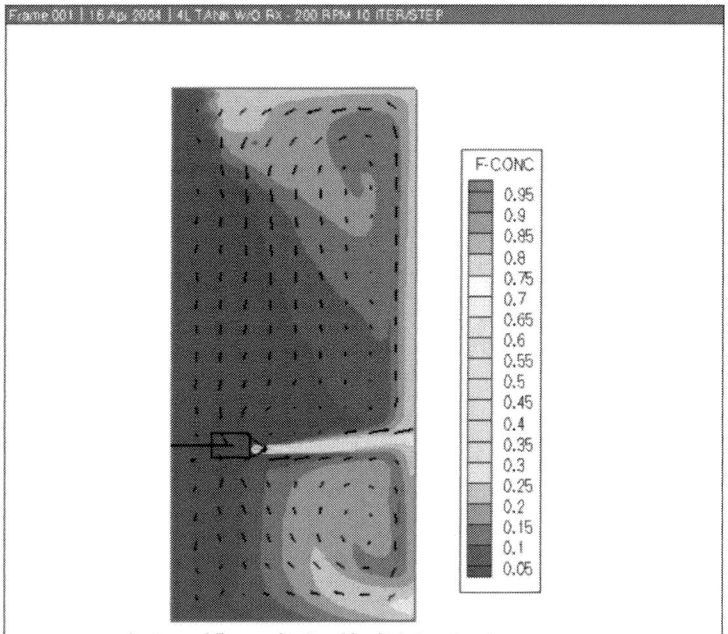

Contours of F-conc after 5 s of feed into impeller stream, 200 rpm

CONCENTRATION DISTRIBUTION FOR PRODUCT-C DURING FEED OF F FOR SINGLE REACTION F + O → C; C-D USED TO MODEL COMPONENT FLOWS AND REACTION

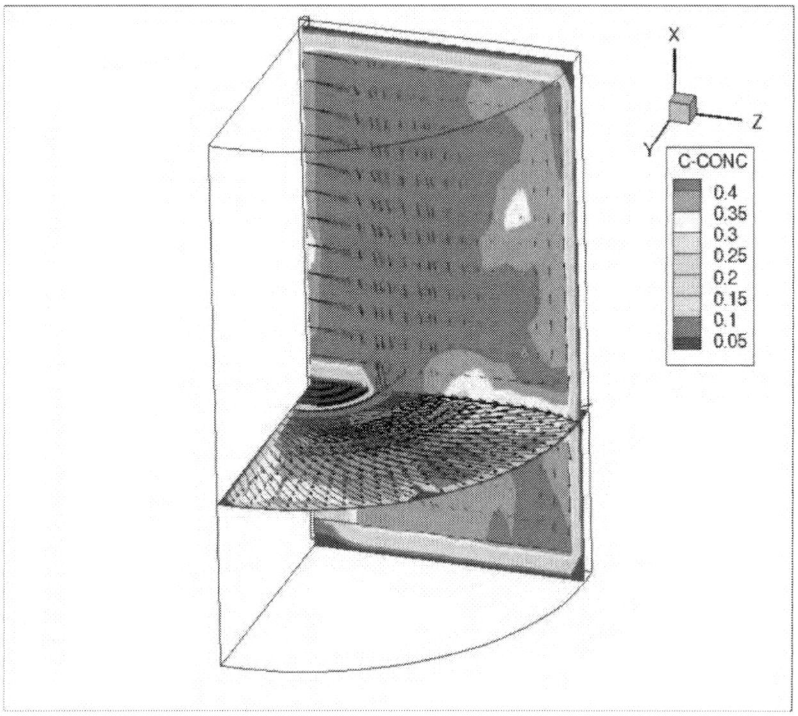

APPLICATION OF C-D TO 2-REACTION SYSTEM: F + O→ C;
O + C→ B; K_1=34 L/M S; K_2=3.8 L/M S
FAST FEED OF F INTO IMPELLER STREAM; MONITORING POINT NEAR VESSEL TOP; FEED DIA. ~6.2 mm; FEED VEL. 10 m/s

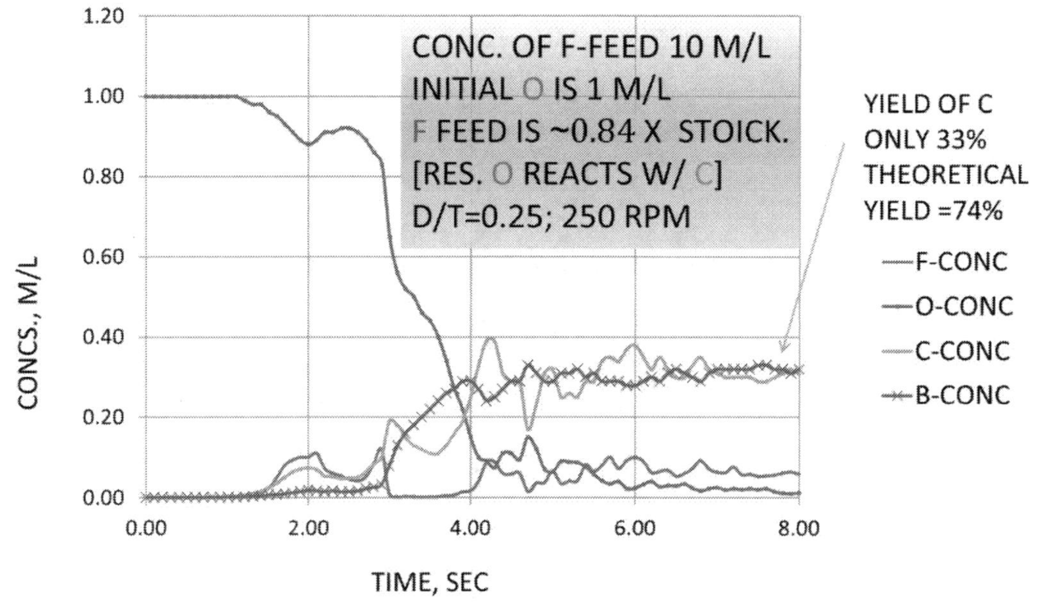

APPLICATION OF C-D TO 2-REACTION SYSTEM: F + O→ C;
F + C→ B; K_1=34 L/M S; K_2=3.8 L/M S
SLOW FEED OF F INTO IMPELLER STREAM; MONITORING POINT NEAR VESSEL TOP; FEED JET DIA. ~3.5 mm; FEED VEL. 1 m/s

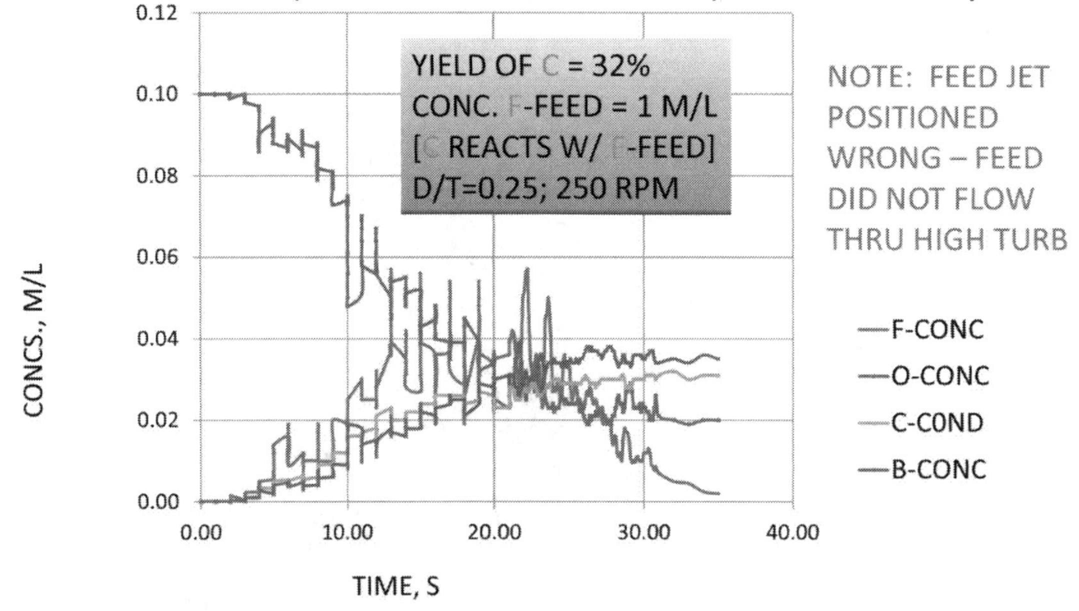

APPLICATION OF C-D TO 2-REACTION SYSTEM: F + O → C;
F + C → B; K_1=34 L/M S; K_2=3.8 L/M S
SLOW FEED OF F INTO IMPELLER STREAM; MONITORING POINT NEAR VESSEL TOP; FEED JET DIA. ~3.5 mm; FEED VEL. 1 m/s

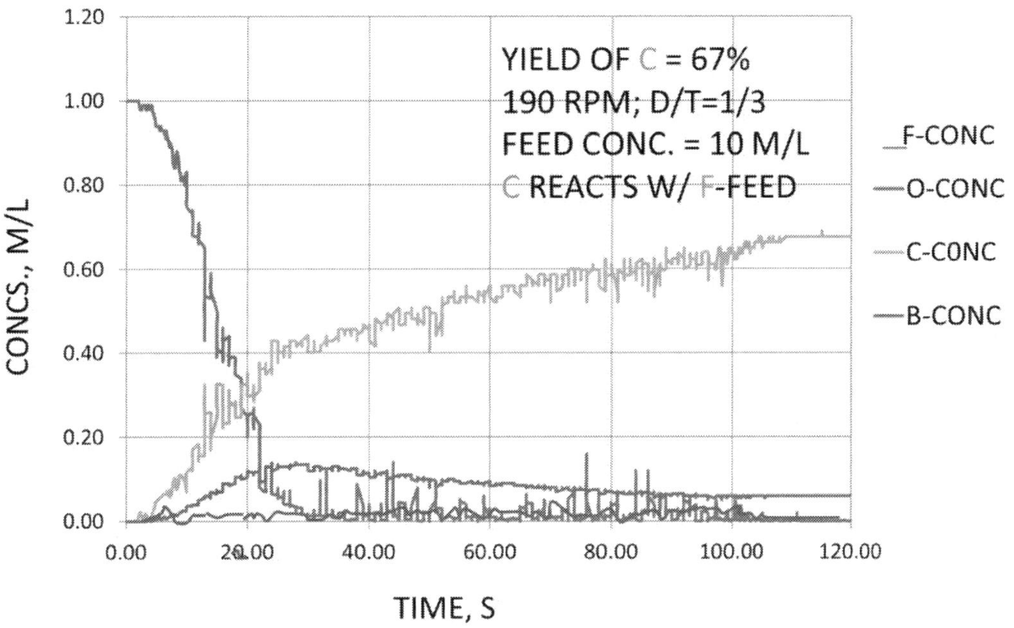

PITFALLS TO AVOID IN C-D SIMULATION

1. POOR POSITIONING OF FEED JET
2. TIME INC. TOO LARGE OR NUMBER OF ELEMENTS TOO SMALL FOR SIMULATED FLOW
3. TIME INC. TOO SMALL FOR SIGN. NO. OF C-Ds
4. CFD SIMULATION NOT SUFFICIENTLY ACCURATE, FOR INSTANCE, FLOW PATTERNS INCORRECT OR TURB. ENERGY AND DISS. TOO LOW OR TOO HIGH
5. POOR INTEGRATION OF KINETICS EQUATIONS – STIFF INTEGRATION METHODS GENERALLY NEEDED
6. LACK OF MATERIAL BALANCE – LOST ELEMENTS OR ELEMENTS MISDIRECTED

Results from Paul and Treybal (1971) Experiments Compared with C-D Simulation and Paired-Interaction Closure (Patterson, 1973); F+O→C; F+C→B

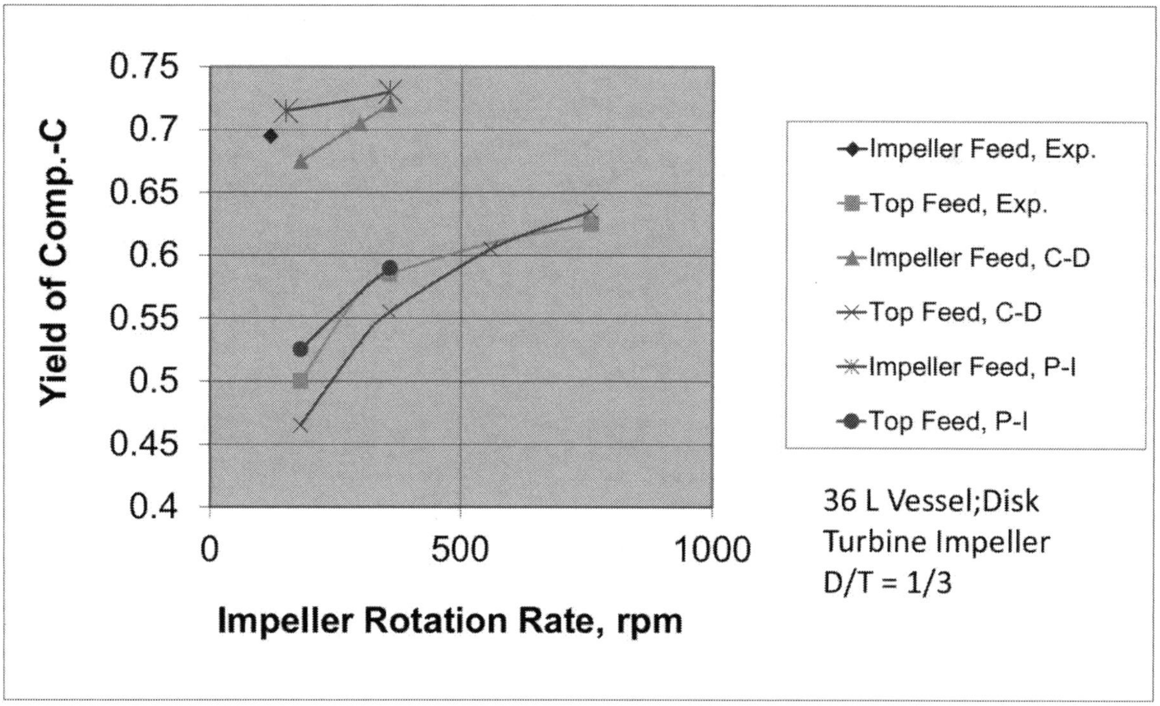

CONCLUSIONS AND PLAN FOR FURTHER WORK

THIS IS A WORK IN PROGRESS:

1. DO MORE TEST SIMULATIONS WITH VALIDATIONS FROM EXP. WORK BY MIDDLETON, ET AL., BOURNE, ET AL., DOSHI, AND PENNEY, ET AL.
2. EXTEND TO MORE COMPLEX REACTIONS: THREE OR MORE AND THOSE WITH A SECOND PHASE
3. CLEAN UP CFD – CD PROGRAM FOR USER FRIEND-LINESS; ADD BETTER INTEGRATION ROUTINE; IMPROVE PREDICTION OF VOLUMETRIC FEED RATE
4. RE-WRITE C-D SUBROUTINE FOR USE IN ANY CFD PROGRAM

379927 Development and Implementation of a Technical and Professional Communication Course for Chemical Engineers

Abstract:

Communication is an inherent component of any career path. For engineers, the ability to communicate technical information to a variety of audiences is critical, but underemphasized in many programs. In many United States institutions, engineering students are required, or given the option, to take a technical communication course early in their curriculum. While this course often provides the necessary base for successful technical communication, in most cases it is taught by someone from an English/communication discipline, and the course content is not integrated into the rest of the curriculum nor applied in an engineering specific context. While technical courses may incorporate writing into their course design through short answer responses or lab reports, evaluation of these materials focus on technical content as opposed to presentation. It is, however, clear and engaging presentation that can set the young engineer apart from others in their career after graduation. For this reason, we have developed a pilot course, Technical and Professional Communication for Chemical Engineers at The Ohio State University, offered for the first time in the Spring 2014 semester. This course is a technical elective that builds upon the required English/communication courses and covers technical and professional communication in a discipline-specific context. The course focuses on personal development of communication skills through a variety of writing exercises, reports, presentations (oral and poster), and development of professional materials (e.g., resume, cover letter, executive summaries). Emphasis of the course is on analyzing the audience and evaluating the purpose of communication in order to choose appropriate content and format. The course structure is mixed-format, combining short lectures with activities that provide opportunities for immediate practice. Focus is on inductive learning, allowing students to experience communication and professionalism principles firsthand. Because the nature of the subject matter makes it highly likely that students will enter the course with strongly developed schemata for many topics, inductive learning allows them to begin developing and adjusting their existing schemata based on their own conclusions. We believe this method is more effective in creating long-term impact on the students' communication and professional skills, as well as exposes them to techniques that reinforce life-long learning. While originally listed as a one-time offering, the course is slated to be offered again in Spring 2015 and we hope to make it an established component of our department and/or the college. Student feedback from the pilot offering indicates that there is a great deal of value for the students in this course. There continues to be a perception of engineers being poor communicators, and recognition that young engineers are in need of further developed communication skills. We believe this course is the beginning to developing a curriculum that better integrates an emphasis on technical and professional communication for engineers, a topic that continues to grow in importance as our means of communication evolve.

381934 Developing Professional Skills in Engineering Graduates through the Use of Scenario-Based Elearning

Abstract:

Recently, there has been a growing demand for Australian Engineering Graduates with better employability attributes. From this, there has been a movement towards implementing learning tasks based on a process of enquiry in order to develop these attributes at an undergraduate level. Scenario-Based learning (SBL) is a learning method based on the same principles as Problem-Based Learning (PBL) but also incorporates eLearning with realistic work-based challenges and is therefore a useful tool in developing employability skills, such as practical knowledge, communication and teamwork, in engineering students.

This gives students the opportunity to develop their problem solving skills and knowledge whilst bridging the gap between a discipline and its practical applications in industry. Scenario based e-learning can also address the challenge of motivating students as interactive scenarios are far more involving than tutorial problems and can show the practical applications of knowledge.

A tool for implementing SBL is the software, SBLi that allows students to interact within an online scenario. This work looks to implement SBLi in to a third level Process Design unit at Monash University. This will be achieved by creating a series of realistic industrial scenarios, which are then designed and developed using SBLi, before being delivered to the students (a cohort of approximately 130 students). A preliminary evaluation of the use of the scenarios will be carried out through a focus group. This will provide feedback to refine the scenarios before they are formally incorporated in the delivery of the unit Process Design. Results indicating the level of student engagement through monitoring the use of the scenarios and the students' perception of their effectiveness for skill developmen by using SBLi will be presented.

386254 Peer Assessment As a Means of Improving Scientific Writing

Abstract:

Writing is one of the marks of human civilisation as a means by which humans communicate with each other. Besides communication of a finalised piece of research, written work is the basis for further opinions, views and critiques from fellow professionals and academics separated by time and distance. Most importantly, it represents a permanent record of scientific work that has been completed. While keeping records is quite easy to do, writing scientific reports or papers can be quite difficult, and especially for undergraduate students who have never done so before. The importance of good scientific communication is indisputable, and good writing skills are a key attribute for careers both in academia and in industry. It scarcely matters how good the technical work is, if its significance is not communicated clearly and accurately.

We train our students in scientific writing by having them write reports regularly as part of their undergraduate and graduate programmes. Students normally write draft reports for their supervisors, and receive feedback on these. They do not always properly understand the feedback, however, as they rarely have a clear appreciation of what a good report should look like. Even if they are shown examples of previous reports they still struggle to see their own work in relation to the work of others. They also find it very challenging to comment on the writing of others, let alone their own.

This work focuses on the use of peer assessment to improve scientific writing in the 4th undergraduate year, or Senior year, at the Department of Chemical Engineering, UCL in London, UK. The peer assessment has been running for 4 years so far and has proven extremely valuable in improving the students' writing skills. The students take two compulsory modules which are mainly assess based on scientific reports, each worth 25% of the final year, focusing on design and research, respectively. The design module is predominantly based on group work, whilst the research project is individual. Both modules run over two terms, and for the same cohort of between 35 and 50 students. In the design module, each group of six students will regularly provide feedback on a draft report from another group. The draft reports are typically 25-30 pages long and cover different aspects of the design and assessment of a control system of a given chemical plant. The feedback is a one page written commentary to the draft report and should cover both technical and presentation aspects. The students do not provide a mark for their peers, and their assessment is hence formative. The students are given the choice at the start of the academic year of whether or not they should also allocate a mark, but have so far every year unanimously decided not to do so. The one page commentary is assessed by the module tutors, and contributing to the peer assessment is therefore necessary to complete the module.

In the research project module, the students work independently on a piece of research in collaboration with an academic supervisor and sometimes also a PhD student or postdoctoral researcher. The student will meet regularly with the supervisor, but is otherwise working on his or her own. The students have to submit an interim report after the first term (4,000 words), outlining the existing literature and the research aims and objectives of their project. The report is not formally assessed but will typically become the first two sections of their final report (10,000 words) which is submitted at the end of term 2. The draft reports are reviewed by a group of four or five students, of which the author is one, together with a supervisor. The students all read each other's reports beforehand, and in the meeting, discuss the relative merits of each draft and together highlight best practice in terms of writing. They also all summarise which changes they will be making to their own drafts following the feedback from their peers.

In this contribution, the detailed module structures, and the motivations behind, will be presented, and examples will be given of the different modes of assessment. The results from end-of-module student surveys will also be included which show that the students find the peer assessment in both modules extremely helpful, and highly relevant to their studies and their future careers in terms of improving their abilities to write clear and concise technical reports.

Innovation and Internationalization in Chemical Engineering: Preparing 21st Century Engineers though Skill Development Modules

21st century engineering education increasingly aims to teach students to be innovators, rather than consumers, of knowledge. Unfortunately, this aspirational language is rarely manifest in courses that have been consciously adapted to these shifting objectives in engineering education. To prepare Carnegie Mellon University undergraduate chemical engineering students as innovators required a course that presented innovation as an outgrowth of engineering fundamentals, while cultivating the professional skills inherent in successful innovators, and while contextualizing innovation within the gobal context of the 21st century. This talk will describe a course entitled "Water Technology Innovation and Policy" designed to blend this dual need for foundational knowledge and professional skills development.

Course Description
Innovation in water technologies is necessary to confront profound water resource challenges facing countries around world. Students successfully completing this course were able to discuss the factors and conditions that drive innovation in the water sector. Students began by describing and classifying the historical drivers for innovation in water treatment, including technical, economic, and regulatory drivers. After an introduction to the fundamental principles of water treatment technologies, students identified present day technology shortcomings and distilled these into discrete design objectives. Students then formulated and answered quantitative and qualitative questions that respond to these design objectives by leveraging their knowledge of engineering fundamentals, regulatory tools, and pricing policies. Comparing their own solutions with those proposed in the peer-reviewed academic literature in engineering and the social sciences, students evaluated the technical feasibility, usability, and social desirability of proposed water innovations in developed and developing countries and summarized their findings through technology pitches and policy briefs.

Course Objectives: The objectives of this course were three-fold.

1) Connect engineering fundamentals to the invention and innovation process.

Fundamental concepts in kinetics, mass transfer, and unit operations, topics familiar to most junior and senior engineering students, were reviewed in the context of their contributions to the invention and innovation process in water treatment. The course covered a series of five major innovations in water treatment, including filtration, disinfection, desalination, water reuse, and technologies to support water efficiency.

Each module began with a single class period, lecture-based introduction to the process fundamentals underlying the water technology innovation. Class time in the proceeding 4-5 classes of the module combined A) a professional development module and B) lecture and case study techniques, where students worked in small groups to formulate potential engineering design solutions to historical water treatment challenges. For example, the first 20 minutes of class might have reviewed forward osmosis membrane separation processes presented in homework readings and examples. The next 30 minutes, students would have been asked to work in assigned groups of 3 to 5 to identify fundamental properties limiting mass transport across forward osmosis membranes and propose modifications to membrane design that would promote the operational feasibility of this water desalination technology. The remaining 20 minutes consisted of guided discussion between groups as they compared perspectives on technological constraints and the innovations that might circumvent them. This process of discovery or investigational learning was aimed at teaching students how to translate engineering knowledge into innovative solutions.

2) Contextualize the role of engineering innovation in relation to economics, social analysis, decision making, and culture as factors in technology development and implementation.

This class also explored technology, market, and regulatory drivers for innovation in the water sector, including water scarcity, water pricing discrepancies, urbanization, public risk perception, and legislation. Basic frameworks for analyzing the range of available policy tools and technology innovation systems were presented in a series of professional skills modules. These modules included:

A) Critically Reviewing Existing Research and Design Paradigms in the Published Literature, a module focused on how to dissect and absorb academic journal articles and patents

B) Implementing the Design Learning Process, a module focused on the innovation process and how to structure innovation within a team setting

C) Pitching Your Idea, a module focused on verbally communicating innovation to corporate and governmental stakeholders

D) Incorporating Economic Analysis into Design Evaluation, a module focused on critically evaluating the costs and benefits of technology deployment

E) Making a Policy Recommendation, a module focused on how to structure a written analysis of tradeoffs in the policy decision-making space via a structured policy brief.

F) Implementing Inventions in Complex Systems, a module addressing unforeseen consequences of technology

G) Working with Actors and Institutions in Developing Economies, a module focused on the unique international context of 21st century problems.

3) Model multi-disciplinary innovation in a project setting that leverages the unique skill sets of students across majors.

This course combined theoretical, quantitative, and comparative approaches. While students began the course with vastly different levels of comfort in employing these approaches, the team-based learning structure accelerated knowledge transfer among class participants. Students worked in a rotating series of small groups during the module component of the course to promote the development of cross-disciplinary dialog.

375099 The Senior Chemical Engineering Laboratory Course As a Communications Course

Abstract:

A common "buzz" word today is "innovation", as is the phrase, "idea generation". However, little is said by those using these buzzes about how idea generation becomes innovation. Communication is the first step in converting idea generation into innovation. It matters little how good your idea is: if you cannot convince others of its value, it has no value; that is --- your idea will not become an innovation.

This point is not stressed in Chemical Engineering programs. Chemical Engineering students receive little training in writing or oral presentations, both of which are required to convert an idea into an innovation. The writing courses Chemical Engineering students take are Freshman English and Technical Writing, neither of which prepares the student for selling ideas. Selling ideas requires clear, succinct writing. Freshman English and Technical Writing teach long, convoluted sentence structures, sentences not conducive to selling ideas. Besides, the longer, more convoluted the sentence, the more likely the writer will violate English grammar. With regard to oral communication, Chemical Engineering students receive scant training.

The Senior Laboratory Practices course provides an opportunity to teach Chemical Engineering students the communication skills they will need to convert their ideas into innovations.

This talk discusses how to use the Senior Laboratory Practices course as a communications course.

379987 The Impact of Team Composition with Respect to Learning Styles on Team Dynamics in a Chemical Engineering Unit Operations Course

Abstract:

Merriam-Webster's dictionary gives one definition of team as "two or more draft animals harnessed to the same vehicle or implement." While we may more traditionally consider another definition, "a number of persons associated together in work or activity," the first definition better captures the essence of a team. Strapped to the same vehicle, so to speak, team members must work together to progress (move the vehicle). Engineering is a highly team-oriented discipline. Any lack of communication or cooperation among team members can seriously deter the team's efforts. Education, on the other hand, is focused in large part on the individual: individual learning, individual performance, and individual evaluation. Attempts to include a teamwork aspect in the curriculum are often deterred by students taking the "divide and conquer" method of splitting the work amongst themselves in order to avoid the need to work side-by-side. Some engineering courses have been designed to focus on teamwork as an inherent part of the coursework in an effort to prepare students for the careers they are about to embark upon. Given the opportunity to self-select teams in these courses, students will often form groups that are homogenous with respect to several indicators, including personality and learning style. While homogeneous groups may be preferred for specific tasks, heterogeneous groups can perform better over a more broad range of tasks. Specifically, heterogeneous teams show greater success in tasks that require creativity and innovation, two essential qualities in engineering teams. We are interested in the relationship between team learning style preferences, team performance, and the individual team member evaluation of collective efficacy in the context of a chemical engineering unit operations laboratory course. While the specific execution may differ, this course is a common component among chemical engineering programs and thus offers results we anticipate will be translatable to chemical engineering departments at other universities.

At The Ohio State University, unit operations begins with a weekly lecture during the spring semester that prepares students for the five-week laboratory course in the May term (Maymester). We are interested in the team dynamics and performance in the Maymester course. Enrollment is consistently 150 to 200 students annually. Teams are assigned by the faculty instructor at the end of the spring term lecture course, based on student preference for experimental track (students may choose from classical, biological, or environmental) and GPA. Teams typically consist of four students that rotate through specific roles (team leader, design engineer, operating engineer, and development engineer), allowing each student to experience each role once.

Our study strives to categorize the team composition with respect to learning styles (homogeneous or heterogeneous) and explore correlations between composition, performance, and team dynamics with no interference in the course. We are using the Felder-Soloman Index of Learning Styles (ILS) to evaluate student learning styles and are interested in separately analyzing each dimension of the related Felder-Silverman learning styles model (active/reflective, sensing/intuitive, visual/verbal, sequential/global) to better understand if team composition with respect to any one dimension is more significant in team dynamics. Team dynamics are evaluated through peer performance evaluations, as well as an individually administered perceptions/collective efficacy survey. The survey includes items such as: group members contributed equally to writing the lab report; I would have preferred to write the lab report myself; I believe in my group's ability to perform unit operations tasks. By administering this survey with each experiment, we are able to control for variations by unit operation laboratory protocol performed (e.g., distillation column, plug-flow reactor, etc.), as well as identify any changes in team dynamics that may occur as the students become more accustomed to working together.

Results indicate there is a high deal of variability in team dynamics among the unit operations students. By evaluating the traditionally run laboratory course without intervention (the only items added for study only purposes are the ILS and perception surveys), we are able to establish a baseline for understanding the role of learning styles and influences of team dynamics in our curriculum. An understanding of the variability in performance based on learning style and team composition can help elucidate the complex nature of team dynamics. Armed with such information, instructors can make decisions regarding how to create teams (in courses at all levels) in a manner most beneficial to the students, and students will have a better understanding of teams as they progress through/complete their degrees.

Graduate Professional Development: Experiences from Three Structured Seminar Courses

Michael A. Matthews, Professor, Associate Dean for Research and Graduate Education

Department of Chemical Engineering

College of Engineering and Computing

University of South Carolina, Columbia SC 29208

Graduate education should produce new scholars who are highly productive, scholarly in their approach to their work, and able to think independently. Toward this end, we have developed a set of three, one-credit structured seminar courses for graduate students in the College of Engineering and Computing at the University of South Carolina. These have been designed to provide "just-in-time" instruction (actually mentoring) for graduate students, particularly for PhD students. The courses are intended to improve their skills in critical reading, writing, and thinking, and to provide instruction in the ethical conduct of research. The courses are generic to students in the engineering fields, because the broad skills described above are applicable to any young scholar. It is our belief that such structured mentoring will increase research productivity on the part of the students, and provide a valuable service for research-active faculty. We have now offered these courses for several semesters, and will report on our approach and our observations. We have in hand for each a sample syllabus, reading materials, a set of possible assignments, and a weekly schedule of activities for a nominal 14-week course. We believe that this approach can be easily adopted elsewhere, as we have paid particular attention on how to "institutionalize" such instruction at a low cost.

389653 Transforming a Basic Engineering Skills Course into a Freshman Design Course with Business Simulation Elements

Abstract:

Foundations in Engineering is a first-year course taught in the second semester at the University of Connecticut. All Engineering students must take this class, regardless of major. Individual sections are taught by a member of each department, traditionally in a lecture format, with the purpose of developing skills that are transferable to other engineering disciplines. The Chemical Engineering section of Foundations in Engineering became the first section to be taught as a project-based design course with business elements in the Spring 2014 semester.

Students were divided into teams (called "companies") and were asked to complete three projects, each with a different set of constraints and design goals in mind. The first project had students build a thermos that was both effective at keeping a vial of water cool, and minimized the weight as these thermoses would be used in shipping. The second project tasked the students to build a water filter that could remove color and particulates while being easy for a hypothetical camper to use and assemble. Finally, students build a baking soda and vinegar powered Chem-E-Car, which needed to be accurate, fuel efficient, and transport passengers (simulated by packing peanuts) without injury.

During each project, companies were given a budget of class-specific currency, which they used to buy materials, machine their devices, and test devices. Deliverables for the class took the form of a short proposal with an expected timetable of major design milestones, a progress report explaining early findings and considerations to be made in the final design, and a final presentation of their device that included a budget for their project. Each company's final device was also tested and compared amongst the performance of the other companies' devices. Rather than have device performance factor into a student's grade, building the best device would contribute to a student's extra credit score.

Based on pre- and post-course student attitude surveys, students seemed to somewhat enjoy the new approach to teaching the course overall. Most of the class enjoyed purchasing materials and the monetary budgeting aspects of the class. Students felt more receptive to working in groups and more comfortable presenting data to their peers at the completion of the course. Other student feedback provided insight into modifications to be made in future semesters.

379684 Developing Experiments Safely in an Undergraduate ChE Laboratory

Abstract:

Every chemical engineer will design experiments and evaluate experimental data at some point during their career. Thus, it is important for chemical engineering students to have a meaninful experience in their Senior laboratory course. Unfortunately, during a one semester course, there is hardly time to design, build, and test an experiment. At the University of Houston, we encourate chemical engineering students to take an Independent Study course during their rising Senior summer. For that course, they are given an experiment to research, design, build, punch, and commission. They must also perform safety audits and procedural audits for their experiment. Safety is stressed during each phase of the development of their experiment.

If successful, the experiment is incorporated into the Junior or Senior laboratory sessions. Not only do these summer students learn experimentation at the fundamental level, but they get a feel for process development, since what they do during the summer is a miniature process development project.

This presentation discusses the positive and negative aspect of this concept, a concept that not only provides evergreen experiments for the Junior and Senior laboratory sessions, but also provides chemical engineering students with experience at experimental design and process development.

387141 One Professor's Journey Toward Implementation of Design Build

Abstract:

This paper will describe efforts by one tenured faculty member to introduce prototyping into a class (Heat Transfer) and capstone design. First, the presentation will review the history of prototyping in chemical engineering education. Next, the challenges and outcomes of a paper design and prototype of a heater with a resistive heating element will be discussed. Following discussion of the heater prototype, the idea of prototyping a battery system for a capstone design project will be presented. The presentation will conclude with a summary of the two design build efforts.

379403 Incorporation of 3-D Printers & Laser Cutters into Chemical Engineering Curriculum

Abstract:

Prototyping tools are common in the laboratory experience of our colleagues in mechanical engineering, but have yet to find widespread use in chemical engineering curriculum. We have recently begun emphasizing design and hands-on projects throughout our curriculum, and have found these projects to correlate with remarkable student satisfaction and engagement. To aid our students in such projects, we have added a laser cutter and multiple 3D printers to our laboratory. As opposed to traditional cutting and manufacturing tools, we have found these pieces of equipment have a low activation barrier to use and low risk of harm.

Our laser cutter has found extensive use in our first-year design laboratory. Students have used the laser to produce parts for unique spectrophotometer designs and flow cells, and it has been used frequently in their final design projects. Some of these design projects have included bench-top heat exchangers, microfluidic devices, and viscometers. The laser cutter has also become an effective part of our department's outreach mission, in the construction of outreach teaching modules, such as cloud chambers, and souvenirs for outreach events.

Our 3D printers have been used primarily within our senior projects laboratory course. They have been used to create novel impeller designs to study the effectiveness of mixing. Students have also used them to create new designs for photobioreactors to minimize algae settling. In other core chemical engineering courses, we have used 3D printers to create Venturi meters and 3D plots of thermodynamic data to help our students better understand complex concepts.

In this work we describe the most effective means we have found to introduce students to the operation of such equipment. We also describe practical resources for putting such equipment to use in chemical engineering courses from introductory to senior-level laboratory courses. We also present results from student surveys showing their positive regard for the use of such manufacturing equipment within our laboratory courses.

19 Nov 2014

WEST POINT
CHEMICAL ENGINEERING

Winemaking at West Point

Robert G. Bozic, Ph.D.
LTC U.S. Army(Retired)
Lecturer in Discipline
Department of Chemical Engineering
Columbia University in the City of New York
rb2335@columbia.edu

COLUMBIA | ENGINEERING
The Fu Foundation School of Engineering and Applied Science

Outline

Background/Motivation
Winemaking Process
Results, Wine Rating & Awards, Cost Analysis, Materials and Lessons Learned
Tailgate Tasting Event
Acknowledgements and References
Conclusion and Future Work

Background and Motivation

Project Goal: Produce homemade wine from raw ingredients for responsible consumption in order to learn more chemical engineering and chemistry of the winemaking process.

Korta Katerina Winery
Orebic, Croatia

http://www.kortakaterinawinery.com/ accessed 28 Sep 2014

West Point Chemical Engineering Club

George M. Taber, *Judgment of Paris*

Grgich Hills Estates

http://www.grgich.com/ accessed 28 Sep 2014

Soil – *"Character"*

Climate – *"Personality"*

Man – *"Spirit"*

http://www.imdb.com/[illegible] accessed 4 Oct 2014

http://www.[illegible] 1 Oct 2014

Cabernet Wine Making Process

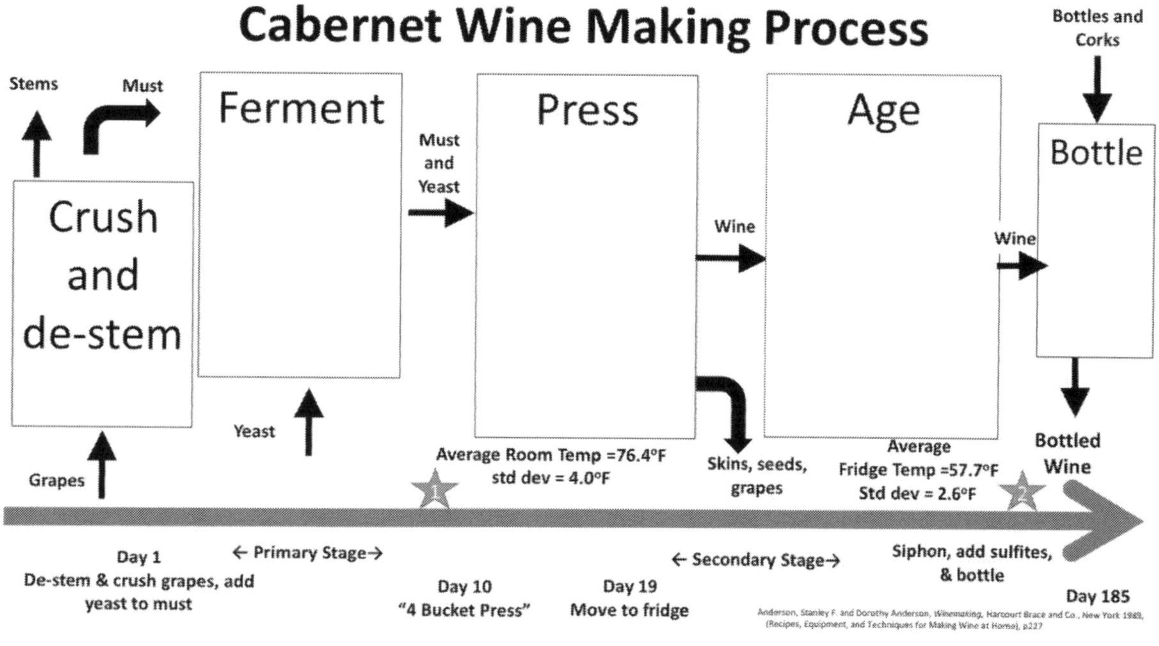

Anderson, Stanley F. and Dorothy Anderson, *Winemaking*, Harcourt Brace and Co., New York 1989, (Recipes, Equipment, and Techniques for Making Wine at Home), p227

Cabernet Wine Making Process

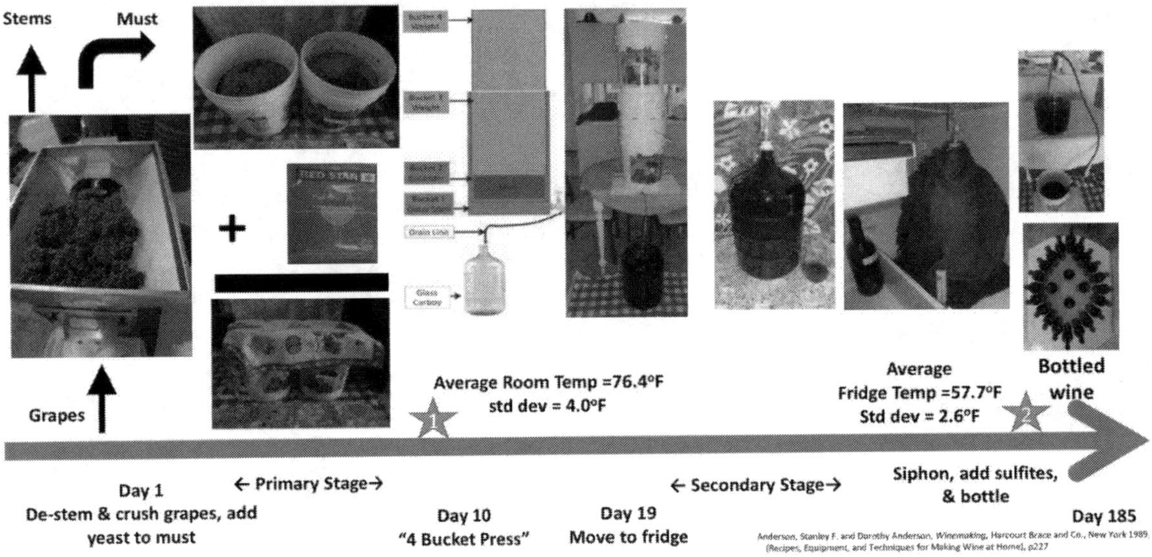

Cabernet & Pinot Results

Volume Percent Alcohol vs. Time (days)

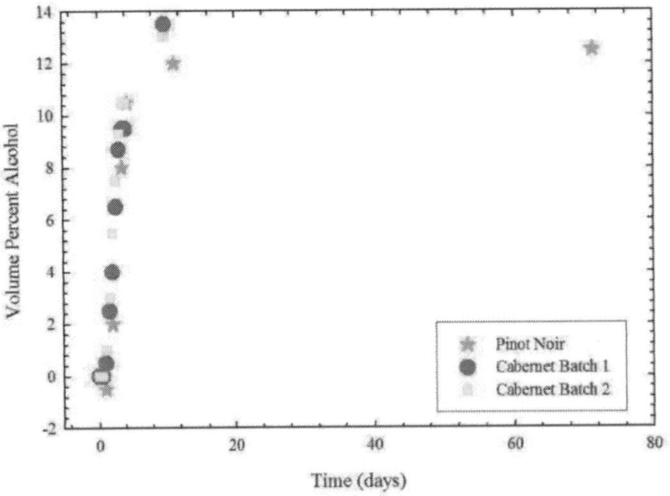

Wine Rating Chart

Appearance
3- Excellent -Brilliant with outstanding characteristic color
2- Good -Clear with characteristic color
1- Poor -Slight haze and/or slight off color
0- Objectionable -Cloudy and/or off color

Aroma and Bouquet
6- Extraordinary - Unmistakable characteristic aroma of grape variety or wine-type. Outstanding and complex bouquet. Exceptional balance of aroma and bouquet
5- Excellent -Characteristic aroma. Complex bouquet. Well balanced.
4- Good -Characteristic aroma. Distinguishable bouquet.
3- Acceptable -- Slight aroma and bouquet.
2- Deficient - No perceptible aroma or bouquet or slight off flavors.
1- Poor - Off odors
0- objectionable -Objectionable or offensive odors

Taste and Texture
6- Extraordinary- Unmistakable characteristic flavor of grape-variety or wine type. Extraordinary balance. Smooth, full-bodies and overwhelming.
5- Excellent All of the above but a little less, Excellent, but not overwhelming.
4- Good Characteristic grape-variety or wine-type flavor. Good balance. Smooth. May have minor imperfections.
3- Acceptable- Undistinguishable wine, but pleasant. May have minor off flavors.
2- Deficient -Undistinguishable wine with more pronounced faults than above.
1- Poor -Disagreeable wine with more pronounced faults than above
0- objectionable -Objectionable or offensive flavors and/or texture

Aftertaste
3- Excellent -Lingering outstanding aftertaste

2- Good -Pleasant aftertaste
1- Poor -Little or no distinguishable aftertaste
0-Objectionable -Unpleasant aftertaste

Overall Impression
2 -Excellent
1 -Good
0 -Poor

Total Scores
18-20 Extraordinary
15-17 Excellent
12-14 Good
9-11 Commercially Acceptable
6-8 Deficient
0-5 Poor and Objectionable

(http://www.americanwinesociety.org accessed 8 Oct 2013)

Category (Possible Points)	Points Awarded
Appearance (3 max)	________
Aroma/Bouquet (6 max)	________
Taste/Texture (6 Max)	________
Aftertaste (3Max)	________
Overall Impression (2 max)	________
Total Score (20 max)	________

Comments:________________________________

(http://www.americanwinesociety.org accessed 8 Oct 2013)
http://cen.acs.org/articles/92/i38/Taste-Wine-Science.html

Wine Colors and Descriptive Words

Clarity/Appearance

Star Bight	Brilliant	Clear
Cloudy	Dull	Watery
Deep Purple	Ruby Red	Red
Red Brown	Deep Brown	Amber

*blue tints in red wine suggest youth
*brown tints in red wine suggest age

Aroma and Bouquet

Acetic	Acidic	Beery
Blackcurrant	Clean	Corky
Dominant	Flinty	Flowery
Foxy	Fruity	Geranium
Green	Honey	Madeira
Musty	Peppery	Powerful
Smoky	Spicy	Sulfury
Woody	Yeasty	Rubbery

Taste

Astringent	Sharp	Green
Mellow	Flat	Thin
Bitter	Sound	Insipid
Peppery	Fruity	Yeasty
Corky	Sweet	Madeira
Balanced	Tart	Smooch
Oxidized	Honey	Flinty
Coarse	Woody	Spicy
Rounded	Baked	Full
Earthy	Metallic	Tannic
Soft	Clean	Harsh
Foxy	Ripe	Velvety
Sulfury	Dry	Light
Grapey		

This wine is brought to you by Cadet Matt Lacey and LTC Robert G. Bozic, Ph.D.

All Grapes and yeast were purchased at Corrado's Market, Clifton, NJ

Anderson, Stanley F. and Dorothy Anderson, *Winemaking*, Harcourt Brace and Co., New York 1989, (Recipes, Equipment, and Techniques for Making Wine at Home), p227

Award Winning Wines

Sep '13, Nouveau Syrah

scale up, wine tasting event with department tailgate & Bronze Medal Winner American Wine Society Contest Nov 2013

http://americanwinesociety.site-ym.com/?page=82 accessed 29 Sep 2014

Christmas 2013, Cabernet Sauvignon

Bronze Medal Winner Corrado's Wine Making Contest Jan 2014

http://www.corradosmarket.com/home/winecontest.html accessed 29 Sep 2014

Cabernet Cost Analysis

$$\text{Annuity} = \text{Principal}\,\frac{i(1+i)^n}{(1+i)^n - 1}$$

Annuity = uniform periodic payments made during n periods in an ordinary annuity, \$ /period

Principal = present worth of capital on which interest is paid, \$

i = interest rate based on length of one interest period, percent /100

n = number of time units or interest periods

The total cost of all supplies and equipment for this project was $177.67.

At a 6% interest rate over 10 periods, a $24.14 annuity was calculated using of the capital-recovery factor. (Peters and Timmerhaus, 1991)

New equipment cost = $69.93 (39%), Supply cost = $101.98 (57%), Sales tax = $5.76 (3%).

The Kicking Mule Brewing Company produced 24 ea bottles of red Cabernet Sauvignon wine.

Some items were on hand and thus not purchased for this project.

Peters, Max S. and Klaus D. Timmerhaus, *Plant Design and Economics for Chemical Engineers*, 4th Ed. McGraw-Hill, New York 1980. p216-251

Materials and Lessons learned

Initial Fermentation Temperature

Colander Pore size and Sediment

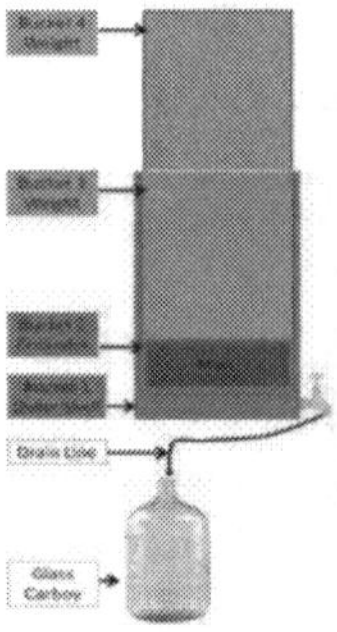

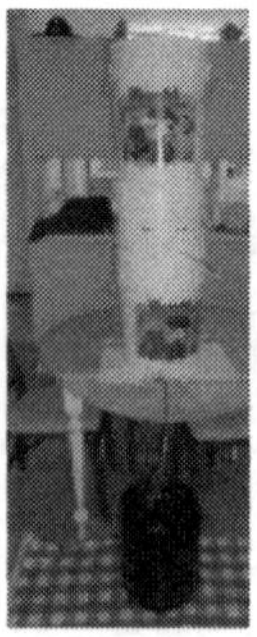

Aging

Syphoning Racking

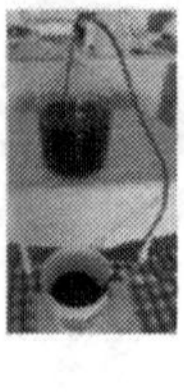

Bottling

Materials

~34 kg (75lbs) of Chilean Cabernet Sauvignon Grapes from Maule Talca, Chile, San Sebastian, SA
5 grams Pasteur Red yeast from Red Star/5 gallon Carboy
4 ea buckets for "4 bucket press", 5 Gallon Glass Carboy
1 ea Large Cylinder Airlock with rubber bung
3/8"x 10 ft siphon hose with plastic tip dispensing rod
1 ea spigot, Thread tape
Black and Decker power drill with drill bits
Racking cane with plastic clamp
Thermometer, Paper towels, Scissors, Measuring cup
Hydrometer with glass cylinder, Wine glass for tasting
Measuring spoons, Beach Towel
Potassium Metabisulfite
(Crosby and Baker LTD) (1/4 tsp /5 gallon ~ 40 ppm),
wine bottles, cork caps

http://www.corradosmarket.com/home/winecontest.html accessed 29 Sep 2014

Tailgate Tasting Event

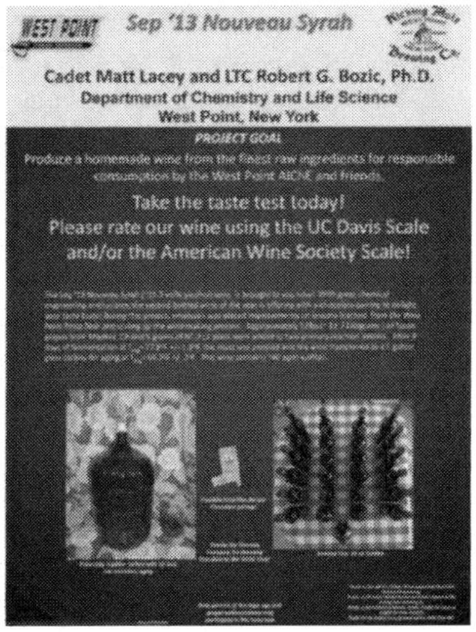

- "Taste and Rate" event
- American Wine Society Scale and UC Davis Scale
- Parings with Chocolate
- Risk Assessment
  Field Manual 5-19, Composite Risk Management, U.S. Government, July 2006

http://americanwinesociety.site-ym.com/?page=82 accessed 29 Sep 2014

http://www.hersheys.com/bliss/experience/pairings/drink-pairings.aspx#/HERSHEY'S-BLISS-Milk-Chocolate accessed 29 Sep 2014

Acknowledgements and References

Thanks to Corrado's in Clifton, NJ for supporting West Point Chemical Engineering!

Thanks to LTC Adam Walton for his continued support to the Kicking Mule Brewing Co.

Thanks to Korta Katarina Winery, Orebic, Croatia for years of support to wine research.

Thanks to Cadet Matt Lacey for his work on the 4 bucket press.

Funds for this project were donated by Robert Bozic.

Eisenmen, Lum. *The Home WineMaker's Manual*, Chapter 12 (1999) available on line www.winebook.webs.com/chapt12.html accessed 25 Jan 2013.

Anderson, Stanley F. and Dorothy Anderson, *Winemaking*, Harcourt Brace and Co., New York 1989, (Recipes, Equipment, and Techniques for Making Wine at Home), p227

Umali, Alona P. et al, "Discrimination of flavonoids and red wine varietals by arrays of differential peptidic sensors" *Chemical Science* 2 2011 p439-445 www.rsc.org/chemicalscience accessed 8 Apr 2013.

Peters, Max S. and Klaus D. Timmerhaus, *Plant Design and Economics for Chemical Engineers*, 4th Ed. McGraw-Hill, New York 1980. p216-251

Fundamentals of Engineering Supplied Reference Handbook, 7th Edition National Council of Examiners for Engineering and Surveying, 2005, Clemson, South Carolina 29631, second printing 2006. p92-98

http://www.kortakatarinawinery.com

http://www.grgich.com/

http://www.emeritusvineyards.com/

http://www.corradosmarket.com/home/winecontest.html

http://americanwinesociety.site-ym.com/?page=82

http://www.hersheys.com/bliss/experience/pairings/drink-pairings.aspx#/HERSHEY'S-BLISS-Milk-Chocolate

http://www.imdb.com/title/tt2419284

Conclusion and Future Work

25 ea bottles of Syrah (Madera , CA grapes)
and 24 each bottles of Cabernet Sauvignon using Chilean Wine Grapes

Increased production capacity from 10 ea bottles of West Point Pinot Noir Wine

Bronze medal winners at amateur wine making contests

**Blending wine, Racking vs Filtering, Higher initial fermentation temperature*

384515 The River Project: Engineering Challenge-Based Learning from Bench-Scale to Pilot-Scale

Abstract:

Northeastern University has combined both deductive and inductive educational approaches to develop a Transport/Separations curriculum sequence over two semesters with an experiential strategy that introduces different modes of learning through **D**iscovery, **D**evelopment and **D**esign; (3D approach). The sequence includes a 4 SH lecture and a 2 SH unit operations laboratory for each of the semesters: the second semester of the second year, and the first semester of the third year. Through the 3D experiential approach, students explore the natural engineering world and discover the underlying concepts (discovery). They are engaged through improving and innovating an experiment (development), and then use their newly acquired knowledge to design an experiment or unit operation (design). The culmination of this sequence is a four-week challenge on water remediation. For this project, students design and construct a small circulating river system which must include a stagnant area and a rapids area. They characterize their clean river and research clean-water standards for rivers as determined by the Environmental Protection Agency of the USA. Students then choose a pollution scenario, deciding type (industrial or sewage), source (point or non-point), and location of the river (urban or rural). Simulated pollutants with an unknown composition are provided to the students. Each pollutant provided includes both chemical and physical contaminants. Groups then use bench scale testing of the pollutant and various chemical and physical modification experiments to design a remediation treatment for their river that solves the requirements to address river water purity. This bench-scale data-based design is then constructed, implemented and tested on their river. Students collect data on what worked and did not work in their remediation solution. Based on this data, the students redesign their remediation solution and test again. Using basic chemicals and novel filtration designs within their treatment systems, students implement their understanding of fluid mechanics, mass transfer, separations, thermodynamics, and kinetics, while also reinforcing writing, presentation, and teamwork skills. The project finishes with a town hall-style presentation in which groups have to defend their decisions and recommendations, as well as respond to critical and probing questions from the audience acting as local citizens. Our presentation will describe the methods, details, and assessment of a unique and rewarding laboratory module.

388984 Desalinisation Process As Integrator Project in Chemical, Environmental and Mechanical Engineering Teaching

Abstract:

DESALINISATION PROCESS AS INTEGRATOR PROJECT IN CHEMICAL, ENVIRONMENTAL AND MECHANICAL ENGINEERING TEACHING

The challenge of potable water supplies involves several branches of the engineering profession and, of course, education. Courses such as, chemistry, physic, material balance, thermodynamics, numerical methods, simulation processes, properties of materials, heat transfer and phase equilibrium, among other, can be used for a specific project in the preparation of new engineers and young researchers.

In the Universidad Libre's Engineering Faculty exists a program of Young Researcher and is possible to develop common project between several departments, working with other Universities. In this case a project for Develop and Design a system of low cost for water desolation using solar energy has served for to begin a series of investigations and experiments to obtain data useful to formulate energetic and exergetics analysis in the solar energy application in several prototypes of solar still. In the design step diverse polymeric materials has been tested. Properties such as viscosity and materials resistance have been measured. For the design and performance of heat transfer of the dome, in the solar still, SolidWorks and Ansys have been used. The weather register has been recorder and the Colombia solar map was used in the early stage of the investigation.

The solar process desolation will be compare with the reverse osmosis for the energetic and exergetics analysis.

This specific project may to use for teaching applied chemistry for mechanical engineering students without courses of this asignature in their curriculum and reinforce several thermodynamics concepts in other courses.

The main objective of this investigation is to supply potable water to a small community in the Colombia Coastal

385682 A Reactor Design Experiment Using Rapid Prototyping

Abstract:

As 3D printers become more commonplace, there is a push to integrate elements of rapid prototyping into engineering curricula. With the ability to produce custom parts with relative speed and minimal waste, 3D printers offer a unique opportunity for students to experience the iterative design process. In this experiment, designed for senior chemical engineering undergraduates, students design a small-scale continuous flow reactor in a CAD software package (in this case, SolidWorks). Students then print the reactor using a FormLabs Form1 3D printer and test it using a saponification reaction of isopropyl acetate and sodium hydroxide. Once students have tested several flow rates and residence times for this reactor, students are encouraged to design a new reactor that they think will allow for better conversion. Students print and test this new reactor as well. Students are also encouraged to model the reactor in COMSOL to verify their results.

Student learning opportunities from this experiment include examination of laminar flow reactors, consideration of variables that impact mixing in laminar flow regimes, Arrhenius reaction kinetics, use of CAD software, and use of predictive modeling software. Students who have completed this experiment are generally excited to be working with the 3D printer. Generally, students with no previous CAD experience are able to produce a reactor channel by the end of the first laboratory period. Students also demonstrate an understanding of reaction kinetics and reactor design when redesigning their device to improve conversion.

385828 Open-Source Hardware and Software in Chemical Engineering Education

Abstract:

Advances in hardware and software have fundamentally impacted every aspect of chemical engineering, but the academic curriculum has been slow to adapt. This presentation will feature case studies that demonstrate how open-source hardware and software can allow instructors to easily and inexpensively introduce students to the powerful tools of the 21st-century engineer.

372786 The Teaching of Transport Phenomena and Related Courses: Survey Results

Abstract:

The authors will present the results of the sixth survey in the resumption of the Chemical Engineering Education Special Projects Committee series of annual surveys on how chemical engineering courses are taught. The surveys are now conducted under the auspices of the Education Division of AIChE. This year's survey covers transport phenomena and related courses, and seeks to determine what concepts are being taught, who is teaching the course(s), and what tools are being used. Teaching methods and novel approaches are, as always, focus elements of this year's survey. A comparison with previous studies on the subject will be included.

Following the presentation of the survey results, a panel consisting of selected participants in the survey will lead a discussion with the audience on topics such as a innovative approaches and common issues observed while teaching transport phenomena. Ideas and approaches discussed during the session will be incorporated into the final survey report.

373250 No Exams, No Graded Homework, No Lectures: Adventures in Active Learning, Peer Instruction, and Flipping the Classroom

Abstract:

In an effort to completely engage students during a semester-long course on "Heat and Mass Transfer", multiple active learning and peer instruction pedagogical approaches were implemented. Examples of these approaches include: online textbook annotations, reflection activities, and student response systems. By minimizing lectures and encouraging learning through peer instruction, a quasi-flipped classroom environment was achieved. In this talk, I will outline how I transformed my classroom from a passive environment to one that actively engages students through peer-based instruction. Suggestions and recommendations will be provided to those instructors seeking to transform their classrooms into dynamic learning environments.

Resources to Implement Flipped Chemical Engineering Classrooms: Course Packages

One of the primary obstacles to the use of best practices in engineering education is overcoming implementation barriers for faculty. These barriers may include a lack of resources, pedagogical reasoning, time, and ease of using the materials. To reduce these difficulties, we have created structured course packages for two chemical engineering courses: thermodynamics and material and energy balances. They contain easy-to-use and effective active learning resources, thereby facilitating the adoption of educational best practices and improving student learning. A course package contains a set of teaching materials that is organized according to learning goals, is easy to navigate, and provides all the necessary resources for active learning and effective assessment. The course packages are built using Microsoft OneNote, allowing users a dynamic and shareable central location for all course materials. They include links to screencasts (short videos with narration) that can be viewed before class as part of a flipped classroom as well as a complete set of class notes that will utilize active learning (ConcepTests with peer instruction, in-class group exercises, etc.). In addition, the course packages contain measureable learning objectives for each class period, a database of solved example problems, homework assignments, and exams. Instructions for the use of the course materials as well as how to adapt them for other courses is an integral part of the package.

An essential element of these course packages is explanations of the underlying theories for these teaching approaches and links to additional references. These course packages will aid new faculty in adopting more effective teaching methods and provide resources to encourage experienced faculty to incorporate active learning techniques in their courses. The materials are also being mapped to popular textbooks, thus aiding novice instructors to most effectively use these textbooks to promote student learning.

Based on summer school sessions where the authors presented on screencasts and ConcepTests, this talk will share detailed information on using the course package, how the specific materials within the package can be modified to suit faculty needs, as well as pedagogical support for implementing active learning methods and a flipped classroom approach.

366063 Active Learning in Small Classroom and Large Classroom of Fundamental Chemical Engineering Courses

Abstract:

Active learning has been implemented after 2012 chemical engineering summer school in sophomore class of Mass and Energy Balances (I and II). In small classroom (30-40), this approach works very well and students like to interact with the instructor and the peers. Active learning keeps the students focused and involved during the instruction. The instruction outcome is very effective. When translating the active learning into the large classroom (more than 60), some difficulties showed up such as keeping the students more focused. Active learning takes longer time in large classroom and managing the students is more time consuming. More creative and structured approaches in active learning need to be developed. The group project is one example of implementing active learning in large classroom. The instruction outcome was assessed by the student survey at the conclusion of the course both for small classroom and large classroom over three semesters. The results indicated that active leaning approach needs to be adjusted based on the class size and more effective active learning skills need to be developed for the classroom that has a large amount of students.

373346 Engineers at Play: Utilization of Games As Teaching Tools for Undergraduate Engineering Students

Abstract:

Students currently entering into our universities have not known a time when there were no computers or internet. These students have high expectations for their education and may not respond strongly to instruction that they do not perceive as appealing. Games and gamification are one pedagogical tool that can be used to help engage these students in a classroom environment. Games can incorporate many aspects of active learning, such as teamwork, immediate feedback, and simulation of different scenarios. As such, games in the classroom or gamified aspects of classes can cater directly to the preferences of these students, which can help them be more engaged in their learning process. A significant open question is whether games in an educational context can meaningfully improve student learning in addition to the apparent benefit to engagement, and if so, how can this gain in learning be quantitatively measured.

Two of the three presenters of this submission met as a direct result of the Chemical Engineering Summer School (specifically the poster session) at UMaine in Orono, in the summer of 2012, and expanded to include the third when presenting their work on games and gamification in the chemical engineering curriculum at the AIChE conference in Pittsburgh in 2012. Since that time, they have all experimented with different types of games and gamified aspects within their classrooms at their respective institutions spanning across all levels of the curriculum. This presentation will present the challenges, lessons learned, and provide ideas for faculty who are interested in introducing games or gamification into their classes. Since meeting nearly two years ago, the presenters have collaborated to write a comprehensive review on games in engineering education, which is currently under review at the *Journal of Engineering Education.*

388417 Going Beyond Demonstrations to "Chose Your Own Adventure" Engineering Experiences

Abstract:

The enthusiasm of STEM teachers and early exposure to STEM fields has been shown to strongly influence a student's selection of engineering as a major/career. Due to the somewhat abstract and often hazardous nature of chemical engineering compared to other engineering fields, it is often difficult to create engaging chemical engineering experiences for K-12 students. To help address this issue, my students and I have created over 50 demonstrations which we have been presented to over 3,000 K-12 students. We have focused on hands-on interactive demonstrations that use as many senses as possible; however, feedback from our experiences has inspired us to modify our approach to involve a higher level of active learning. Here we present our efforts to create open-ended, hypothesis-driven, "chose your own adventure" chemical engineering experiences that are safe, fun, inexpensive, and easily tailored to elementary, middle, or high school students.

387038 Does Time-to-Degree Matter? Survival Analysis and Students' Scores Evaluation

Abstract:

Does time-to-degree matter? Survival analysis and students' scores evaluation

The undergraduation dropout and the long duration are remarkable features of the current teaching of Chemical Engineering in Brazil. In general, the course proposes the formation of a multidisciplinary professional, capable of acting in the management, operation and control of chemical and petrochemical industries. The Chemical Engineering Undergraduation Course at Federal University of Bahia is over 50 years old and is among the 10 state colleges with major prestige in the labor market in Brazil. The last few years have been particularly important for the course because of the change of the curriculum in 2008 and the change of the approval average from 7 to 5 on a scale of 0 to 10. In this context, this research attempts to make a statistical analysis of the students' scores from 1971 to the present with a focus on modeling the time of conclusion or abandonment from the course through the survival analysis.

Questions like, if there were changes in the length of time needed to complete the Chemical Engineering or on their score over time or whether it would be possible to predict the likelihood of the graduation at the present time based on the students' score and semester sought be understood. In the analysis of survival, the time until an event of interest is adjusted by means of asymmetric probability curves with the possibility of using censored data (length of stay of students who have not yet completed or dropped out of the course). Based on the interpretation of the shape of the distribution and the value of its parameters is possible to obtain important information about the phenomenon of interest. The statistical analysis was developed with graphs of correlation between the variables (year of entry, score and time of conclusion/ dropout of the course), boxplot charts and histograms to verify initial assumptions that are important for survival analysis and to verify the hypothesis of randomness of the variables along the time. The entire study was conducted through open language R through the R-Studio GUI with the function package called Survival. The database had approximately 3300 students' data classified by the way they entered and left the college. There are a few outliers or inconsistent information, which was eliminated in the process. The behavior of the Chemical Engineering students' score has shown to be superior to the consensus with median and average values equal to 6.9 and 6.7, respectively, and with right asymmetry distribution. The application of the Fisher test for non-randomness in time series rejected the null hypothesis against a p-value less than 0.05, which indicates a periodic behavior. When the same test was applied to the time series of the score, the results were similar, indicating the frequency of this variable over time. The correlation between the score and time-to-degree shows a triangular shape. These variables are negatively correlated when the students' score are over five and are positively correlated when students' score are below five.

Survival analysis was conducted with two groups of data: (1) students who graduated and (2) students who dropout the course. The data sets were modeled by using exponential, Weibull, log-normal, Gaussian, log-logistic and logistic distributions. Kaplan-Meier estimator was used to compare the parametric distribution with the non-parametric survival curve, and to select the best model.. The first analysis indicated a significant fit of the data with the Weibull distribution (scale = 1.16 and shape = 0.0688). The shape's value indicates a distribution with "Child mortality" that means in our terms a well-expressed capacity of the students in finish the course on less time. The median was equal five, value that correspond to the time expected by the course administration for the course conclusion. The second group of data obtained best fit with the exponential distribution (scale = 0.937). The histogram indicated that the greater number of dropouts happens during 3rd and 4th semesters of the course. At this stage of the course, the students start to deal with chemical engineering aspects in chemical process on industry, statistics to engineers and computational methods for engineering courses. The other courses are related with theoretical math and physics.

It is expected that the obtained results contribute to uncover the student's profiles involving the student's time-to-degree and scores, and base the development of long term enrollment plans at the Polytechnic School and central campus levels.

384354 Optimizing the Design and Implementation of the Upper Level Curriculum: Making the Case for Scheduling Analysis of an Undergraduate Program

Abstract:

Over the past decade, undergraduate programs in chemical engineering have experienced a groundswell of enrollment, coinciding with renewed interest in reducing the time to graduation and the associated costs for undergraduate education in general. At Texas A&M University, undergraduate graduating classes have increased from an average of 40 students (~1995 – 2000) to over 150 students since 2008. Most recently, the Dwight Look College of Engineering made headlines announcing the "25 by 25 Initiative" which aims to double engineering enrollment over the next decade. This significant increase in student enrollment has reinvigorated several debates regarding how a department may best utilize its limited resources (the number of available teaching faculty) while facilitating student graduation in a timely fashion. Several departments have followed a strategy of offering courses multiple times per year to facilitate 4-5 year graduation rates; however in doing so departments reduce the number of faculty available to team-teach courses with large enrollment. This ultimately leads to a debate as to what scheduling strategy is 'best,' which in turn raises the question as to whether the same techniques industry employs to ensure the most efficient and effective process (i.e., scheduling analysis) can by employed in academia to identify more efficient and effective implementation of an undergraduate curriculum.

This talk details a recent analysis performed during my tenure as director of the undergraduate program in Chemical Engineering at Texas A&M, and serves to illustrate the potential for scheduling analysis to provide a firm footing for weighing the impact of curriculum planning upon faculty teaching loads and student graduation rates. Alternately, for a given departmental vision regarding curricular implementation, such analysis has the potential to identify the resources needed to meet said vision. The scheduling analysis consists of (i) construction of a curricular structure matrix detailing course sequencing via pre-requisites, and its use for generating a course node structure divided into multiple sub-structures for subsequent scheduling analysis; (ii) construction of a course scheduling matrix for identifying the availability of each course on a per-semester basis, which combines with the curricular structure matrix to enable construction of a scheduling node structure; and (iii) use of this scheduling node structure to predict student graduation rates and per-semester course enrollments. As will be shown, scheduling analysis has the potential to be a valuable tool for any department wishing to quantitatively and accurately study the impact of student enrollment, curriculum design and course offering strategies upon classroom size, faculty demands and graduate rates.

Engineering (and other) curricula are based on a logical progression of courses enforced with prerequisite requirements. Effectiveness of this structure is enhanced if students can be made to remember the concepts, methods and skills they acquire in each course for future application, and also to appreciate the concrete usefulness of what they learn and connections between different learned elements. An appreciation of future usefulness provides motivation in the present. Ideally, students might say, for example,

> "I am learning and will remember how to solve differential equations in my computational methods course this year so that (among other things) I will be able solve for the time evolution of a given reaction network in my kinetics course next year"

looking forward in time, and

> "I am able solve for the time evolution of this reaction network in my kinetics course this year because I learned and remembered how to solve differential equations in my computational methods course last year"

looking backward in time. A major challenge in engineering education is fostering this kind of retention, and appreciation of future usefulness and connectivity, in students' minds.

This talk describes a new initiative toward this end being undertaken at the University at Buffalo based on the spiral learning concept. The capstone senior plant design course is decided three years in advance for a given entering class, and used as a focal point for the curriculum experienced by that group of students. Specifically targeted "spiral problems" are developed for instructors in different courses that introduce an element of theory, a problem solving method or a skill that will be needed for the design project. Each spiral problem comprises a formal problem statement, a solution, and lecture material that explains how it fits into the broader context of the objective of the capstone project. In this way, students are introduced to the underlying technical theme, and return to it ~ten times throughout their sophomore, junior and senior years, from various perspectives including thermodynamics, kinetics, transport, unit operations, materials, bioengineering, mathematical and computational methods, safety and ethics. The process is illustrated using spiral problems that have been used to reinforce design projects relating to bioethanol and poly-lactic acid production, and the manner in which the various problems reinforce the ultimate senior-year plant design objective is mapped onto a flow chart of the student curriculum. We also describe the information technology aspects needed to make the process work, and how elements of ceremony are used to make students remember the spiral moments in their education.

368020 A Unit Operations V-Mixing Experiment to Integrate Particle Technology, Experimental Design and Advanced Data Analysis with Pharmaceutical Applications in the Undergraduate Curriculum

Abstract:

V-mixers are used to mix powders and particulates in a wide range of applications in the pharmaceutical and food industries. This work involves the development of a unit operations v-mixing experiment using a sugar and salt particulate system (80wt% sugar/20wt% salt). This particulate system simulates the excipient and active ingredient formulation in standard aspirin (acetylsalicylic acid). Students were introduced to and applied v-mixing and particulate and powder technology, design of experiments (DOE) and engineering statistical data analysis techniques. They developed a mixing quality index and studied the effects of mixing process variables on the mixing quality of the salt/sugar system. Mixing speeds ranging from 10 to 30 RPM were investigated along with mixing times of 40 to 200 sec. A 50% loading was used in the mixer. The electroconductivity of sample water solutions was used to measure the salt concentration in samples and ascertain mixing quality. A factorial experimental design was used to fully describe the system mixing characteristics and develop predictive models for mixing efficacy as a function of v-mixer process variables. Detailed experimental results and data analysis are highlighted in this work. The v-mixer system and experimental protocol can be adapted to a wide range of applications ranging from K-12 demonstrations to research investigations.

372840 How to Teach Best a Basic Understanding of Molecular Interactions and Fluid Phase Thermodynamics

Abstract:

A basic understanding of the interactions of molecules and the resulting fluid phase thermodynamics is essential for practicing good chemical engineering. Nevertheless, it is not so simple to teach such understanding to students. Most simple concepts of polarity and dispersion are not able to explain the problems faced in modern chemical engineering. Group contribution methods are just frameworks of numbers, which do not provide a qualitative understanding of the molecules and their interactions, and molecular modelling also is too indirect and to slow for being useful in a chemical engineering course.

Being based on quantum chemistry, the COSMO-RS method provides vivid graphical pictures of the surface polarity of molecules, making use of the polarization charge density σ of the molecules virtually swimming in a conductor. The electrostatic and hydrogen bond interactions of molecules, as well hydrophobic interactions are quantified based on this surface polarity by simple expressions. As a result the interactions of the molecular surfaces can be easily and intuitively comprehended and students can simply understand why some molecules like each others and others do not, and why it gets warm when you mix acetone and chloroform. The methods leads straight to chemical potentials and activity coefficients, and hence to the entire fluid phase thermodynamics of pure liquids and mixtures.

Using the free educational version of the COSMOtherm program, the students can start to play around with ~300 diverse molecules, check mutual activity coefficients and binary phase diagrams, and do many other educative examples and exercises. They will keep in mind fluid phase thermodynamics as a vivid and colorful discipline.

375803 A Laboratory Tool for Distributed Solar PV Systems Education

Abstract:

Solar Photovoltaic (PV) systems have been used extensively for residential applications. There are also several programs that promote the installation of solar PV systems in the US, such as the California Solar Initiative (CSI) program which provides cash back for solar energy systems for existing homes. As a result, educators have an opportunity to provide a broad based understanding of solar PV systems and promote greater awareness of the technology as it spreads. Some universities are providing master's degree courses in solar energy. However at the undergraduate level, solar energy is mainly the subject of a curriculum of engineering and energy courses. We seek to augment available educational materials with practical hands-on experience in the nature and design of solar PV systems. We believe that hands-on education is critical to the evolution of energy systems, so that a broad base of people can understand the basics of how energy systems work. Widespread understanding is important for developing rational policy and building the consensus needed for the large capital expenditures required for energy systems.

A laboratory simulation tool was developed for distributed solar PV systems based on an energy system research model. The simulation tool was organized to (1) explore electricity usage patterns, (2) study the time and regional availability of solar energy, (3) design a solar PV system, and (4) compare business cases for different designs. To maximize applicability and interest, the simulation tool allows students to customize electricity demand to match household characteristics, change weather assumptions, select system location, and vary the solar module area and energy storage capacity. The simulation tool was applied to one graduate course and two undergraduate courses as a lab project. A questionnaire was created and applied for undergraduate students both before and after the lab project. Comparison of the pre and post survey results demonstrates that the tool successfully helped students learn about solar PV systems: (1) 65% of the students believe that their knowledge of solar PV systems after finishing the lab project is equal to or above medium level and (2) 91% of the students claimed that they are more likely to accept solar PV systems based on what they have learned in this project. Finally, in order to greatly increase availability and potential use, a refined user interface has been created and the tool has been published online using Purdue University Hub Technology as part of the NanoHub. The tool is named as "SolarPV" and can be accessed at: https://nanohub.org/tools/solarpv .

362686 Illustrative and Easy Topics for Undergraduate Thesis, and Their Influence on Students to Pursue Graduate Studies

Abstract:

The insertion of undergraduate students in scientific research and academic work is significantly enhanced through the following activities: 1) engaging in summer research stays, 2) undertaking an undergraduate thesis, 3) having the opportunity to be included as co - authors in publications, memories, proceedings, as well as 4) forging relations with students and professors of other universities. All of the aforementioned tend to encourage students to pursue graduate studies, which is absolutely necessary for the formation of high level human resources. During more than fifteen years I have promoted these activities in my university and I can state that due those facts, more than 18 undergraduate students have graduated under my supervision since 2006. Currently, some of them are working in the industry, while five of them are doctors, six are masters, and others are currently graduate students at universities in Canada, USA, Germany and Mexico.

A crucial aspect that favors students be attracted by scientific research, is to offer them executable thesis with relatively easy topics and guaranteed success, such that they can realize the importance of focusing on the details, while applying the theory in the experimental work, but most importantly, so that their scientific curiosity is awakened. If professors include students in complicated thesis topics, such that they are very likely to fail, instead of generating enthusiasm, the failure, frustration, and other complications, will disappoint them and move them away from science, in some cases, even resulting in quitting the thesis project. More important aspects that contribute to the success in the elaboration of the undergraduate thesis are the good communication between the student and advisor, to have available all the facilities to do the thesis, to take to the students to local and domestic conferences and to give them enough attention.

It is very important to take advantage of graduate thesis details, gaps and even unexplained inconsistencies, specifically those that can be solved by undergraduate students. The following are some examples of undergraduate thesis completed under my supervision that have lead my students to pursue graduate studies.

1.- "Study of irradiance field, energy balance, phenol degradation and kinetic modeling of the photocatalytic reactor Photo CREC Water II".

In this case, the topic was water purification, specifically phenol degradation using photocatalysis in an annular reactor operating in batch mode in recirculation, the Photo CREC Water II, with catalyst in aqueous suspension is illuminated by a UV lamp to activate the catalyst. Samples of the aqueous suspension are taken and analyzed in the HPLC, TOC and UV-Vis to monitor the model compound and intermediates. As a result a complicated reaction mechanism and network is developed in this process.

One of the most important contribution of this thesis was the measurement of temperature profiles during the time of the experiments, 1) when the water was just in recirculation with the lamp turned off, 2) then turned on, and finally 3) with the phenol degradation. This information was used to perform the energy balance for the photocatalytic process taking place during the experiments. The earlier temperature measurements, allowed calculating the heat added to the water due to pumping, by using the equation Q =mcpDT. Then, by use of the temperature profiles with the lamp turned on, the heat added to the water due to the effect of both pumping and lamp irradiance was accounted, and the energy absorbed due to the lamp emission was calculated, subtracting the heat determined when lamp was turned off. Part of this energy is known to activate the catalyst, to achieve chemical reaction, while some is going to be delivered as heat to the water due to both the reactions exothermic nature and inefficiencies arising from charge recombination during the photoelectric effect. Then, the heat calculated using the later temperature profiles, i.e. when the phenol degradation was carried out, allowed to calculate energy consumption for photocatalyst activation while properly subtracting all other relevant effects.

In this way, by use of simple temperature measurements, several problems are posed to the students: 1) the calculation of the energy to activate the catalyst, 2) the calculation of the enthalpies of the reaction and formation of the free radical OH, 3) approximation of the irradiance field in the reaction annular section, 4) the number of active sites estimation, as well as 5) proposal of the reaction network and mechanism, with its respective kinetic and adsorption parameter estimation. Skills and notions are developed through the project, such as usage of Math Lab, to face the challenge of reproducibility of experimental results or to know the apparatus detection limits.

These challenges wake up their scientific curiosity. The involved students were very enthusiastic during the elaboration of their thesis and I am sure, due to this, they now are graduate students. Once they gathered intermediate and model compound concentration profiles, they immediately realize there is a need to propose a logical structure that relates them, i.e. a reaction network. Additionally, the reaction network indicates them that parameter estimation comes next, to determine whether model hypotheses are rejected or validated, and refine through an iterative process a model that arises from both experiments and adventuring to propose new ideas based on solid application of organic chemistry, mathematics, physics and scientific research principles in chemical engineering.

Inasmuch as they know that different catalysts lead to different results, they feel the need to understand the nature of the materials, to characterize them, and they get eventually the ability to handle and operate the pieces of equipment, the interpretation of results, among many other things. In summary we are taking advantage of the details of graduate thesis to structure undergraduate thesis.

2.- Impact of different boundary conditions in the angular momentum of the molecules inside a spinning cylinder.

The angular momentum of water inside a rotating cylinder is analyzed for several cases where different boundary conditions are applied. In the first case the cylinder is spinning with water inside, as a result a parabolic surface profile is generated. In the second case the cylinder is fixed, with a central bar rotating. As expected the surface profile has changed, due the different boundary conditions.

In both cases, it is necessary to analyze the three components of the Navier Stokes equations (momentum equations), and to solve the resulting equations, once the appropriate terms are discarded. From the solutions of these equations, the surface profile equation is deducted. The theoretical results are evaluated y compared with the experimental ones, obtained through photographs of the devices in rotation. In this way the student understand the utility of the mathematical modeling, the importance of the correct deduction of the mathematical models, and how they can allow the prediction of the behavior of the physical system, avoiding experimentation, if the model is already validated.

3.- Non stationary axial temperature profiles from heat conduction in a cylindrical bar.

A cylindrical bar is warmed up in one end and the heat is transmitted axially through the bar by conduction, the experiments are performed for both an insulated and a non-insulated bar, while the other end is exposed to the environment.

The bar has holes in the axial, angular and radial directions, such that it is possible to place thermocouples and measure temperature profiles for all directions. The corresponding mathematical models are solved and the theoretical results are then compared with the experimental profiles. Again, as in former cases, the importance of mathematical modeling, and the acquired notion that, if the model is correctly deducted, heat transfer predictions will avoid unnecessary experimentation. This work can be extended in several ways , such as 1) using a heat source to account for fed heat, 2) putting two bars of different material in series, to verify the continuity boundary condition in the interphase between materials, among others possibilities.

There are of course more examples. So far, all the involved students are doctors, masters, and the elaboration of these theses gave them the scientific curiosity, but more importantly, because they liked the scientific research and academic activities. It is evident that the presented examples are simple enough for undergraduate level studies and the achieved success was one of the reasons why these students are now graduate students and and/or active and highly motivated scientists.

390220 The I-REU: Integrating Innovation, Entrepreneurship, and Industrially Relevant Research into an REU Site

Abstract:

Most REU sites are designed around a scientific theme which draws student interest from a single field or discipline. These may include topics such as "Energy", "Water", "Process Design", or other similar themes. Our site differs from this by offering broader scientific options and instead focusing on the commercialization potential of the technology. Each project has both an academic and an industrial mentor that provides two unique perspectives for the student participants. The industrial mentors range in size from small startups to large multinational corporations. The academic mentors are mostly anchored in the chemical engineering department, though faculty in Chemistry, Materials Science, and other closely related departments have participated. A key part of our program is the business oriented short course that all students take. This course, taught by the former Dean of the UConn School of Business, teaches young engineers the mechanics of business, enabling them to at the very least communicate better with the business community and possibly inspire them to start their own businesses someday.

372545 REU Experience at the University of Alabama: The Benefits of Engaging Students in the Experimental and Computational Aspects of Research Projects

Abstract:

We are currently in our second round of funding for an NSF REU Site program at The University of Alabama. Our first program was driven by an intellectual focus on developing more environmentally-benign "clean" energy technologies. In order to engage and train the next generation of scientists and engineers in providing clean energy solutions, we assembled a set of projects that addressed various technical aspects of future energy solutions. These projects included: technologies for oil spill cleanup; ionic liquids for CO_2 separation and capture; CO_2 emissions monitoring from sequestration formations; fuel cell, supercapacitor, and electrochemistry modeling; biogas and biodiesel production; energy storage in clathrate hydrates; and environmental catalysis. While these research projects spanned a number of different technical areas, the projects that involved both experimental and computational components resulted in elevated scholarly productivity. In addition, the combined experimental and computational perspectives enriched the knowledge-base of the students (as judged by student feedback responses). In order to leverage the benefits observed from our first REU Site program (during three years of operation), we expanded the computational aspects of our projects by refocusing our theme to: "Leveraging Computational Tools for Enhancing Engineering Innovation". This program primarily involves faculty members from the Department of Chemical and Biological Engineering, but it also includes faculty from Aerospace Engineering and Metallurgical and Materials Engineering. Each project has a clearly defined hypothesis, and the hands-on research experiences involve a varying balance of experiments and computational work across a variety of cross-cutting topics including 3-D printing, molecular/materials design, coding/simulations and biology.

Within the context of our previous REU Site and our more recent expansion of computational projects, we will highlight some of the primary challenges and successes that we have experienced. In particular, we will outline our strategy for: (a) initial student recruitment; (b) successfully acclimating students to campus and their research environments; (c) unconventional seminars for professional development; and (d) long-term student mentoring.

IMPLEMENTATION STRATEGIES FOR A LARGE, MULTI-INSTITUTIONAL REU PROGRAM AND KEY ACTIONS OF SUCCESSFUL SUMMER RESEARCH MENTORS

Paper #386517

D. Raj Raman[1,2], Brandi N. Geisinger[2,3], Mari R. Kemis[2,3], Arlene de la Mora[4] and MaryAnn Moore[1,2], (1) Ag & Biosystems Engineering, Iowa State University, Ames, IA, (2) NSF Engineering Research Center for Biorenewable Chemicals (CBiRC), Ames, IA, (3) RISE, Iowa State University, Ames, IA, (4) School of Education, Iowa State University, Ames, IA

Abstract

Summer research opportunities for undergraduates, such as those supported by the National Science Foundation's Research Experience for Undergraduates (REU) program, can be critical experiences that help persuade students to pursue graduate work. Our team has been involved in the administration and assessment of a large, multi-year REU associated with one of NSF's Engineering Research Centers (ERC) with a significant emphasis on Chemical Engineering. Because ERC's are 10-year projects, we have thus far six project years of experience with running a large, interdisciplinary, multi-institutional REU program that has graduated 78 students since its inception. Here, we report on approaches to recruitment, project definition, placing and advising students, and assessing programs.

We also discuss our unique approaches to training REU mentors. Because of the relatively small cohort size associated with single programs (ca. 8 – 12 students typically), the existing literature regarding mentoring in these programs is fairly anecdotal in nature; we are unaware of any examples of hypothesis driven research exploring the factors leading to positive outcomes for REU students, nor of any regarding the qualities associated with successful undergraduate research mentoring. In 2013, we conducted a survey of over 100 participants in multiple REU programs at our institution, to explore the validity of several hypotheses we had regarding the importance of mentoring and the qualities that characterize good mentoring. We will report preliminary findings from that effort, which include evidence-based recommendations on factors that lead to participant satisfaction, and on approaches to successful mentoring.

Introduction & Overview of Best Practices

Since its inception in 2008, The National Science Foundation Engineering Research Center for Biorenewable Chemicals (CBiRC) has had associated with it a dedicated Research Experience for Undergraduates (REU) site. With CBiRC's formal launch occurring in the fall of 2008, the first cohort of REUs interns were hosted by CBiRC labs in the summer of 2009, and the total number of program alumni now stands at 78. Contingent upon continuing funding, up to an additional 36 students might graduate from the program before project year 10. Furthermore, during the span of the CBiRC REU, the majority of our author team have been involved in the direction, administration, and/or assessment of two other summer internship programs at Iowa State University – one USDA funded, the other a separate NSF site – together serving an additional 64 students since 2011. The significant load of students through our programs offered ample opportunities to refine our approaches to serving participants, and many of the challenges and successes in the CBiRC REU were echoed in the other two programs.

Over the years, we have developed a series of best practices that have improved the quality of the student experience of our program participants, as well as of the faculty principal investigators (PIs) and graduate students who do the bulk of the day-to-day mentoring of the REU's. This paper describes some of the most important of those practices, and also presents results from our efforts to understand how mentoring impacts REU experiences.

Our best practices can be subdivided broadly into the following areas, here presented roughly chronologically through one year of the REU cycle: Do project level recruiting; Use a web-based application portal; Implement a hybrid selection approach; Conduct phone interviews prior to making offers; Mind the logistical details; Deploy a podcast-based mentor training program; Provide context, expectations, and opportunities to build community; and finally, Implement a robust assessment and evaluation program.

CBiRC's faculty and research transcend the discipline of chemical engineering, but chemical engineering is at CBiRC's core, and the other disciplines associated with CBiRC are closely allied to chemical engineering, e.g., chemistry, biochemistry and molecular biology, bioinformatics, agricultural and biosystems engineering. The centrality of chemical engineering to CBiRC motivated the presentation of this work at the AIChE meeting.

Best Practices: Do Project Level Recruiting

As with any educational endeavor, the abilities and motivational levels of incoming students is a critical factor in the program's success. To attract a diverse pool of qualified and motivated applicants, we begin by soliciting potential research project opportunities from CBiRC labs. These opportunities are communicated to us by principal investigators (PIs), in the form of a short project title and two-sentence description. This step is critical in the operation of any REU (or other summer research) program that offers options that reflect a wide range of disciplines, even when these disciplines share some degree of commonality. This is because it enables students to have an inkling of the nature of the work that they may be doing, and allows us to ask that they articulate, in their applications, how they see themselves contributing to the project that they've identified as most interesting to them. This in turn assists with both evaluation and final placement, and increases the likelihood that each participating student is highly motivated when they arrive for the program.

Participants are recruited nationally through a variety of channels, including: NSF's 'Search for an REU' website, our own REU Project website; lists of students developed by ISU's graduate college during their visits to multiple fairs and or colleges that serve underrepresented populations; and direct communication with targeted departments across the nation including regional community colleges and other non-research-intensive higher-education institutions.

Best Practices: Use a Web-Based Application Portal

Having populated the program website with descriptions of specific projects, we require applicants to list their top three projects. That information is on their application form, which accompanies a personal narrative (that includes past research experience, current research interests, and long-term educational/career plans), undergraduate transcript, and two letters of reference, all of which are submitted electronically through a system we have developed in conjunction with the ISU College of Engineering Information Technology Services unit. Using this system avoids the challenge of tracking potentially hundreds of e-mail messages containing various portions of each application. The system we use allows recommenders to directly provide such letters, thus avoiding the awkward and undesirable situation of applicants handling (and potentially reading) their own letters of recommendation.

Best Practices: Implement a Hybrid Selection Approach

Once the application deadline has passed, we do a central (programmatic level) screening to identify 3 – 6 strong candidates for each open research project. Lab PIs then receive full packets for this subgroup of students and review and rank those candidates. The project director then makes the final selection for each lab based on PI ranking and programmatic needs (e.g., diversity of school of origin, etc.). It should be noted that PIs can rank any candidate as unacceptable, and if that occurs we will not place that student in that particular lab. We appreciate that labs are making a significant resource commitment to host an REU student, and a placement that is not desired by the PI serves neither the student nor the overall goals of the project.

Best Practices: Conduct Phone Interviews Prior to Making Offers

With the short list of candidates selected, it is important to begin making offers rapidly, for multiple reasons: First, there is a high demand for well-qualified REU candidates, and programs that delay the decision process risk losing good candidates to other opportunities. Second, but equally important, is that finalizing the cohort precedes making travel and housing plans – logistical details which for a large REU cohort are non-trivial! We have instituted a short phone-interview of all prospective participants prior to making a formal offer. This brief (ca. 10 minute, 4-question) interview is conducted by the program director, typically in the evening, and is not a surprise – we first e-mail candidates to let them know they are finalists but that we would like to phone interview them prior to making formal offers. The goal of the interview is to get a sense of personality of the candidates, and more importantly to establish the program expectations before a formal offer is made. In particular, the phone interview emphasizes the importance of collegiality with the cohort and lab members, as well as commitment to helping the lab achieve its goals. Since instituting this step, we have had far fewer instances of personality conflicts within the cohorts. *Note – because this article will be publically available, and because we will be continuing to run this program for several more years, we do not wish to include the questions herein. However, any PI interested in the list of questions is welcome to contact the lead author who will provide them via e-mail.*

Best Practices: Mind the Logistical Details

Although perhaps fitting into the category of "too obvious to mention" we have found that it is easy to underestimate the volume of logistical details associated with a 12 – 15 student REU cohort, especially one where four or five students will spend most of their 10-week program at an institution far from the director and administrator of the program. And it is hard to overestimate the importance of getting the logistical details correct, from a student perspective. If not attended to, issues such as housing being ready when students arrive, travel plans being straightforward and negotiable by inexperienced travelers (as are many students from lower-resource backgrounds), and stipend checks arriving in a timely manner, can severely detract from student overall experiences. And if these problems are widespread (e.g., not just one or two students who perhaps have complex travel arrangements due to starting locations), it is possible for an entire cohort to sour on project leadership – we had a close call with this in one year and have heard from students in other projects where this has occurred – it is a situation best avoided through judicious planning! In our experience, having a faculty member (director) in charge of logistics does not work well at all. Typical faculty members associated with REU (and similar) programs have a multitude of other tasks as part of their core job responsibilities, and the heavy lifting of logistical details occurs throughout spring semester and particularly heavily in the latter half of that semester, which at many academic institutions is a time packed with other departmental, college, and university level activities in which faculty members are expected to participate. We have always found ways to fund – either through the program funds or through salary saving – approximately one quarter-time staff member per REU program, and this has been central to our success in keeping up with the logistical efforts required for these programs.

Best Practices: Deploy a Podcast-Based Mentor Training Program

Although mentoring, like teaching, is a skill that some seem to possess innately, it is too critical a function to leave to chance. This is especially true in highly research-active labs in which mentoring is most often done by a graduate student, post-doc, or other non-faculty member, and where that person may only spend one year in that role: it is often persons with low-experience mentoring to whom the REU student is assigned. We realized, after several years of operation, that our program would benefit from some form of formalized mentor training. CBiRC-affiliated labs, like many labs associated with strong NSF-funded Engineering Research Centers (ERCs), are highly research productive and are working to meet project-based targets. To respect the time constraints on these labs – which have already committed the time needed to take an REU – and to handle the multi-institution nature of our REU program, we wished to develop mentor training that could be delivered rapidly and at a distance. To that end, in 2012 we developed and delivered a podcast for REU mentor training (this was updated in 2013, and used in both 2013 and 2014). We requested that mentors, who are generally graduate students and post-docs generally, view the 13-minute podcast and then attend a face-to-face and electronic (for offsite mentors) discussion about mentoring approaches. Our mentoring approach is described in detail in a following section, but judging by the nature of the conversations that occur in the follow-up discussion session, this approach is increasing the awareness of mentoring key-actions, and is causing mentors to reflect far more on their role in the REU.

Best Practices: Provide Context, Expectations, and Opportunities to Build Community

As mentioned previously, we set some expectations during the brief phone interview process. Participants typically arrive on campus on the afternoon of Wednesday immediately after Memorial Day. This schedule works well for all except some quarter-system academic institutions. We typically identify one or two participants who are from our institution (the vast majority are from other institutions) to serve as local hosts with the understanding that they will help orient the visitors to our campus. These local hosts are normally embedded in the same housing with the other students, and are there on the day that the other students arrive. We also have the director and administrative lead visit the dormitory in which the students are staying to welcome them on the night that they arrive. The following morning, we begin with a breakfast followed immediately by the intake assessment – we do not want to contaminate our pre-experience data with any of the conversations that come later on the first day. After the intake assessment, we formally welcome students to the REU site and introduce them to the PIs, other faculty mentors, graduate students, and fellow REU participants. An introductory presentation describes the overall theme of the REU Project Site and its relevance, providing context for their work. The goals and logistics of the REU program are presented, as are our hopes and expectations for the program – this includes a discussion of the importance of contributing to the lab and to building community in the cohort – we are great believers in providing unambiguous expectations. Students participate in general safety training during orientation also, although all labs provide additional specific training as needed. We also leverage non-federal funds to allow participants to go on a team-building event, typically a canoe trip down a local river, on the first Saturday of the program. On the next day (Sunday), we typically have the off-site students travel to their labs at other CBiRC member institutions ranging from 300 to 2000 miles away; the formal lab experiences begin on Monday.

Throughout the summer the entire cohort participates in weekly meeting with the program director. We use web-conferencing software to allow off-site students to participate, and these 30 – 45 minutes meetings allow an opportunity for progress updates, questions, and concerns to be discussed amongst all. We provide a poster template during week 1 and encourage students to begin building their poster immediately – for example, it's often possible to provide a working title and authors, along with rationale, by the second week. Materials and methods can often be drafted by the fourth week, and so on. In fact, we have found that students can use development of the posters as a catalyst to move their projects forward or to clarify key concepts. We have also found that the weekly meetings – and the 3-day orientation – seem to be sufficient for the cohort to maintain a reasonable level of cohesion throughout the summer. And after complaints in the second and third program years about the off-site students missing out on the camps-wide REU poster session (which attracts well over 100 posters typically), we have budgeted for and had all participants come back to campus two days before the end of the program, so that the entire group is reunited and participates as a single cohort in the campus-wide poster session.

Best Practices: Robust Assessment and Evaluation

From the program outset with have contracted, and worked closely with a third-party educational assessment and evaluation team (the Research Institute for Studies in Education – RISE) based at our home campus. The assessment instruments were developed collaboratively, and include a written pre-program assessment given first thing on the morning of the program start, a written post-program assessment given immediately after the campus-wide REU session, a cohort-wide focus group conducted after the campus-wide REU session, and a written 6-month post program survey. In early years, we also had a mid-program written survey, but concerns about survey burnout and the limited value of those data in such a brief program caused us to drop that portion of the assessment and evaluation program. The anonymous results of the surveys and focus groups are summarized by RISE, and we have used these results to better understand and improve the program, and well as to illustrate some of the program's impacts to NSF and to our peers.

Mentoring Impacts

Based on our research team's experience in overseeing and assessing multiple cohorts of summer undergraduate REU participants, we developed a set of hypotheses regarding key actions that typify good mentoring. In some cases, the hypothesis were driven by negative examples; that is, we considered cases where an internship had gone poorly and reflected on the mentor actions (or lack thereof) that contributed to the problem. The six key components of effective mentoring hypothesized were as follow:

- *Safe*—effective mentors teach students about laboratory procedures and safety protocol.
- *Prepared*—effective mentors are ready for their REU students – they have the necessary supplies and equipment and have ideas or a plan for the student's research project.
- *Proactive*—effective mentors monitor the student's progress and make meaningful changes when problems arise.
- *Patient*—effective mentors are patient and understanding with students.
- *Present*—effective mentors are readily available for students to approach, seek advice, and ask questions.
- *Positive*—effective mentors maintain a positive attitude in the lab and giving the student positive feedback on his or her work.

To examine the validity and relative importance of these six components of mentoring, we conducted a survey of over 100 participants in REU programs at our institution in the summer of 2012. The survey completed by each student consisted of four sections. In the first section, students were asked who served as their primary mentor (e.g., faculty member, graduate student, postdoctoral research associate, etc.), how many contact hours they had with their primary mentor, lab PIs, or other members in the lab group, and to indicate the value of various REU program activities (e.g., hands-on experience in lab, mentoring by faculty members, poster development, etc.) using a 7-point Likert-type scale anchored by 1 = *not valuable at all* and 7 = *very valuable.*

The second section of the survey focused on an assessment of students' REU experience. Specifically, students were asked to report their level of agreement ranging from 1 (*strongly disagree)* to 5 (*strongly agree*) with statements indicating whether their experience influenced their decisions to pursue graduate school and/or feeling that they made intellectual contributions to research in their respective host labs. Two questions asked whether students are or will be coauthors on a published paper and/or paper presented at a professional meeting. Using a 5-point Likert-type scale ranging from 1 (*poor*) to 5 (*excellent*), students were asked to rate the following REU experiences: laboratory experience, mentoring, social experience, seminars/lectures, tours, and overall experience. Students were also asked to rate the quality of their relationships with their primary mentor, lab PI (if different than primary mentor), and others in their lab, using a 5-point scale ranging from 1 (*very poor*) to 5 (*very good*).

The final section of the survey was comprised of the 29-item Dimensions of Mentoring Scale (not shown here for reasons of space). The items on the scale were developed to assess students' attitudes towards mentoring along the six dimensions previously described. For example, questions were related to safety (e.g., my mentor emphasized the importance of lab safety), preparedness (e.g., clear laboratory protocols and techniques were provided to me), presence (e.g., my mentor helped me become familiar with the tools and equipment available in the lab), positivity (e.g., my mentor gave me positive feedback on aspects of my work at least once a week), proactive behavior (e.g., my mentor stayed engaged in my project through the duration of the program), and patience (e.g., my mentor was patient and understanding when I asked questions). Students were asked to indicate the extent with which they agreed with each statement on a 5-point Likert-type scale ranging from 1(*strongly disagree*) to 5 (*strongly agree*).

The 70+ responses from the survey were then analyzed, with particular attention paid to the correlation between specific responses and the overall quality of internship experience. To summarize, all the dimensions of mentoring that were hypothesized were highly correlated with the overall REU experience at p values less than or equal 0.01 (table below). Furthermore, second to the hands-on experience in the lab itself, mentoring was the most important single factor in predicting overall quality of experience.

Correlations (r) of Dimensions of Mentoring and Overall REU Experience

Dimension	Overall Experience
Safe	**.48****
Prepared	**.53****
Present	**.47****
Positive	**.36****
Proactive	**.48****
Patient	**.31****

Note: *N*=66; Bold indicates correlations with significant *p*-values, $** = p \leq .01$

Conclusions

We have presented here a series of best practices for running summer research internship programs, developed through experience with over 140 summer interns during the past six years. These practices address REU programs from the stage at which projects are sought from participating labs, to the stage at which program evaluation occurs. They address recruitment, application, participant selection, logistics, mentor training, setting expectations, and conducting assessment and evaluation. We hope that readers find these ideas useful in their own programs, because the deployment of successful REU programs (and similar) is a tremendous opportunity to build human capital and to further human knowledge through research – the practices here seek to increase the likelihood that these overarching goals are in fact met by the program.

Polymer REU at the Colorado School of Mines

Polymer science and engineering provides jobs and produces technologies that advance our quality of life. Therefore, research into novel polymer materials is needed, especially related to cutting edge technologies like fuel cells or solar materials. Here, a unique interdisciplinary research experience for undergraduates combines polymer science (with emphasis on chemistry) and polymer engineering (physical and mechanical property optimization). The overall objectives for the REU site are: 1. Engage and educate undergraduate students in polymer science and engineering with emphasis on technologies needs in the 21st century (e.g., energy), 2. Emphasize diversity in research teams with respect to experience (from faculty to freshmen, non-Ph.D. granting institutions, etc.), gender, and ethnicity, and 3. Quantify successes of the program via thorough assessment, including number of participants earning their Bachelor's degrees, presentations at national and regional meetings, and publications in peer reviewed journals. A close collaboration with NREL includes laboratory tours, a diverse group of mentors and students, a mentored exchange program between chemistry and chemical engineering students, and a vibrant social program highlight the beauty of a Colorado summer.

379938 Successful Operation of a Multi-Institution NSF REU Site

Abstract:

The Colorado Center for Biorefining and Biofuels, C2B2 (www.c2b2web.org) was founded in 2006 by the Colorado Energy Research Collaboratory, CERC (www.coloradocollaboratory.org). Funds came from the State of Colorado to support energy research among the three premier universities in the State – University of Colorado, Colorado State University, Colorado School of Mines and the National Renewable Energy Laboratory; with the objective being that the institutions would partner in renewable energy research and education. A multi-institution REU program was implemented on a trial basis during the summers of 2008 and 2009 (28 students served) using State of Colorado funds, and then later expanded with the addition of NSF award funding since 2010. The C2B2/NSF REU program offers undergraduate students the opportunity to engage in meaningful research alongside faculty, post-doctoral researchers, and graduate students at each of our four partnering institutions. Along with research duties, all students participate in educational seminars, C2B2 institution site visits, industry site visits, planned social activities, and cultural experiences. Students learn how researchers of varying disciplines at all four sites collaborate to solve biofuel and biorefining challenges. This program provides the students a rich and diverse ten week experience within the beautiful backdrop of the Colorado Rocky Mountains. On average, 15 to 18 students are served annually depending on the funds available from NSF, C2B2, and related PI research grants. The program has been overwhelmingly successful and typically considers > 300 applications/yr. The C2B2/NSF REU Program is a favorite of faculty and NREL PIs for summer research. This presentation will discuss how a multi-institutional REU program successfully operates and will provide a framework by which others might consider developing such a program.

392500 REU on Engineered Bioactive Interfaces and Devices at the University of Kentucky

Abstract:

For the past six year, the University of Kentucky has offered an NSF REU Program focusing on Engineered Bioactive Interfaces and Devices. This multidisciplinary program provides rising juniors and seniors with the opportunity to work on multidisciplinary research with faculty from the College of Pharmacy, Department of Chemistry and Engineering. To ensure a multidisciplinary experience, each REU student is assigned two advisors from separate disciplines. The program goes beyond research in the laboratory and includes a number of technical workshops, panel discussions and seminars related to teamwork, intellectual property development, communication skills and what to expect from graduate school. The students are assigned a graduate student who serves as a mentor to the REU students during the 10-week program. The program ends with a poster competition and awards banquet. Best practices and assessment results over the past 6 years will be presented.

Sustainable Energy REU
at Kansas State University

Larry Erickson, Keith Hohn and Wendy Griswold
Kansas State University

National Science Foundation
WHERE DISCOVERIES BEGIN

"Sustainable Energy in the 21st Century" REU

- Funding to start an REU at K-State in the area of sustainable energy was obtained in 2009
- Specific goals for this REU are:

  1. To develop and apply knowledge and skills in the areas of mathematics, science, engineering, and research methods
  2. To develop an understanding of sustainability as an applied multidisciplinary field encompassing economics, science, and technology, and the social, political, and ethical issues facing contemporary societies
  3. To develop a recognition of the need for, and an ability to engage in life-long learning by increasing the capacity for self directed learning and original investigation

Changes to REU Program

- NSF REU grant was renewed for three more years (2012-2015)
- Program maintains its general structure
- Increased emphasis on outreach
  - Outreach to K-12 students (Planetkeepers, EXCITE! Programs)
  - Developing a hands-on activity to demonstrate sustainable energy to be used in the Discovery Center in Manhattan

Added Objective:
To develop skills for communicating scientific and engineering concepts to the public and to K-12 students

"Sustainable Energy in the 21st Century" REU

- REU consists of multiple parts:
  - Research on sustainable energy under a faculty advisor
  - Sustainability project
  - Sustainability seminar
  - Field trips
  - Social events
  - Research meetings, poster session
  - Reflective activities – journals, focus group

Sustainability Seminar

June 4 - Sustainability - What is it and How do we Quantify it? Keith Hohn and Larry Erickson

June 11 - Sustainable Energy: Overview of Opportunities and Challenges - Judy Wu

June 18 - Energy Conservation and Efficiency - David Carter and Ben Champion

June 25 - Photocatalysis for Production of Fuels and Chemicals - Ken Klabunde and Keith Hohn

July 2 - Solar Energy - Ryszard Jankowiak

July 9 - Nanotech Batteries - Jun Li and Clinton Whiteley

July 16 - Renewable and Distributed Power Generation - Ruth Douglas Miller

July 23 - Biofuel Production and Economics - Praveen Vadlani and Donghai Wang

July 30 - Production of Algae for Sustainable Energy and Other Useful Products - Wayne Yuan and Larry Erickson

August 6 - Final REU and EPSCoR Presentations

Integration of REU/ESPCOR

- EPSCOR undergraduates were integrated into the NSF REU site in the following ways:
  - All students viewed sustainability seminars
  - All students attended two field trips, EPSCOR 2011 annual meeting, and final poster session
  - KSU EPSCOR students attended social events and SUROP seminars, and participated in sustainability projects

Participants

- This site engaged 77 students (58 supported by the REU site, 27 by the NSF EPSCOR project, and 2 supported by other research grants) in cutting-edge research on sustainable energy from 2009-2014

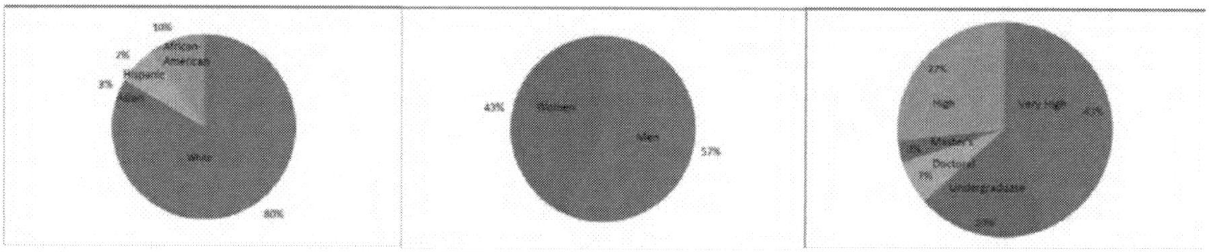

Figure 2. Demographics of "Earth, Wind, and Fire: Sustainability in the 21st Century" program (2012-2014): ethnic/racial on left, gender in middle, and institution type (Carnegie classification) on right.

Field Trips

- Field trips were taken to:
  - Meridian Way Wind Farm
  - Nesika Energy ethanol plant
  - Bowersock Mills & Power Company hydroelectric plant
  - Algal reactor, Lawrence, KS wastewater treatment plant

Sustainability Project

- Two group projects:
  - Solar parking lot
    - White paper was prepared that investigated the potential for implementing solar panels in a parking lot to generate electricity
  - Outreach projects: student-led project to communicate with the public on sustainable energy
    - Develop activity for Discovery Center, discuss sustainable energy with K-12 students in Planetkeepers and EXCITE! programs

Social Events

- Social events included:
  - Welcoming picnic
  - Picnic, tour of Konza Prairie
  - Trip to Cosmosphere
  - Brown Bag lunches
  - Pizza and movie night

Final Poster Session

- At end of ten weeks, students prepared posters on their research
- For last three years, a joint poster session has been held with six different REU programs (over 50 posters in 2014). Awards, sponsored by Phillips 66 were provided for the top five posters as judged by faculty and graduate student judges and to the top three posters as voted on by the presenters themselves.

Evaluation

- Parallel mixed data analysis
- Qualitative data collection is a reflective program element

Variable	Mean (1-10 scale)
Skills in technical area	8.51
Understanding sustainability	7.91
Attend graduate school	6.85
Research career	7.03
REU career benefit	8.66

Sample 2011 participant summative comments:

"The internship I had last summer was absolutely one of the best experiences I have had. It helped me to solidify my decision to pursue sustainability as a major part of the rest of my education and future career. I learned so much and it has proven useful in a lot of my classes (both natural and social science based classes). Thanks so much for giving me the opportunity!"

"This experience has assured me that I would like to continue my education and career in academics for the foreseeable future. I have also been heavily impacted personally by the REU. I have developed several friendships that will last much longer than the summer did. It has furthered my advancement in my field. My resume has become much more impressive due to the experience and the conference participation it enabled. I feel more confident in my ability to succeed and excel in academics."

"I find it easier to find examples of sustainability inside and outside of class now. Before, whenever I was asked to give an example of sustainability, it would take me a while. Now, when I am posed with the same question, it takes me seconds to come up with one since I have participated in so many sustainability-related discussions during my summer."

More recent summative comments

- *In the past I viewed the issue [sustainability] as more of an abstract idea with non-concrete issues and solutions involved. Instead, I now have a solid grounding in the obvious problems we face and some of the things we must do to even hope to resolve these serious problems.* **Student 223, 2013**
- *Education is a very important thing on the road to sustainability. Many people do not really understand what it means or how we can accomplish it. By practicing with the kids [outreach project] we learned more about how to take our research and put it into simpler terms for the common person to understand. If we want people to support our work, we have to be able to let them know what we're doing and why it's important, and I think this helped us learn that.* **Student 211, 2012**
- *Most of the time in my [prior] research, I've been given a specific task with specific instructions, but here I was pretty much left to my own devices and I had to become a better problem solver and think more in an advanced state to interpret my own data. Essentially I had to move up from my regurgitation comfort zone and start creating. While this was scary at first, I have come to see how COOL it is. The poster I just made and will present is MY OWN NOVEL RESEARCH. No one else has done it or gotten my results. That just blows my mind!* **Student 214, 2013**

Two manuscripts have been published:

Goldin, E., L. Erickson, B. Natarajan, G. Brase, and A. Pahwa. 2014. "Solar Powered Charge Stations for Electric Vehicles," Environmental Progress and Sustainable Energy 33: 1298-1308.

Robinson, J., G. Brase, W. Griswold, C. Jackson, and L. Erickson. 2014. "Business Models for Solar Powered Charging Stations to Develop Infrastructure for Electric Vehicles," Sustainability 6: 7358-7387.

Internet Site:

http://www.che.ksu.edu/reu/

Speaker Contact Information:

Larry E. Erickson
Professor of Chemical Engineering and
Director, Center for Hazardous Substance Research
Kansas State University, Manhattan, KS 66506
lerick@ksu.edu
Phone: (785) 532-4313

Publications

1. Walker, K., Vadlani, P., Brijwani, K., Madl, R., Ugorowski, P., and Hohn, K.L. "Ethanol fermentation from food processing waste", *Environmental Progress and Sustainable Energy*, 32:1280-1283 (2013).
2. Roy, S., Aguirre, A., Higgins, D., and Chikan, V., "Investigation of Charge Transfer Interactions in CdSe Nanorod P3HT/PMMA Blends by Optical Microscopy", *Journal of Physical Chemistry C*, 116:3153-3160 (2012).
3. Y. Kuo, C.D. Frye, M. Ikenberry, K.J. Klabunde, "Titanium-indium oxy(nitride) with and without RuO2 loading as photocatalysts for hydrogen production under visible light from water", *Catalysis Today* **199** (2013) 15-21.
4. Reinhard, E., Champion, B., Schulz, N.N., Gould, R., Perez, E., and Brown, N., "Computer power consumption and management: "Earth, wind, and fire: Sustainable energy for the 21st century" research experience for undergraduates," *Power Systems Conference and Exposition (PSCE), 2011 IEEE/PES* , pp.1-8, 2011, doi: 10.1109/PSCE.2011.5772621.
5. Ko, T.-W. and Pahwa, A., "Investigating Small Renewable Energy Light Sources to Meet the Needs of Off-Grid People," *Proceedings of International Conference on Power Systems Operation and Planning*, Abuja, Nigeria, Jan. 2010.
6. Erickson, L.E., Griswold, W., Hohn, K.L., and Saulters, O. "Enriching and Evaluating Sustainability Education", *Journal of Sustainability Education* **1** (2010) 1.
7. Reed, R., Erickson, L.E., and Yuan, W., "Growth of Algae Using Nitrogen and Phosphorous in Waste Water", Proceedings of the 39th Annual Biochemical Engineering Symposium, Kansas State University, pp. 16-24.
8. J. Aronoff, B. Champion, C. Lauer, A. Pahwa, "The Effectiveness of Community-Based Social Marketing on Energy Conservation for Sustainable University Campuses", *IEEE Power and Energy* **11** (2013) 30-38.
9. E.L. Goldin, L.E. Erickson, B. Natarajan, G. Brase, A. Pahwa, "Solar Powered Charge Stations for Electric Vehicles", *Environmental Progress and Sustainable Energy,* 33:1298-1308 (2014).

10. K. Hudek, L.C. Davis, J. Ibbini, L.E. Erickson, “Commercial Products from Algae”, in: R. Bajpai, A. Prokop and M. Zappi (Eds.), Algal Biorefineries, Springer, New York, NY, p. 275-295 (2013).
11. M. Ikenberry, L. Pena, D. Wei, H. Wang, S. Bossmann, T. Wilke, D. Wang, V.R. Komreddy, D.P. Rillema, K.L. Hohn , "Acid monolayer functionalized iron oxide nanoparticles as catalysts for carbohydrate hydrolysis", *Green Chem.* **16** (2014) 836-843.
12. W.W. McNeary, L.E. Erickson, "Sustainable Management of Algae in Eutrophic Ecosystems", *Journal of Environmental Protection* **4** (2013) 9-19.
13. S. Pahwa, A. Hodges, C. Scoglie, S. Wood, “Topological Characteristics of the Power Grid and Mitigation Strategies Against Cascading Failures”, IEEE Systems Conference, 2010; DOI 10.1109/SYSTEMS.2010.5482329.
14. J. Robinson, G. Brase, W. Griswold, C. Jackson, L.E. Erickson, "Business Models for Solar Powered Charging Stations to Develop Infrastructure for Electric Vehicles", *Sustainability* 6:7358-7367 (2014).
15. B. Ronnebaum, L.E. Erickson, A. Pahwa, G. Brase, M. Babcock, "Economics and External Costs of Solar Powered Charge Stations for Electric Vehicles", *Renewable and Sustainable Energy Reviews* (2014), submitted.
16. T. Wagner, L.E. Erickson, "Sustainable Management for Eutrophic Lakes and Reservoirs", *Ecological Engineering* (2014), submitted.
17. Griswold, W.M. (2012). Journals as evaluative tools. In Glowacki-Dudka, M. (Ed.), Proceedings of the 31st Annual Midwest Research-to-Practice Conference in Adult, Continuing, Community and Extension Education, (pp. 59-64). Accessible at http://www.uco.edu/ceps/events/mwr2p.asp

The Importance of Educating and Mentoring High School Students on the Selection of the Chemical Engineering Bachelor's Degree

*Contributing Authors: Holly Robillard (holly.robillard@uconn.edu)
and Teresa Piliouras, PhD (piliour@optonline.net)*

Students that are highly successful in their mathematics and science courses are often told by family, teachers, and guidance counselors to pursue college degrees in related fields. However, the breadth of STEM fields is wide ranging and can be challenging to narrow without further advice.

Students that have family members and/or friends in chemical engineering are inclined to choose it as their major in college. After speaking to undergraduate students enrolled in the chemical engineering program and alumni at the University of Connecticut, it has been found that their selection of the major is frequently the result of exposure and encouragement from mentors in the field, second to an interest in chemistry. In contrast, high school students who are uneducated about chemical engineering as a career are less likely to adopt the major in their undergraduate years. Thus, it is highly likely that many students that have the potential to be successful in the field do not pursue the degree because of this lack of education and mentorship. Such as can be seen most clearly amidst the population of underrepresented students in engineering (i.e. women and minorities).

In the hope of remedying these issues, undergraduate chemical engineering students are working on providing information and support to high school students around their university. Organizations such as Engineering Ambassadors visit local schools to "engage a diverse student population... in activities that inspire them to explore a variety of creative solutions to the problems facing humanity" [1]. Through developing and giving presentations regarding current engineering feats and by "educating students on the importance of engineering to the health, happiness, and safety of our world," the group hopes to inspire students to consider college and a major in engineering [1]. Furthermore, the Engineering Ambassadors at the University of Connecticut have worked to provide students with continuing education and mentorship through recurring outreach visits, engineering events for middle and high school students, and website content suited towards their questions.

On a national scale, organizations and companies are working to provide all high school students with the education regarding career paths in STEM and mentoring through online presentations and webinars. One such program is Best We Can Be™, a part of Technical Consulting & Research, Inc., that "cultivates life skills needed to succeed, promotes college & career readiness and develops proficiency using sophisticated technologies" [2]. Through online, interactive presentations made by undergraduates, graduates, and young professionals, high school students are not only informed about the presenters' respective fields but also develop a connection with these mentors.

This project will show that, by providing high school students with valuable information about the field and effective mentorship opportunities, they will be more inclined to select chemical engineering as a major in college prior to or during their first two years as an undergraduate. An analysis of the issues as well as the motivation, execution, and impact of these programs will be performed.

Works Cited:

[1] *Engineering Ambassadors*. University of Connecticut, 12 May 2014. Web. 12 May 2014.

[2] *Best We Can Be*^{TM}. Technical Consulting & Research, Inc., 12 May 2014. Web. 12 May 2014.

azengineering

Leading Institutional Change Through Faculty Support and Engagement with Administrators: Use of Evidence Based Teaching Practices

Gail Burd[a,b], Debra Tomanek[b,c], **Paul Blowers[d]**, Molly Bolger[b], Jonathan Cox[b], Lisa Elfring[b,e], Elmer Grubbs[f], Jane Hunter[c], Ken Johns[g], Loukas Lazos[h], Roman Lysecky[h], Drew Milsom[g], Ingrid Novodvorsky[c], John Pollard[e], Edward Prather[i], Vicente Talanquer[e], Kay Thamvichai[h], Hal Tharp[h], and Colin Wallace[j]

[a] Senior Vice Provost for Academic Affairs, University of Arizona
[b] Department of Molecular and Cellular Biology, University of Arizona
[c] Office of Instruction and Assessment, University of Arizona
[d] Department of Chemical and Environmental Engineering, University of Arizona
[e] Department of Chemistry and Biochemistry, University of Arizona
[f] Department of Electrical and Computer Engineering, Northern Arizona University
[g] Department of Physics, University of Arizona
[h] Department of Electrical and Computer Engineering, University of Arizona
[i] Department of Astronomy, University of Arizona
[j] Department of Physics and Astronomy, University of North Carolina at Chapel Hill

1

THE UNIVERSITY OF ARIZONA.

azengineering

A Question for Faculty…

What is the purpose of having students come together to meet several times per week for an hour or more?

What do we want students to be able to do after a class meeting?

2

THE UNIVERSITY OF ARIZONA.

azengineering

In the Literature…

Active learning is crucial

8410–8415 | PNAS | June 10, 2014 | vol. 111 | no. 23

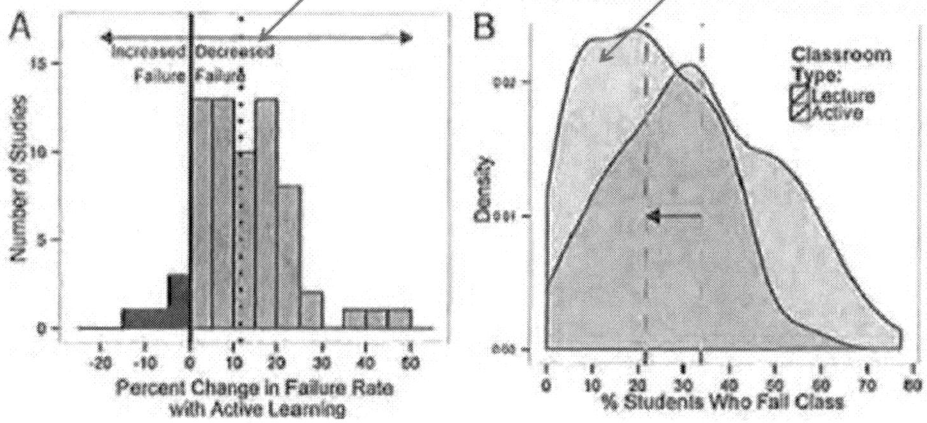

Active learning increases student performance in science, engineering, and mathematics

Scott Freeman[a,1], Sarah L. Eddy[a], Miles McDonough[a], Michelle K. Smith[b], Nnadozie Okoroafor[a], Hannah Jordt[a], and Mary Pat Wenderoth[a]

THE UNIVERSITY OF ARIZONA.

azengineering

More Literature Informing the Work

Course structure matters

Increased Structure and Active Learning Reduce the Achievement Gap in Introductory Biology

David C. Haak,[1*] Janneke HilleRisLambers,[1] Emile Pitre,[2] Scott Freeman[1]†

Active Learning and Peer Instruction

Peer Instruction: Ten years of experience and results

Catherine H. Crouch and Eric Mazur[a)]
Department of Physics, Harvard University, Cambridge, Massachusetts 02138

Am. J. Phys. 69 (9), September 2001

Innovation in large lectures--teaching for active learning

Diane Ebert-May; Carol Brewer; Sylvester Allred

Bioscience; Oct 1997; 47, 9; Research Library Core pg. 601

Transforming the lecture-hall environment: The fully interactive physics lecture

David E. Meltzer[a)]
Department of Physics and Astronomy, Iowa State University, Ames, Iowa 50011
Kandiah Manivannan
Department of Physics, Astronomy, and Materials Science, Southwest Missouri State University, 901 South National Avenue, Springfield, Missouri 65804

Am. J. Phys. 70 (6), June 2002

4

THE UNIVERSITY OF ARIZONA.

azengineering

How We Moved from Individuals Working Alone to A Larger Group Getting Things Done

What were the Boundary Conditions:

Had a coalition of the willing: 10-15 random faculty spread across 7 STEM departments who were "doing stuff"

Have many administrators who came from STEM fields and know the challenges of educating students in this area:

- Provost
- Vice Provost for Instruction
- Interim Vice President of Research
- Vice President of the Office of Instruction and Assessment

Opportunity presented itself through the AAU-STEM grant application – which we received, and that was the catalyst that carried the effort forward

5

THE UNIVERSITY OF ARIZONA

azengineering

Our Model for Influencing Faculty Teaching Practices: Year 1

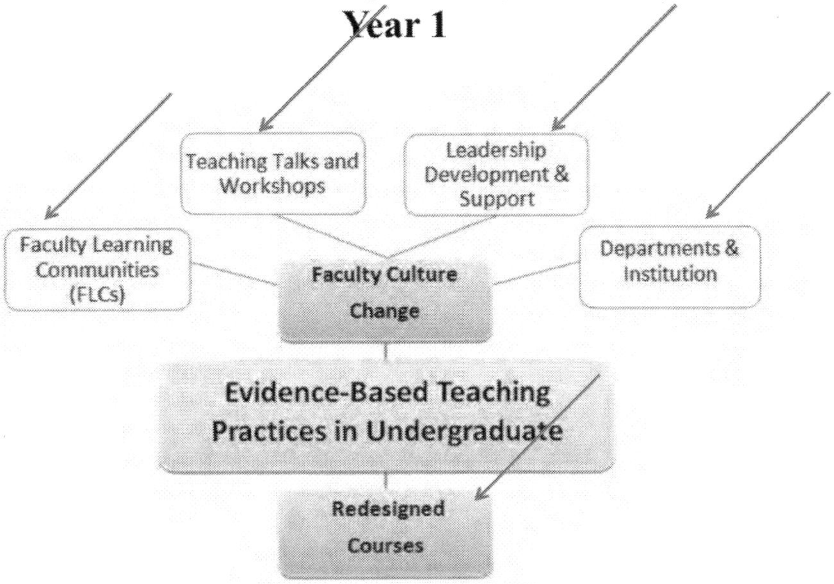

THE UNIVERSITY OF ARIZONA

azengineering

Faculty Learning Communities: Year 1

Sent open invitation to STEM instructors and faculty. Optimistically hoped for 8-10 responses.

Had 36 faculty join: we formed 4 groups.

AAU STEM FLC SPRING 2014
Draft Curriculum

Week	Meeting	Topic
1. Jan 20-24	FCL Meeting 1	The Learning Cycle: Exploration and Concept Introduction Reading: • Lawson, A. E. (2001). Using the learning cycle to teach biology concepts and reasoning patterns. *Journal of Biological Education*, 35(4), 165-169 Notes: Present learning-cycle graphic in group after doing a learning-cycle exploration phase activity. In addition to the reading, ask participants to come with some ideas about skills/concepts their students struggle with around the time of the two implementation weeks (see Projects 1 & 2 Implementation weeks below). Introduce the principle of backwards design.
2. Jan 27-31		
3. Feb 3-7	FLC Meeting 2	The Learning Cycle: Concept Application Notes: Complete the learning cycle with an application activity involving skills/concepts brought to the "table" at Meeting 1. Perhaps a discussion of a reading on backwards design (Lisa has identified one that we can review). Plan and provide guidelines for Project 1 to be presented by each participant at Meeting 3.
4. Feb 10-14		
5. Feb 17-21	FLC Meeting 3	Presentation and Discussion of Individual Project Plans (Project 1)
6. Feb 24-28	Project 1 Implementation in FLC participants' classes	
7. Mar 3-7	Notes: Use FLC mentors, other AAU instructors, CCT graduate students to observe and provide feedback.	
8. Mar 10-14	FLC Meeting 4	Reflection on Project 1 Implementation Notes: Debrief on the implementation. Target assessment in particular. Find an assessment article appropriate for the group to read in preparation for this meeting.
9. Mar 17-21	**SPRING BREAK**	

THE UNIVERSITY OF ARIZONA

azengineering

Teaching Talks: Year 1

The AAU STEM Education Project
Teaching-Talks

ARIZONA

The UA Science-Technology-Engineering-Math (STEM) community values excellent teaching. The AAU STEM Education Project was developed to support our pursuit of this excellence. If you are a STEM faculty member, please join us in our *Teaching-Talks*, a series of discussions about best practices in STEM teaching.

RSVP with your intention to join us Jan. 27 to kick off a new series of great discussions for Spring 2014. Send your RSVP to Becky Pérez at rperez@email.arizona.edu by Friday, Jan. 24th. Invite your STEM friends. See you soon, STEM colleagues!

Teaching-Talks are sponsored by the AAU STEM Education Project and the Office of Instruction and Assessment.

Please join us:

Monday, January 27th, 1-2pm in the Ventana Room, SUMC
Topic: Pre-Post Course Assessment of Learning

Thursday, February 27th, 9-10am in the Ventana Room, SUMC
Topic: Effective Practices for Student Collaborative Learning

Tuesday, April 1st, 3-4pm in the Agave Room, SUMC
Topic: The Role of Student Predictions in Science Learning

Tuesday, April 29th, 9-10am in the Ventana Room, SUMC
Topic: Daily Class Reflections that Enhance Learning

Idea was to catch people with quick "one-offs" that were bursts of information

Attendance was 6-40 faculty members per meeting

Most received high evaluations for ability to help faculty

Did not seem fulfilling for leaders who gave these…unfocused as a whole.

THE UNIVERSITY OF ARIZONA

azengineering

Leadership Development: Year 1

Internal Leadership Faculty Learning Community Strengths:

Broad discussions across discipline boundaries about what learning means and how it occurs: biology, chemical engineering, chemistry, electrical and computer engineering, physics, astronomy.

Areas explored:

- modeling
- predictions
- quantitative reasoning
- computations and reaching end answers
- information literacy
- assessment of learning gains

Weaknesses: We were not as focused as we could have been to get more done

THE UNIVERSITY OF ARIZONA.

azengineering

Departments and the Institution : Year 1

Weaknesses: We were not as focused as we could have been to get more done

This did not happen. We had plans to host department heads and engage them in discussions

Class Redesigns: Year 1

Phys compared best lecturer to new lecturer who did active engagement – AE led to +10% on each exam question consistently after first exam

Chem voted to implement large scale transformation of 6,000 person per year gen chem with support for faculty training

ENGR, including chem. Eng., implemented redesigns with some hitches

Biology started a series of discussions that would normalize biology experiences for freshmen students, much like chem has now done

THE UNIVERSITY OF ARIZONA.

azengineering

Year 2: Faculty Learning Communities

Week	Meeting	Topic
1. Sept 15-19	FCL Meeting 1	*Active Learning: Student Performance and Instructional Models* Readings: • Freeman, S. et al. (2013). Active learning increases student performance in science, engineering, and mathematics. PNAS, 111(23), 8410-8415. • Lawson, A. E. (2001). Using the learning cycle to teach biology concepts and reasoning patterns. *Journal of Biological Education*, 35(4), 165-169. Participants should complete these two readings before the meeting and come prepared to share their ideas, opinions, and beliefs about the information presented in these papers. They should also come prepared to share their expectations and goals for their participation in the FLC this semester.
2. Sept 22-26	FLC Meeting 2	*Active Learning: Some Examples* Readings: • Exemplars of different learning cycles (Exploration, Concept Introduction, and Application Phases) Participants should read and analyze the exemplars provided before the meeting. They should also identify some conceptual challenge frequently faced by the students in the classes they teach (my students often struggle to….), and think of potential instructional projects they may want to implement in their course this semester to improve student learning in that area. Participants should analyze the different exemplars of learning cycles and use them as a guide to create a first draft of a learning cycle they plan to implement in their class this semester. Participants are expected to share their initial ideas at the meeting.
3. Sept 29-Oct 3	FLC Meeting 3	*Sharing and discussing final plans for Project 1* Participants will share and discuss their final plans for Project 1 implementation
4. Oct 6-10		**Project 1 Implementation in FLC participants' classes**
5. Oct 13-17		

Tightened up curriculum to be done in one semester.
Asked faculty what tools they were comfortable with.

Some returning members – some departments have an "advocate" who is sending faculty to join the effort. The coalition of the willing is growing.

THE UNIVERSITY OF ARIZONA.

azengineering

Year 2: Faculty Learning Communities 2

"In order for the next phase of our discussion to have some basis on what in in the literature for active learning techniques, please circle your level of exposure from 0 (have never heard of this technique) to 5 (I am very familiar and use this technique very comfortably) for each of the techniques named below:"

	Count of Scores					
Technique	**0**	**1**	**2**	**3**	**4**	**5**
Animations	0	1	1	0	1	3
Case Studies	0	1	0	0	4	2
Clickers	0	4	0	0	0	3
Collaborative learning	1	0	0	2	2	2
Computer simulations	0	1	2	0	3	1
Concept inventories	4	0	3	0	0	0
Concept maps	2	1	3	0	1	0
Cooperative learning	0	1	1	0	2	3
Formative assessment	4	1	0	1	1	0
Interactive demos	0	1	1	1	0	4
Just-In-Time teaching	6	0	1	0	0	0
Peer instruction	0	0	2	2	2	1
POGIL	4	0	1	2	0	0
Problem-based learning	1	1	0	1	2	2
SALG	6	1	0	0	0	0
Think-Pair-Share	2	1	0	1	0	3

We are finding a need for assessment training and discussions.

12

THE UNIVERSITY OF ARIZONA.

azengineering

Teaching Talk Topics

Designing Classes for Student Thinking: The Role of Predictions

When you plan for your classes, how is your time spent? Outlining your talking points? Selecting diagrams to structure your explanations? Creating student thinking exercises? Research on learning suggests that the last planning activity has the greatest potential for advancing student learning. In this session, you will consider examples of stimulating student thinking with student predictions.

Changing Students from Passive Note-takers (or Daydreamers!) to Active Learners

Research on learning tells us that when students grapple with concepts and problems during class, they are more likely to learn. Strategies to foster this engagement can be easily incorporated. In this session, you will try several strategies that you can implement in your own classes.

Classroom Assessment Techniques: How to Determine What Your Students are Learning

Assessing what your students are learning during your class sessions can be a powerful guide for your teaching and their learning. And these assessment ... ou ... ly to apply them in your course.

In fall 2014, 3 departments requested 1 or more talks

THE UNIVERSITY OF ARIZONA

azengineering

Yea

Expanded team and are looking outside of STE

Designed a Collaborativ done and we are in the m

Invited Chairs from across UA to observe: 10 or more observations that are generating discussions in just 3 weeks…The importance of chairs.

A Different Kind of Space Launch

By La Monica Everett-Haynes, University Relations - Communications | October 29, 2014

The UA Collaborative Learning Spaces Project was launched to revolutionize the way faculty teach and the way students engage in their own learning. (Photo: Paul O'Mara)

The UA is enhancing classroom technology, restructuring the physical space of classroom and moving away from tradition to improve instruction and learning.

THE UNIVERSITY OF ARIZONA

azengineering

Year 2: Course Redesigns

More faculty from Phys have moved from FLCs to leadership positions

Chemical engineering has a new chair who is shifting instructors around to get active engagement faculty to saturate the sophomore courses, which was the hitch in the prior year.

Chemistry is feeling growing pains as they try to get faculty who have not been AE faculty to use the AE materials that have been created…Student pushback is occuring

Biology is testing new curricula and comparing to traditional offerings while trying to build consensus about what THE biology course should contain.

More redesigns are cascading from these into higher level courses through discussions in FLCs and growing coalition.

THE UNIVERSITY OF ARIZONA.

azengineering

Unexpected Gestalt – Classroom Workshop: How Do People Learn?

16

THE UNIVERSITY OF ARIZONA.

azengineering

Unexpected Gestalt – Classroom Redesigns

UA is gearing up to transform 2 or more spaces to be up and fully running in Fall 2015, using what we learned in the CLS this fall.

17

THE UNIVERSITY OF ARIZONA.

azengineering

Conclusions and Future Work

Even at a research I institution, an open call can generate discussions among faculty about how to teach with evidence based methods

Faculty learning communities with implementation plans and trials are more effective than theoretical discussions about teaching

A goal oriented project manager is critical when building a grass-roots effort, even with the support of upper administration

Having participation from many departments strengthens efforts to achieve culture change

Undergraduate STEM Education Initiative
Association of American Universities

18

THE UNIVERSITY OF ARIZONA.

377120 What Is the Value of a Campus Classroom Experience Anyway?

Abstract:

Innovative learning environments such as massively online open courses (MOOCs) present interesting questions about traditional in-class, on-campus learning experiences. We previously demonstrated that in-person social networks can significantly influence student performance and retention. For example, women who know the most connected or the top students in class experience 13-15% rise in their grades relative to those who did not. Yet, key questions remain about how these networks develop and why they are effective. For example, how do in-person student networks form over the course of a semester? What do students gain from these networks? Do students cluster based on their Felder-Soloman learning styles, grades, or connectivity? How does classroom seating correlate with student performance and course evaluations? This talk will discuss our findings for each of these questions.

385130 Impact of Learning Styles on Student Performance and Self-Efficacy in Material Balances Course

Abstract:

The popular American children's story, *The Little Engine that Could*, is most renowned for the engine saying "I think I can, I think I can" and instilling the power of self-belief in the young. The importance of this self-belief, otherwise termed self-efficacy, has been studied by many researchers at the university level. Ties between self-efficacy and learning style, learning-style and performance, and self-efficacy and performance have been independently investigated by others in a variety of contexts. We are studying the link between learning style preferences, self-efficacy, and student performance in the context of an introductory chemical engineering course (process fundamentals, *i.e.*, material balances). We will be using the natural bias of homework and exam problems to evaluate student self-efficacy, performance, and whether this correlates with their learning style preferences. Using the Felder-Silverman model of learning styles, and the subsequent Felder-Soloman Index of Learning Styles (ILS) indicator, we are individually analyzing each learning style dimension (active/reflective, sensing/intuitive, visual/verbal, and sequential global) to better understand the subtleties of learning style effects. We are interested in how student preferences may influence their performance, and perceptions/self-efficacy when presented with tasks demonstrating learning style bias. By exploiting the naturally occurring biases in the regularly assigned homework and exam problems, we are able to study correlations between learning styles, performance, and self-efficacy without impacting the course. By evaluating the traditionally run lecture course as it is, we are able to establish a baseline for understanding the role of learning styles in our curriculum. The sample size of this study is approximately 100 students per semester. Data collected include ILS results, student grades by specific task/problem and self-efficacy survey responses, as well as final grade. Surveys are administered with specific homework assignments and exam problems, with questions & statements evaluating student responses to problems based on a Likert scale. Surveys include items such as: I am confident in my answer for problem 5; I am uncertain I solved problem 5 correctly; I am confident in solving problems similar to problem 5. Results provide evidence regarding whether learning styles can affect student performance or self-efficacy, as well as identify naturally occurring biases in chemical engineering problems that are translatable to other fields. We believe that this integrated study of learning styles, student performance, and self-efficacy using the Felder-Silverman model in the context of a traditional, lecture-based engineering classroom will provide valuable information for engineering education. It is commonly accepted that every field attracts a certain "type" of individual, but a correlation between learning styles and success that demonstrates similarities among students who do not succeed in introductory chemical engineering would suggest that our curriculum repels students otherwise interested in the field. A better understanding of student background and learning preferences is invaluable in developing improved curricula, especially in the ever-changing modern classroom.

379441 Building Chemical Engineering Students from Miscellaneous Parts & Vague Instructions: A Hands-on First-Year Design Laboratory

Abstract:

Lab courses and open-ended problem solving offer benefits in conceptual learning and student satisfaction that are often underutilized early in our curriculum. Chemical engineering departments do not typically offer their first-year students a laboratory experience that is focused on core chemical engineering concepts and the design process. Instead, first-year laboratory courses primarily occur within chemistry and physics departments, and these courses do not typically exercise the type of creativity and innovation that attract many students to engineering.

We created a freshman chemical engineering design course and laboratory, *Chemical Engineering Design & Innovation*, and have refined it over three semesters. This course is situated in the second semester of our curriculum, after a more traditional lecture-based introduction to chemical engineering course.

For each of the course projects, individual students first use browser-based simulations to familiarize themselves with the relevant theory and data analysis. They are then put into teams and given project design goals and access to a wide range of inexpensive materials and tools. Such a project may include the design and construction of a spectrophotometer and flow cell to be used to track a reaction, or of a photobioreactor designed to grow algae effectively. Students then validate their designs and compare their data to that predicted by theory. Students are trained on the use of Arduino microcontrollers and MATLAB to enable them to collect data from a variety of sensors. Students finish this course with a collaborative assignment on our seniors' final project, andthen a final project of their own design.

We have compiled results from pre- and post-tests and student surveys from this course. The data seems to indicate that this course has been greatly successfully in achieving its goals and has improved with each iteration. Students report remarkably high levels of satisfaction with the course compared to more traditional lecture-based courses, and indicate that it has helped them make social connections within the department. The prototyping and data-analysis skills that students have gained early in the curriculum have also greatly increased the value of our freshmen to faculty research programs and those who hire our students as interns.

384077 Analysis of Student Usage on Interactive Simulations

Abstract:

We have developed open-source interactive browser-based simulations that model realistic chemical engineering experiments [1]. In these simulations, students are able to adjust several parameters of a system and observe the results. These simulations can be easily implemented into courses as homework assignments, as they contain randomized unknown parameters that students may be required to determine and submit. We also use these tools to prepare students for hands-on design projects involving the same chemical engineering core concepts and theory.

Our simulation code not only records if students correctly determine unknown parameters, but also records the time and location of all mouse movements, clicks, and field alterations. The resulting database of student usage data may be used to determine trends inherent in successful and unsuccessful results, allowing for iterative development of simulations that optimally benefit student learning.

We have incorporated several of these simulations into our department's newly developed freshman laboratory course, such as a spectrophotometer with batch reaction, a diffusion experiment, and a bioreactor. In this course, the students are required to determine an unknown parameter in the simulations for several of their homework assignments, with the goal of developing a student's engineering intuition. For example, in one homework assignment, students are required to deduce mass transfer principles from a simulation modeling diffusion of a dye out of a sphere [2]; namely, they are to determine how the rate of diffusion is affected by both the diameter of the sphere and the diffusivity of the dye. This simulation mimics a student lab assignment in which a number of alginate beads containing a high concentration of dye are put into a beaker of pure water. The concentration of the dye in the water outside of the bead is plotted over time, and later the students use this simulation with real-world data to estimate the diffusivity of their dye.

This process of simulation usage followed by hands-on experimentation has proven to be an effective way to develop a foundation for core chemical engineering concepts within freshmen. The simulations allowed students to discover on their own, using any internet-enabled device, how specific physical parameters can be determined, while going at their own pace, and without wasting valuable lab resources or in-class time.

Preliminary analysis of the student usage data has revealed usage patterns and students' apparent assumptions about simulation functionality that may be used to improve that functionality. Furthermore, differences in behavior between successful and unsuccessful student attempts may be used to aid in student learning and identify and intervene with struggling students while they are still in the process of working through a problem.

[1] A. Butterfield, "Online Teaching Material," 2011. [Online]. Available: http://www.che.utah.edu/~tony/OTM/. [Accessed: 10-May-2014].

[2] A. Butterfield, "1-D Mass Transfer in a Sphere," 2011. [Online]. Available: http://www.che.utah.edu/~tony/OTM/DiffusionSphere/. [Accessed: 10-May-2014].

Assessment of the impact of customer focused activities within a virtual internship environment on students' product design process

Matthew R Markovetz [1], Renee M Clark[2], Cheryl A Bodnar [1],[2]
[1] Department of Chemical and Petroleum Engineering, Swanson School of Engineering, University of Pittsburgh, Pittsburgh, PA, USA.
[2] Engineering Education Research Center, Swanson School of Engineering, University of Pittsburgh, Pittsburgh, PA, USA.

Engineering design courses are traditionally focused towards generating a product that maximizes performance while providing a solution to the design problem. In this paradigm, the needs of the end-user (or customer) are not necessarily taken into direct consideration. A better approach may be to incorporate customer-desired attributes from the onset of the design process and iterate on the design to ensure the final product will meet the end-user needs. Integration of customer-focused experiences within design courses may then ultimately provide a model for this process that will easily be translatable to industry.

In order to determine whether the use of customer focused activities within a model of the product design process can impact the students' product design process, two separate sections of a chemical engineering, sophomore level product design course participated in the epistemic game Nephrotex. In the Nephrotex environment, the ultimate goal is for teams of 5 or 6 students to design a hemodialysis membrane as virtual interns for the company Nephrotex. One class section conducted focus groups with virtual customers wherein desirable attributes for the end-user were discussed; a second (control) section played through the same environment, but without focus groups. Both groups had access to otherwise identical virtual resources and live mentorship from graduate students playing the role of design advisors at the company. Additionally, students in both sections were given the same design constraints from virtual internal consultants for Nephrotex. Outcome measures of interest for the study were the number of students in each class that reported spending most of their time analyzing customer needs and feedback as well as the accessibility of the product to the consumer in terms of final market cost.

Students from each section were asked each week through a survey what three tasks they had spent the most time addressing. Of those students that reported "customer needs analysis and feedback" at any point in the project, 30 were in the section with customer focus groups, and 12 were in the section without. This difference was statistically significant based on a z test of proportions ($p<0.0005$). An important issue to the virtual customers was that of controlling product cost. Upon completion of the internship, each group presented a final design along with its sale price. The average price of the membrane designs generated by groups that had participated in focus groups ($n=10$) was $112, whereas the average price of membrane designs generated in the other section ($n=10$) was $125, which is also a significant difference ($p=0.0318$, unpaired t-test; $p=0.0261$, Wilcoxon rank sum test). Combined, these results make a case for the somewhat intuitive, but highly important, notion that customer-focused design can be fostered early in the engineering curriculum in a way that also satisfies traditional performance criteria. Ultimately, establishing the consumer as an important part of the design process for inexperienced students may set the stage for increased success upon entrance into the workforce.

370867 Exposing Chemical Engineering Students to Pharmaceutical Concepts through Introductory-Level Experiments and Illustrative Exercises

Abstract:

Over the past several years we have explored ways to incorporate concepts of pharmaceutical engineering within the chemical engineering curriculum. Our philosophy is to integrate these concepts in freshman and sophomore level courses and provide an experience that reinforces core educational objectives and increases student interest in the pharmaceutical field. This paper is a continuation of our educational methods development, and will describe several pharmaceutical and consumer product experiments. The pharmaceutical and consumer product concepts include: drug delivery, drug formulations, quality control, pharmacology, active ingredient function, and pharmaceutical processing. These experimental methods and supporting exercises were created to be used in the lower level engineering curriculum and in outreach activities. One of these experiments involves the manufacture and evaluation of drug delivery strip films, used in consumer products (breath fresheners and energy supplements) as well as analgesic drugs. Their production and consumer use provide an opportunity to engage engineering students in learning the concepts of phase change, material balances, diffusional mass transfer, and statistical analysis. In addition, methods used to measure strip film efficacy (dissolution and API delivered) and aesthetics (color and elasticity) are integrated. A hypothetical exercise is provided to help students understand how manufacturing requirements translate to the amount of raw materials used. An experiment in pharmacodynamics/pharmacokinetics with Alka-Seltzer® helps students understand how to develop a RICE (Reagent, Initial concentration, Change in concentration, Equilibrium concentration) table to determine the equilibrium constant. The stoichiometry of the "plop, plop fizz, fizz"™ is also explored. Crystallization experiments using dextromethorphan illustrate API solubility curves for different solvents. This work is part of the educational outreach efforts of the NSF ERC for Structured Organic Particulate Systems.

377212 The Engineering and Computing Community at USC

Abstract:

The Engineering and Computing Community (ECC) is a residential living and learning community at the University of South Carolina. The ECC is designed to enrich the educational environment for students in the College of Engineering and Computing, focusing mostly on entering freshmen. Various activities are organized throughout the year to help these students develop personally, professionally, academically, and socially. A one-hour professional development course has been developed and offered for these students, focusing on various related activities. Additionally, a fellows program has been implemented to recognize and reward students that take part in the relevant activities. While still facing many challenges, the ECC activities appear to be helping students with their initial progression in the College of Engineering and Computing at USC.

387639 Where Education and Industry Meet: Lessons from a Trip to India

Abstract:

The focus of Cornell University's, NSF-funded Grass Roots GK-12 Program is to advance research and education in renewable and alternative energy. Through a two-week program in India, in cooperation with the Mumbai-based Somaiya Group, a group of Cornell chemical engineering graduate students and New York State high school teachers explored these ideas at an international level. The trip focused on educating students through inexpensive hands-on activities, visiting energy efficient schools, and learning about the Somaiya Group's use of biomass as a renewable energy source. Hands-on classroom activities involving physics (hand made centrifuge), chemistry (surface tension lab), and biology (DNA extraction) were presented to eighth and ninth grade students at four schools spread across two Indian states. The Somaiya schools and factories incorporated various renewable energy sources. For example, one rural school was installing a solar concentrator to generate heat for cooking and designing a biodigestor to produce electricity. The biorefineries utilized renewable energy at a larger scale. From sugar cane production, the company produces not only sugar, but also ethanol, specialty chemicals, wax, and electricity from burning bagasse. The processes were interconnected, with waste from sugar production serving as inputs for wax and electricity production. This talk will focus on what we learned from the trip, as well as the benefits of international experiences to education at all levels.

International educational opportunities for engineering students include, but are not limited to, internships, study abroad programs, and humanitarian programs such as Engineers Without Borders (EWB). Since 2007, a course at xxx University involving engineering students from many disciplines has focused on developing and implementing small-scale engineering projects in developing communities located in Tonga, Ghana, and Peru. The two-semester, three-credit Global Engineering Outreach (GEO) course has recently focused on projects in communities located in northern and southern Peru. Most importantly, the course focuses on integrating both engineering and social aspects that are critical for developing sustainable products. Students in the course, who speak Spanish, interact on a near-weekly basis with the communities in their native language during project development.

This work provides insights into the development of sustainable international programs in a developing community context. Highlights of this work will present student survey results associated with the social and engineering context of development projects. Recent results of implemented projects and the associated impacts will also be discussed. Particularly, recent projects involving pumps, washing machines, water filters, and ovens implemented in the Uros Islands of Southern Peru will be addressed. Feedback from those living on the islands have shown that these projects have had an impact on their health and economic conditions.

Fuel me up!

Van der Borght, K., Toch, K., Marin G.B., Thybaut, J.W.
Ghent University, Laboratory for Chemical Technology, Ghent, B-9052, Belgium.

Introduction

The training of chemical engineering students is of multi-disciplinary nature which ensures that a master of science in chemical engineering is readily employable in highly demanding industrial as well as academic environments. Numerous efforts are put in by the educational community to prepare these students at their best capabilities [1-4]. At Ghent University, Belgium (Europe), this learning trajectory comprises 5 years, i.e., 3 years bachelor of science in chemical engineering followed by 2 years master of science in chemical engineering. During the bachelor years, a fundamental basis of mathematics and sciences is laid upon which a solid engineering training can be constructed. Learning via integrated projects starts already in the 1st semester of the first bachelor year and is followed by a project in the 2nd bachelor in chemical technology and material science and culminates in a cross course project in the 3rd bachelor year focusing on chemical engineering which combines knowledge gained during the bachelor years. After obtaining the bachelor degree, students enter the master years in which chemical engineering is approached in a multi-scale ideology, see Figure 1, which corresponds to the one followed at the Laboratory for Chemical Technology [5].

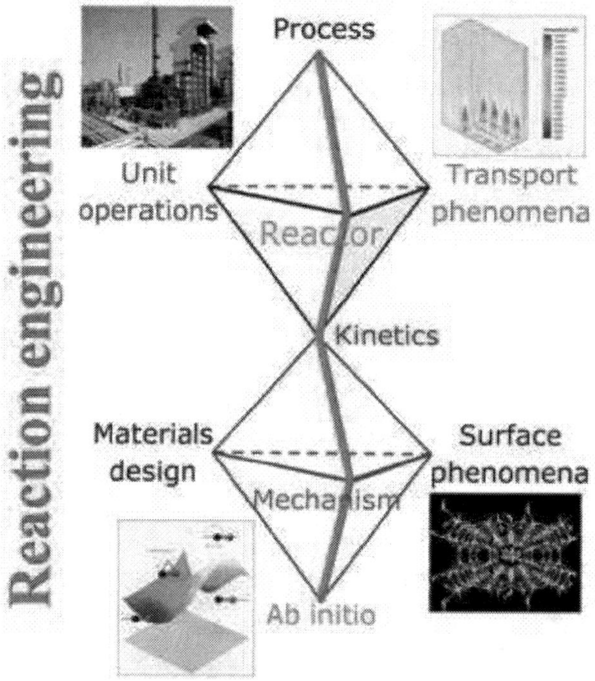

Figure 1: Multi-scale approach to chemical reaction engineering as envisioned by the Laboratory of Chemical Technology, Ghent University [5].

This work presents a case study of the cross course project for students in the 3rd bachelor year. This project already incorporates a wide variety of (chemical) engineering topics, indicated by Figure 1, among others chemical reactors, unit operations, analysis techniques, kinetic modeling and transport phenomena.

The starting point for this project was to produce a fuel for a radio controlled (RC) model car equipped with a glow engine, which is of the internal combustion type, see Figure 2. Typical RC fuels for such a two-stroke engine consist of a mixture of 10 – 40 % nitromethane mixed with methanol and additives for lubrication. In the framework of evolving towards a more sustainable society, the students' assignment was to produce a bio based fuel and have the car running. Because of the open-ended character of the assignment, an overview of possible renewable alternatives was composed and resulted in the selection of bioethanol and biodiesel for an in-depth study. Bioethanol was selected to be a possible 'green' substituent for the methanol while biodiesel would be used as a lubricant. A multi scale approach was used for a critical evaluation of the two biofuel candidates: an experimental investigation, kinetic modeling and an industrial plant simulation.

Figure 2: RC car used for the cross course project

Biofuel production: experimental investigation

For bioethanol production, a lab scale fermentation unit was implemented which started from sugar and baker's yeast, see

Figure 3, left. Special attention was given to the necessity for nutrient addition and the effect of ethanol on the performance of the yeast. The effect of different initial sugar and nutrition concentration was investigated, see

Figure 3, left. The progress of the fermentation was followed as function of time via sampling in which the unconverted sucrose and fructose is dehydrated to hydroxymethylfurfural by addition of H_2SO_4 and followed by the complexion of HMF with phenol. The formed complex can be detected by UV/VIS.

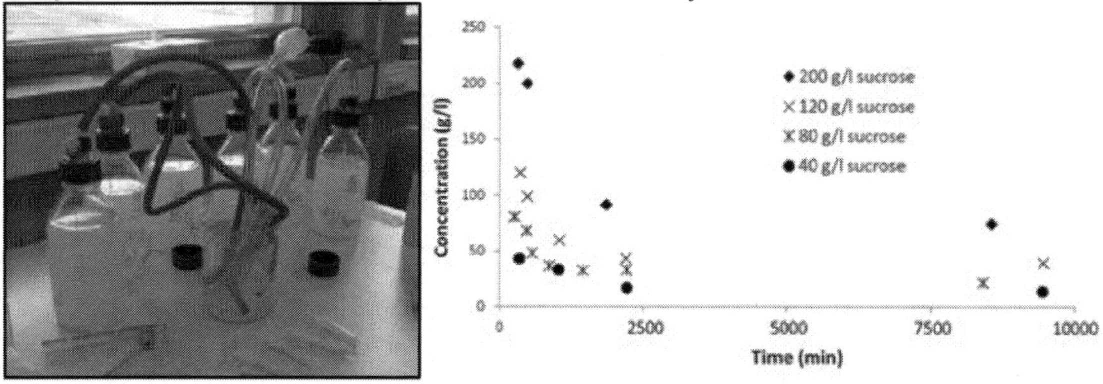

Figure 3: left: Lab scale fermentation of sugar; right: Sucrose concentration as function of initial sucrose concentrations (◆, 200 g/l sucrose; X: 120 g/l sucrose; ✳ : 80 g/l sucrose; ●, 40 g/l sucrose

For fuel application purposes, the ethanol mixture obtained after fermentation needs to be enriched in ethanol via distillation. De distillation unit and Txy diagram is shown in Figure 4. Due to the azeotrope at 96 wt.% ethanol in water, not all water could be removed. The ultimately obtained ethanol concentration was determined using a density measurement and amounted to ± 93%.

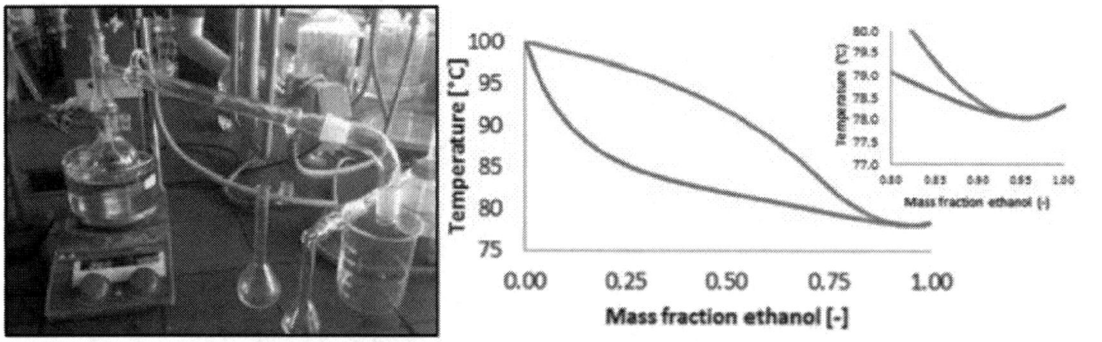

Figure 4: left: Destillation unit; right: Txy diagram of water/ethanol. The red curve is the liquid line and the blue curve is the vapor line.

The biodiesel production occurred through the transesterification of a triglyceride, e.g., sunflower oil, with an alcohol, i.e., methanol, using a homogenous catalyst, i.e., NaOH, in a lab scale experimental setup, see Figure 5, left. A hands-on approach was required to acquire reliable lab scale information on the reaction kinetics: ensuring that all reagents are dry prior to reaction, phase mixing during reaction and purification of the product by catalyst removal and separation of the organic biodiesel phase from the polar glycerol-methanol phase. The product analysis for the biodiesel experiments was performed via ATR-FTIR to follow up the yield of biodiesel as function of time, see Figure 5, right. The effect of reaction temperature (298 to 340 K), molar oil to methanol ratio (1:3 to 1:9), and catalyst concentration (0.1 to 0.6 wt%) on the reaction rates was studied.

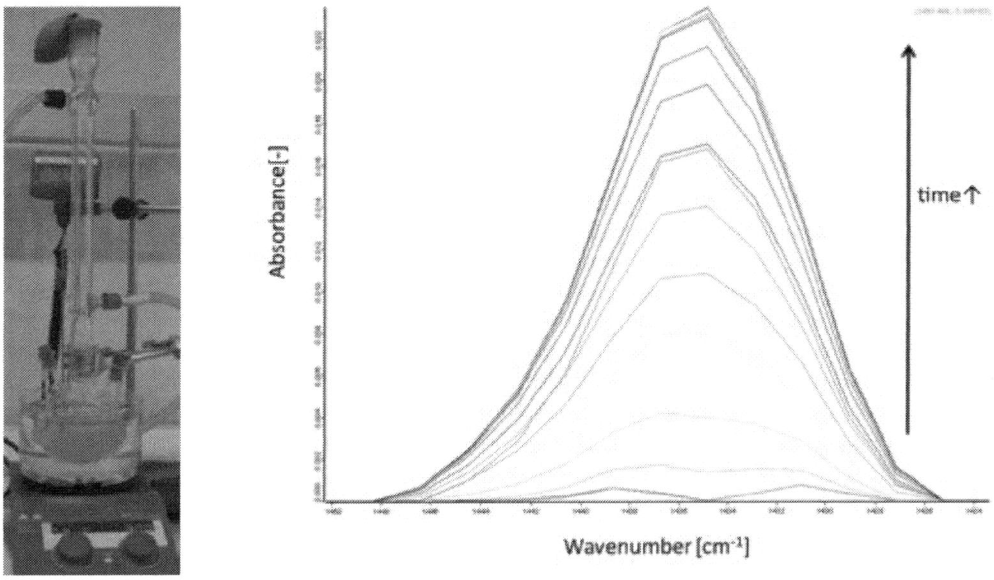

Figure 5: left: lab scale transesterification of sunflower oil, right: ATR-FTIR spectra of the product from sunflower oil transesterification as function of time

Biodiesel production: kinetic modeling

A user-friendly regression package was used for the kinetic modeling of the biodiesel production, i.e., Athena Visual Studio [6]. Based upon the experimental observations for the transesterification of sunflower oil, a kinetic model was constructed and regressed to the experimental data. Initially, a simple power law model was applied but due to some model shortcoming, it was quickly replaced by more fundamental, reaction mechanism based, kinetics. The final kinetic model was able to adequately describe the experimental data as shown in Figure 6.

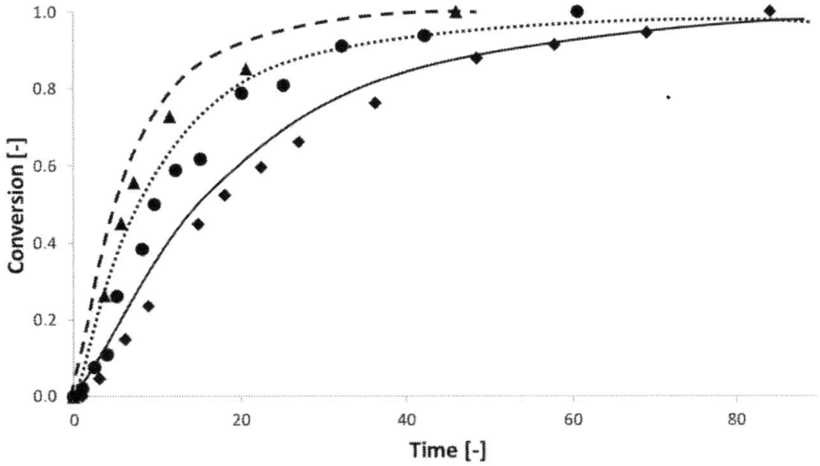

Figure 6: Sunflower oil conversion as function of time at different molar sunflower oil to methanol ratio. Symbols represent experimental data, lines represent model simulations. ◆, full line: molar ratio of sunflower oil to methanol of 1:12; ●, dotted line: : molar ratio of sunflower oil to methanol of 1:9; ▲, dashed line: molar ratio of sunflower oil to methanol of 1:6

Biofuel production: industrial plant simulation

The industrial plant simulations for the bioethanol and biodiesel production based upon typical production capacities were performed using ASPEN® software [7]. Inspired by literature and visits to bioethanol and biodiesel production plants, several plant layouts were benchmarked in ASPEN®. For the reactor, where the actual biofuel production is happening, kinetics were used based upon the experimental observations and kinetic modeling efforts. An example of such a simulation for an industrial biodiesel production plant is given in Figure 7. Similar simulations were performed for the production and purification for the production of fuel-grade ethanol where different separation methods for ethanol and water were investigated.

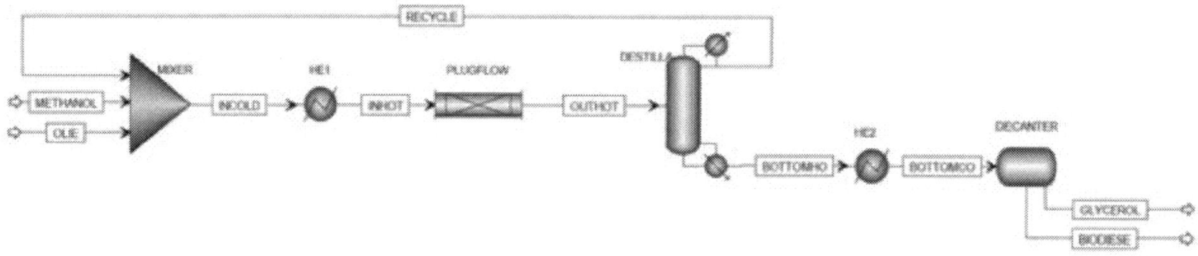

Figure 7: Example for an industrial plant simulation for biodiesel production in ASPEN®

Green RC fuel production

Finally, a fuel was composed using the produced ethanol and biodiesel, to substitute resp. the methanol and lubrication components in the original RC car fuel, and 20 wt.% of Nitromethane (Sigma Aldrich). Nitromethane still needed to be added to this mixture to ensure the proper ignition of the fuel. The fuel was successfully tested in a RC car by the students, see Figure 8. The white smoke produced during testing can be related to the remaining 7% water in the bioethanol.

Figure 8: Testing the green RC fuel

Conclusions

Overall, the project resulted in a critical evaluation of biofuels indicating both the opportunities and challenges of a bio-based society. The students had to follow a multi-disciplinary approach in which chemical reactors, unit operations, analysis techniques, kinetic modeling and transport phenomena were used. Using lab scale equipment, common reactants and easy analysis techniques, the students were able to produce and test a green alternative for RC fuel. Additionally, they gained experience in experimental investigation and kinetic modeling which were used for an industrial plant simulation.

References

[1] C.L. Dym, A.M. Agogino, O. Eris, D.D. Frey, L.J. Leifer, Engineering Design Thinking, Teaching, and Learning, Journal of Engineering Education, 94 (2005).

[2] J.G. Mackenzie, W.B. Earl, R.M. Allen, I.A. Gilmour, Amoco Computer Simulation in Chemical Engineering Education, Journal of Engineering Education, 90 (2001) 331-334.

[3] D.J. Schneck, Integrated Learning: Paradigm for a Unified Approach, Journal of Engineering Education, 90 (2001) 213-217.

[4] B.U. Sherrett, E.J. Nefcy, E.S. Gummer, M.D. Keretsky, An Expert Solution to Assess an Industrially Situated, Computer-Enabled Design Project, Journal of Engineering Education, 102 (2013) 541-576.

[5] Laboratory for Chemical Technology, Web page http://www.lct.ugent.be/.

[6] Athena Visual Studio, Web page http://www.athenavisual.com/.

[7] Aspen Plus(r) , Web page http://www.aspentech.com/products/aspen-plus.aspx.

382114 What Can We Learn from Students Using Facebook?

Abstract:

Over the past two decades, we have witnessed a remarkable growth in the internet and its usage across the globe. Furthermore, rapid advances in digital technology have allowed the internet to become more readily accessible, becoming a part of our everyday life. In recent years, social network sites (e.g. *Facebook*) have provided a new means of interaction and communication between individuals and groups. This has percolated through the education system mainly via the students. We are now starting to realise the potential for *Facebook* to be used for educational purposes; largely due to its desirable qualities which replicate other existing educational tools that allow peer feedback, possess good interaction tools, and fit into the social context of learning environments. This work investigates the extent to which *Facebook* is currently used in engineering education, the factors that motivate educators and students to use *Facebook* for study purposes, and whether *Facebook* is used as 'communities of practise' or mostly as social spaces.

The extent to which Facebook is used in the Chemical Engineering course at Monash University is presented in here. Results from a student survey carried out across different years are presented and ideas of how facebook could be further used as an education tool discussed. Similarly, limitations of using Facebook and the expectations raised by students will be presented. The outcomes of this work will serve as basis to determine if *Facebook* has a place within engineering education, and what of its features could prove to have a positive impact in the students' learning experience.

379025 Internationalizing Access to Chemical Engineering Education By a Lifelong Language Learning Environment

Abstract:

Both chemical engineering and engineering education are rapidly obtaining international dimension. The U.S. has served as home for leading chemical engineers from all over the world for the last 50 years. The challenges of finding best candidates for chemical engineering career and best chemical engineering education will shape the chemical engineering future in the USA in the 21 Century. In turn, the ability to internationalize best practices in chemical engineering (both at the education and industrial level) is necessary for the successful development of the chemical industry in the global perspective.

Attracting promising students to chemical engineering education system and their transforming into chemical engineering professionals for the US and world chemical industry requires a continuous flow of talents. Successful enrollment of talents by top world universities specialized in chemical engineering is an excellent proof of successful internationalization of chemical industry and education. These universities are, however, facing a rapidly growing challenge of dropping demand for chemical engineering programs both by domestic and international students. Chemical engineering education and engineering education in general are becoming less popular. This fact is confirmed by the statistics of the institute of International Education.

The global demand for world class chemical engineering programs taught in U.S. and other top world chemical engineering universities can be increased if we attract more potential students from Asia, Russia and post-Soviet countries. These countries are proud of talented school graduates and offer government funds programs to sponsor their education in best world chemical education centers. The problem of reduction of the influx of talented school graduates from these countries is that academic mobility in these states is closely linked to the language learning problem (in other words, the ability of involved faculty and students to speak world languages such as English at a level of fluency ensuring their integration into the global community). For example, several percent of Russian secondary school graduates and undergraduate students demonstrate TOEFL results sufficient for enrollment by chemical engineering universities in the USA. These students can be, however, excellent chemists and scientists and be motivated to obtain a good-paid position at world-class chemical industry companies requiring a degree from top universities offering chemical engineering programs.

A new approach is offered to involve more students into the U.S.-offered chemical engineering programs – a language learner focused approach. This strategy is called a "Lifelong Language Learning Environment" (a "Triple-L E") meaning that English learning in the universities from above-mentioned countries should be treated as a truly life-long process with the focus on chemical engineering linguistic aspects. Starting at the secondary school level, engineering English learning continues at a university to offer Bachelors able to pursue Master's degree in a U.S. university or a secondary school graduate with TOEFL scores allowing entering a chemical engineering program. Another component of this "Environment" is that faculty in involved in language learning together with students. An indispensible prerequisite for a successful Lifelong Language Learning Environment is involvement of U.S. universities into its development and functioning at their partnering higher education centers abroad (in other words, it requires strong international partnership that is successfully achieved by all top chemical engineering universities in the USA).

This paper also provides a case study of development and implementing of a "Triple-L E" initiative at one of Russian National Research universities specialized in chemical engineering. A peculiar feature of the Lifelong Language Learning Environment implementation in that university is that mostly technical English training is provided, English for Specific Purposes (ESP). English for Specific Purposes is aimed at developing specific skills of communicating in English on the topics of professional interest, depending on the background and career prospects. English teaching is focused on the engineering field of study with the application of computer aided language learning solutions. The core of the process if formed by professors having both engineering (major) and linguistic (minor) degrees or the linguists with 5-10 years of experience in technical English. Successful involvement of U.S. universities is achieved through the development and implementation of joint short-term and dual degree programs with the focus on the spread of disciplines taught in English from these programs to other university chemical engineering programs. This strategy allowed training of faculty, administrators and students for the internationalized chemical engineering programs and attracting English-speaking students from other countries as well as stimulating outgoing academic mobility.

Student Handbook for Process Safety

Robert Forest
PhD Candidate, Department of Chemical and Bimolecular Engineering
University of Delaware
150 Academy St.
(302) 831-2326
rforest@udel.edu

Jerry Forest
Global Process Safety Manager
Celanese
225 E. John Carpenter Freeway Suite 1200
Dallas, TX 75234
(214) 417-1920
Jerry.forest@celanese.com

Keywords: ABET, process safety education, student, handbook

Abstract

In its investigation report of the T2 Laboratory explosion, the U.S. Chemical Safety Board recommended that AIChE and the Accreditation Board for Engineering and Technology (ABET) modify the university chemical engineering curricula to increase awareness of chemical reactivity hazards. In 2011, ABET approved new program criteria for accredited chemical engineering programs. In part, the criteria includes "the engineering application of these basic sciences to the design, analysis, and control of chemical, physical, and/or biological processes, including the hazards associated with these processes." This criterion has been in effect since the fall of 2012. All accredited chemical engineering programs must now include content addressing the hazards and control of the hazards of processes.

The Center for Chemical Process Safety (CCPS) has responded in part to this recommendation by initiating project #242, Student Handbook for Process Safety. The scope of the book includes the development and evaluation of tools for application of process safety principles throughout the chemical engineering curricula. It may also be used as a resource for a stand-alone process safety course. The target audience is primarily chemical engineers at the junior and senior level and recent graduates new to industry. It however can be applied to all engineering disciplines. This paper will give an overview of the need to include process safety basics in university curricula and give details on the content and use of the book.

1. Introduction

AIChE's Center for Chemical Process Safety (CCPS) was formed in 1985 and has always sought to provide resources to bridge process safety knowledge gaps for engineers entering and practicing in industry. In 1992 the Safety in Chemical Engineering Education (SaChE) program was formed to provide material that chemical engineering professors can incorporate into their curricula [1].

Despite these efforts to provide process safety resources for both engineers and chemical engineering professors, causes of process safety incidents in industry have been linked to lack of education on recognition and control of process safety hazards. This has led the U.S. Chemical Safety and Hazard Investigation Board (CSB) to make recommendations to both AIChE and the Accreditation Board for Engineering and Technology (ABET) to add reactive hazard awareness to baccalaureate chemical engineering curricula. The CCPS has responded by making the change to ABET requirements. In addition, CCPS continues to bridge the knowledge gap with a comprehensive improvement to the SaChE program and by developing a process safety student handbook for use at the university level by professors, student engineers, and relatively inexperienced engineers already in industry.

2. The Need for New ABET Requirements

In 2002, the CSB published *Improving Reactive Chemical Hazard Management* with objectives that included determination of how industry addresses reactive hazards and recommendations to reduce the number and severity of incidents [2]. The report identifies 167 serious accidents between 1980 and 2001 that resulted in 108 deaths. Among other conclusions, this report notes that reactive incidents are often caused by inadequate recognition and evaluation of reactive hazards. Evidence of this conclusion is reinforced with industry incidents such as the explosion that occurred at T2 Laboratories Inc., in December of 2007. This incident killed 4 people, seriously injured 13, and had significant community impact.

The incident that occurred at T2 Laboratories was investigated by the CSB and it was determined that there was a probable cooling system failure that led to a runaway reaction, system overpressure, and explosion [3]. The CSB identified the root cause: "T2 did not recognize the runaway reaction hazard associated with the MCMT it was producing." A key finding from the incident is "Most baccalaureate chemical engineering curricula in the U.S. do not specifically address reactive hazard recognition or management." The specific recommendation to AIChE is: "Work with the Accreditation Board for Engineering and Technology, Inc. to add reactive hazard awareness to baccalaureate chemical engineering curricula requirements."

In addition to industrial incidents caused by reactive chemical hazards, industry has increasing requirements that new hire engineers have a better grasp of process safety knowledge in general.

This can either be introduced through the existing engineering curricula, as a separate course that covers the material in depth, or both [4][5].

AIChE, through CCPS worked with ABET and in October 2011, ABET issued the following statement for accredited chemical engineering programs [6]:

> *The curriculum must provide a thorough grounding in the basic sciences, including chemistry, physics, and biology, with some content at an advanced level, as appropriate to the objectives of the program. The curriculum must include the engineering application of these basic sciences to the design, analysis, and control of chemical, physical, and/or biological processes, **including the hazards associated with these processes.***

The need for this ABET requirement is that many U.S. universities do not include process safety principles in the engineering curricula [7]. This in turn leads to knowledge gaps that could result in process safety incidents as students enter the workplace and begin their practice.

3. Student Handbook for Process Safety

In 2012, the CSB formally commended AIChE for exceeding the recommended action to work with ABET to include reactive hazards in undergraduate chemical engineering curricula [8]. However, the CCPS continues to go beyond this recommendation by working to close the process safety knowledge gap that exists with some chemical engineering professors, students, and new engineers. This includes improvement to the SaChE program and development of a Student Handbook for Process Safety.

The member companies of the CCPS technical steering committee define which projects progress. The fact that the Student Handbook for Process Safety project progressed indicates the importance that industry places on improving the process safety competency of new hire engineers.

3.1 Scope of Book

The CCPS project steering team identified that the scope of the project include the development and evaluation of tools for application of process safety principles throughout the standard chemical engineering curricula in addition to, or as an alternative to adding a new specific process safety course. The ABET requirement did not specifically state that a stand-alone process safety course should be developed, nor did it require that process safety be included in specific elements of the curricula. The project team therefore felt that the outline could be structured to do both; serve as a standalone course or be used as a reference in specific engineering courses. The book is cross-referenced throughout for ease of use by the student and the professor. As Dickson and Crowl point out, engineering faculty do not always have and adequate background in process safety [7]. This book will therefore serve as both a reference and a resource for engineering professors.

3.2 Primary Target Audience

Given the scope of the book as either supplemental process safety reference material or a standalone course, the intended audience is broader than students alone. While it is primarily written for chemical engineering undergraduate students, it can be used as a resource for graduate students, recent graduates, non-chemical engineering majors, engineers with little industrial experience, and engineering professors.

3.3 Development of Outline

In order to determine the book outline, the project team completed an analysis of existing process safety books and resources at the university level, and completed an analysis of a 14 element and the CCPS 20 element process safety model with common engineering courses to determine which elements fit with each course. In addition, a similar matrix was developed for IChemE chartered requirements and the common courses. Existing SaChE material was also reviewed to determine how that material could supplement the book, and if there were any gaps that could be filled.

The project team surfaced only one process safety book in use at several U.S. universities: *Chemical Process Safety Fundamentals with Applications* by Crowl and Louvar [7]. This excellent resource covers: 1. Toxicology, 2. Industrial hygiene, 3. Source models, 4. Toxic release and dispersion modelling, 5. Fires and explosions, 6. Preventing fires and explosions, 7. Introduction to reliefs, 8. Relief sizing, 9. Hazards identification, and 10. Risk Assessment. It is not our intent to recreate this material. We therefore took the approach of providing course material specific for common engineering curricula and common to industry using the CCPS 20 element model of process safety.

The project team made 3 matrices. The first compared a 14 element process safety model from the OSHA PSM regulation to common engineering courses. The matrix categorized if material should be covered in the subject, has a potential to upgrade process safety knowledge of the student, or if it is a fundamental topic that might be covered by all. Table 1 is an example 14 element matrix. The second matrix compared the 20 element CCPS process safety model while the third matrix compared iChemE chartered requirements. Each matrix produced similar results.

The project team felt that there was not a significant overlap in existing SaChE materials the outline planned. We therefore decided to reference SaChE materials as an additional reference for both the student and the professor.

	Employee Participation	PHA	Training	PSSR	Hot Work Permit	Incident Investigation	Compliance Audit	PSI	Operating Procedures	Contractors	MI	MOC	Emergency Planning	Trade Secrets
Fundamentals														
Engineering Orientation								o	o				o	
Statics/Dynamics														
Circuits											o			
Thermodynamics														
Fluid Dynamics											o			
Math Sequence														
Non-Engineering Lab Courses														
Lab Chemistry Sequence		o		o		o		o	x				x	
Physics		o		o		o		o	x				x	
Biology/Geology/etc.		o		o		o		o	x				x	
ChE Core														
Process Analysis		o		o									o	
Multicomponent Thermo								o						
Heat Transfer								o			o			
Mass Transfer								o						
Unit Operations		o		o				o			x		o	
Process Economics											x			
Reaction Kinetics/Engieering		o						o			o		o	
ChE/Unit Ops Lab		x		x		o	o		x		x		x	
Process Dynamics/Controls		o						x			x		x	
Process/Plant Design		x		o							x		x	
Computing														
Simulation		o		o				o			x		o	

x — Inherently covered or should be covered in this subject
o — potential opportunity to upgrade PSM knowledge
PHA and MOC highlighted as a backbone subject

Table 1 – 14 point Process Safety Matrix

3.4 Outline

From the book review, process safety element and IChemE verses course gap analysis and SaChE review, the project team thought that it was appropriate to develop an outline that includes a review of risk based process safety, establishing the need for process safety and touching on process safety risk encountered in plant design. Table 2 list the general outline of the Student Handbook for Process Safety.

Chapter	Title
1	Introduction
2	Process Safety Basics
3	The Need for Process Safety
4	Process Safety for Engineering Disciplines
5	Process Safety in Design
6	Course Materials
7	Process Safety in the Workplace

Table 2 – General Outline

Section 4 gives more detail of each chapter.

3.5 Format of Book

This book will be published in hardback. In addition, it will be published as an e-book. The vision is to include hyperlinks for cross reference among chapters. For example, Chapter 3 describes several classic process safety case studies. Each case study is referenced to a Chapter 2 process safety element. Similarly, design elements are linked to both case studies and process safety elements. It is our intention to also provide links to resource material such as CSB videos and SaChE materials. The book will also make use of a process safety dictionary for the student reference.

4. Content

CCPS has an extensive library of process safety resources. It is not our intent to recreate that content, but rather present it in such a way as to introduce the target audience to the concepts and provide a reference to the CCPS materials. More detail on selected chapters follows.

4.1 Process Safety Basics

This chapter introduces the reader to CCPS's risk based process safety pillars:

- Commit to process safety
- Understand hazards and risks
- Manage risks
- Learn from experience

The 20 process safety elements associated with each pillar are explained. Incident examples that have causes related to each process safety element are given. Note, the CCPS process safety elements go beyond the traditional OSHA regulatory elements.

Key concepts such as culture, hazard and risk are defined. The pillars and elements discussed are shown in table 3.

Pillar	Element
Commit to Process Safety	Process Safety Culture
	Compliance with Standards
	Process Safety Competency
	Workforce Involvement
	Stakeholder Outreach
Understand Hazards and Risks	Process Knowledge Management
	Hazard Identification and Risk Analysis
Manage Risk	Operating Procedures
	Safe Work Practices
	Asset Integrity and Reliability
	Contractor Management
	Training for Performance Assurance
	Management of Change
	Operational Readiness
	Conduct of Operations
	Emergency Management
Learn from Experience	Incident Investigation
	Measurements and Metrics
	Auditing

	Management Review and Continuous Improvement

Table 3 – CCPS Pillars and Elements

A process safety foundation is laid by introducing the reader to the fundamentals of process safety. This approach differentiates this book from Crowl's technical analysis of process safety as young engineers will need to be conversant in the elements as they enter industry [9].

4.2 The Need for Process Safety

AIChE created the CCPS in response to the 1984 Bhopal tragedy. Many of the incidents that have defined process safety occurred before today's engineering students were born. As industrial risk is reduced, we expect that young engineers will be involved in and exposed to fewer incidents. While this is obviously a good thing, it also takes away an element of experience from the young engineer that has been afforded to more seasoned practitioners.

This chapter provides reference materials to familiarize the reader with incidents that have defined the current state of process safety. The incidents are categorized per CCPS process safety pillar in order to highlight the importance of all process safety elements. Table 4 details the incidents covered in this chapter.

	Pillars of Risk Based Process Safety			
Incident	**Commit to Process Safety**	**Understand Hazards and Risk**	**Managing Risk**	**Learning and Experience**
BP Explosion, Texas City, 2005	Process Safety Culture	Process Knowledge Management Training & Performance Assurance	Management of Change Asset Integrity & Reliability	
Arco Channelview Explosion, 1990		Process Knowledge Management		
Space Shuttle Columbia, 2003	Process Safety Culture			
Concept Sciences Explosion, 1999		Process Knowledge Management Hazard Identification & Risk Management		
Esso Longford Gas Plant Explosion, 2001	Process Safety Competency	Hazard Identification & Risk Management		
Port Neal, Ammonia Nitrate Explosion, 1994		Hazard Identification & Risk Management	Operating Procedures	

	Pillars of Risk Based Process Safety			
Incident	**Commit to Process Safety**	**Understand Hazards and Risk**	**Managing Risk**	**Learning and Experience**
Piper Alpha, 1988			Safe Work Practices Emergency Management	
Partridge Raleigh Oilfield Explosion, 2006	Compliance with Standards		Safe Work Practices Contractor Management	
Texaco, Milford Haven Explosion, 1994			Asset Integrity and Reliability	
Formosa Plastics VCM Explosion, 2004			Conduct of Operations Emergency Management	
Flixborough Explosion, 1974			Management of Change	
Sandoz Warehouse Fire, 1986			Emergency Management	
Exxon Valdez, 1989			Conduct of Operations Management of Change	
PEMEX LPG Explosion, 1984	Compliance with Standards			
Bhopal Methyl Isocyanate Release, 1984		Hazard Identification & Risk Assessment	Management of Change	
Deepwater Horizon, 2010	Process Safety Culture			Incident Investigation

Table 4 – Incidents that define process safety

4.3 Process Safety for Engineering Disciplines

While each traditional engineering discipline may be involved in all aspects of process safety, certain disciplines tend to participate in those elements matching their educational experience. For example, mechanical engineers might be called to participate in mechanical integrity programs. Chemical engineers might be involved with consequence analysis for process hazard analysis, and all disciplines might participate in management of change.

This chapter focuses on the activities of a new engineer related to the 20 process safety elements. The chapter differentiates typical engineering disciplined activities for each element. Table 5 is a summary of activities by element and discipline.

Element	Activity	Discipline
Commit to Process Safety		
Process Safety Culture	Learn responsibilities for process safety roles	All
Compliance with Standards	Learn and apply standards and regulations that apply to processes and equipment	All
Process Safety Competency	Take advantage of training opportunities	All
Workforce Involvement	Contribute Listen to the input from operators, technicians, etc.	All
Stakeholder Outreach	Become aware of the organization outreach efforts	All
Understand Hazards and Risk		
Process Knowledge Management	Ensure accuracy of PSI Be familiar with Safety Data Sheets Develop reactivity matrix Develop and update design basis calculations Develop and update Piping and Instrumentation Diagrams (P&IDs) Verify electrical classification Maintain equipment files (design and fabrication information)	All All Chemical Chemical Chemical Electrical Mechanical
HIRA	Participate in PHAs and other studies Assemble required PSI Review applicable RAGAGEP Compile industry incidents	All Chemical All Chemical & Safety
Manage Risk		
Operating Procedures	Write new operating procedures Update operating procedures	Chemical
Safe Work Practices	Write Permits Approve permits	Safety Safety & Chemical
Asset Integrity and Reliability	Identify equipment covered by ITPM program Write ITPM procedures Analyze inspection results Approve and monitor repairs Maintain ITPM records Testing of Safety Instrumented Systems	Chemical & Mechanical Mechanical Mechanical Mechanical Mechanical Mechanical & Electrical
Contractor Management	Approve contractors with respect to safety Train contractors	Safety All
Training and Performance Assurance	Not applicable for new engineers	
Management of Change	Identify changes Participate in MOCs Assemble required PSI Review applicable RAGAGEP	All All Chemical, Mechanical & Electrical All
Operational Readiness	Participate in readiness reviews Assemble required information	All All
Conduct of Operations	Perform all tasks reliably	All
Emergency Management	Learn emergency response plans for your area	All
Learn from Experience		

Element	Activity	Discipline
Incidents	Participate in investigations Recognize and identify near misses	All All
Metrics	Record and maintain data Analyze data	All All
Auditing	Assemble requested information for audits	All
Management Review	Not applicable for new engineers	

Table 5 – Typical engineering activity by process safety element and discipline

4.4 Process Safety in Design

Many engineering courses teach the student how to design equipment. Kinetics teaches reactor design, fluids may teach pump design, and in unit operations we learn much about how to specify equipment.

This chapter is written for the engineering professor and student to think about the process safety incidents that could possibly result from common unit operations. A good process engineer always has the process safety in mind when designing and specifying equipment. Examples of process safety incidents that have occurred with each equipment type are given. The chapter also includes several diagrams and pictures to better familiarize the reader with the topic. A complete section is devoted to process safety in refining, as the team's experience is that this information is often not included in engineering curricula. Process safety concerns for the following types of equipment are detailed:

- Pumps, compressors and fans
- Heat exchange equipment
- Mass transfer; distillation, leaching and extraction, and adsorption
- Mechanical separation
- Reactors, reactive hazards
- Incinerators, flares, and fired equipment
- Storage
- Petroleum
    - Crude handling & separation
    - Light hydrocarbon handling and separation
    - Hydrotreating
    - Catalytic Cracking
    - Reforming
    - Alkylation
    - Coking
- Transient Operating States

4.5 Course Materials

The purpose of this chapter is to show how process safety can fit into chemical engineering courses. This chapter is a guide for teachers as well as students. A list of process safety incidents and of modules from SACHE are described and the courses that are relevant to support are provided. Access to SACHE modules is available through the SCAHE website, www.sache.org.

The SACHE courses are grouped by topic; Inherently Safer Design, Process Safety Management, Hazards, Hazard Identification and Risk Assessment, Protection Systems, Case Histories and Other.

4.6 Process Safety in the Workplace

This chapter describes what process safety duties a young engineer may encounter when first entering the workplace. The purpose is to introduce the student to work activities and skills that are not presently part of any engineering curricula. Some of these duties are described below.

- Interacting with operators, craftsmen, contractors, and unions
- Action item management
- New skills training such as meeting facilitation, e-mail etiquette and leading MOC reviews
- Consequence assessments such as gas models and techniques to estimate releases
- Operating discipline – review of operator long books, shift instructions, evaluation sheet monitoring, and leading toolbox meetings,
- Engineering discipline - safe operating limit tracking, and tracking safety system challenges
- Management discipline – communication boards, bypass boards, taking action on data
- Other Conduct of Operations topics – lighting, hose management, equipment labeling, and two-way communication
- Lead and develop process safety training for operators
- PI system basics, taking command of the unit material and energy balance
- Process safety in turnarounds
- Commissioning and how it relates to PSSR
- Other miscellaneous topics

5. Conclusions

In summary, the need for university chemical engineering ciricula to include an increased awareness of chemical reactivity hazards has led ABET to include new process safety criteria for accredited programs. In response to this requirement, the CCPS initiated project #242, Student Handbook for Process Safety. This handbook, which primarily targets junior and senior chemical engineering students, aims to teach process safety principles in both stand-alone classes as well as in existing curricula. Additionally it will serve as a reference to professors who seek to include process safety principles in their classes. The ultimate goal is that chemical engineering programs will exceed the new ABET requirement with the assistance of this handbook.

6. References

[1] SaChE. *Welcome to SaChE*, http://www.sache.org/ (accessed September 16, 2014)

[2] Chemical Safety Board, Improving Reactive Hazard Management, U.S. Chemical Safety and Hazard Investigation Board, Washington, DC (2002)

[3] Chemical Safety Board, T2 Laboratories Inc. Runaway Reaction, U.S. Chemical Safety and Hazard Investigation Board, Washington, DC (2007)

[4] Dee, Sean J., Cox, B.L. and Ogle, R. A., "Process Safety in the Classroom: The Current State of Chemical Engineering Programs at US Universities." AIChE 2014 Spring Meeting, 10th Global Congress on Process Safety, New Orleans, LA (2014)

[5] Crowl, Daniel, A. "Process Safety Education: Meeting the new ABET Requirements.", CEP, April 2012, 19.

[6] ABET. *Criteria for Accrediting Engineering Programs, 2014-2015*, http://www.abet.org (accessed September 16, 2014)

[7] Dickson, Brian, and Crowl, D.A. Is There an Acceptable Level of Competence in Process Safety for a Graduate Chemical Engineer?, AIChE 2014 Spring Meeting, 10th Global Congress on Process Safety, New Orleans, LA (2014)

[8] Chemical Safety board, CSB Commends center for Chemical Process Safety response to Reactive Hazards Study Recommendations, Says CCPS "Exceeds Recommended Action." http://www.csb.gov (accessed September 16, 2014)

[9] Crowl, Daniel, A. and Louvar, J. F., *Chemical Process Safety: Fundamentals with Applications*, 1st ed., Prentice Hall, Englewood Cliffs, NJ (1990)

384539 Development of a Course to Address Safety in the Undergraduate Chemical Engineering Curricula at Mississippi State University

Abstract:

Safety is a very important issue for chemical engineers. Feedback from the chemical engineering alumni as well as input from the advisory board of the Dave C. Swalm school of Chemical Engineering at Mississippi State University (MSU) indicated that safety should be taught at the undergraduate level. As a result of this feedback an undergraduate course was developed and was initially offered as a technical elective in the Spring of 2009. This course has been offered annually and now is a required course as part of the chemical engineering degree requirement at MSU.

Currently this course uses "Chemical Process Safety" by Daniel Crowl and Joseph Louvar, as the text book and covers many aspects of safety that an entry level chemical engineer will face as part of their first job. Details of the course offering, the course evolution, as well as the impact on the students and entry level engineers will be presented.

Using the SAChE Website to Enhance Undergraduate Safety Education

AIChE National Meeting

Thomas O. Spicer, III
Professor
Ralph E. Martin Department of Chemical Engineering

20 November 2014

Background of SAChE

Safety and Chemical Engineering Education Program

- Started in 1992
- Cooperative effort
  - Center for Chemical Process Safety (CCPS) of the American Institute of Chemical Engineers (AIChE)
  - Engineering educators
- Objective: enhance the educational experience of undergraduate and graduate students studying chemical and biochemical products and processes
- Committee membership
  - Academics
  - Industry
  - Government
  - AIChE Staff
- Safety Scale-Up Partnership between CCPS and corporate sponsors
  - All universities now have no charge access to the SAChE site.

SAChE Web Site - www.sache.org

- Originally intended to distribute SAChE Products
  - SAChE Products are classroom materials and aids designed to help faculty bring safety issues to the attention of students
    - Overcome the "activation energy" barrier of faculty including new material in a course
    - Easily modified to fit an instructor's style
    - Originally a single copy sent to departments
    - Problems: costly to distribute; easily misplaced in the academic institution

SAChE Web Site

- Web distribution could address these problems
  - Reduction of distribution costs
  - Ease of updating materials
  - Materials readily available to multiple faculty (and students) in participating departments
- Web host: Ralph E. Martin Department of Chemical Engineering (2003)
  - SAChE Product distribution
  - Access provided free of charge to all academic users

SAChE Web Site Today

- SAChE Products
- Process Safety Beacon Archive
- Student Safety Certificate Program
- SAChE Faculty Workshop Archive
- Links to CSB Video Room and Dow Lab Safety Academy
- SAChE Newsletter archive
- SAChE Committee Listing
- SAChE Membership including:
  - Higher education
  - Corporate members
  - Beacon Archive subscribers
- Links to relevant safety information
- Account and messaging services

SAChE Web Site Accounts

- Group Manager/Point of Contact structure to distribute the workload of maintaining accounts
  - Web site Administrators set up Groups and Group Managers (also identify users as SAChE Committee members)
  - Group Managers are responsible for setting up Group Members
  - All accounts based on email address; web site passwords stored in an encrypted data base (no passwords can be retrieved)
  - Higher education Group Managers responsible for setting their student group password (one password per institution)

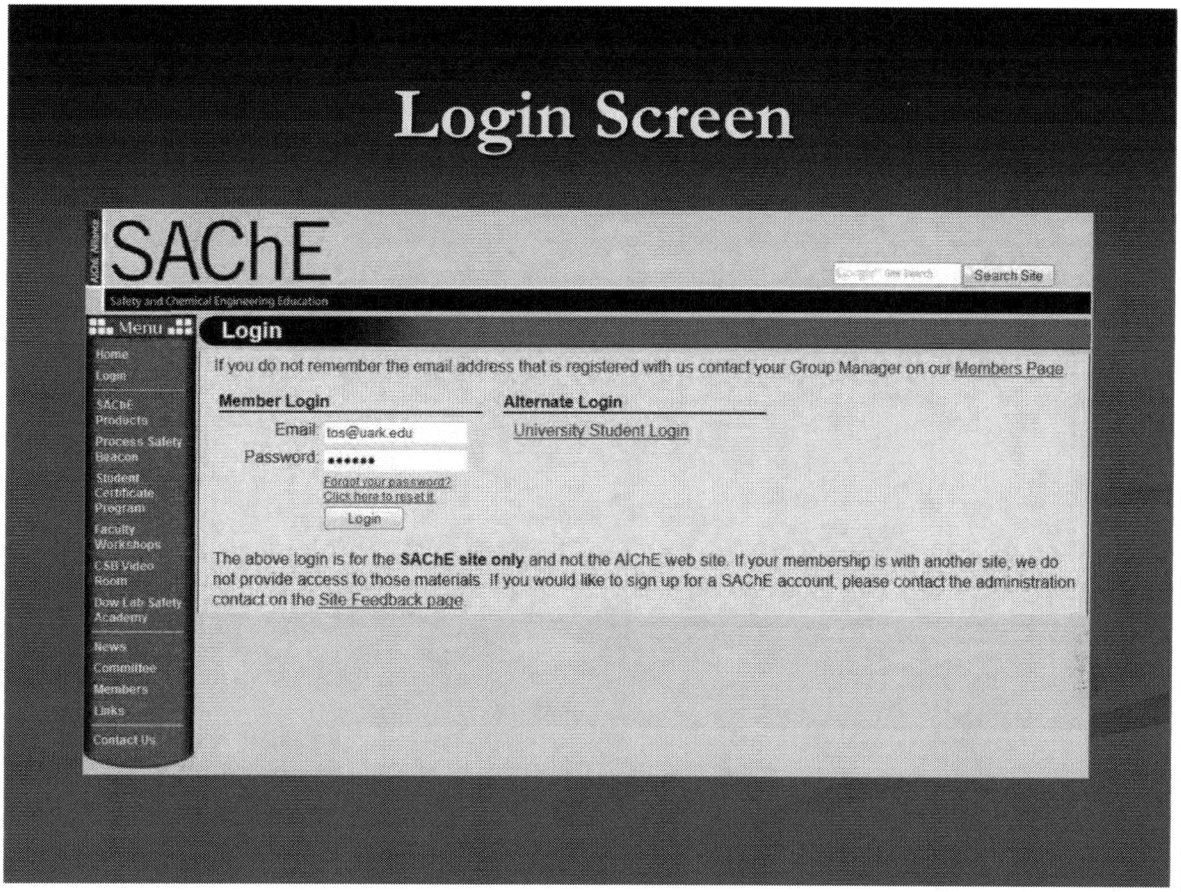
Login Screen
SAChE
Safety and Chemical Engineering Education
Search Site
Menu
Home
Login
SAChE Products
Process Safety Beacon
Student Certificate Program
Faculty Workshops
CSB Video Room
Dow Lab Safety Academy
News
Committee
Members
Links
Contact Us
Login
If you do not remember the email address that is registered with us contact your Group Manager on our Members Page.
Member Login
Email: tos@uark.edu
Password: ******
Forgot your password? Click here to reset it.
Login
Alternate Login
University Student Login
The above login is for the SAChE site only and not the AIChE web site. If your membership is with another site, we do not provide access to those materials. If you would like to sign up for a SAChE account, please contact the administration contact on the Site Feedback page.

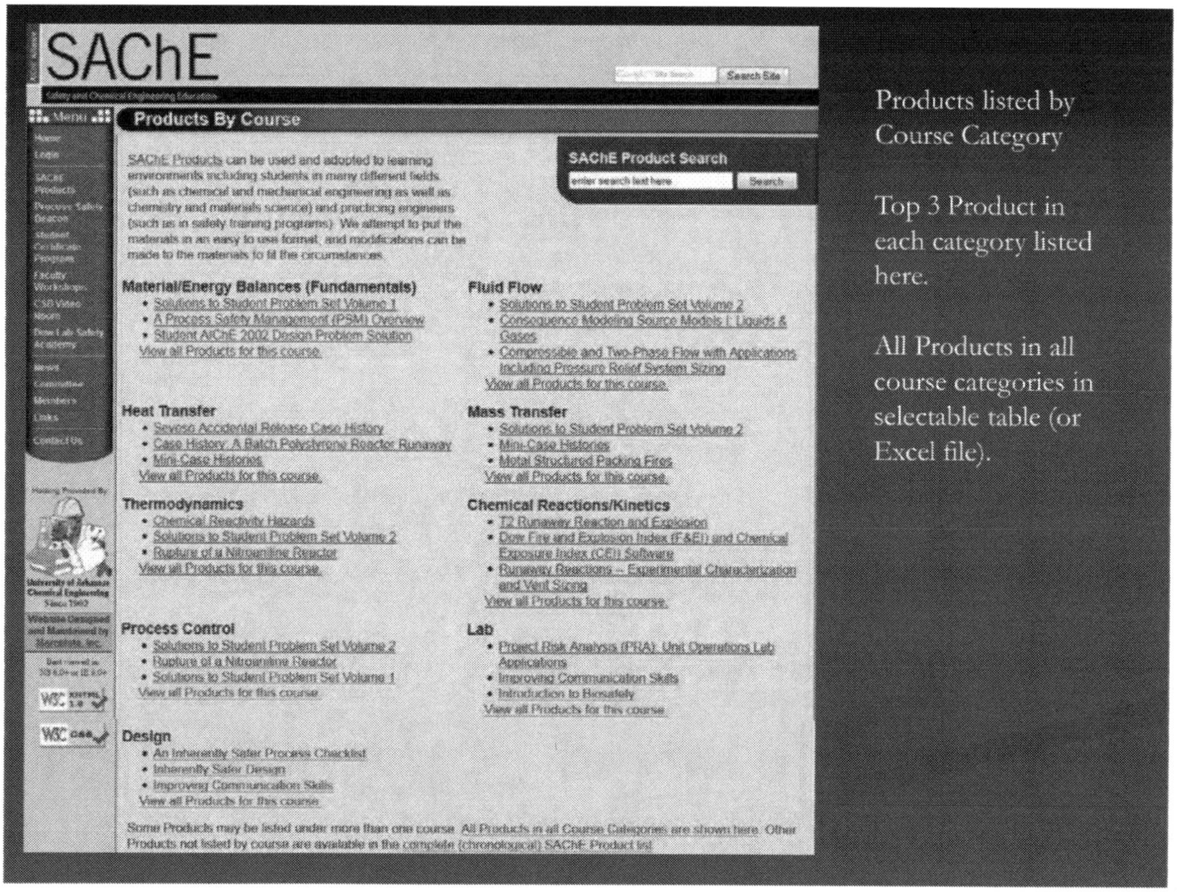
SAChE
Search Site
Products By Course
SAChE Product Search
Search
Material/Energy Balances (Fundamentals)
Fluid Flow
Heat Transfer
Mass Transfer
Thermodynamics
Chemical Reactions/Kinetics
Process Control
Lab
Design
Products listed by Course Category
Top 3 Product in each category listed here.
All Products in all course categories in selectable table (or Excel file).

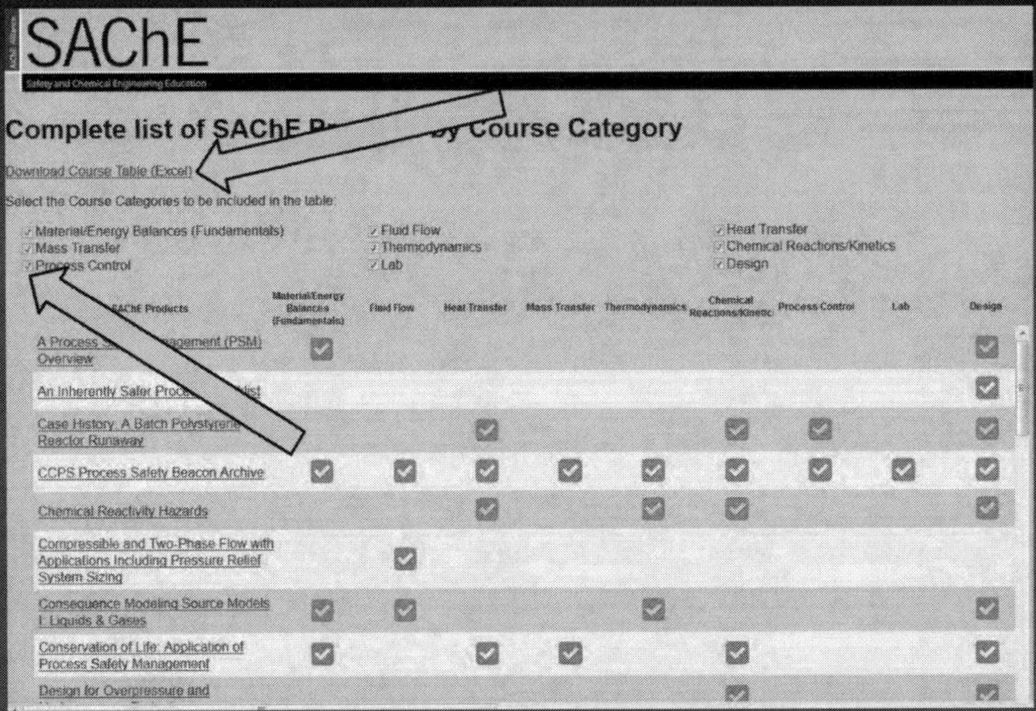

Select the course category to be included in the table.

All of the information can be downloaded to an Excel spreadsheet.

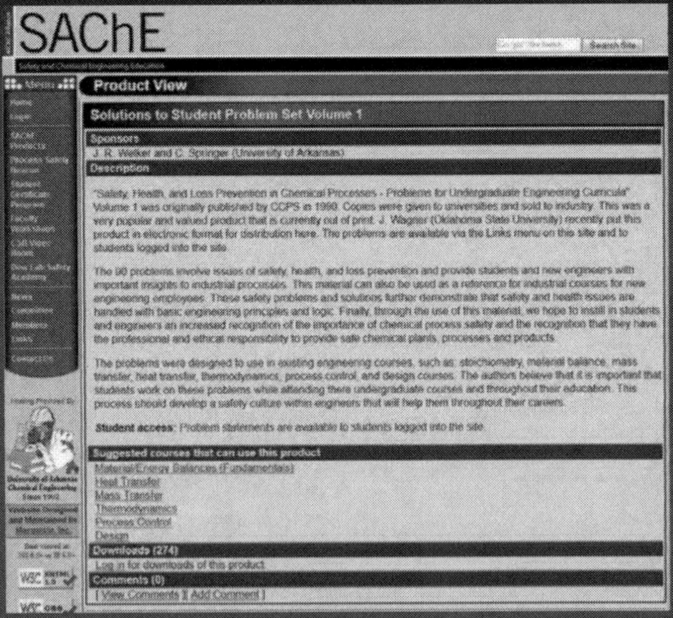

Before logging in, Product description is visible to all including suggested courses.

Comments on the product can be posted and viewed by all. (Comments approved by site administrator.)

Note on which files are available to students once logged in.

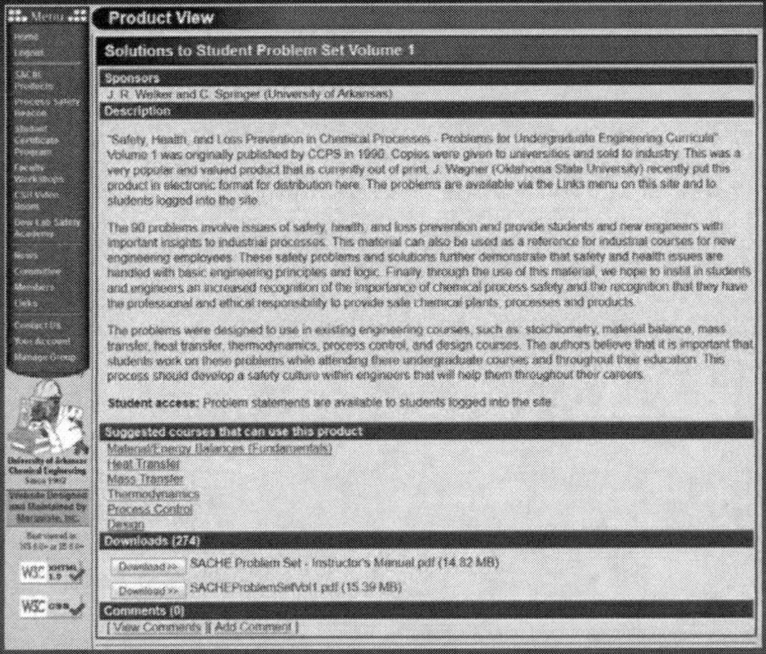

After logging in, Product files are available for downloading.

Streaming video services are provided by the University's Global Campus Media Services.

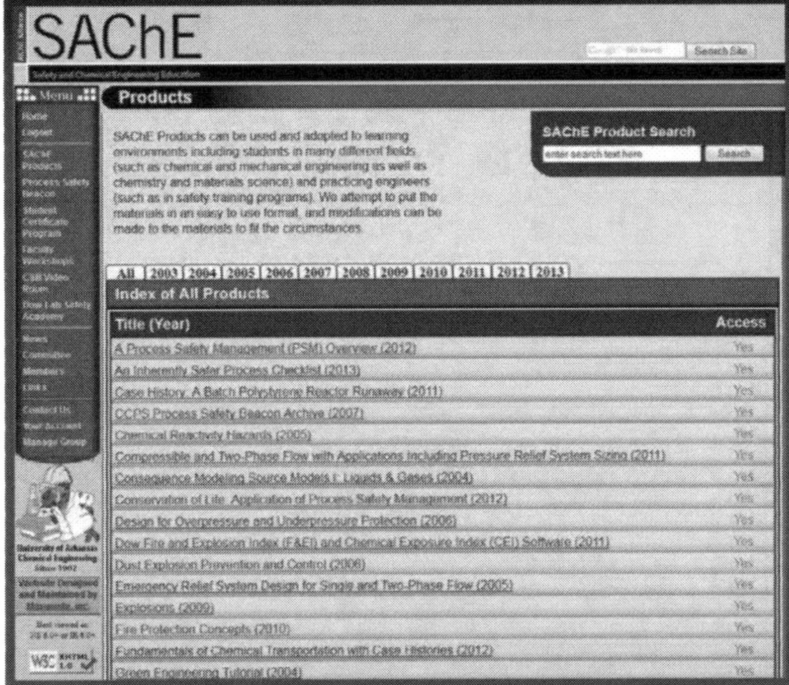

All tab provides an alphabetic listing of all products available for all years.

The SAChE products can be searched (simple text search) for words in the Product titles, sponsors, and descriptions.

Google ™ Custom Search for the site has been added in the past year.

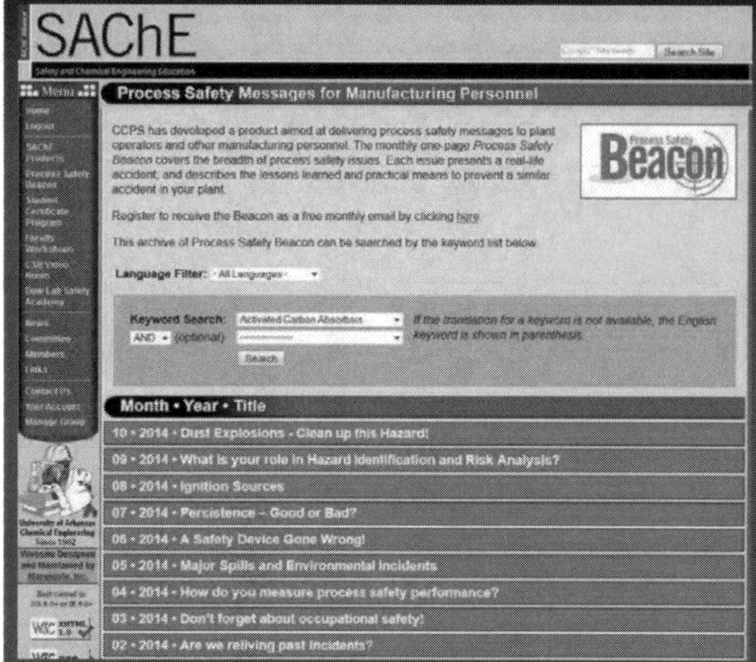

Process Safety Beacon Archive

Archive can be viewed by all. Printable versions are restricted to registered users.

Available translations are listed for each issue.

The archive can be searched by available translation or keyword.

Keywords are available in a few languages other than English – including French. If the keyword translation is unavailable, the English word is listed.

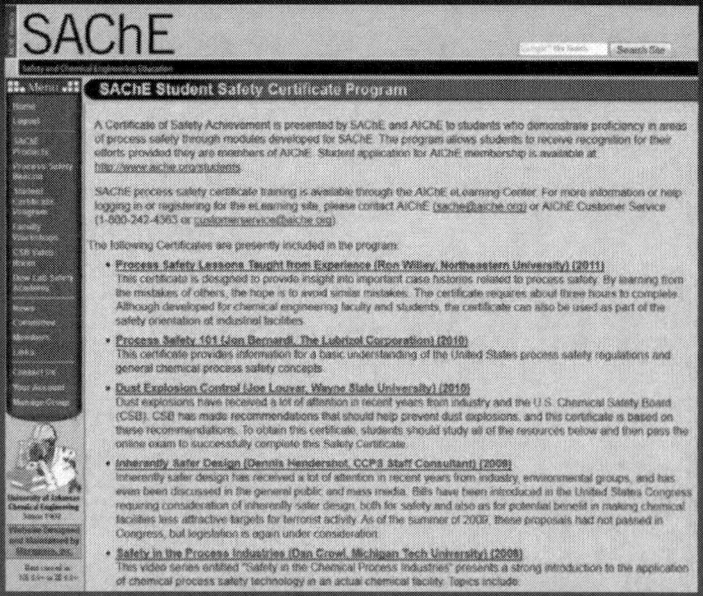

Student Safety Certificate Program

Nine certificate modules developed since 2008.

Students must be members of AIChE.

Basic procedure is to study the material provided then complete the quiz available through the AIChE web site. All material is now on the AIChE web site.

A list of students who successfully complete a module is available for all.

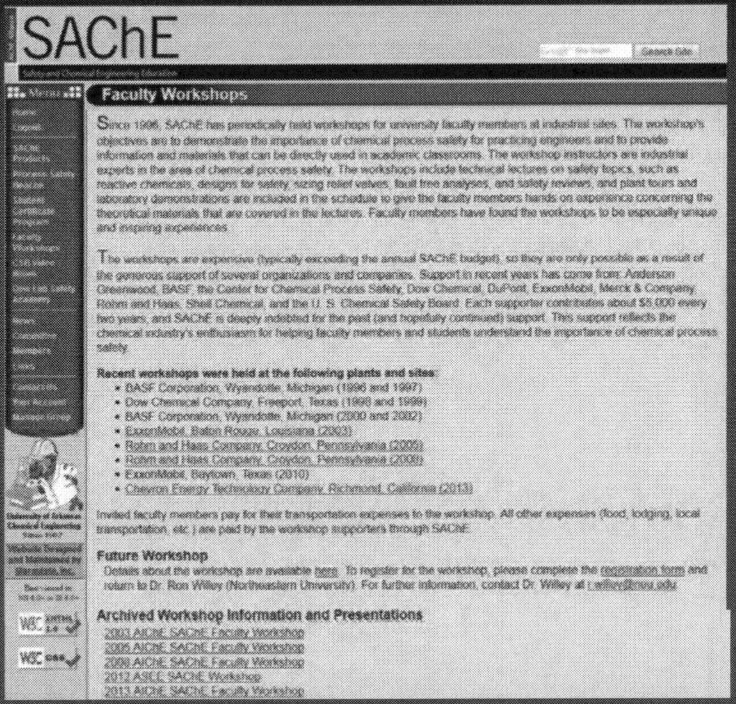

SAChE faculty workshops offer an excellent opportunity for faculty to see the importance of chemical process safety training.

Materials are made available to participants some of which is available through the SAChE site.

The Faculty Workshop was held in 2014, but the materials from that workshop have not been uploaded to the site as yet.

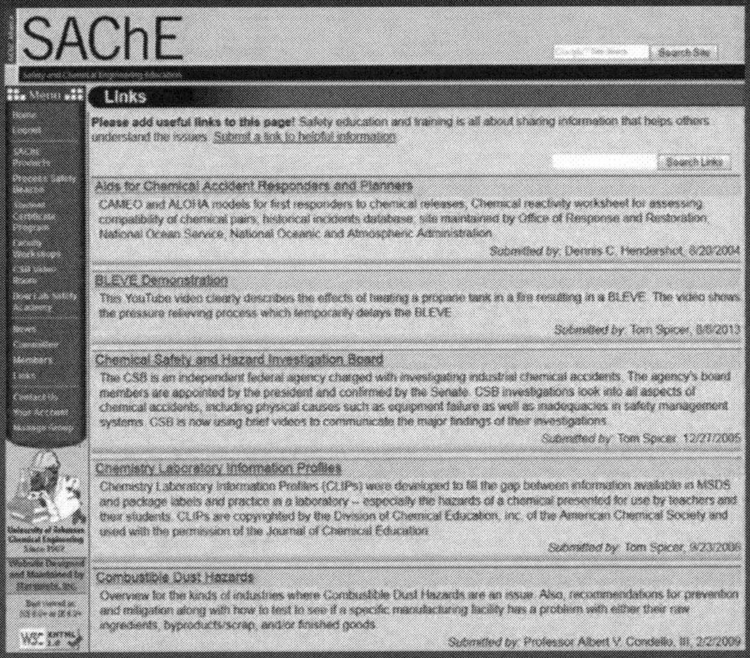

Many resources are available on the web, but this Links page provides a mechanism for vetting resources for students.

The links can be searched by title, text, or contributor.

Link contributions are approved by the site Administrator based on general content only.

Contributions are welcomed!

Submit a link today!

Common sense is not common – it must be passed on to others.

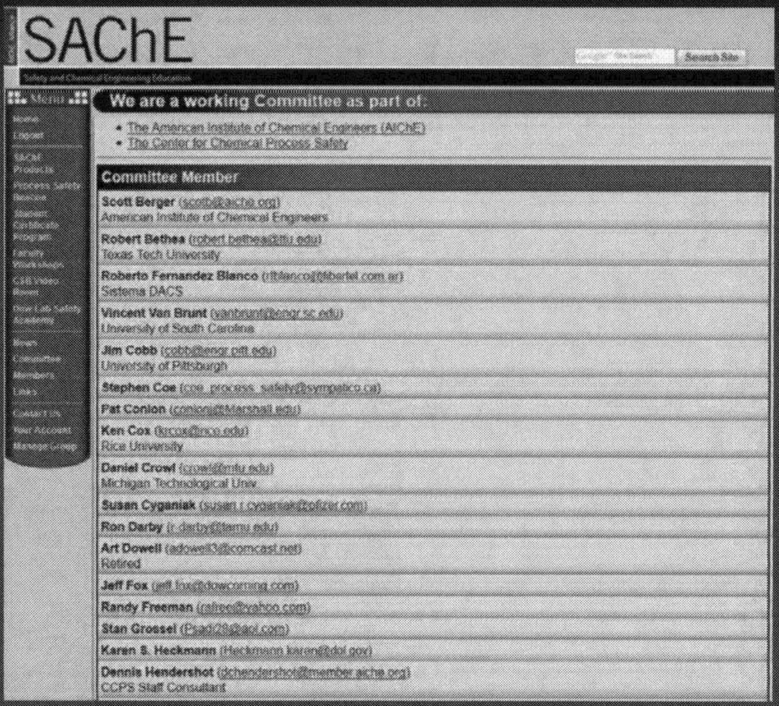

SAChE Committee Members

The complete list of Committee members is available at the SAChE site (partial list at left).

Primary use is in the US, but some use world wide.

Statistics for the past year.

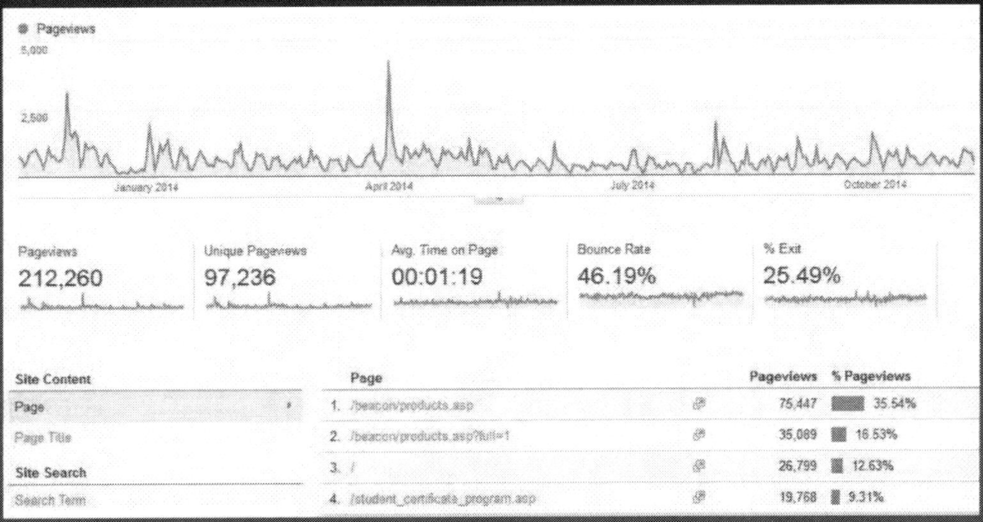

Top use attributed to the Beacon archive and the Student Safety Certificate Program.

Conclusions

- Many resources are available through the site to:
  - Enhance the educational experience of students in an academic environment
  - Provide training resources for organizations
- Many of the resources on the site are available to all site visitors.
- All SAChE Products available to academic institutions without charge because of the Safety Scale-Up Partnership with AIChE.

Conclusions

- Why has the site been successful?
  - Quality content
  - Good technical support (Marquiste, Inc.)
  - Prompt action when problems are identified

Conclusions

- What should you do?
  - Tell your colleagues about the information available on the SAChE site.
  - Use SAChE materials in training and educational activities.
  - Provide feedback about the materials you use – submit a comment about the Products.
  - Submit information about links you have found helpful on issues related to safety education (Links Page).

Questions?

http://www.sache.org/

Tom Spicer, PhD, PE
Professor and Barker Chair in Chemical Engineering
Ralph E. Martin Department of Chemical Engineering
3202 BEC/CHEG
University of Arkansas
Fayetteville, AR 72701

479-575-6516
479-575-7926 fax
http://www.cheg.uark.edu

SAFETY ANALYSIS WITH MODEL-BASED DYNAMIC SIMULATION ON MOBILE DEVICES

Mordechai Shacham
and
Michael Elly
Ben Gurion University of the Negev
Beer-Sheva, Israel

Michael B. Cutlip
University of Connecticut
Storrs, CT

The Motivation

➢Prediction and prevention of chemical process hazards are essential parts of the chemical engineer's education.

➢An unconventional and intriguing approach for safety instruction is to let the students discover potential safety hazards and the ways to prevent them by model-based dynamic simulation (MDBS) of processes.

➢Process modeling, simulation and optimization programs, such as Aspen Plus, Hysys, GAMS, gProms etc. are routinely used for dynamic simulation.

➢Such large scale programs can be used only in locations where computers and the required programs are available. Internet access is often also essential.

➢In this presentation the use of the PolyMathLite app for Android-based devices for MDBS of potentially dangerous processes is demonstrated.

The PolyMathLite App*

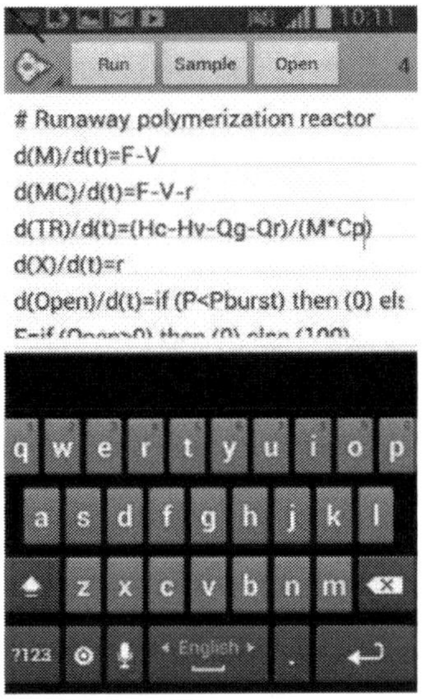

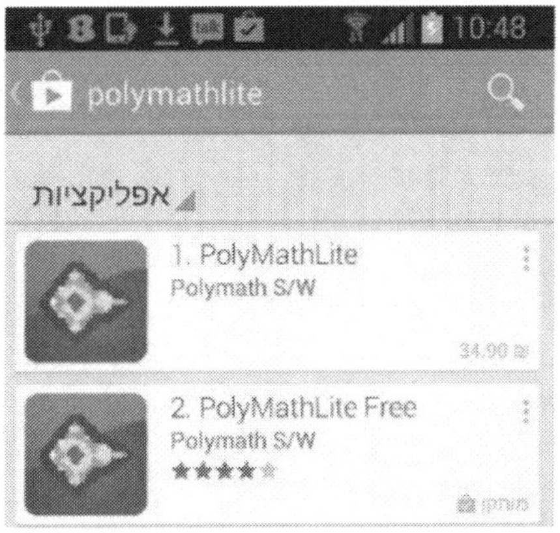

*PolyMathLite is an Android app produced by PolyMath Software. (http://www.polymathlite.com)

Example 1 – Temp. Runaway in a Polymerization Reaction

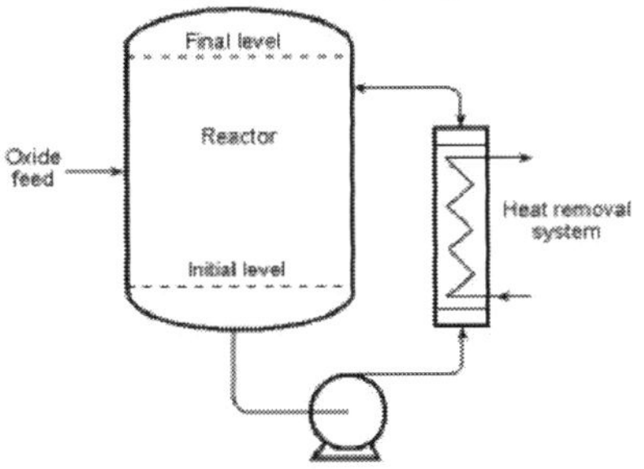

Consider the manufacture of a polyol lubricant by step-wise condensation of *propylene oxide* with *butanol*:

$C_4H_9OH + (n+1)\ C_3H_6O \rightarrow$ heat + $C_4H_9(OC_3H_6)nOCH_2CHOHCH_3$

- The catalyzed alcohol is initially charged into the reactor, up to the "initial" level.
- The oxide is fed into the reactor at a constant rate, until the batch is ready and the reactor is full.
- Excess heat of the reaction is removed via an external heat removal system.

Example 1 – Temp. Runaway in a Polymerization Reaction*

➢Economical considerations dictate that the reaction should be completed at the highest possible rate.

➢The reaction rate is a function of the temperature, catalyst concentration and liquid phase oxide concentration (which is function of the pressure).

➢The limits on the reactor temperature and catalyst concentration are set by considerations of thermal degradation and purification difficulties.

➢To maximize the reaction rate, the pressure must be kept as high as possible for the entire duration of the batch.

*Shacham et al. Computers & Chemical Engineering, 24, 415-421 (2000)

Example 1 –Typical Student Assignment

a. Simulate the ***normal operation*** of the semi-batch reactor (F_c = 3300 kg/min).

b. Check the effects of the ***reduction of the cooling re-circulation*** mass flow rate by reducing it to F_c = 2500 kg/min.

c. Check the effects of ***cooling failure*** with 10 minutes duration. Introduce such a failure 12 hrs after the startup.

d. Rework (c), but **increase** F_c to 5000 kg/min ***after the failure.***

e. Check the effects of the ***failure of the bursting disk*** to open at the specified rupture pressure (= 8 atm).

PolyMathLite Model of the Polymerization Reactor

Run Sample Open 16

```
# Safety related parameters
T0 = 80 # Feed temperature (deg. C)
Fc=3300 #Recycle mass flowrate (kg.
M0 = 4400 # Initial mass in the react
M(0)=4400 # Initial mass in the react
MC(0)=0 # Initial oxide component m
TR(0)=80 # Initial reactor temp. (deg
# Additional parameters
Lamda = 670 # Heat of vaporization (
Cp = 3.5 # Specific heat of feed react
HR = -1660 # Heat of reaction (kJ/ k
Pburst = 8# Disk rupture pressure (ba
R = 1.987 # Gas constant
E = 21000 # Activation Energy
Kv = 100
t(0)=0
```

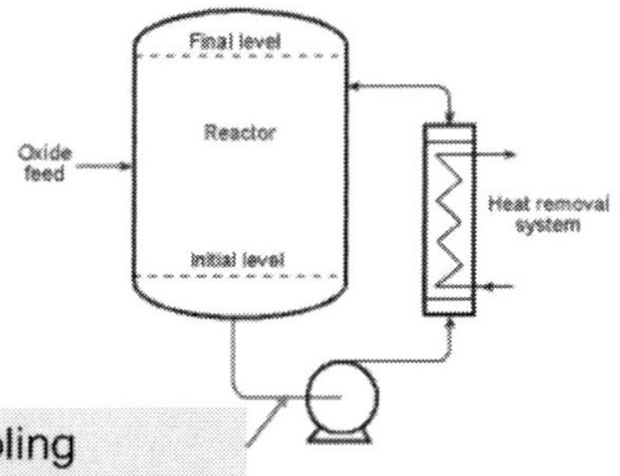

Normal cooling recirculation flow-rate

Highest pressure allowed

PolyMathLite Model of the Polymerization Reactor – Equation's Section

```
# Equations
d(M)/d(t) = F-V # Total mass in the reactor (kg)
d(MC)/d(t) = F-V-r # Oxide component mass in the reactor (kg)
d(TR)/d(t) =(Hc-Hv-Qg-Qr )/(M*Cp # Temperature in the reactor (deg C)
d(X)/d(t) = r # The mass of oxide reacted (kg)
d(Open)/d(t) = if (P<Pburst) then (0) else (0.001) # Status of the bursting disk: 0 closed, >0 open
F = if (Open>0) then (0) else (100) # Oxide feed rate (kg/min)
V = if ((P<=1) or (Open==0)) then (0) else (V1) # Vapor discharge rate (kg/min)
V1 = if (P<1.9) then (Vsubs) else (Vs) # Vapor discharge rate (kg/min)
Vs=0.85*Kv*P/sqrt(TR+273) # Sonic vapor discharge rate (kg/min)
Vsubs=Kv*P/sqrt((TR+273))*sqrt(1+1/P^2) # Sub-sonic - vapor discharge rate (kg/min)
r = k*MC # Reaction rate (kg oxide/min)
Hc= F*Cp*(T0-TR) # Feed enthalpy change (kJ/min)
Hv=V*Lamda # Latent heat of vapor discharge (kJ/min)
Qg=r*HR # Heat of reaction (kJ/min)
Qr=Fc*Cp*(TR-T0) # Heat removal rate ( kJ/min)
P = if (P1<1) then (1) else (P1) # Oxide vapor press
P1 = (exp(-3430/(TR+273)+11.7)+1.45e-3*MW)*C #
k = 9e9*exp(-E/(R*(TR+273))) # Reaction rate coeffic
C = MC/M # Oxide conncentration (kg/kg)
MW = (M0+X)/(M0/74) # Molecular weight of the polymer (kg/mol)
```

If the bursting disk is open the oxide feed stops and the vapor discharged

The model contains complete documentation, including the description of the variables and their units.

Polymerization Reactor Results – Normal Operating Conditions

	Variable	Initial value	Final value	Minimal value	Maximal value
1	C	0	0.1578	0	0.744704
6	Hc	0	-3955.89	-1.11E+04	0
9	k	0.000895	0.002265	0.000895	0.010554
12	M	4400	2.04E+05	4400	2.04E+05
14	MC	0	3.23E+04	0	3.99E+04
15	MW	74	2895.18	74	2895.18
16	Open	0	0	0	0
17	P	1	2.21255	1	6.35029
18	P1	0	2.21255	0	6.35029
20	Qg	0	-1.21E+05	-3.98E+05	0
21	Qr	0	1.31E+05	0	3.66E+05
22	r	0	73.0491	0	239.664
23	R	1.987	1.987	1.987	1.987
24	t	0	2000	0	2000
25	T0	80	80	80	80
26	TR	80	91.3025	80	111.897
27	V	0	0	0	0
28	V1	7.5271	9.8533	7.5271	28.4331
29	Vs	4.52409	9.8533	4.52409	28.4331
30	Vsubs	7.5271	12.7211	7.5271	33.863
31	X	0	1.68E+05	0	1.68E+05

Molecular weight of the polymer (kg/mol)

Maximal pressure (bar)

Maximal temp. (°C) in the reactor

Reactor's Operation with Cooling Media Failure

Fc=if (*t*<700) then (3300) else (if (*t*>710) then (3300) else (0))

	Variable	Initial value	Final value	Minimal value	Maximal value
1	C	0	7.82E-25	0	0.744704
6	Hc	0	0	-5663.84	0
9	k	0.000895	0.000895	0.000895	4.08745
12	M	4400	7.55E+04	4400	7.59E+04
14	MC	0	5.90E-20	0	3.81E+04
15	MW	74	1269.51	74	1269.51
16	Open	0	0.01324	0	0.01324
17	P	1	1	1	8.30799
18	P1	0	7.12E-24	0	8.30799
20	Qg	0	-8.76E-20	-7.85E+05	0
21	Qr	0	1.64E-10	0	1.60E+06
22	r	0	5.28E-23	0	472.828
23	R	1.987	1.987	1.987	1.987
24	t	0	2000	0	2000
26	TR	80	80	80	215.563
28	V1	7.5271	7.5271	6.38044	35.8127
29	Vs	4.52409	4.52409	3.8349	35.8127
30	Vsubs	7.5271	7.5271	6.38044	42.4368
31	X	0	7.11E+04	0	7.11E+04

Molecular weight of the polymer (kg/mol)

Maximal pressure (bar)

Maximal temp. (° C) in the reactor

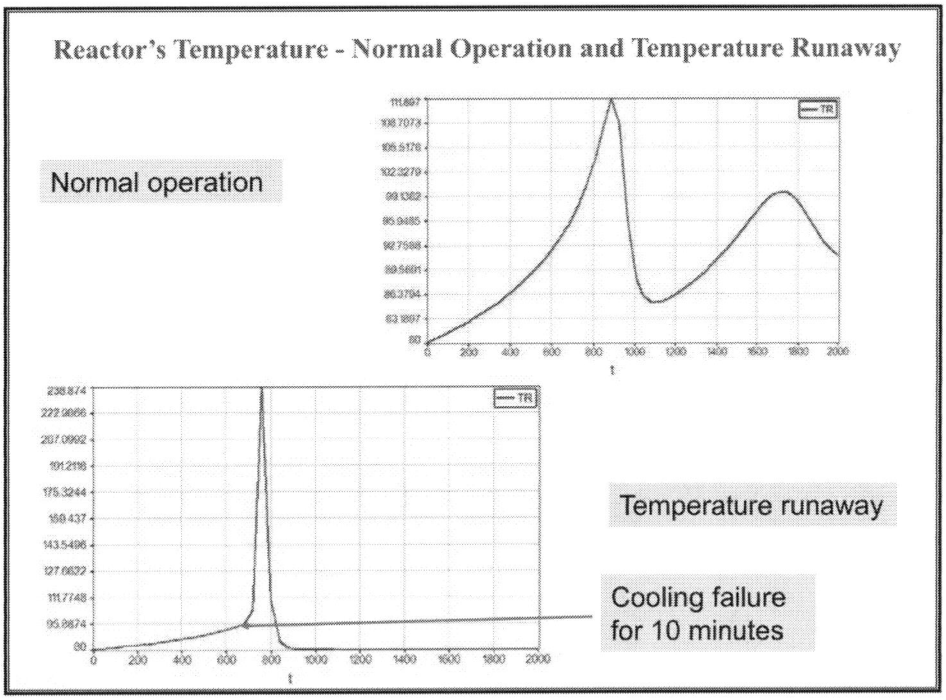
Reactor's Temperature - Normal Operation and Temperature Runaway
Normal operation
Temperature runaway
Cooling failure
for 10 minutes
TR
t

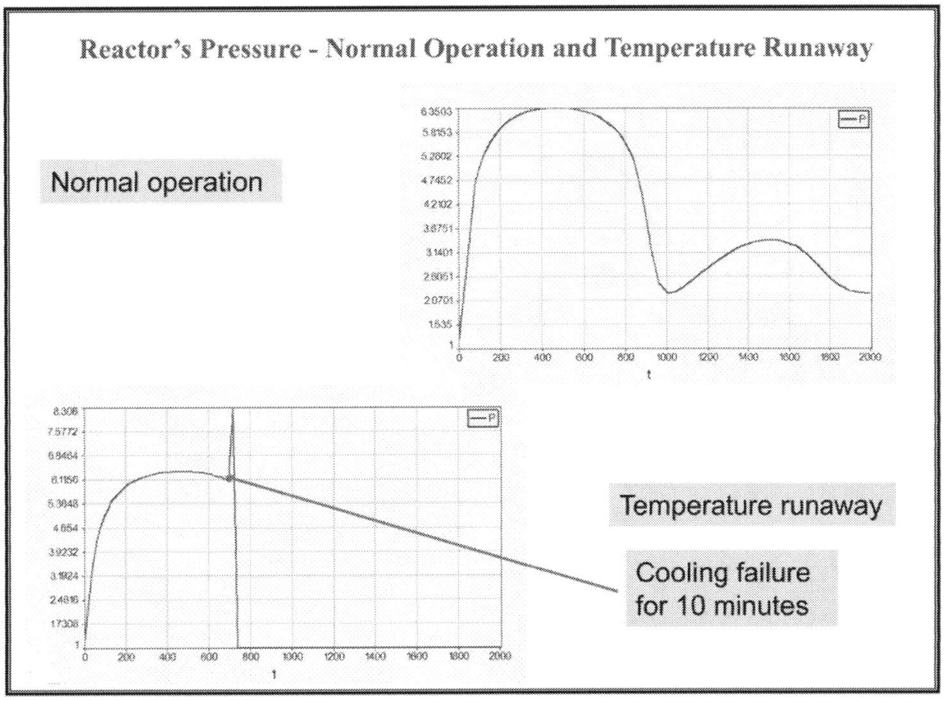
Reactor's Pressure - Normal Operation and Temperature Runaway
Normal operation
Temperature runaway
Cooling failure
for 10 minutes
P
t

Results of the Safety Analysis for Example 1

a. Simulate the normal operation of the semi-batch reactor

T_{max} = 111.8 °C; P_{max} = 6.35 bar and MW_{final} = 2895 kg/mol;

b. Check the effects of the reduction of the cooling re-circulation mass flow rate

T_{max} = 303 °C; P_{max} = 8.15 bar and MW_{final} = 893 kg/mol;

c. Check the effects of cooling failure with 10 minutes duration

T_{max} = 215.6 °C; P_{max} = 8.3 bar and MW_{final} = 1269 kg/mol;

d. Rework (c), but increase F_c to 5000 kg/min after the failure.

T_{max} = 105.6 °C; P_{max} = 6.9 bar and MW_{final} = 2565 kg/mol;

e. Check the effects of the failure of the bursting disk to open

The model cannot handle properly such a scenario

Export of the Polymerization Reactor Model to MATLAB

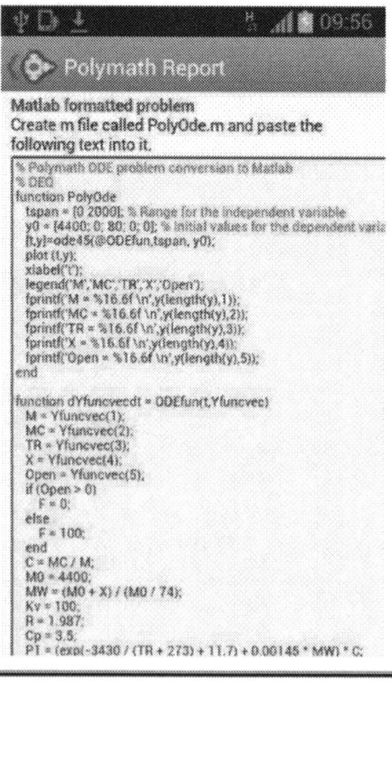

Polymath Report

Matlab formatted problem
Create m file called PolyOde.m and paste the following text into it.

```
% Polymath ODE problem conversion to Matlab
% DEQ
function PolyOde
  tspan = [0 2000]; % Range for the independent variable
  y0 = [4400; 0; 80; 0; 0]; % Initial values for the dependent varia
  [t,y]=ode45(@ODEfun,tspan, y0);
  plot (t,y);
  xlabel('t');
  legend('M','MC','TR','X','Open');
  fprintf('M = %16.6f \n',y(length(y),1));
  fprintf('MC = %16.6f \n',y(length(y),2));
  fprintf('TR = %16.6f \n',y(length(y),3));
  fprintf('X = %16.6f \n',y(length(y),4));
  fprintf('Open = %16.6f \n',y(length(y),5));
end

function dYfuncvecdt = ODEfun(t,Yfuncvec)
  M = Yfuncvec(1);
  MC = Yfuncvec(2);
  TR = Yfuncvec(3);
  X = Yfuncvec(4);
  Open = Yfuncvec(5);
  if (Open > 0)
    F = 0;
  else
    F = 100;
  end
  C = MC / M;
  M0 = 4400;
  MW = (M0 + X) / (M0 / 74);
  Kv = 100;
  R = 1.987;
  Cp = 3.5;
  P1 = (exp(-3430 / (TR + 273) + 11.7) + 0.00145 * MW) * C;
```

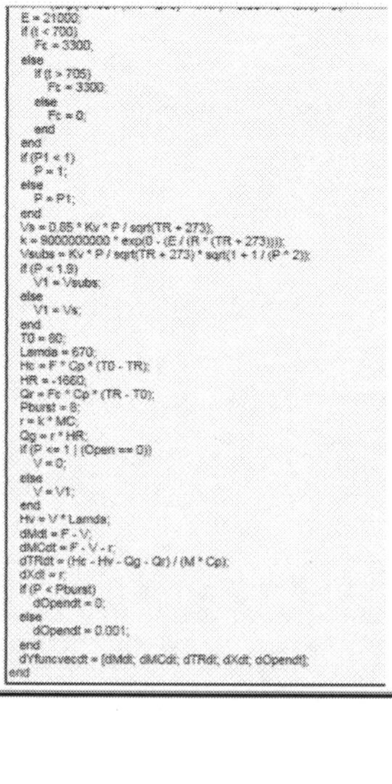

```
  E = 21000;
  if (t < 700)
    Fc = 3300;
  else
    if (t > 705)
      Fc = 3300;
    else
      Fc = 0;
    end
  end
  if (P1 < 1)
    P = 1;
  else
    P = P1;
  end
  Vs = 0.85 * Kv * P / sqrt(TR + 273);
  k = 9000000000 * exp(0 - (E / (R * (TR + 273))));
  Vsubs = Kv * P / sqrt(TR + 273) * sqrt(1 + 1 / (P ^ 2));
  if (P < 1.9)
    V1 = Vsubs;
  else
    V1 = Vs;
  end
  T0 = 80;
  Lamda = 670;
  Hc = F * Cp * (T0 - TR);
  HR = -1660;
  Qr = Fc * Cp * (TR - T0);
  Pburst = 8;
  r = k * MC;
  Qg = r * HR;
  if (P <= 1 | (Open == 0))
    V = 0;
  else
    V = V1;
  end
  Hv = V * Lamda;
  dMdt = F - V;
  dMCdt = F - V - r;
  dTRdt = (Hc - Hv - Qg - Qr) / (M * Cp);
  dXdt = r;
  if (P < Pburst)
    dOpendt = 0;
  else
    dOpendt = 0.001;
  end
  dYfuncvecdt = [dMdt; dMCdt; dTRdt; dXdt; dOpendt];
end
```

Example 2 -Nitric Acid oxidation of 2-octanol in a Semi-Batch Reactor

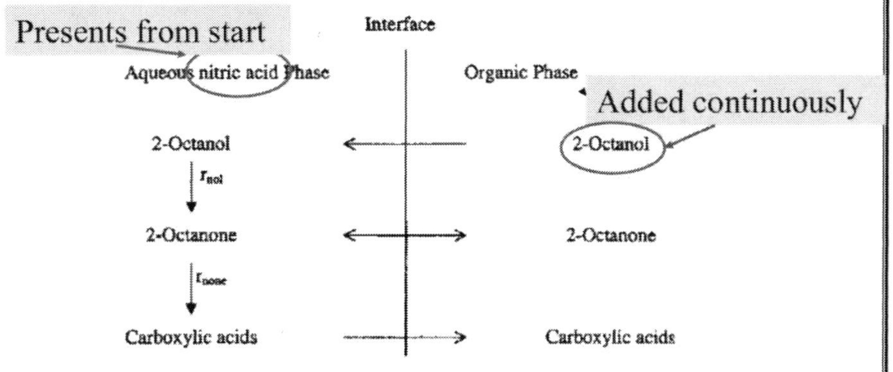

$$A + B \xrightarrow{r_1} P + 2B$$
$$P + B \xrightarrow{r_2} X$$

A - 2- octanol, P - 2-octanone , X – undesired oxidation products B - nitrosonium ion, causes an autocatalytic behavior

Example 2 -Nitric Acid oxidation of 2-octanol in a Semi-Batch Reactor (van Woezik and Westerterp*)

- In the semi-batch reactor aqueous nitric acid is initially present, and 2- octanol (A) is added at a constant rate until a desired molar ratio of the reactants is reached..
- The 2-octanol reacts to form 2-octanone and carboxylic acid.
- Under normal operating conditions, the temperature in the reactor does not exceed the limit of ~ 0 °C.
- If the temperature at any time exceeds ~ 5 °C, runaway conditions develop, which may lead to a maximal temperature of over 200 °C, and conversion of all of the 2-octanone to carboxylic acid.

**Chemical Engineering and Processing*, *39*, 521(2000) and *41*, 59(2001).

Example 2 – Important Safety Related Parameters

```
Run   Sample   Open
# Semi - batch reactor - 2-octanol ox
#Safety related parameters
Tcool_IN = 260 # Inlet coolant temp. [
Fwl = 100 # Flow rate of coolant [L/m
Tdos = 293.15 # Temp. of feed dose [
tdos = 36000 # Dosing time [s], 10h
 Vdos1=0.6 #Final volume of the dose
Percent=0.6 #Initial mass concentrat
UA0=1500 # Initial cooling surface he
UA1=2100 # Final cooling surface hea
t(f) = 72000 # Process duration(s)
Tr(0) = 260 # Temp. in the reactor at t
t(0) = 0.0001 # Starting time (s)
# Equations
d(Np)/d(t) = (r1 - r2) * Vr0 / (1 - Epsd)
Np(0) = 0 # Number of moles of 2-oct
```

Potential Causes of Temperature Runaway

- Increase of coolant inlet temp.
- Reduction of coolant flow rate
- Shortening of dosing time
- Increase of dose volume
- Increase of nitric acid concentration
- Reduction of effective heat transfer rate

PolyMAthLite Model of the of the 2-octanol Oxidation Reactor– Mole Balance Equations

```
# Equations
d(Np)/d(t) = (r1 - r2) * Vr0 / (1 - Epsd) # Mole balance for 2-octanone (P)
Np(0) = 0 # Number of moles of 2-octanone (P) at t = t0
d(Nx)/d(t) = r2 * Vr0 / (1 - Epsd) # Mole balance for carboxylic acids (X)
Nx(0) = 0 # Number of moles of carboxylic acids (X) at t = t0
r1 = k1 * CaOrg * CbAq * (1 - Epsd) # Reaction rate of a and b to p[kmol/m3/s]
r2 = k2 * CpOrg * CbAq * (1 - Epsd) # Reaction rate of p and b to x[kmol/m3/s]
Vr0 = 1.5 # Initial volume in a reactor [m3]
Epsd = Vdos1 / (Vdos1 + Vr0) # Volume fraction of dispersed phase
k1 = maA1 * exp(-E1perR / Tr - m1 * H) # Specific reaction rate 1
k2 = mpA2 * exp(-E2perR / Tr - m2 * H) # Specific reaction rate 2
CaOrg = (Theta * NaF - Np - Nx) / (Vdos1 * Theta) # Concentr of a in org phase [kmole/m3]
CpOrg = Np / (Vdos1 * Theta) # Concentr. of (P) in org phase [kmol/m3]
CbAq = (Np + Y * NaF) / Vr0 # Concentr. of (B) in aq. phase [kmole/m3]
maA1 = 10 ^ 5 # Pre-exponential factor reaction 1 [m3/kmol/s]
mpA2 = 10 ^ 10 # Pre-exponential factor reaction 2[m3/kmol/s]
E1perR = 11300 # Activation temperature reaction 1 [K]
E2perR = 12000 # Activation tempetature reaction 2 [K]
m1 = 6.6 # Hammett's reaction rate coeff. reaction 1
m2 = 2.2 # Hammett's reaction rate coeff. reaction 2
H = -0.6221 - 3.7214 * wt - 1.5714 * wt ^ 2 # Hammett's acidity function
Theta = If (t <= tdos) Then (t / tdos) Else (1) # Dimensionless time up to t=tdos
NaF = Vdos1 * RhoOctan / MwOctan # Total amount of 2-octanol (a) fed [kmol]
Y = 0.035 # Initial concentr. of nitrosonium ion Y=Nb0/NaF
wt = Nn * Mw / (Vr0 * RhoAcid) # Mass concentr. of nitric acid sol [%/100%]
RhoOctan = 820.7 # Density of 2-octanol [kg/m3]
MwOctan = 130.23 # Molar mass of 2-octanol [kg/kmol]
Nn = CnAq * Vr0 # Number of moles of HNO3 [kmol]
Mw = 63 # Molar mass of HNO3 [kg/kmol]
RhoAcid = 1500 # Density of pure nitric acid [kg/m3]
CnAq = (NnO - Y * NaF - Np - 2 * Nx) / Vr0 # Concentr. of HNO3 in the aq. phase [kmol/m3]
```

PolyMAthLite Model of the of the 2-octanol Oxidation Reactor– Energy Balance Equations

```
d(Tr)/d(t) = (Qr + Qdos + Qcool) / Gamma # Reactor energy balance (Tr in K)
Qr = Qnol + Qnone # Sum of the heat of reaction the reactions [W)
Qdos = Phi * RhoCPdos * (Tdos - Tr) # Heat input due to reactant addition [W]
Qcool = UAcool * (Tcool - Tr) # Heat removed by the cooling jacket [W]
Gamma = Gamma0 + RhoCPdos * Phi * t # Total heat capacity of the system [J/K]
Qnol = r1 * Vr0 * Hnol / (1 - Epsd) # Heat of reaction, 1 [W]
Qnone = r2 * Vr0 * Hnone / (1 - Epsd) # Heat of reaction, 2 [W]
Phi = Vdos1 / tdos # Volumetric flow rate of the feed [m3/s]
RhoCPdos = 2 * 10 ^ 6 # Heat capacity of dose [J/m3/K]
UAcool = UA0 + (UA1 - UA0) * Theta # Cooling surface heat transfer coefficient [W/K]
Gamma0 = 5.4 * 10 ^ 6 # Initial heat capacity of the system [J/K]
Hnol = 160 * 10 ^ 6 # Specific heat of reaction 1 [J/kmol]
Hnone = 520 * 10 ^ 6 # Specific heat of reaction 2 (J/kmol)
Fw = Fwl / 60 * 10 ^ (-3) # Flow rate of coolant [m3/s]
d(Tcool)/d(t) = (Fw * (Tcool_IN - Tcool) - Qcool / (RhoCoolant * CpCoolant)) / Vj # Jacket energy balance (T in K)
Tcool(0) = 273.15 # Coolant exit temp. at t = t0 (K)
RhoCoolant = 1000 # The density of coolant [kg/m3]
CpCoolant = 4183 # Heat capacity of coolant [J/kg/K]
Vj = 1.5 # Volume of the jacket [m3]
```

2-octanol Oxidation Reactor– Simulation Results

Normal operating conditions

	Variable	Initial value	Final value	Maximal value
Cooling Temp. in (K)	Tcool_IN	260	260	260
Cooling Temp. out (K)	Tcool	273.15	260.38	273.15
Reactors temp. (K)	Tr	260	261.64	274.57
2-octanone (mol)	Np	0	3.1224	3.1224
Carboxylic acid (mol)	Nx	0	0.2521	0.2521

Desired product

Temperature runaway

Maximal temp. (K)

	Variable	Initial value	Final value	Maximal value
Cooling Temp. in (K)	Tcool_IN	265	265	265
Cooling Temp. out (K)	Tcool	273.15	265.16	288.12
Reactors temp. (K)	Tr	260	265.67	392.98
2-octanone (mol)	Np	0	0.3353	1.3596
Carboxylic acid (mol)	Nx	0	2.4050	2.4050

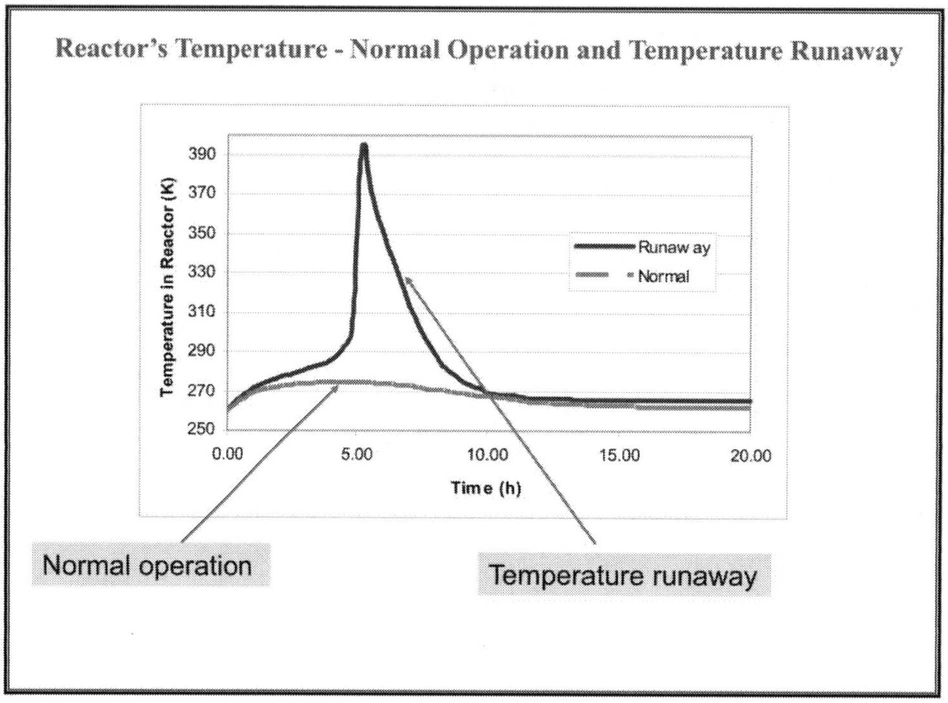
Reactor's Temperature - Normal Operation and Temperature Runaway
Temperature in Reactor (K)
390
370
350
330
310
290
270
250
0.00
5.00
10.00
15.00
20.00
Time (h)
Runaway
Normal
Normal operation
Temperature runaway

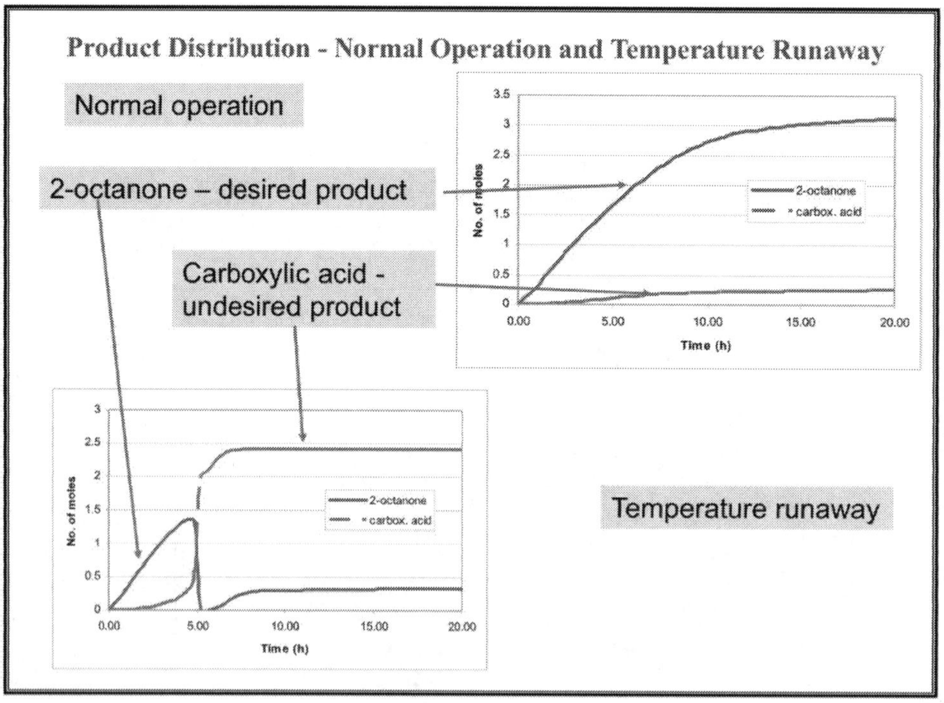
Product Distribution - Normal Operation and Temperature Runaway
Normal operation
2-octanone – desired product
Carboxylic acid - undesired product
Temperature runaway
No. of moles
Time (h)
2-octanone
carbox. acid

Example 2 - Additional Causes of Temperature Runaway

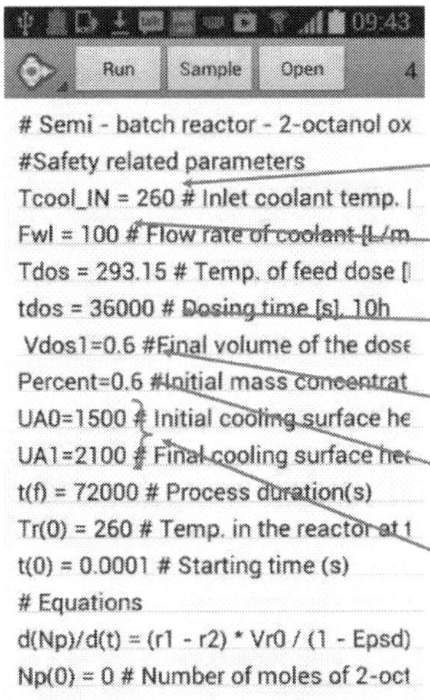

Increase of coolant inlet temp. ≥ 266 K

Reduction of coolant flow rate ≤ 12

Shortening of dosing time ≤ 7.5 h

Increase of dose volume ≥ 0.81 m^3

Increase of nitric acid concentration ≥ 0.64

Reduction of effective heat transfer rate ≤ 81%

Conclusions

- It has been shown that model-based dynamic simulation of processes enable students to *learn* safety related issues *by discovering potential safety hazards and trying out methods for their prevention*.

- This method of instruction makes it possible to *incorporate process safety into various chemical engineering courses*, such as process design, process simulation and reaction engineering.

- The availability of PolyMathLite (or other mobile device based mathematical modeling programs) can considerably extend the use of MDBS for safety analysis, *beyond the classroom, eliminating the need to use large scale simulation programs.*

ABSTRACT

AIChE Annual Meeting

November, 2014

Education Division

Title: *PSM eBook Material – Class Outcome Metrics*

Authors: Deborah L. Grubbe, PE, Co-Author, DuPont (retired)

Ronald L. Cutshall, Sr, PE, Co-Author, Amoco (retired)

Steven J. Swanson, PhD, Co-Author, Shell (retired)

Abstract:

Due to ABET requirements, many universities are including more safety information in their curricula. This paper will outline the outcome metrics and comments from a 3 credit Process Safety Management Course developed for ChE and ME students at a large, Midwestern University. Over a two year period, there were a number of questions asked and data collected about the course, so as to ensure appropriate perception around value. Results from the student surveys and comments from other key constituents will also be discussed: teaching assistants, guest lecturers, advisory board members and employer comments. Students and employers were so adamant about the value of this course, that the faculty moved from concept to required course in 18 months, and the Engineering Faculty is now evaluating a safety review course for all cooperative education students that must be completed before the first co-op work period. The paper will also outline the safety material recommended for inclusion in the pre-cooperative education review.

AUTHOR INDEX

AUTHOR INDEX

AUTHOR INDEX

AUTHOR INDEX

AUTHOR INDEX